Studies in Fuzziness and Soft Computing

Volume 361

Series editor

Janusz Kacprzyk, Polish Academy of Sciences, Warsaw, Poland
e-mail: kacprzyk@ibspan.waw.pl

The series "Studies in Fuzziness and Soft Computing" contains publications on various topics in the area of soft computing, which include fuzzy sets, rough sets, neural networks, evolutionary computation, probabilistic and evidential reasoning, multi-valued logic, and related fields. The publications within "Studies in Fuzziness and Soft Computing" are primarily monographs and edited volumes. They cover significant recent developments in the field, both of a foundational and applicable character. An important feature of the series is its short publication time and world-wide distribution. This permits a rapid and broad dissemination of research results.

More information about this series at http://www.springer.com/series/2941

Lotfi A. Zadeh · Ronald R. Yager
Shahnaz N. Shahbazova
Marek Z. Reformat · Vladik Kreinovich
Editors

Recent Developments and the New Direction in Soft-Computing Foundations and Applications

Selected Papers from the 6th World Conference on Soft Computing, May 22–25, 2016, Berkeley, USA

 Springer

Editors
Lotfi A. Zadeh
Department of Electrical Engineering
    and Computer Sciences
University of California
Berkeley, CA
USA

Ronald R. Yager
Machine Intelligence Institute
Iona College
New Rochelle, NY
USA

Shahnaz N. Shahbazova
Department of Information Technology
    and Programming
Azerbaijan Technical University
Baku
Azerbaijan

Marek Z. Reformat
Electrical and Computer Engineering
University of Alberta
Edmonton, AB
Canada

Vladik Kreinovich
Department of Computer Science
University of Texas at El Paso
El Paso, TX
USA

ISSN 1434-9922   ISSN 1860-0808   (electronic)
Studies in Fuzziness and Soft Computing
ISBN 978-3-030-09223-8   ISBN 978-3-319-75408-6   (eBook)
https://doi.org/10.1007/978-3-319-75408-6

Contents

Part I
Information and Data Analysis

Big Data Analytics and Fuzzy Technology: Extracting Information from Social Data

Shahnaz N. Shahbazova and Sabina Shahbazzade

Abstract Data becomes overwhelming present in almost all aspects of manufacturing, finance, commerce and entertainment. Today's world seems to generate tons of data related to all aspect of human activities every minute. A lot of hope and expectations are linked to benefits that analysis of such data could bring. Among many sources of data, social networks start to play a very important role. Indications what individuals think about almost anything related to their lives, what they like and dislike are embedded in posts and notes they leave on the social media platforms. Therefore, discovering the users' opinions and needs is very critical for industries as well as governments. Analysis of such data—recognized as a big data due to its tremendous size—is of critical importance. The theory of fuzzy sets and systems, introduced in 1965, provides the researchers with techniques that are able to cope with imprecise information expressed linguistically. This theory constitutes a basis for designing and developing methodologies of processing data that are able to identify and understand views and judgments expressed in a unique, human way —the core of information generated by the users of social networks. The paper tries to recognize a few important example of extracting value from social network data. Attention is put on application of fuzzy set and systems based methodologies in processing such data.

S. N. Shahbazova (✉)
Department of Information Technology & Programming,
Azerbaijan Technical University, Baku, Azerbaijan
e-mail: shahbazova@gmail.com

S. Shahbazzade
EECS Department, University of California, Berkeley, USA
e-mail: shahbazzade@berkeley.edu

© Springer International Publishing AG, part of Springer Nature 2018
L. A. Zadeh et al. (eds.), *Recent Developments and the New Direction in Soft-Computing Foundations and Applications*, Studies in Fuzziness and Soft Computing 361, https://doi.org/10.1007/978-3-319-75408-6_1

3

1 Introduction

More and more often social networks, or shall we say data generated by its users, is an object of research activities conducted by variety of organizations and corporations in order to extract information and knowledge about multiple aspects characterizing activities, behaviour, as well as interests and likes of users and group of users. We can find multiple examples of corporations and agencies putting enormous effort to analyse and understand data that is generated by actions, interactions, and conversations involving users, as well as by the users' views and opinions on almost everything what happens in their lives and surroundings.

Varieties of techniques are used to process data generated by users in social networks: statistical approaches, graph based approaches, and many others [1–6]. However, a human nature is present in the social networks. This means that the networks are human-like—full of imprecise relations and connections between individuals, vague terms, groups and individuals with indefinite descriptions and characteristics of interests. It seems that many aspects of social networks resembles the ones of their users.

In the light of these statements, we would like to state that techniques of processing social networks of users and groups should reflect such human facets. Therefor, techniques that are based on a human-like methodology such as the theory of fuzzy sets and systems [7] are the most suitable for such a purpose. The abilities of fuzzy methods to deal with ambiguous data and facts, to describe things in a human manner, and to handle imprecision and ambiguity make fuzzy based technologies and methods [8] one of the best tools for analysing social network [9].

2 Analysis of Social Networks

The nature of data generated by users of social networks allows us to identify multiple areas of human life, as well as industrial, corporate and governmental activities that could benefit from analysing social network data. In the following subsections we try to identify a few important aspects contributing to improvements in: quality of human life, political life, shopping and entertainment, manufacturing, and corporate visions and goals, Fig. 1.

2.1 Quality of Life

The users' ability to observe and quickly react to different events—would they be positive or negative—means that analysis of data can be an important element of sophisticated and intelligent **Disaster Management** systems. An early detection of disasters, for example earthquakes, floods or wild fires, would enable quick

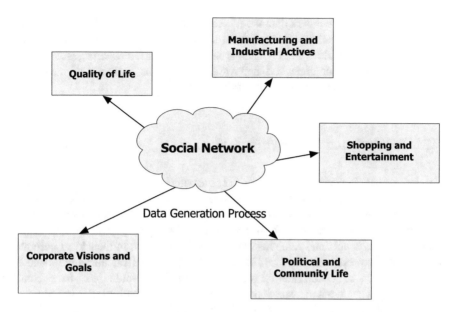

Fig. 1 Social network as data source for multiple types of information and knowledge

interventions that increase chances of minimizing or even mitigating the effects of disasters. Tools that allow visualization of social media data would lead to better understanding of dynamics of investigated calamities. They would provide close, real-time monitoring of disasters and their progress. They would enable initial analysis and estimation of effectiveness of **prevention actions**, as well as determine the most adequate forms of help. All that is closely related to another very important area that could benefit form analysis of social network data: **Health Care**. In this case, application of big data techniques would have an enormous impact on coping with diseases and health problems in the contexts of **interactions human-animal-environment**. In general, data analysis would provide better identification and recognition of origins of problems, better knowledge of mechanisms of disease spreading, and better understanding of impact of variety of health problems on human behaviour and actions. To sum up, systems able to detect and track any health related problems related to humans and animals are of great importance, especially in the time of increased human and animal migration activities, as well as potential **climate changes**.

Education is yet another domain where preforming big data analysis on social network data could play an important role. For example, it is already known that analysis of posts generated by pupils and students can lead to detecting bullying and even preventing tragedies caused by such behaviour. Development of systems monitoring and analysing variety of issues related to ways of study as well as problems and issues encountered by students would mean creation of a better environment for education. Such an environment would increase **effectiveness of education systems** and contribute to better-educated societies.

Analysis of social network data would lead to a more effective **Criminal Justice** system. More information extracted from social data related to actions violating laws and by-laws would definitely change the way law enforcement agencies work. The abilities to detect undesired behaviours as well as presence of dangerous individuals would create s**afer communities**.

2.2 Political and Community Life

Facebook [10], Pinterest [11], Twitter [12] and other social sites continue to generate and provide an uninterrupted stream of data. Users publish variety of messages and notes concerning multiple aspects of their lives. One of aspects of high concern is their **community life and national affairs**. They provide uncensored opinions, assessments and thoughts about what is going on in their local communities. They express their likes and dislikes regarding **political decisions** made at different levels of government. Processing such data would lead to a better understating what types of decisions are popular among individuals—would that relate to simple local things, or national important issues. It seems that it will be difficult for any government to ignore what citizens have to say about its **governing practices**, introduced laws and its solutions of 'small' and 'big' problems and concerns. Fuzzy processing could provide here a special analysis based on linguistic summarization of data. That would provide easy to understand descriptions and generation of **people's opinions**.

Elections are big events that trigger a lot of reactions and emotions among citizens. Opinions regarding individual candidates are subjects of many posted statements. Importance and relevance of issues they raise and cover can be analysed in order to better address needs and requirements of voters. Social networks already play a significant role in political life, and analysis of data generated by the users during an election time would provide evaluation of **voters' emotions and reactions**. That would shape and influence a whole election process. Overall that would lead to election of governments addressing important issues in a way majority of population would like to see.

2.3 Shopping and Entertainment

The social media is a platform for exchange of any types of information among users. This means that a lot of posted notes, statements and texts are related to the users' opinions and thoughts about things and activities they do every day. Quite a portion of that is related to shopping: what they buy, what they like and want to buy, what they should not buy, and what is their shopping—including **on-line shopping**—experience. Also entertainment plays a significant role in the users' lives and it is also reflected in network data generated by individuals.

They comment on multiple things from what **music** they listen to and what **movies** they watch to look, behaviour and likes/dislikes of actors, directors, and singers.

All of this is an enormous source of data and potentially information that is invaluable for **entertainment and retail industries**. The ability to understand a customer, its desires and need is simply essential. Such knowledge would allow manufacturers, on-line stores, and entrainment conglomerates to prepare their products to better satisfy customers' needs.

2.4 Manufacturing and Industrial Activities

The mentioned above social networks, such as Facebook and Twitter, continue to generate and provide an uninterrupted stream of data. Development of methods and approaches that analyse this data from a perspective of the **users' opinions on different products and services** will allow companies to identify the customers' preferences, as well as their needs and likes. This would lead to a so-called **data-driven manufacturing**—a scenario where existing and potential customers influence what is being manufactured. Analysis of the users' data would also help identifying weaknesses and strengths of manufactured goods. Social media **sentiment investigations** could determine if users intent to purchase specific products or if they dislike these products. All this would provide insight that can be explored and acted upon. Such analyses could also assess consumers' interest in a product before it is launched.

Introduction of tools and systems analysing social network data would affect variety of **service and utility** companies. Such systems could change the way companies provide services to the users, and how these systems could react to changes in the demand and needs. For example, General Electric (GE) is about to release a system that uses analysis of social media to support estimation of **potential problems** in an electrical grid—it is called Grid IQ [13]. Many aspects of individuals' lives are being discussed on the forum of social networks therefore any company that provides services would benefit a lot from social network data. Another interesting aspect is related to a so-called **geo-tagging** where many aspects of collected data is location sensitive—and this alone will bring valuable information to be analysed.

2.5 Corporate Visions and Goals

Understanding the users' needs and their attitude to multiple services and products offered by companies are key elements of building corporate strategies and plans for future. Also here, vast amounts of data collected during the users' social network activities could provide corporations with valuable information. In such areas like **advertising and marketing** any indications regarding the users' moods, attitudes and opinions expressed directly or indirectly would be able to change policies and strategies of companies. The analysis would provide companies with indicators of their social presence, their ranking, and popularity among users.

One of the most interesting and intriguing aspects of analysis of social network data could be related to strategies and operations of **insurance and financial** companies. Information—in a form of opinions, facts or evaluations—that describe the users' behaviours, patterns and rules of actions would be of great interest to insurance companies. They could adopt and customize their policies and offer insurance packages to variety of customers trying to fit their specific needs. Financial companies could use social media data to improve returns on investment —analysis of the users' discussions and posts from the point of few of **sentiments**

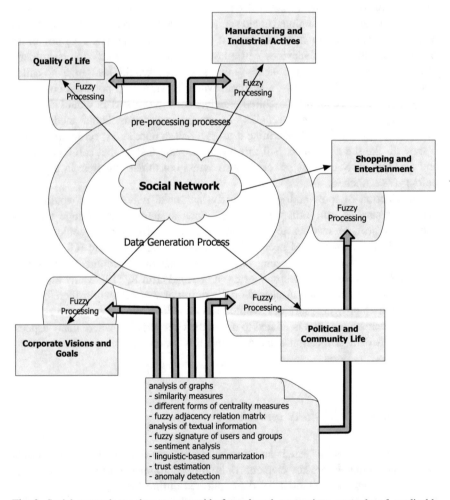

Fig. 2 Social network as data source with fuzzy-based processing: example of applicable approaches

and attitudes towards investments and associated with it expectations can lead to identification of trends, patterns and motivations of the users. All this would translate into **better strategies** and policies of investment polices of companies (Fig. 2).

3 Fuzzy-Based Data Processing

In the light of importance of analysing social data it is critical to use variety of methods and approaches that prepare and statically process data, as well as methodologies that are suitable to 'see' data in a more human-like way. The latter ones come under the name of soft computing methods. Among them fuzzy sets and systems are of special interest. They allow for treating data in a 'relaxed' form— they do not anticipate crisp boundaries between like and dislike, good and bad, adequate and impropriate—all that is very important from the perspective of analysis and understanding of human emotions, attitudes and behaviours related to variety of things and issues a single individual deal with every day. Some examples of application of fuzzy based techniques for processing of social network data are illustrated below.

One of the first steps in analysing social network is building an adjacency relation matrix based on relations between users. Those users are nodes in a social network, therefore, it is reasonable to assume that a centrality measure can be used to identify important features of the network, for example users that are most influential during a decision making process [14–16]. The measure of centrality of directed social fuzzy networks is a topic of investigations presented in [17]. A few new centrality measures have been proposed: fuzzy in-degree centrality, fuzzy out-degree centrality, fuzzy in-closeness centrality and fuzzy out-closeness centrality.

In general, a process of finding leaders and communities in a social network is a very interesting research topic in the network analysis. The works presented in [18–23] are examples of a fuzzy-based approach, fuzzy clustering-based to be exact, leading to detecting communities in the network. Another, linguistic based, approach to analysis of networks and detecting leaders is described in [24]. The pivotal point of the work is related to different measures describing nodes in graphs. Based on analysis of these measures the authors provide a linguistic-based definition of a leader. This definition describes many of the requirements and criteria represented with linguistic terms and expressions that a leader should satisfy. This work is an example of "Intelligent Social Network Analysis ISNA" introduced in [25].

Social networks are treated as a source of multiple types of information. Some of this information comes directly from the structure of networks and is associated with users and relations between them, while some is embedded in textual statements left by the users on networks. In such cases, methods and techniques of analysis of social networks should be equipped with the abilities to process this information—attributes, labels, text—and fuse it with "structure-based" information.

In [26] the authors describe a methodology for building a signature that represents the user's interests and opinions. This signature is built based on items as well as labels used by the users when expressing their opinions and thought about those items. Such signatures are constructed based on two fuzzy sets: fuzzy sets that represent items' attractiveness, and fuzzy sets representing popularity of different descriptions. The usefulness of the users' signatures is demonstrated in a process of evaluating similarity between users. The work also provides a description of a method for constructing signatures of groups of users. Firstly, the signatures of members of the group are built; secondly, these signatures are aggregated using ordered weighted aggregation [27] operator using different linguistic quantifiers. That allows for construct the group's description in a multiple different ways.

A few mentioned above examples of applications of fuzzy technologies to analysis of social networks focus on the users and relations between them. Their main objective is to find out as much as possible about the users and the roles they play in the network. Those example show how fuzzy methods can be used to detect different relations, identify communities and their leaders, find friends, and describe groups of users. There is also a body of work that looks at applications of social networks for 'higher-order' purposes. For example, a social network can be treated as a communication media for exchanging data and information between actors. Such applications of social network have a different sort of problems. Also here, there is a place for fuzzy technologies enabling identification of sources and sink of information, activities of users and types of important information—all that in a linguistic form without worrying about splitting everything into crisp sets.

In the context of advanced analysis of social network data, we would like to state that application of fuzzy-based approaches will enhance the capabilities of data analytical methods, will enable deeper and more semantic oriented analysis, and what is also essential will make the obtained outcomes more human-like.

4 Discussion and Conclusion

The above section provides just a few examples of areas that can benefit from processing, analysing and modelling social network data. At this stage we would like to foresee how fusion of big data methods with fuzzy technologies could contribute to analysis of social networks.

Most of the work dedicate to analysis of network data is targeting structural information of networks. The interconnection and relations between users have been the main source of information [15, 24, 28–31]. Some works [25, 26, 32–37] show an attempt to use additional information that brings different aspects of analysis. It is well known that each of the nodes/users as well as connections/relation is associated with supplementary information. It seems very reasonable that including that information in analysis procedure and methods would increase the scope of analysis and allow for looking and discovering new insights regarding the users and their behaviour.

A very important aspect of analysis of network data is to look into posts, comments, tweets, or any form of text generated by the users. They contain enormous amounts of information about users: what is being currently discussed, what are moods among people, what they like, what types of things invoke positive responses, what type of things are perceived as negative. It seems critical to be armed with the ability to find answers to these questions—this would result in detecting trends, popular and/or important topics, determine users' requirements and expectations regarding variety of items and events. In such a context, we would like to state that application of fuzzy-based approaches will enhance the capabilities of data analytical methods, will enable deeper and more semantic oriented analysis, and what is also essential will make the obtained outcomes more human-like. Fuzzy methods targeting **summarization** of texts [38], **sentiment analysis** [39, 40], **trust** inference and propagation [19, 41], and **event detection** [42] are just a few examples of research topics that constitute important issues that can be addressed with, and will benefit from application of fuzzy-based techniques and methods.

References

1. L. Freeman, *The Development of Social Network Analysis* (Empirical Press, Vancouver, 2006)
2. P.N. Krivitsky, M.S. Handcock, A.E. Raftery, P.D. Hoff, Representing degree distributions, clustering, and homophily in social networks with latent cluster random effects models. Soc. Netw. **31**, 204–213 (2009)
3. J. Scott, *Social Network Analysis*. A Handbook, London, Sage (2000)
4. T.A.B. Snijders, C. Baerveldt, A multilevel network study of the effects of delinquent behavior on friendship evolution. J. Math. Sociol. **27**, 123–151 (2003)
5. T.A.B. Snijders, Statistical models for social networks. Annu. Rev. Soc. (2011)
6. F. Vega-Redondo, Complex Social Networks (Cambridge University Press, 2007)
7. L.A. Zadeh, Fuzzy sets. Inf. Control **8**, 338–353 (1965)
8. G.J. Klir, B. Yuan, Fuzzy Sets and Fuzzy Logic: Theory and Applications (Pretience Hall, 1995)
9. W. Pedrycz, *Social Networks: A Framework of Computational Intelligence, Studies in Computational Intelligence*, ed. by S.-M. Chen, vol. 526 (Springer International Publishing Switzerland, 2014)
10. https://www.facebook.com. Accessed 4 May 2016
11. https://www.pinterest.com. Accessed 4 May 2016
12. https://twitter.com. Accessed 4 May 2016
13. http://www.gegridsolutions.com/demandopt/Catalog/GridIQ.htm. Accessed 4 May 2016
14. P. Bonacich, Power and centrality: a family of measures. Am. J. Sociol. **92**(5), 1170–1182 (1987)
15. S.P. Borgatti, Centrality and network flow. Soc. Netw. **27**(1), 55–71 (2005)
16. S.P. Borgatti, M.G. Everett, A graph-theoretic perspective on centrality. Soc. Netw. **28**(4), 466–484 (2006)

17. R.-J. Hu, Q. Li, G.-Y. Zhang, W.-C. Ma, Centrality measures in directed fuzzy social networks. Fuzzy Inf. Eng. **7**, 115–128 (2015)
18. T.C. Havens, J.C. Bezdek, C. Leckie, K. Ramamohanarao, M. Palaniswami, A soft modularity function for detecting fuzzy communities in social network. IEEE Trans. Fuzzy Sets **21**(6), 1170–1175 (2013)
19. S. Kim, S.Han, The method of inferring trust in web-based social network using fuzzy logic, in *Proceedings of the International Workshop on Machine Intelligence Research* (2009), pp. 140–144
20. J. Su, T.C. Havens, A generalized fuzzy t-norm formulation of fuzzy modularity for community detection in social networks, in *Advance Trends in Soft Computing WCSC 2013, Studies in Fuzziness and Soft Computing*, ed. by M. Jamshidi et al., vol. 312 (Springer International Publishing Switzerland, 2014), pp. 65–76
21. J. Su, T.C. Havens, Fuzzy community detection in social networks using a genetic algorithm, in *Proceedings of 2014 IEEE International Conference on Fuzzy Systems (FUZZ-IEEE)* (2014), pp. 2039–2046
22. J. Su, T.C. Havens, Quadratic program-based modularity maximization for fuzzy community detection in social networks. IEEE Trans. Fuzzy Syst. (in press)
23. S. Zhang, R. Wang, X. Zhang, Identification of overlapping community structure in complex networks using fuzzy c-means clustering. Phys. A Stat. Mech. Appl. **374**, 483–490 (2007)
24. M. Brunelli, M. Fedrizzi, M. Fedrizzi, Fuzzy m-ary adjacency relations in social network analysis: Optimization and consensus evaluation. Inf. Fusion **17**, 36–45 (2014)
25. R.R. Yager, Intelligent social network analysis using granular computing. Int. J. Intell. Syst. **23**, 1196–1219 (2008)
26. R.R. Yager, M.Z. Reformat, Looking for like-minded individuals in social networks using tagging and fuzzy sets. IEEE Trans. Fuzzy Sets **21**(4), 672–687 (2013)
27. R.R. Yager, On ordered weighted averaging aggregation operators in multi-criteria decision making. IEEE Trans. Syst. Man Cybern. **18**, 183–190 (1988)
28. J. Boyd, M. Everett, Relations, residuals, regular interiors, and relative regular equivalence. Soc. Netw. **21**(2), 147–165 (1999)
29. T. Casasús-Estellés, R.R. Yager, Fuzzy concepts in small worlds and the identification of leaders in social networks, in *IPMU 2014, Part II, CCIS* vol. 443 (Springer International Publishing Switzerland, 2014), pp. 37–45
30. R. Hannemanand, M. Riddle, *Introduction to Social Network Methods* (University of California, Riverside, 2005)
31. J. Liu, Fuzzy modularity and fuzzy community structure in networks. Eur. Phys. J. B **77**(4), 547–557 (2010)
32. Y. Cao, J. Cao, M. Li, Distributed data distribution mechanism in social network based on fuzzy clustering, in *Foundations and Applications of Intelligent Systems, Advances in Intelligent Systems and Computing*, vol. 213, ed. by F. Sun et al. (Springer, Berlin, Heidelberg, 2014), pp. 603–620
33. S. Elkosantini, D. Gien, A dynamic model for the behavior of an operator in a company, in *Proceedings of the 12th IFAC Symposium on Information Control Problems in Manufacturing, France*, vol. 2 (2006), pp. 187–192
34. S. Elkosantini, D. Gien, Human behavior and social network simulation: fuzzy sets/logic and agents-based approach, in *Proceedings of the 2007 Spring Simulation Multi-conference SpringSim '07*, vol. 1 (2007), pp. 102–109
35. M.J. Lanham, G.P. Morgan, K.M. Carley, Social network modeling and agent-based simulation in support of crisis de-escalation. IEEE Trans. Syst. Man Cybern. **44**(1), 103–110 (2014)
36. M.Z. Reformat, R.R. Yager, Using tagging in social networks to find groups of compatible users, in *Proceedings of Join IFSA World Congress and NAFIPS Annual Meeting (IFSA/NAFIPS), Edmonton, Canada*, June 24–28, 2013, pp. 697–702

37. G. Stakias, M. Psoras, M. Glykas, Fuzzy cognitive maps in social and business network analysis, in *Business Process Management, SCI*, ed. by M. Glykas, vol. 444 (Springer, Berlin, Heidelberg, 2013), pp. 241–279
38. X.H. Liu, Y.T. Li, F.R. Wei, M. Zhou, Graph-based multi-tweet summarization using social signals, in *Proceedings of COLING 2012* (2012), pp. 1699–1714
39. D.N. Trung, J.J. Jung, L.A. Vu, A. Kiss, Towards modeling fuzzy propagation for sentiment analysis in online social networks: a case study on TweetScope, in *Proceedings of 4th IEEE International Conference on Cognitive Info-communications* (2013), pp. 331–337
40. D.N. Trung, J.J. Jung, Sentiment analysis based on fuzzy propagation in online social networks: a case study on TweetScop. Comput. Sci. Inf. Syst. **11**(1), 215–228 (2014)
41. F. Hao, G. Min, M. Lin, C. Luo, L.T. Yang, IEEE mobi fuzzy trust: an efficient fuzzy trust inference mechanism in mobile social networks. IEEE Trans. Parallel Distrib. Syst. **25**(11), 2944–2955 (2014)
42. T. Matuszka, Z. Vincellér, S. Laki, On a keyword-lifecycle model for real-time event detection in social network data, in *Proceedings of 4th IEEE International Conference on Cognitive Info-communications* (2013), pp. 453–458

Personalization and Optimization of Information Retrieval: Adaptive Semantic Layer Approach

Alexander Ryjov

Abstract This work describes the idea of an adaptive semantic layer for large-scale databases, allowing to effectively handling a large amount of information. This effect is reached by providing an opportunity to search information on the basis of generalized concepts, or in other words, linguistic descriptions. These concepts are formulated by the user in natural language, and modelled by fuzzy sets, defined on the universum of the significances of the attributes of the database. After adjustment of user's concepts based on search results, we have "personalized semantics" for all terms which particular person uses for communications with database (for example, "young person" will be different for teenager and for old person; "good restaurant" will be different for people with different income, age, etc.).

1 Introduction

Why people spend time, efforts, money, etc. for collecting the data? Why Big Data, Internet of Things are the most disruptive technologies for modern society and economics [1]? We can find explanations in a number of analytical reports and white papers—for example [2–5]. In general sense, the summary is: it is because the data is a model of real world. Using the data, we can have more adequate picture of the interesting for us part of a real world, make more well-founded and quick decisions, and make a number of important things by more effective and efficient manner (see cases in the references above).

So, database is an information model of the real world (Fig. 1). All important from user's point of view objects from a real world are presented in database as an "information image". We describe the real objects, search their images in database, and use the results for operation in a real world.

A. Ryjov (✉)
Department of Mechanics and Mathematics,
Lomonosov Moscow State University, 119899 Moscow, Russia
e-mail: alexander.ryjov@gmail.com

© Springer International Publishing AG, part of Springer Nature 2018
L. A. Zadeh et al. (eds.), *Recent Developments and the New Direction in Soft-Computing Foundations and Applications*, Studies in Fuzziness and Soft Computing 361, https://doi.org/10.1007/978-3-319-75408-6_2

Fig. 1 Database as an
information model of real
world

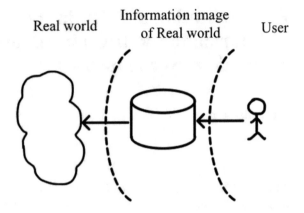

The present data processing technologies only allows us to search information using concepts (words, symbols, figures), which are present in the data base descriptions of objects. This leads to difficulties in those situations where the information we need is not expressed unequivocally in the language of significances of attributes of object descriptions. The translation of such queries towards the latter search language tends to deform their meaning, and, hence, to reduce the efficiency and the quality of using these data bases.

One of the important properties of the information we need for solving different tasks, and which distinguishes it from the information in the database, is the fuzziness of the concepts of the user. The user, like any human being, thinks in qualitative categories [6–8], whereas the database information is basically clear (sharp, non-fuzzy). This is of one of the main problems in the "translation" of user's needs of information towards the database query.

We can illustrate this with the following example.

Example 1 Choosing a car.

We consider the situation of a customer choosing a car from a database (electronic catalogue), which contains the following information:

- Price
- Model
- Year of issue
- Fuel consumption

For the formulation of a request in such a database, the user is forced to present his query in the language of concrete, definite numbers and models. If our user has in mind a very specific car, for which all the above-mentioned attributes have definite values, and if he has decided that other cars (similar, or for instance a little more economical) are not suitable, the standard technology of databases solves all his problems. But, he wants to buy a car, which is "not very expensive", "prestigious", "economical", and "not old", the formulation of such a query to the database is not a simple task.

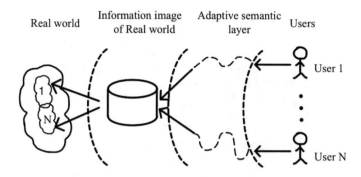

Fig. 2 Adaptive semantic layer

A similar example can be given for the task of choosing an apartment or other real estate property using a database, containing the description of particular flats in a city. The "correct" processing of queries of the type "comfortable, not very expensive flat, close to a park, and not very far from an underground station" depends on the parameters of the user—his/her age, income, habits, and other personal features.

In this case Fig. 1 could be transformed to Fig. 2—different users/classes of users can have different (similar, but not equivalent) conception of correct presentation of data from database. We need in additional layer for implementation this feature.

Applications of this approach for political problems (for example, monitoring and evaluation of State's nuclear activities [9], department of safeguards, International Atomic Energy Agency) have been described in [10].

In general, it can be stated that the problem described above is very important when using automated (electronic) catalogues of goods and services, i.e., data bases, containing particular information about particular objects of the same general type. Especially this way could be effective for several tasks in big data analysis—the next frontier for innovation, competition, and productivity which was introduced by McKinsey Global Institute in May 2011 [2] and was supported by leading companies [3–5].

2 The Concept of an Adaptive Semantic Layer

The structure of an adaptive semantic layer is shown in Fig. 3. The idea of the adaptive semantic layer is to provide by user an interface, which allows:

- define user's concepts;
- search an information by this concepts;
- adjustment of user's concepts based on search results.

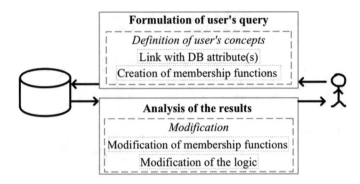

Fig. 3 The structure of an adaptive semantic layer

2.1   Definition of the User's Concepts

The work of user begins with a linguistic description of the objects he wants to find in the data base. If the system does not recognize or know this linguistic description, control is transmitted to the program block for the construction of membership functions[1]. If, on the other hand, the system recognizes the description, it will retrieve the membership function, associated with it. The user can then, in case of disagreement, edit the membership function, and a new membership function will now be associated with this user, reflecting his "view" (interpretation) of the description. The membership function editor is based on the principle of cognitive graphics and does not require a specialized knowledge of computers.

We can mark out two situations: metric ($U \subseteq R^I$) and non-metric ($U \not\subseteq R^I$) universum. For both cases we have well-defined methods which were tested in a number of applications of fuzzy models. We can provide the following continuation of example 1:

Example 2

(a) Metric universum: $U = $ [\$10000, \$50000]; user's concept $A = $ "not-very-expensive car" is presented in Fig. 4. We can also use fuzzy clustering methods (for instance, Fuzzy C-Means) for building membership functions.
(b) Non-metric universum: U is a set of all cars' models from the database; user's concept $B = $ "sport cars for city" is presented in Fig. 5. Using the same way, we can define "cars for hunting", "cars for farmers", "cars for young girls", etc.

Here in right column we have all the cars; green box is collection of cars definitely belongs to user's concept; red box is collection of cars definitely not belongs to user's concept; yellow box is a set of cars which partially belongs to user's concept. We start from empty boxes (all models are in the right column) and

---

[1]Following [6], we associate semantics of the terms (words) with membership functions.

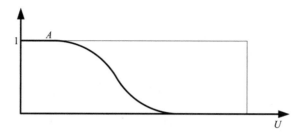

Fig. 4 Semantic of user's concept A

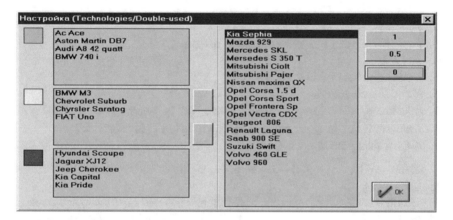

Fig. 5 Semantic of user's concept B

split all the models to these three boxes. We can order elements from yellow box according to belongings to user's concept and define membership function as linear one.

After the formulation of the linguistic description, and the association of the membership functions with this description, an information search in the database can be effected.

2.2   The Information Retrieval Algorithm

The information retrieval algorithm consists in calculating for each record in data base the degree of satisfaction to the formulated request: from 1 (total satisfaction) to 0 (total non-satisfaction). The result of the search is an ordering of the records in the data base on the basis of the degree of satisfaction to the request.

Notations:

- $i(1 \leq i \leq N)$—index of database attributes;
- U_i—domain of attribute i;
- $A^i = \{A_1^i, \ldots, A_{ni}^i\}$—user's concepts defined on i—attribute ($ni \geq 0$);
- $a_{nk}^i(u) = \mu_{A_{nk}^i}(u_i)$—membership function of nk—concept of i—attribute (nk ni, $u_i \in U_i$);
- $Q = \langle A_{k1}^1 \circ A_{k2}^2 \circ \cdots \circ A_{kN}^N \rangle$, where $ki \leq ni$, $A_{ki}^i \in A^i \cup \emptyset (1 \leq i \leq N)$; $\circ \in \{$and, or, not$\}$—is users concept's based database query.

Algorithm:

1. $r = 0$ (r—index of records in database; $r \leq R$);
2. $r = r + 1$;
3. $i = 0$ (i—index of record's attribute in database);
4. $i = i + 1$;
5. if $A^i(.) \in Q$ then calculate $a_{nk}^i(u_i)$;
6. if $i < N$ then goto 4;
7. calculate $\mu_Q(r) = a_{k1}^1 \bullet a_{k2}^2 \bullet \cdots \bullet a_{kN}^N$, where "$\bullet$" is: t-norm if "$\circ$" in Q is "and"; t-conorm if "$\circ$" in Q is "or"; $1 - a_{nk}^i(u_i)$ if "$\circ$" in Q is "not";
8. if $r < R$ then goto 2.

Result: $\mu_Q(1), \ldots, \mu_Q(R)$—degree of belonging of each records from database to user's query.

As different classes of users can have different membership functions, the results of the search for the same query can be different for different users, or classes of users (Fig. 2). This allows us to have different "views" on the same data base.

2.3 Adjustment of User's Concepts Based on Search Results

It is obvious enough that different users (classes of users) can have different formalization of the concepts (different membership functions). For example, concept "expensive" for student and for businessman can be different. How can we make our interface "personalized"?

In general terms, if we allow using uncertainty at the point of "entry" of the system, we have to provide tools for manipulation of uncertainty at the "output".

We can propose two ways to adjust or tune the interface. First way is adjustment of membership function, second one is tuning of t-notms and t-conorms. The following example can explain this idea.

Fig. 6 Adjustment of the semantic of user's concept A

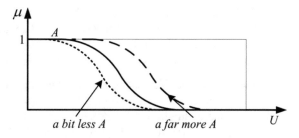

Example 3

(a) Adjustment of the membership functions is shown in Fig. 6.

Here *more*, *less*—directions of modification; *a bit*, *not so far*, *a far*, ...—volume ("power") of the modificators.

This approach is described in [11].

(b) Adjustment of the logic (*t*-norms and *t*-conorms).

We can use parametric representation of a family of *t*-norms and *t*-conorms like

$$T_\lambda(a,b) = \frac{\mu_a \times \mu_b}{\lambda + (1-\lambda)(\mu_a + \mu_b - \mu_a \times \mu_b)}$$

and use genetic algorithms for choosing the best value of λ.

This approach is described in detail in [12].

3   Optimization of Semantic Layer

It is assumed that the person describes the properties of real objects in the form of linguistic values. The subjective degree of convenience of such a description depends on the selection and the composition of such linguistic values. Let us explain this on a model example.

Example 4 Let it be required to evaluate the height of a man. Let us consider two extreme situations.

Situation 1. It is permitted to use only two values: "small" and "high".
Situation 2. It is permitted to use many values: "very small", "not very high", ..., "not small and not high", ..., "very high".

Situation 1 is inconvenient. In fact, for many men both the permitted values may be unsuitable and, in describing them, we select between two "bad" values.
Situation 2 is also inconvenient. In fact, in describing height of men, several of the permitted values may be suitable. We again experience a problem but now due to the fact that we are forced to select between two or more "good" values. Could a set of linguistic values be optimal in this case?

Different persons may describe one object for database. Therefore, it is desirable to have assurances that the different sources of information for the database describe one and the same object in the most "uniform" way.

On the basis of the above we may formulate the first problem as follows:

Problem 1 Is it possible, taking into account certain features of the man's perception of objects of the real world and their description, to formulate a rule for selection of the optimum set of values of characteristics on the basis of which these objects may be described? Two optimality criteria are possible:

Criterion 1. We regard as optimum those sets of values through whose use man experiences the minimum uncertainty in describing objects.
Criterion 2. If the object is described by a certain number of users, then we regard as optimum those sets of values which provide the minimum degree of divergence of the descriptions.

This problem is studied in [13, 14]. It is shown that we can formulate a method of selecting the optimum set of values of qualitative indications. Moreover, it is shown that such a method is robust, i.e. the natural small errors that may occur in constructing the membership functions do not have a significant influence on the selection of the optimum set of values. The sets which are optimal according to criteria 1 and 2 coincide.

What gives us the optimal set of values of qualitative attributes for information retrieval in a database? In this connection the following problem arises.

Problem 2 Is it possible to define the indices of quality of information retrieval in fuzzy (linguistic) databases and to formulate a rule for the selection of such a set of linguistic values, use of which would provide the maximum indices of quality of information retrieval?

This problem is studied in [11, 15]. It is shown that it is possible to introduce indices of the quality of information retrieval in fuzzy (linguistic) databases and to formalize them. It is shown that it is possible to formulate a method of selecting the optimum set of values of qualitative indications which provides the maximum quality indices of information retrieval. Moreover, it is shown that such a method is also robust.

4   Conclusion Remarks

Databases are essential part of information environment we have in modern life. We use databases for solving a number of day-to-day tasks: buying food, clothes, tickets, etc. in e-shops, planning trips, communicate with friend using social networks, and many others.

Users of databases are different, because they are human beings. It means that we have to take into consideration human's perception of objects of the real world and manner of their description for databases.

Here we have focused on two important from our point of view issues:

- How we can make databases more personal, i.e. different for different users?
- How we can make databases more optimal, i.e. more adequate as a model of a real world we would like to operate in?

This article describes only general properties of database as a model of real world. I do hope that these ideas and results will allow building a maximal comfortable information environment for the participants.

References

1. *A Gallery of Disruptive Technologies* (McKinsey Global Institute), http://www.mckinsey.com/assets/dotcom/mgi/slideshows/disruptive_tech/index.html
2. *Big Data: The Next Frontier for Innovation, Competition, and Productivity* (McKinsey Global Institute, 2011), http://www.mckinsey.com/insights/mgi/research/technology_and_innovation/big_data_the_next_frontier_for_innovation
3. *IBM. Bringing Big Data to the enterprise,* http://www-01.ibm.com/software/data/bigdata/
4. *Microsoft   Big   Data,* http://www.microsoft.com/sqlserver/en/us/solutions-technologies/business-intelligence/big-data.aspx
5. *Oracle and Big Data. Big Data for the enterprise,* http://www.oracle.com/us/technologies/big-data/index.html
6. L.A. Zadeh, The concept of a linguistic variable and its application to approximate reasoning. Part 1, 2, 3. Inform. Sci. **8**, 199–249; 8,301–8,357, 943–980 (1975)
7. L.A. Zadeh, *Computing with Words—Principal Concepts and Ideas.* Studies in Fuzziness and Soft Computing, vol. 277 (Springer, 2012), ISBN 978-3-642-27472-5
8. L.A. Zadeh, From computing with numbers to computing with words—from manipulation of measurements to manipulation of perceptions. Int. J. Appl. Math. Comput. Sci. **12**(3), 307–324 (2002)
9. A. Ryjov, A. Belenki, R. Hooper, V. Pouchkarev, A. Fattah, L.A. Zadeh, *Development of an Intelligent System for Monitoring and Evaluation of Peaceful Nuclear Activities (DISNA),* IAEA, STR-310, Vienna (1998), 122 p.
10. A. Ryjov, On application of a linguistic modeling approach in information collection for future evaluation, in *Book of Extended Synopses, International Seminar on Integrated Information Systems*, Vienna, Austria, IAEA—SR April 2000, vol. 212, pp. 30–34
11. A. Ryjov, Models of information retrieval in fuzzy environment, in *Publishing House of Center of Applied Research, Department of Mechanics and Mathematics* (Moscow University Publishing, Moscow, 2004), 96 p.
12. A. Ryjov, M. Feodorova, Genetic algorithms in selection of adequate aggregation operators for information monitoring systems, in *Proceedings of the V Russian Conference "Neurocomputers and Its Applications"*, Moscow, Feb. 1999, pp. 267–270
13. A. Ryjov, Fuzzy linguistic scales: definition. properties and applications, in *Soft Computing in Measurement and Information Acquisition*, ed. by L. Reznik, V. Kreinovich (Springer, 2003), pp. 23–38

14. A. Ryjov, *The Principles of Fuzzy Set Theory and Measurement of Fuzziness* (Dialog-MSU, Moscow, 1998), 116 p.
15. A. Ryjov, Modeling and optimization of information retrieval for perception-based information, in *Brain Informatics. International Conference, BI 2012, Proceedings*, ed. by F. Zanzotto, S. Tsumoto, N. Taatgen, Y.Y Yao, Dec. 2012, http://link.springer.com/chapter/ 10.1007/978-3-642-35139-6_14

Frequent Itemset Mining for a Combination of Certain and Uncertain Databases

Samar Wazir, Tanvir Ahmad and M. M. Sufyan Beg

Abstract Modern industries and business firms are widely using data mining applications in which the problem of Frequent Itemset Mining (FIM) has a major role. FIM problem can be solved by standard traditional algorithms like Apriori in certain transactional database and can also be solved by different exact (UApriori, UFP Growth) and approximate (Poisson Distribution based UApriori, Normal Distribution based UApriori) probabilistic frequent itemset mining algorithm in uncertain transactional database (database in which each item has its existential probability). In our algorithm it is considered that database is distributed among different locations of globe in which one location has certain transactional database, we call this location as *main site* and all other locations have uncertain transactional databases, we call these locations as *remote sites*. To the best of our knowledge no algorithm is developed yet which can calculate frequent itemsets on the combination of certain and uncertain transactional database. We introduced a novel approach for finding itemsets which are globally frequent among the combination of all uncertain transactional databases on remote site with certain database at main site.

Keywords Data mining · Uncertain transactional database · KDD
Probabilistic data · Expected support · Frequent itemset · Probabilistic frequent itemset

S. Wazir (✉) · T. Ahmad
Department of Computer Engineering, Jamia Millia Islamia, New Delhi, India
e-mail: samar.wazir786@gmail.com

T. Ahmad
e-mail: tahmad2@jmi.ac.in

M. M. Sufyan Beg
Department of Computer Engineering, Aligarh Muslim University, Aligarh, India
e-mail: mmsbeg@eecs.berkeley.edu

© Springer International Publishing AG, part of Springer Nature 2018
L. A. Zadeh et al. (eds.), *Recent Developments and the New Direction in Soft-Computing Foundations and Applications*, Studies in Fuzziness and Soft Computing 361, https://doi.org/10.1007/978-3-319-75408-6_3

1 Introduction

The size of data is growing exponentially day by day and we cannot stop this but we can develop the strategies to capture this data and develop tool to analyze it. These tools give us the results in exact or predictable format. This process is well known as KDD (Knowledge Discovery in Databases) or in general term Data Mining. Data Mining used in various application like marketing, airlines, surveillance, medical science, pattern recognition.

In data mining, Frequent Itemset Mining problem is one of the most discussed issue in which a database called transactional database is used. This database consists of some transactions, each transaction consists of some items and group of items is called itemset. So if count of an item or itemset in transactional database is greater than a user specified threshold (i.e. Minimum Support) than that itemset is called Frequent Itemset and this problem is called Frequent Itemset Mining (FIM) problem [1].

In FIM two types of databases are used:

- First is certain transactional database in which each transaction has some items that are associated with each other. For example transaction T1 *{Bread, Butter, Milk}* means that the customer who purchased *Bread* also bought *Butter* and *Milk*. All such type of transactions capture doubtless presence of an item in a transaction.
- Second is Uncertain Transactional Database in which each item in a transaction has its existential probability associated with it. In many applications, the presence of an item in a transaction is doubtful and is best captured by its probability of existence or likelihood measure. For example T2 *{Bread: 0.23, Butter: 0.45, Milk: 0.78}* means that there is 23% chance that a customer will purchase *Bread*. *Bread, Butter* and *Milk* present in same transaction T2, so they are associated with each other. Therefore if customer will purchase *Bread* then there is 45% chance that it will purchase *Butter* and 78% chance that it will purchase *Milk*. This uncertain purchase behavior of customer for each item in a transaction is also called Attribute Uncertainty [2, 3].

In case of certain transactional database, for computing frequent itemsets two categories of algorithms are used called sequential and parallel algorithms. Agrawal et al. proposed Apriori which is the best example of sequential algorithm for FIM [1]. After that several researcher proposed different sequential and parallel algorithms for FIM, e.g. two parallel formulations of the *Apriori* algorithm were proposed in [4], *Count Distribution (CD)* and *Data Distribution (DD)*. To improve the performance of Data Distribution, Intelligent Data Distribution was proposed in [5–7]. After this Hybrid Distribution was developed by combining the advantages of both CD and DD [5]. Two more parallel algorithms were subsequently proposed to improve the performance of previous work. These algorithms was Fast Distributed Mining (FDM) [8] and Fast Parallel Mining (FPM) [9] Algorithms.

In case of Uncertain Transactional Database the first algorithm of FIM was based on expected support, which follow the Apriori sequential pattern of FIM and given by Chui et al. called UApriori [10]. After UApriori several algorithms were proposed for FIM over Uncertain Transactional Database [2, 3, 10–17]. These algorithms cab be divided into two different categories that is Expected Support based FIM algorithms (UFP-Growth, UH-Mine) and Probabilistic FIM algorithms (Dynamic Programming based algorithms, Divide and Conquer based algorithm, PDU Apriori, NPDU Apriori) [16].

This paper proposed a model in which the combination of Apriori, UApriori and CD is used to find the Global Frequent Itemset on the combination of certain and uncertain transactional databases. In sequential or parallel association mining, for computing frequent itemsets either certain transactional database is used or uncertain transactional database is used. To the best of our knowledge this is the first approach in which frequent itemsets are discovered on the combination of certain and uncertain transactional databases. For example, an "Accidents" dataset is provided by Frequent Itemset Mining [18] dataset repository. National Institute of Statistics (NIS) collects this dataset from Belgium region in 1991–2000. "Accidents" dataset is a certain transactional database and computing frequent itemset on this database gives us an idea about the most favorable condition for an accident and how a preventive action can be taken. Later in this dataset some uncertainty issues also considered like chances of accidents in case of rain, heavy fog and huge traffic on Christmas holidays. Therefore "Accidents" dataset is modified and existential probability of each item is added in the dataset hence it become "Accidents" uncertain transactional database. This dataset is used by Tong et al. [16] for computing frequent itemset over "Accidents" uncertain transactional database. Let assume that we want to calculate global frequent itemset for more than one country, in which some countries have rainy and cold weather and some have dry and hot weather. Now we need a mechanism for combining certain and uncertain database collected from different countries. After combining the databases we compute items which are globally frequent, the location at which we calculate global frequent items is called main site and all other locations are remote sites.

In this case we may have the following solutions:

- First solution of the problem is to bring all databases at one location and then use FIM algorithm like Apriori to calculate frequent items. This method create communication overhead between main site and remote sites and also not time efficient.
- Second solution is we can compute frequent items at main site by Apriori and at remote sites by UApriori and then bring the results at main site to calculate global frequent itemsets.
- We proposed a more effective solution in which we compute frequent items at remote sites by UApriori and send these itemsets to main site. In main site we do not compute frequent itemsets of certain database in advance, like we did in second solution. As shown in Fig. 1, We first calculate size-1 frequent items

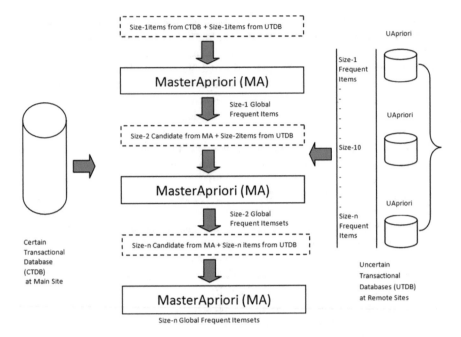

Fig. 1 MasterApriori

then combine these items to the same size-1 items we got from the remote sites for calculating size-1 global frequent items. These size-1 global frequent items are to be sent to the next stage for generating size-2 candidate itemsets and after this we perform pruning operation. We repeat this procedure until we got all global frequent itemsets. We call this algorithm as MasterApriori. MasterApriori decreases the execution time dramatically.

2   Related Work

Mining frequent items in certain and uncertain databases and discovering association rules is often regarded as an important research topic among researchers. Many efficient algorithms have been proposed for sequential and parallel Frequent Itemset Mining. Apriori is well known algorithm for sequential frequent itemset mining proposed by [1] in certain transactional database. Similarly for uncertain database we use UApriori by [10] and for parallel frequent itemset mining we use count distribution proposed by [4, 19]. Here we discuss these three algorithms than we focus on our algorithm for finding global frequent itemset.

2.1   Apriori Algorithm [1]

Apriori uses multiple passes for computing frequent itemset. In this algorithm we first calculate all Size-1 items then calculate their counts. If the count is greater than user defined threshold called Minimum Support than that item is called Frequent Item or Large Item. In next pass we combine all these Size-1 large items in form of Size-2 Itemset and repeat the same procedure. After joining and before counting the occurrence of itemset in database we perform one more operation called pruning. Pruning use the concept that if an itemset is frequent than all its subset must also be frequent. After pruning we check itemset count with minimum support for calculating frequent itemset. The structure of Apriori algorithm is given below.

Apriori Algorithm

1. Input (certain database D, minimum support minSup)
2. L1 = {large-1 itemset};
3. for (k=2; $L_{K-1} \neq \emptyset$; k++) do begin
       C_K = **apriori-gen (L_K-1)** // New Candidates
       for all transactions t ϵ D do begin
           C_t = subset (C_K, t);
           for all candidates c ϵ C_t do
               c.count++;
     End
     L_K = {c ϵ C_K | c.count >= minSup}
     End
   Answer = $U_K L_K$

2.2   UApriori Algorithm [10]

This algorithm extends the Apriori algorithm on Uncertain Transactional Database (UTDB) and this is the first FIM algorithm on Uncertain Transactional Database based on Expected Support.

Let T = $\{t_1, t_2, ..., t_i\}$ be a set of transactions in UTDB and X = $\{x_1, x_2, ..., x_j\}$ be a set of distinct items in UTDB. Let |D| is the no of transactions and |X| is the no of items then according to UApriori the expected support of an item or itemset X can be calculated as

$$Expected\,Support(X) = \sum_{i=1}^{|D|} \prod_{j=1}^{|X|} P_{t_i}(x_j)$$

where $P_{t_i}(x_j)$ is the existential probability of item x_j in transaction t_i

Table 1 Uncertain transactional database

T_1	I_1 (0.5)	I_2 (0.7)
T_2	I_1 (0.3)	I_2 (0.1)

For example in Table 1 UTDB is given

So in this case

Expected Support $(I_1) = 0.5 + 0.3 = 0.8$

Expected Support $(I_1, I_2) = 0.5 \times 0.7 + 0.3 \times 0.1 = 0.38$

In Sect. 3.1 we discuss UApriori in detail.

2.3 Count Distribution [4]

In case of certain transaction database count distribution is used to compute frequent itemset in parallel when database is distributed among different nodes e.g. in case of Fig. 2, if our database is distributed among 3 processors. Then we calculate the count of each itemset separately on each processor.

After that using a Global Reduction Operation we compute the sum of these individual counts. The database is distributed among P processors so each processor handles N/P transactions. However this algorithm can only compute the count of itemset in parallel. It cannot parallelize the computation of building candidate itemsets.

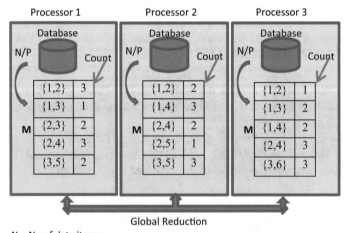

N = No of data items

M = Size of candidate set

P = No of processors/Nodes

Fig. 2 Count distribution (CD) algorithm

3 Problem Definition and Paper Organization

The problem of computing Global Frequent Items over the combination of certain and uncertain transactional database can be decomposed into two subproblems:

1. Find all itemsets which have expected support equal or greater than the lowest probability (P_L) at remote site (P_L is to be discussed later in Sect. 3.1 of this part). The expected support of an itemset can be calculated by adding the existential probabilities of that itemset in all transaction. In Sect. 3.1 we use UApriori algorithm for solving this problem and use lowest probability (P_L) as minimum support.
2. In Sect. 3.2 we use the frequent items from remote sites and combine them with items of main site and check their total expected support with global minimum support to calculate Global Frequent Itemset for all sites. We use MasterApriori algorithm to solve this problem.

3.1 Discovering Frequent Itemset at Remote Sites

For discovering frequent itemset at remote sites, UApriori algorithm is used [10, 15–17]. For calculating expected support of an itemset we first multiply the existential probabilities of each item of itemset in each transaction than we add the existential probabilities of all transactions for getting the expected support of an itemset (*Like in* Sect. 2.2). We know that multiplying probabilities results in lower values so if e.g. size of an itemset is 10 or it have 10 items. In this case we are multiplying 10 values which are lower than one. So we can conclude that when itemset size become large than the expected support become negligible. Here we call this value as lowest probability represented by P_L. The UApriori algorithm we use for calculating frequent itemset at remote site is given below:

UApriori	**//Executes at remote sites**
1. Input (UDB, P_L)	
2. L1 = {large-1 itemset};	
for all items in UDB where	
$P_e >= P_L$	
3. for (k=2; $L_{K-1} \neq \emptyset$; k++) do begin	
C_K = **UApriori-gen (L_K-1)** // New Candidates	
L_K = {c ϵ C_K	c.expectedSupport >= P_L}	
End	
Answer = $\cup_K L_K$	

In the above algorithm P_e is Existential Probability L_K is large K frequent itemset and C_K is size k candidate itemset.

UApriori-gen () function perform join and prune operation for generating size k candidate itemset by taking $k - 1$ large frequent itemset as input. For example let's consider the uncertain transactional database given in Table 2 [20].

We choose P_L as 0.5, ES is the Expected Support. So by UApriori we get Size-1, 2, 3 frequent itemset based on their expected support ES greater than lowest probability P_L. The procedure is given in Tables 3, 4, and 5.

Table 2 Uncertain transactional database (UTDB)

TID	Items			
0	1(0.5)	2(0.4)	4(0.3)	5(0.7)
1	2(0.5)	3(0.4)	5(0.4)	
2	1(0.6)	2(0.5)	4(0.1)	5(0.5)
3	1(0.7)	2(0.4)	3(0.3)	5(0.9)

Table 3 Calculating Size-1 frequent items

C_1	ES_1	$L_1(ES_1 >= P_L)$
1	$0.5 + 0.6 + 0.7 = 1.8$	1
2	1.8	2
3	0.7	3
4	0.4	
5	2.5	5

Table 4 Calculating Size-2 expected frequent itemsets

C_2	ES_2	$L_2(ES_2 >= P_L)$
1, 2	$0.5 \times 0.4 + 0.6 \times 0.5 +$ $0.7 \times 0.4 = 0.78$	1, 2
1, 3	0.21	
1, 5	1.28	1, 5
2, 3	0.32	
2, 5	1.09	2, 5
3, 5	0.43	

Table 5 Calculating Size-3 expected frequent itemsets

C_3	ES_3	$L_3(ES_3 >= P_L)$
1, 2, 5	0.542	1, 2, 5

3.2 Calculating Global Frequent Itemset at Main Site

At main site we use MasterApriori as follows

MasterApriori   **//Executed at main site**

---

1. Input(CTDB, UTDB, minSup)
2. Generate size-1 items from CTDB and prepare (itemset, mean) pair   // e.g. ABCD(itemset) 0.16(mean)
3. Generate L_1 = {Large -1 itemset};
   for (i=0; i<size_of_CTDB || i<size_of_UTDB) do begin
            if item.CTDB is present in UTDB
            than Item.CTDB.mean += item.UTDB.ES
            if item.UTDB is not present in CTDB
            add item.UTDB in CTDB
       end
       L_1 = { l ∈ L_K | l.mean >= minSup}
       Add to L_1
4. for (k=2; L_{K-1} ≠ ø ; k++) do begin
            C_K = **apriori-gen (L_{K-1})** // New Candidates
            for all transactions t ∈ CTDB do begin
               C_t = subset(C_K, t) ; //candidates contained t
               for all candidates c ∈ C_t do
                  c.count++;
               C_t.mean = (C_t.count ÷ NoOfTrans _in_CTDB)+
                            C_t.mean_in_UTDB
           end
           L_K = {c ∈ C_K | c.mean >= minSup}
       end
       Answer = $U_K L_K$ // frequent itemsets

In the given algorithm we use the following abbreviation.
CTDB Certain Transactional Database (Table 6)
UTDB Uncertain Transactional Database (Table 2)
minSup global minimum support
mean = item_count/No_Of_Transactions_in_CTDB;
apriori-gen(L_{K-1}) apriori join and pruning stage

Table 6 Certain transactional database (CTDB)

TID	Items				
0	1	2	3	4	5
1	1	3	4	5	6
2	1	3	4	6	7
3	1	4	5	6	7
4	1	4	6	7	8

At main site we get the expected frequent itemsets from the remote sites. Here we generate size-1 items and their counts from certain transactional database. These size-1 items counts is divided by no of transaction in CTDB for getting the mean of each size-1 items. This mean is added with the expected support of same item of UTDB. Now we check this total mean with global minimum support to get Global Size-1 Frequent Itemset. In the next stage we generate candidate itemsets by apriori join and prune method. Again we add the mean of the candidates at CTDB with the expected support of same candidates at UTDB and check the result with global minimum support to get Global Frequent Itemset. We repeat this procedure until we get all Global Frequent Itemsets at main site. To understand the complete procedure let's take the following example. In Table 6 we use the certain database at main site. We choose global minimum support (minSup) 0.5. In each stage of candidate itemset generation we check the total expected support (local + remote) of an itemset with minSup for getting Global Frequent Itemsets for all sites. For example let itemset {1, 2, 5}, so first step is to check count of {1, 2, 5} in CTDB which is 1 and no of transactions in CTDB is 5 so mean for {1, 2, 5} is 1/5 = 0.2. Now expected support of {1, 2, 5} in UTDB can be calculated as:

$$\text{Expected Support}(1, 2, 5) = 0.5 \times 0.4 \times 0.7 + 0.6 \times 0.5 \times 0.5 + 0.7 \times 0.4 \times 0.9 = 0.542$$

Tables 7, 8, 9, and 10 shows the generation of Global Frequent Itemsets.

4   Experiments and Result Analysis

For performance study and checking the efficiency of the proposed MasterApriori algorithm, experiments are divided in two steps.

Table 7 Calculating Size-1 global frequent items

Size-1 CTDB items	Size-1 CTDB items count	Size-1 CTDB items mean	Size-1 UTDB items mean	Size-1 total mean_1	$L_1(\text{mean}_1 >= \text{minSup})$
1	5	1.00	1.8	2.8	1
2	1	0.20	1.8	2.00	2
3	3	0.60	0.7	1.30	3
4	5	1.00		1.00	4
5	3	0.60	2.5	3.1	5
6	4	0.80		0.80	6
7	3	0.60		0.60	7
8	1	0.20		0.20	

Table 8 Calculating Size-2 global frequent items

Size-2 CTDB items	Size-2 CTDB items count	Size-2 CTDB items mean	Size-2 UTDB items mean	Size-2 total mean$_2$	$L_2(\text{mean}_2 \geq \text{minSup})$
1, 2	1	0.20	0.78	0.98	1, 2
1, 3	3	0.60		0.60	1, 3
1, 4	5	1.00		1.00	1, 4
1, 5	3	0.60	1.28	1.88	1, 5
1, 6	4	0.80		0.80	1, 6
1, 7	3	0.60		0.60	1, 7
2, 3	1	0.20		0.20	
2, 4	1	0.20		0.20	
2, 5	1	0.20	1.09	1.29	2, 5
2, 6	1	0.20		0.20	
2, 7	1	0.20		0.20	
3, 4	3	0.60		0.60	3, 4
3, 5	2	0.40		0.40	
3, 6	2	0.40		0.40	
3, 7	1	0.20		0.20	
4, 5	3	0.60		0.60	4, 5
4, 6	4	0.80		0.80	4, 6
4, 7	3	0.60		0.60	4, 7
5, 6	2	0.40		0.40	
5, 7	1	0.20		0.20	
6, 7	3	0.60		0.60	6, 7

1. First Apriori algorithm was implemented on certain transactional database and execution time was recorded. Then UApriori algorithm was implemented on uncertain transactional database and execution time was recorded. After that Execution time of Apriori and UApriori was added to calculate total time.
2. In second step global frequent items were calculated by MasterApriori and execution time was recorded. This time was combined with the time of UApriori to calculate total time.

So in the results we got that for calculating Global Frequent Itemsets if we use Apriori and UApriori separately and then combine the results, it will take more time comparatively to MasterApriori with UApriori. All experiments were performed on Intel(R) Core(TM) i5 2.2 GHz Machine with 4 GB of RAM. Algorithms was implemented on Microsoft's Visual Studio 2013 running on Microsoft Windows 10Pro.

For certain transactional database we generate synthetic database by using synthetic database generation program given by [1]. We set parameters for synthetic database as T20.I6.D100K.N100.L200, defined as

Table 9 Calculating Size-3 global frequent items

Size-3 CTDB items	Size-3 CTDB items count	Size-3 CTDB items mean	Size-3 UTDB items mean	Size-3 total mean$_3$	L$_3$(mean$_3$ >= minSup)
1, 2, 3	1	0.20		0.20	
1, 2, 4	1	0.20		0.20	
1, 2, 5	1	0.20	0.542	0.742	1, 2, 5
1, 2, 6	1	0.20		0.20	
1, 2, 7	1	0.20		0.20	
1, 3, 4	3	0.60		0.60	1, 3, 4
1, 3, 5	2	0.40		0.40	
1, 3, 6	2	0.40		0.40	
1, 3, 7	1	0.20		0.20	
1, 4, 5	3	0.60		0.60	1, 4, 5
1, 4, 6	4	0.80		0.80	1, 4, 6
1, 4, 7	3	0.60		0.60	1, 4, 7
1, 5, 6	2	0.40		0.40	
1, 5, 7	1	0.20		0.20	
1, 6, 7	3	0.60		0.60	1, 6, 7
4, 5, 6	2	0.40		0.40	
4, 5, 7	1	0.20		0.20	
4, 6, 7	3	0.60		0.60	4, 6, 7

Table 10 Calculating Size-4 global frequent items

Size-4 CTDB items	Size-4 CTDB items count	Size-4 CTDB items mean	Size-4 UTDB items mean	Size-4 total mean$_4$	L$_4$(mean$_4$ >= minSup)
1, 4, 5, 6	2	0.40		0.40	
1, 4, 5, 7	1	0.20		0.20	
1, 4, 6, 7	3	0.60		0.60	1, 4, 6, 7

T20 = average size of transactions is 20
I6 = Average size of Maximal Potential Large Itemset is 6
D100K = Number of Transactions in Database are 100,000
N100 = Number of items in Database are 100
L200 = Number of Maximal Potential Large itemsets are 200.

For uncertain transactional database we use datasets used in [16].

We use Gazelle dataset as uncertain database in which total no of transactions are 59,601 and there are 498 distinct items in dataset.

The performance of MasterApriori with the combined results of Apriori and UApriori can be compared from Tables 11 and 12 for calculating Global Frequent

Table 11 Execution time of Apriori + UApriori

T20.I6.D100K.N100.L200 (CTDB)		Gazelle dataset (UTDB)	Total time
minSup	Time (Apriori)	Time (UApriroi)	(Apriori + UApriori)
0.1	3178.083	279.489	279.489 + 3178.083 = 3457.572
0.15	1158.304	279.489	1437.739
0.20	411.022	279.489	690.511

Table 12 Execution time of MasterApriori

Master Apriori		Gazelle dataset (UTDB)	Total Time
MinSup	Time	Time (UApriroi)	
0.1	2.059	279.489	279.489 + 2.059 = 281.548
0.01	83.626	279.489	363.115
0.001	180.84	279.489	460.33

Itemsets. The combined time of Apriori and UApriroi decreases when we increase the minimum support for certain transactional database which reflects in losing of significant frequent itemsets and association rules and when the minimum support decreases the execution time increases exponentially (refer Fig. 3). On the other hand the execution time of MasterApriori increases linearly when minimum support decreases. In MasterApriori we check the minimum support of certain database after combining its count with uncertain database so here we are not losing any information. Only we can lose the information when we are choosing minimum probability for UApriori. But we use the same result of UApriori with MasterApriori and Apriori.

So here we get the following improvements:

1. We are taking very low global minimum support for MasterApriori so that we can calculate maximum no of Global Frequent Itemset. If we take the same support for certain transactional database then it will take exponentially high time for computing frequent itemset.

Fig. 3 MasterApriori versus Apriori + UApriori

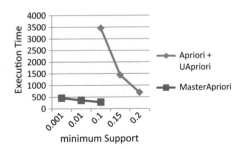

2. In MasterApriori, there is very less communication between local and remote sites. After calculating the frequent itemset by UApriori we need to bring the result at one site and these results are of very less size.
3. According to Fig. 3 it is seen that our algorithm is much faster than combination of Apriori and UApriori for calculating Global Frequent Itemset. We can see that when global minimum support decreases time for Apriori + UApriori increases exponentially but for MasterApriori it is increasing linearly.

For Gazelle dataset by taking Lowest probability $P_L = 0.02$, time taken by UApriori is 279.489 s.

5 Conclusion and Future Directions

The proposed algorithm is useful when we are dealing with uncertain and certain database at the same time. It also very useful when we want to know that a particular item or itemset is frequent all over the globe or not. The main attraction of our algorithm is that there is very less communication between local and remote sites and it is very fast in operation. Producing frequent itemset by using Apriori, UApriori or Count Distribution approach giving amazing results and are milestones in this area but when we compare the performance with time and no of frequent itemset produced then we know that still we need some better strategy. If we count frequent itemset by using Apriori on certain database and we convert the same certain database in uncertain database by assigning probability generated by Gaussian distribution (which is widely accepted by [11, 10, 15]) and then calculate frequent items by UApriori. We get the result that there is huge difference between no of frequent itemset by both approaches. Because one is probabilistic and other is exact so difference may occur but it should not be exceeded beyond a limit. So in future we can work on the strategies that how to assign probabilities on uncertain items and how we calculate our desired frequent items in minimum time. Our approach gives an introduction that how can we manage the gap between certain and uncertain items because some items in database can never be represented as certain and some items in database can never be represented as uncertain. So we need a strategy of dealing with both certain and uncertain databases.

References

1. R. Agrawal, R. Srikant, Fast algorithms for mining association rules, in *Proceedings of the 20th VLDB Conference, Santiago, Chile* (1994), pp. 487–499
2. T. Bernecker, R. Cheng, H.P. Kriegel, M. Renz, F. Verhein, A. Züfle, D.W. Cheung, S.D. Lee, Wang Liang, Model-based probabilistic frequent itemset mining. Knowl. Inf. Syst. **37**, 181–217 (2013)

3. L. Wang, R. Cheng, S.D. Lee, D.W.-L. Cheung, Accelerating probabilistic frequent itemset mining: a model-based approach, in *CIKM* (2010), pp. 429–438
4. R. Agrawal, J.C. Shafer, Parallel mining of association rules. IEEE Trans. Knowl. Data Eng. **8** (6), 962–969 (1996)
5. E. Han, G. Karypis, V. Kumar, Scalable parallel data mining for association rules, in *Proceedings 1997 ACM-SIGMOD International Conferences on Management of Data, Tucson, Arizona* (1997)
6. E. Han, G. Karypis, V. Kumar, Scalable parallel data mining for association rules. IEEE Trans. Knowl. Data Eng. **12**(3), 337–352 (2000)
7. M.V. Joshi, E. Han, G. Karypis, V. Kumar, Efficient parallel algorithms for mining associations, in *Large-scale Parallel and Distributed Data Mining*, ed. by M. Zaki, C.-T. Ho. Lecture Notes in Computer Science/Lecture Notes in Artificial Intelligence (LNCS/LNAI), vol. 1759 (Springer, 2000)
8. D.W. Cheung, J. Han, V.T. Ng, A.W. Fu, Y. Fu, A fast distributed algorithm for mining association rules, in *Proceedings of the 4th International Conferences Parallel and Distributed Information System, IEEE Computer Society Press, Los Alamitos, CA* (1996), pp. 31–42
9. D.W. Cheung, Y. Xiao, Effect of data distribution in parallel mining of associations, in *Data Mining and Knowledge Discovery*, vol. 3 (Kluwer Academic Publishers, 1999), pp. 219–314
10. C.K. Chui, B. Kao, E. Hung, Mining frequent itemsets from uncertain data, in *11th Pacific-Asia Conference on Advances in Knowledge Discovery and Data Mining, PAKDD 2007, Nanjing, China*
11. T. Bernecker, H.-P. Kriegel, M. Renz, F. Verhein, A. Züfle, Probabilistic frequent itemset mining in uncertain databases in *Proceedings of the 15th ACM SIGKDD Conference on Knowledge Discovery and Data Mining (KDD'09), Paris, France*
12. R. Cheng, D. Kalashnikov, S. Prabhakar, Evaluating probabilistic queries over imprecise data, in *SIGMOD* (2003)
13. Q. Zhang, F. Li, K. Yi, Finding frequent items in probabilistic data, in *SIGMOD* (2008)
14. L. Sun, R. Cheng, D.W. Cheung, J. Cheng, Mining uncertain data with probabilistic guarantees, in *SIGKDD* (2010)
15. C.K. Chui, B. Kao, A decremental approach for mining frequent itemsets from uncertain data, in *PAKDD* (2008), pp. 64–75
16. Y. Tong, L. Chen, Y. Cheng, P.S. Yu, Mining frequent itemsets over uncertain databases, in *VLDB'12*
17. Y. Tong, L. Chen, P.S. Yu, UFIMT: an uncertain frequent itemset mining toolbox, in *KDD'12, Beijing, China*. 12–16 August 2012
18. Frequent Itemset Mining Implementations Repository, http://fimi.ua.ac.be/
19. D.W. Cheung, V.T. Ng, A.W. Fu, Y. Fu, Efficient mining of association rules in distributed databases. IEEE Trans. Knowl. Data Eng. **8**(6), 911–922 (1996)
20. SPMF An Open-Source Data Mining Library, http://www.philippe-fournier-viger.com/spmf/index.php?link=datasets.php

New Method Based on Rough Set for Filling Missing Value

R. Çekik and S. Telçeken

Abstract The presence of missing value in a dataset can affect the performance of an analysis system such as classifier. To solve this problem many methods have been proposed in different studies using different theorems, analysis systems and methods such as Neural Network (NN), k-Nearest Neighbor (k-NN), closest fit etc. In this paper, we propose novel method based on RST for solving the problem of missing value that was lost (e.g., was erased). After dataset filling with proposed method, it has been observed improvement the performance of used analysis systems.

Keywords Missing value · Rough set · Data mining

1 Introduction

Missing value is a common problem in areas such as statistical analysis and data mining. Rates of less than 1% missing value are generally considered insignificant, 1–5% governable. However, when more than 15% missing values may effectuate bad an impression and affect the quality of the supervised learning process or the performance of analysis systems such as classifiers, missing value is seen as a significant problem. Find solutions to such problems is to be a reformative effect to performance of analysis system and it is revealed more robust analysis.

There are two main reasons for the missing information: the value was lost (e.g., was erased) and was not important. In the former case the value useful but currently we have no access to it for reasons such as deleted some values so flawed, not saved result of misunderstanding and see insignificant some attributes during data entry.

R. Çekik (✉)
Department of Computer Engineering, Anadolu University, 26555 Eskisehir, Turkey
e-mail: rasimcekik@anadolu.edu.tr

S. Telçeken
Department of Computer Science, Anadolu University, 26555 Eskisehir, Turkey
e-mail: stelceken@gmail.com

© Springer International Publishing AG, part of Springer Nature 2018
L. A. Zadeh et al. (eds.), *Recent Developments and the New Direction in Soft-Computing Foundations and Applications*, Studies in Fuzziness and Soft Computing 361, https://doi.org/10.1007/978-3-319-75408-6_4

41

In the latter case the value does not matter for reasons such as deleted value due to inconsistency with other data records, so such values are called unimportant values (or "do not care" conditions). In this study, it is focused on the problem of missing value in the former case.

Various methods have been developed to solve the problems of missing value. The simple solution is the reduction of data set and elimination of all samples with missing values or replace missing value with mean or with mean for given class [1]. Grzymala-Busse and Hu [2] proposed treating missing values as special values or replace missing value with most common attribute value or concept most common attribute. The closest fit algorithm defined by Grzymala-Busse et al. [3] is based on replace a missing value with an existing value of the same attribute from another case that resembles as much as possible the case with missing values [4]. Greco et al. was defined an analysis method using Rough Set Theory (RST) called ROUSTIDA [5]. Another solution is replacing missing attributes by Adaptive Method in Rough Sets defined by Jeong et al. [6].

In this paper we propose novel method based on RST for solving the problem of missing value that was lost (e.g., was erased).

2 Rough Set Theory: Basic Definitions

Rough set theory proposed by Pawlak is a mathematical tool to with the data analysis and to discover hidden pattern in data. It is popular approach analyzing incompleteness, vagueness, uncertainties data in many disciplines such as data mining, learning machine, pattern recognition etc. It can also be used for process of feature selection and feature extraction in data, data reduction, decision rule generation, and pattern extraction (templates, association rules) etc.

2.1 Information and Decision Systems

An information system is a data set that contains the most comprehensive information. This data set is matrix that represents as a table where each row represents a case (an event, a pattern or simply, and an object etc.) and every column represents an attribute (a variable, a feature etc.). More formally, an information system $I = (U, A)$ where U is a non-empty finite set of objects called Universe Set and A is a non-empty finite set of attributes such that $a: U \rightarrow V_a$ for every $a \in A$. The set V_a is called the value set of a. Decision system is obtained adding a decision attribute by information system. Mathematically a decision systems is any information system of the form $D = (U, AU\{d\})$, where d is called decision attribute.

2.2 Indiscernibility

For any $R \subset A$ there is an associated equivalence relation $IND(R)$:

$$IND(R) = \{(x, y \in U^2) \,|\, \square a \in R, a(x) = a(y)\} \tag{1}$$

IF $(x, y) \in IND(R)$, then x and y are indiscernible by attributes from R. The equivalence classes of the R-indiscernibility relation are denoted $[x]_R$, $x \in U$.

2.3 Reduct and Core

In an information systems, reduct is defined as a subset of minimal (R) of the conditional attribute set (E) such that for attribute set D, $\gamma R(D) = \gamma E(D)$. For $\square a \in R$, where R is a minimal subset $\gamma R - \{a\}(D) \neq \gamma E(D)$. A given data set may have many reduct sets, and all reducts is denoted by

$$\begin{aligned} &\square a \in X, and\ X \subseteq C \\ &Rall = \{X | \gamma X(D) = \gamma C(D); \ \gamma X - \{a\}(D) \neq \gamma X(D)\} \end{aligned} \tag{2}$$

The intersection of all the set in *Rall* is called the core. $CORE(C)$ is denoted by $CORE(C) = \cap RED(C)$, where is the set of all reducts of C.

2.4 Discernibility Matrix and Functions

Core and reduct of attributes and attributes value are computed by using *discernibility functions* $\left(f_A\left(a_1^*, a_2^*, \ldots, a_3^*\right)\right)$ that a Boolean function of m Boolean variables $a_1^*, a_2^*, \ldots, a_3^*$ and a *discernibility matrix* $(Z_{i,j})$ of size $n \times n$, where n denotes the number of objects. $Z_{i,j}$ is defined as the set of all attributes which discern objects x_i and x_j.

3 New Method Using RST for Filling Missing Value

RST pass names more about complete decision tables or information systems studies in literature. However, RST is working very successfully on missing value data. It is very successful to reveal the relationship between attributes and objects. A new method is proposed using this advantage of RST. Therefore, some basic concepts of rough set for new method have been changed or have been used another concepts using related to missing data in literature. For instance, instead of the discernibility relations is used characteristic relations. For decision tables, in which all missing values are lost, a special characteristic relation was defined by Stefanowski and Tsoukias [7]. In this paper, characteristic relations has been denoted by $LV(B)$, where B is a nonempty subset of the set A that is be set all attributes. *For* $x, y \in U$, characteristic relation $LV(B)$ is given as follow:

$$(x; y) \in LV(B) \text{ if and only if } p(x; a) = p(y; a)$$
$$\text{for all } a \in B \text{ such that } p(x; a) \neq ?. \tag{3}$$

Consequently, discernibility matrix is calculated to use characteristic relations. Accordingly, a discernibility matrix defined as below,

$$c_{ij} = \{a \in A \mid a(x_i) = a(x_j), \forall a \in B \text{ such that } a(x_i) \neq ?\} \tag{4}$$

In here, discernibility matrix is reflexive but in general does not need to be symmetric or transitive. Additionally, conventional some concepts of RST is changed for calculated discernibility functions. For example, it is used operators of union of mathematical set approach instead of Boolean expression, which is said to be in the normal form, if it is composed of Boolean variable and constants only, liked by operators of disjunction and conjunction. Mathematically, discernibility functions is denoted by,
Where $c_{ij}^* = \{a^* \mid a \in c_{ij}\}$

$$f_D(a_1^*, a_2^*, \ldots, a_m^*) = \cup \{c_{ij}^* \mid 1 \leq j \leq i \leq |U|\} \tag{5}$$

To calculate discernibility function for each object:

$$f_i = \subseteq \{\cup f_D(a_1^*, a_2^*, \ldots, a_m^*) \mid 1 \leq i \leq |U|\} \tag{6}$$

According to this (7), where all discernibility functions set denoted by,

$$f_{all}(a_1^*, a_2^*, \ldots, a_m^*) = \{f_i \mid 1 \leq i \leq |U|\} \tag{7}$$

An example of fill missing value with new method for a decision table (Table 1) is presented.

Table 1 Decision table

U/A	a_1	a_2	a_3	d
X_1	1	1	1	1
X_2	2	?	2	2
X_3	1	2	?	3
X_4	2	1	2	3
X_5	?	1	1	3

Table 2 Discernibility matrix

	X_1	X_2	X_3	X_4	X_5
X_1	λ	a_2	a_1, a_3	a_2	a_1, a_2, a_3
X_2	λ	λ	a_3	a_1, a_2	a_1, a_2
X_3	a_1	a_2	λ	λ	a_1, a_2
X_4	a_2	a_1, a_2, a_3	a_3	λ	a_1, a_2
X_5	a_2, a_3	a_2	a_3	a_2	λ

The discernibility matrix is constructed in the following way. A discernibility matrix can be used to find the minimal subset of attributes, which leads to expose the indiscernibility relation between objects. To do this, one has to construct the discernibility function. For the discernibility matrix presented in Table 2, the discernibility function has the following form:

$$f_1 = a_1 \cup a_2 \cup a_2 a_3 = a_1 \cup a_2 a_3$$
$$f_2 = a_2 \cup a_2 \cup a_1 a_2 a_3 \cup a_2 = a_1 a_2 a_3$$
$$f_3 = a_1 a_3 \cup a_3 \cup a_3 \cup a_3 = a_1 a_3$$
$$f_4 = a_2 \cup a_1 a_2 \cup a_2 = a_1 a_2$$
$$f_5 = a_1 a_2 a_3 \cup a_1 a_2 \cup a_1 a_2 \cup a_1 a_2 = a_1 a_2 a_3$$

To calculate the final form of f_{all}, the absorption law is applied. According to the absorption law, if a set that separated by operator of union is to be subset any of the other set, this set is discarded. If it is not to be subset, this set remains as it is. For our example f_{all} set is as follows,

$$f_{all} = \{a_1, a_2, a_3, a_1, a_2, a_3, a_1, a_3, a_1, a_2, a_1, a_2, a_3\}$$

After determining the discernibility functions of all objects f_{all}, probability of each attribute is calculated in f_{all}. Accordingly, relationship with other attribute of each attribute is determined. So, it is determined the most commonly using attribute. Probability of each attribute is given as follow:

$$p(a_i \backslash f_{all}) = \frac{Count(a_i, f_{all})}{Count(a_{all}, f_{all})} \tag{8}$$

For our example, if it is calculated to probability of each attribute:

$$p(a_1 \backslash f_{all}) = \frac{Count(a_1, f_{all})}{Count(a_{all}, f_{all})} = \frac{5}{13}$$
$$p(a_2 \backslash f_{all}) = \frac{Count(a_2, f_{all})}{Count(a_{all}, f_{all})} = \frac{4}{13}$$
$$p(a_3 \backslash f_{all}) = \frac{Count(a_3, f_{all})}{Count(a_{all}, f_{all})} = \frac{4}{13}$$

Table 3 Completed decision table

U/A	a_1	a_2	a_3	d
X_1	1	1	1	1
X_2	2	1	2	2
X_3	1	2	2	3
X_4	2	1	2	3
X_5	2	1	1	3

Hereafter step is to select an attribute that is to be the highest probability attribute in f_{all}. For our example, the highest probability attribute is a_1. Latter step is determined the highest probability value of attribute for the highest probability attribute $p(v_i/v_{all})$, where v_i is any value and v_{all} is all values in highest probability attribute. In our example, since for attribute a_1 probability of value 1 and 2 are equal, any one of these values is chosen randomly. We selects value 2 for this example. Missing value fills with value the highest probability value in attribute where the missing values corresponding to attribute value that selected the highest probability value of attribute for the highest probability attribute. In example, there are three missing value in for $m(X_5, a_1)$, $m(X_2, a_2)$ and $m(X_3, a_3)$. Accordingly, $m(X_5, a_1) = 2$ (in here, since probability of value 1 and 2 are equal, this value are chosen randomly.), $m(X_2, a_2) = 1$ and $m(X_3, a_3) = 2$ is to be. Completed table is given (Table 3).

3.1 Experimental Results

In this section we quantitatively evaluate the performance of the proposed method. In experiments used dataset have been taken from the website of UCI Machine Learning Repository (https://archive.ics.uci.edu/ml/datasets.html). It is chosen two dataset as Bands and Horse Collic datasets (See Table 4).

In this paper, it is practiced to classification that an ordered set of related categories used to group data according to its similarities. To do this, it is used some classification methods in a program called WEKA. There are several criterions for the evaluation of a classification method. In this paper, accuracy, f-score (f-measure) and mean absolute error (MAE) criterions have been used. In here, it is make two test. First when datasets are empty, this criterions is evaluated. Second after filling data sets with proposed method, criterions is evaluated. Performance of classifiers also are illustrated in Tables 5, 6, 7 and 8.

Table 4 Using datasets

Name	Attributes	Examples	Classes	Percentage of missing values (%)
Bands	19	539	2	32.28
Horse collic	27	368	2	46.67

Table 5 Performance of classifiers for incomplete bands dataset

Classifier	Accuracy (%)	f-score	MAE
Naïve Bayes	61.4100	0.6100	0.4080
IBk	62.8942	0.6210	0.3717
PART	64.9351	0.6510	0.3766
J48	67.7180	0.6750	0.3587
RandomForst	71.7996	0.7190	0.3889

Table 6 Performance of classifiers for complete bands dataset

Classifier	Accuracy (%)	f-score	MAE
Naïve Bayes	63.4508	0.6360	0.3909
IBk	71.4286	0.7150	0.2867
PART	69.9443	0.6820	0.3377
J48	70.3154	0.7010	0.3253
RandomForst	76.6234	0.7610	0.3512

Table 7 Performance of classifiers for incomplete horse dataset

Classifier	Accuracy (%)	f-score	MAE
Naïve Bayes	64.6667	0.6520	0.3530
IBk	47.6667	0.4620	0.5231
PART	71.3333	0.7120	0.3356
J48	68.3333	0.6780	0.3418
RandomForst	72.6667	0.6750	0.3315

Table 8 Performance of classifiers for complete horse dataset

Classifier	Accuracy (%)	f-score	MAE
Naïve Bayes	69.6667	0.7060	0.3053
IBk	67.3333	0.6700	0.3281
PART	83.6667	0.8340	0.1736
J48	83.3333	0.8300	0.1978
RandomForst	83.6667	0.8260	0.2660

Analyzing results given all table, we see that proposed method improves performances of classifiers on both Bands and Horse dataset. We also can say that proposed method improves performances of classifiers for all three criteria.

4 Conclusions and Future Work

Missing value is important problem in data mining. It can affect the performance of methods that is used to data analysis in data mining. To solve this problem or effect reduce improves the performance of this methods. In this paper, we propose novel method for filling dataset including missing value. We show the performance of proposed method on two dataset. The experimental results say that after dataset filling with proposed method, it has been observed improvement the performance of used classifiers for accuracy, f-score and MAE. For example, proposed method has been provided improvement about 5–6% on Bands dataset and about 8–9% on Horse dataset for accuracy. As future work will be made to improve proposed method.

References

1. M. Kantardzic, *Data Mining: Concepts, Models, Methods, and Algorithms* (2003), pp. 277–277
2. J.W. Grzymala-Busse, H. Ming, A comparison of several approaches to missing attribute values in data mining, in *Rough Sets and Current Trends in Computing* (Springer, Berlin, Heidelberg, 2000)
3. J.W. Grzymala-Busse, L.K. Goodwin, A closest fit approach to missing attribute values in preterm birth data, in *New Directions in Rough Sets, Data Mining, and Granular-Soft Computing* (Springer, Berlin, Heidelberg, 1999), pp. 405–413
4. J. Kaiser, Dealing with missing values in data. J. Syst. Integr. **5**(1), 42 (2014)
5. G. Salvatore, B. Matarazzo, R. Słowinski, Handling missing values in rough set analysis of multi-attribute and multi-criteria decision problems, in *New Directions in Rough Sets, Data Mining, and Granular-Soft Computing* (Springer, Berlin, Heidelberg, 1999), pp. 146–157
6. L. Jeong-Gi, L. Sang-Hyun, M. Kyung-Il, *Replacing Missing Attributes by Adaptive Method in Rough Sets* (2013)
7. J. Stefanowski, A. Tsoukias, On the extension of rough sets under incomplete information, *in Proceedings of the 7th International Workshop on New Directions in Rough Sets, Data Mining, and Granular-Soft Computing, RSFDGrC'1999* (Yamaguchi, Japan), pp. 73–81

A Hierarchy-Aware Approach to the Multiaspect Text Categorization Problem

Sławomir Zadrożny, Janusz Kacprzyk and Marek Gajewski

Abstract We advance our work on a special text categorization problem, the multiaspect text categorization, introduced in our previous works. In general case, it assumes a hierarchy of categories, and documents are assigned to leaves of a category but within categories documents are further structured into sequences of documents, referred to as cases. This is much more complex than the classic text categorization. Previously, we proposed a number of approaches to deal the above problem but we took into account to a limited extent hierarchies occurring in the definition of the problem. Here, we we start with one of our best approaches proposed so far and extend it by assuming that categories are arranged into a hierarchy, and that there is a hierarchical relation between a category and its offspring cases.

1 Introduction

We address an extended version of the classic, very relevant *text categorization* (TC) [1, 2] problem, an example of multiclass classification with documents to be assigned to one of some predefined classes/categories. In our previous work [3–10] we introduced the *multiaspect text categorization* (MTC), a novel problem that adds another dimension to the classic TC problem.

An intuitive example may be: a document concerning a citizen's application for a driving license is submitted to a system. It is assigned (by a clerc, for now) to, e.g., the category of "Social and civic cases", and then to, e.g., its subcategory of "transportation", etc. arriving finally at a specialized descendant subcategory, at the bottom of the hierarchy, "Documentation of a vehicle registration". Within this the document has to be classified to a specific *case*, i.e., a sequence of documents related

S. Zadrożny (✉) · J. Kacprzyk · M. Gajewski
Systems Research Institute, Polish Academy of Sciences, ul. Newelska 6,
01-447 Warszawa, Poland
e-mail: Slawomir.Zadrozny@ibspan.waw.pl

J. Kacprzyk
e-mail: Janusz.Kacprzyk@ibspan.waw.pl

© Springer International Publishing AG, part of Springer Nature 2018
L. A. Zadeh et al. (eds.), *Recent Developments and the New Direction in Soft-Computing Foundations and Applications*, Studies in Fuzziness and Soft Computing 361, https://doi.org/10.1007/978-3-319-75408-6_5

to an appropriate business process instance. This sequence can exist when, e.g., the submitted document is an office request for an extra document which is then put at the end of a specific person's driving license application case. However, the document in question may be an original application of a citizen for the license so that a new case should be formed initiated by his/her application.

The above task is surely non-trivial and is usually dealt with manually, which is costly and time consuming, and our aim is to develop a system automatically generating an advice concerning the proper classification of documents. In MTC, there is a hierarchy of categories and documents are classified only to its lowest level, i.e., to categories represented by leaves of a hierarchy tree; this can be done using a standard text categorization techniques [1], possibly with a hierarchy of categories [2, 11, 12]. However, in MTC documents are to be classified within a category to a sequence of documents concerning some business process. Thus, the essence of belongingness of a document to a sequence, referred to as a *a case*, is different. The list of cases is not known in advance and it has to be decided if the document should be assigned to an existing case or to launch a new case.

So far, we proposed new solutions to MTC for a flat structure of categories. Here, we extend our approach based on a degree of fuzzy sets subsethood used for the assignment of documents to cases [5], adding information on the hierarchy of categories, inspired by [13, 14], and also [2, 11, 12].

2   A Formal Problem Statement

We introduced the multiaspect text categorization (MTC) in [15]. Basically, documents result from various business processes and form a collection $D = \{d_1, \dots, d_n\}$. Each document is assigned to exactly one category $c \in C = \{c_1, \dots, c_m\}$ which corresponds to the set of leaves of a hierarchy H that is, in general, a tree with nodes of varying degrees. Within categories documents are further organized into sequences referred to as *cases*. Hence, each document d belongs to exactly one sequence σ, and the set of all sequences is $\Sigma = \{\sigma_1, \dots, \sigma_p\}$. Each (leaf) category can comprise any number of cases.

The essence of MTC is the classification of a newly arriving document d^* both to an appropriate category $c \in C$ and to an appropriate sequence $\sigma \in \Sigma$. The first task may be solved by one of many supervised TC methods [1]. On the other hand, the classification of a document to a sequence is more challenging as sequences in Σ may be short. Also, a newly arrived document d^* may not belong to any existing sequence and a new sequence should be established initiated with d. As to related problems, the Topic Detection and Tracking (TDT) [16] may be mentioned, cf. our [7].

3 A Subsethood Measure Based Solution and its Extension

In [5] we proposed a compact representation of documents, adopted also here, based on keywords and the tf × IDF, i.e., a classic vector space model weighting scheme. Each document $d \in D$ is initially represented by a vector $d = [w_1, \dots, w_M]$ in the space of keywords from a vocabulary $T = \{t_i\}_{i \in \{1,\dots,M\}}$. The coordinates of the particular w_i^d's in d are computed as functions of the number of occurrences of a given keyword t_j in d, denoted as $freq(d, t_j)$, and of the number of documents in D in which t_j occurs at least once, denoted as n_{t_i}. Thus, the the weight w_i^d of $t_i \in T$ in $d \in D$ is $w_i^d = freq(d, t_i) \times \log \frac{n_{t_i}}{N}$ where N is the number of documents in D.

The w_i^d's are computed for all $t_i \in T$'s but each document is represented only with K keywords with the highest weights; here $K = 5$ [5]. The representation of a document is thus: $d \rightarrow [w_{t_1}^d, \dots, w_{t_N}^d] \rightarrow [w_{t_{k_1}}^d, \dots, w_{t_{k_K}}^d]$ where $T = \{t_1, \dots, t_N\}$ is the set of all keywords used to index documents in the collection, w_{t_i} are their weights computed for d and $k_1, \dots, k_K$ are indexes such that $w_{t_{k_1}} \geq w_{t_{k_2}} \geq \cdots \geq w_{t_{k_K}}$.

The cases are sequences of documents and are represented by the concatenation of vectors. Formally, a case $\sigma = < d_{s_1}, \dots, d_{s_l} >$ is represented as:

$$\sigma \rightarrow [w_{t_{k_1}}^{d_{s_1}}, \dots, w_{t_{k_K}}^{d_{s_1}}, \dots, w_{t_{k_1}}^{d_{s_l}}, \dots, w_{t_{k_K}}^{d_{s_l}}] \tag{1}$$

Now, when a document $d^* \rightarrow [w_{t_{k_1}}^{d^*}, \dots, w_{t_{k_K}}^{d^*}]$ is matched against the case σ represented by (1), then only the keywords in the representation of d^* are taken into account. As for a given t_i there may be many documents in the case σ having this keyword, a decision on if and how they are aggregated to get one weight of t_i for σ with respect to d^* is necessary. We assume that this weight is the weight of t_i in the most recent document in σ. If t_i does not appear in the representation of any document from σ, then this weight is equal 0.

Formally, when a case σ is matched against a document d^*, then σ—for this matching—is represented as $\sigma \rightarrow [w_{t_{k_1}}^\sigma, \dots, w_{t_{k_K}}^\sigma]$ where $w_{t_{k_i}}$ refers to the weight of the i-th keyword in the representation of d^*, $i \in [1, K]$ and is computed as follows:

$$w_{t_{k_i}} = \begin{cases} 0 & \text{if } t_{k_i} \text{ does not occur in any document of } \sigma \\ w_{t_{k_j}}^{d_{s_m}} & \text{otherwise} \end{cases} \tag{2}$$

where $t_{k_j} = t_{k_i}$ and $m \in [1, l]$ is the largest m such that t_{k_i} occurs in the representation of d_{s_m}.

Based on the training documents, the categories are represented as: for each (leaf) category $c \in C$ the mean vector of all vectors representing documents in c is $c \rightarrow [w_{t_1}^c, \dots, w_{t_{M_c}}^c]$ where $w_{t_i}^c = \frac{\sum_{d \in c} w_{t_i}^d}{|d \in c|}$ and $d \in c$ denotes documents assigned to c while $|\cdot|$ denotes the set cardinality. Thus, if a t_i is not in the representation of a d, then

$w_{t_i}^d = 0$. The number M_c of keywords in the representation of c is therefore from $[K, N]$.

The essence of our approach to solve the MTC [5] is to use a weighted sum of the similarities of a document to a given case and to the category it belongs to. Since the representations of both categories and cases may be treated as fuzzy sets in the space of keywords, both similarities may be defined in terms of the fuzzy sets subsethoods. For a document to be classified, d^*, and each case σ, which belongs to $c \in C$, the two similarity degrees are computed denoted, respectively, as:

$$sim(d^*, \sigma) \tag{3}$$

$$sim(d^*, c) \tag{4}$$

So, we compute the similarity of a new document d^* to all ongoing cases σ and choose the most similar (least dissimilar) case. The overall similarity of a d^* to a σ is the weighted sum of two similarity measures:

$$w_{seq} sim(d^*, \sigma) + w_{cat} sim(d^*, c) \tag{5}$$

the first one related to σ itself, and the second to the category to which σ belongs. Clearly [5]: (1) appropriate (dis)similarity measures, and (2) weights controlling the influence of the similarity to a particular case and a particular category, should be chosen.

In Sect. 3.1 we propose an extension of the generic algorithm which makes use of the information concerning the hierarchical relations: between the categories in the hierarchy and between the categories and the cases. In Sect. 3.2 we discuss some fuzzy subsethood measures used in [5] and extend them to include the above hierarchy related information.

3.1 Information Extracted from the Hierarchy of Categories

The essence of the extension of our algorithm [5] given here is that, first, we consider a hierarchy (tree) H of categories C and assign documents to the lowest level categories only, and due to [5] the decision on assigning a d to a category is partly based on whether the fuzzy set of keywords representing d^* is the subset of that representing the categories $c_k \in C$. Moreover, the keywords shared by all categories having the same parent in the hierarchy are assumed as less meaningful. Thus, for sibling leaves of H we find an intersection of their fuzzy sets of keywords. Assuming two leaf categories c_{k1} and c_{k2} which are the only children of a higher level category $c_k \in C$, and treating representations of categories c_{k1} and c_{k2} as fuzzy sets, their intersection is:

$$c_{k1} \cap c_{k2} \to [\min(w_{t_1}^{c_{k1}}, w_{t_1}^{c_{k2}}), \dots, \min(w_{t_M}^{c_{k1}}, w_{t_M}^{c_{k2}})] \tag{6}$$

where, as previously, if a t_i does not appear in the representation of c_l, then $w_{t_i}^{c_l} = 0$, where $l \in \{k1, k2\}$; M belongs therefore to $[0, \min(M_{k1}, M_{k2})]$, where M_{k1} and M_{k2} correspond to M_C, i.e., are the numbers of keywords in the representations of the categories c_{k1} and c_{k2}, respectively.

The higher the degree of a t_i in (6) the less important the its role in distinguishing between c_{k1} and c_{k2}, i.e. its lower influence. In general, the importance weight I of keywords $t_i \in T$ taken into account for computing the match between a d^* and a c_{k_l} whose parent in H is a category c_k, $c_{k_l} \subseteq c_k$, is:

$$I(t_i) = 1 - (\bigcap_{c_{k_l} \subseteq c_k} c_{k_l})(t_i) \tag{7}$$

where I is interpreted as a fuzzy set in T.

3.1.1 Identifying Keywords Shared by Cases Belonging to a Category

Now, we consider how to measure the similarity of a document to be classified d^* w.r.t. the particular cases σ. One should expect that not all keywords are equally important as the keywords which are common to many cases within the same category may be expected to be less useful in making a decision as to which case to assign d^*. That is, a keyword t_i is deemed less important from the perspective of the cases belonging to category c, the higher the truth value of:

Keyword t_i occurs in *most* of the cases $\sigma_j \in \Sigma_c$

belonging to a category c

what may be formally denoted as:

$Q\sigma_j's$ are O_{ij}

where Q is a linguistic quantifier with the membership function $\mu_Q(y) = y$ and O_{ij} represents the property that the keyword t_i occurs in the representation of the case σ_j with a non-zero weight. The linguistic quantifier Q may be replaced by another one, e.g., "most".

Thus, in general, the importance weight I of keywords $t_i \in T$ in the matching between a d^* and a σ from a $c \in C$ is:

$$I(t_i) = 1 - Q_{\sigma_j \in \Sigma_c} \sigma_j(t_i) \tag{8}$$

where I is a fuzzy set in T, Σ_c is the set of all cases belonging to c, $\sigma_j(t_i)$ is a crisp predicate stating that a t_i occurs in the representation of the case σ_j with a non-zero weight; as previously, $A(\cdot)$ is the membership function of A.

Table 1 Subsethood measures considered in the paper (in the second and third row the implication/conjunction operators involved are shown)

No.	Name	General formula	Detailed formula
1.	Kosko's subsethood measure	$\frac{\|A\cap B\|}{\|A\|}$	$sub_1(A,B) =$ $\frac{\sum_i \min(A(x_i),B(x_i))}{\sum_i A(x_i)}$
2.	Implication based (Reichenbach/product)	$\forall x A(x) \rightarrow B(x)$	$sub_2(A,B) =$ $\prod_i (\overline{A(x_i)} + A(x_i)B(x_i))$
3.	Implication based (Łukasiewicz/mean)	$Q_{avg} x A(x) \rightarrow B(x)$	$sub_3(A,B) =$ $\frac{\sum_i \min(\overline{A(x_i)}+B(x_i),1)}{5}$

The lower the degree of a keyword t_i in (8) the less important its role in distinguishing between the cases belonging to a given category c, and while computing the similarity degree (3) these keywords should have a lesser influence. In Sect. 3.2 we discuss in a detail how to do this via an appropriate modification of the formulas shown in Table 1.

3.2   Subsethood Measures and Their Hierarchy-Aware Extensions

In [5] we considered 5 types of fuzzy sets subsethood measures, and here we chose the two most promising ones, from a theoretical and computational viewpoints. The measures used are shown in Table 1. In Table 1 A corresponds to a fuzzy set representing d^* and B corresponds to a fuzzy set representing a case σ or a category c profile, depending on which similarity index (3) or (4) is to be computed; only 5 keywords with the highest weights are used (cf. Sect. 3) to represent a document and then cases and categories. Thus, in all formulas in Table 1 both fuzzy sets A and B are defined over the same space of 5 keywords representing the document, i.e., the set A. The following notation is used in Table 1:

- $\overline{A(x)}$—the complement to 1 of the membership degree of x to A, i.e., $\overline{A(x)} = 1 - A(x)$,
- $|A|$—the ΣCount cardinality of the fuzzy set, i.e., $|A| = \sum_i A(x_i)$,
- Q_{avg}—Zadeh's linguistic quantifier [17], $\mu_{Q_{avg}}(x) = x$.

The first subsethood measure in Table 1 was proposed by Kosko [18] and is interpreted as the truth degree of a linguistically quantified proposition [17]: "$Q_{avg}Ax's$ are B".

The remaining two measures of subsethood are examples of a "fuzzification" of the classical definition stating that A is a subset of B if: "$\forall x \quad x \in A \rightarrow x \in B$". The two versions in rows 2–3 of Table 1 differ w.r.t. different implications and different types of aggregation to model "$\forall$". Hence, in Table 1 we have:

- in row 2: the Reichenbach implication is used, i.e., $a \rightarrow b = 1 - a + ab$, and "$\forall$" is modeled via the algebraic product (t-norm),
- in row 3: the Łukasiewicz implication is used, i.e., $a \rightarrow b = 1 - a + b$ and "$\forall$" is interpreted as the arithmetic mean; this is also known also as Goguen's inclusion grade [19].

Now the problem is how to include the information extracted from the hierarchy of categories, given as some extra importance weights of the keywords, cf. Sect. 3.1, in the formulas for the particular three indexes in Table 1.

3.2.1 Kosko's Subsethood Index sub_1

The inclusion of importance degrees is here quite simple thanks to the interpretation of this index. Namely, if the importance of keywords is given as a fuzzy set I in the space of all keywords, then to account for I using to Zadeh's [17] calculus of linguistically quantified propositions, we should write:

$$Q(A \wedge I)x's \text{ are } B \qquad (9)$$

Thus, the formula from Table 1, for the degree of subsethood of A in B w.r.t. I important keywords, becomes:

$$sub_1^I(A, B) = \frac{\sum_i \min(A(x_i), I(x_i), B(x_i))}{\sum_i \min(A(x_i), I(x_i))} \qquad (10)$$

In a rare case when the denominator of (10) becomes 0 the value of the index is assumed to be 0.5 to represent a lack of information as to the subsethood of A in B when all keywords are not important.

Another possible approach to extend Kosko's subsethood measure with the importance weights is to interpret this index in terms of the conditional probability of a fuzzy event, here assumed in Zadeh's [20], i.e. $P(A) = \sum_{i=1}^{n} A(x_i)p(x_i)$, where A is a fuzzy set defined in some space $X = \{x_i\}_{i=1,...,n}$, $A(\cdot)$ is its membership function and p is a probability distribution over X, i.e., $p : X \rightarrow [0, 1]$, $\sum_{i=1}^{n} p(x_i) = 1$. Then, assuming for simplicity the uniform probability distribution p over X, i.e., $\forall x_i \in X \; p(x_i) = \frac{1}{n}$, we have the conditional probability of the fuzzy event B, conditioned on A, $P(B|A)$ as:

$$P(B|A) = \frac{P(A \cap B)}{P(A)} = \frac{\sum_{i=1}^{n} \min(A(x_i), B(x_i))\frac{1}{n}}{\sum_{i=1}^{n} A(x_i)\frac{1}{n}} \qquad (11)$$

and, thus:

$$P(B|A) = \frac{\sum_{i=1}^{n} \min(A(x_i), B(x_i))}{\sum_{i=1}^{n} A(x_i)} = sub_1(A, B) \qquad (12)$$

where $sub_1(A, B)$ is Kosko's subsethood measure (cf. Table 1).

We can now normalize the importance weights $I(\cdot)$ to obtain $I_{norm}(\cdot)$:

$$I_{norm}(x_i) = \frac{I(x_i)}{\sum_{i=1}^n I(x_i)}$$

and then use them as probability distribution to model Kosko's subsethood measure with importance weights:

$$sub_1^{I_2}(A, B) = \frac{\sum_{i=1}^n \min(A(x_i), B(x_i))I_{norm}(x_i)}{\sum_{i=1}^n A(x_i)I_{norm}(x_i)} \tag{13}$$

3.2.2 Reichenbach's Implication with the Algebraic Product Based Index sub_2

The implication based indexes may be seen as the conjunction of the implications involving particular keywords, i.e., those used to represent the documents, cases and categories. The conjunction corresponds here to the general quantifier $\forall$ and is represented by the algebraic product (or the product t-norm). Thus, with the importance weights of the keywords, we obtain the weighted conjunction of elements $(x_1, \ldots, x_n)$ with assigned weights $(w_1, \ldots, w_n)$ denoted as $T((w_1, x_1), \ldots, (w_n, x_n))$. For the product t-norm the weighted conjunction may be defined as (cf., e.g., [21]):

$$T((w_1, x_1), \ldots, (w_n, x_n)) = \prod_i x_i^{w_i} \tag{14}$$

Based on (14) we employ the following version of the subsethood index sub_2 extended with importance weight, with $I(x_i) = w_i$:

$$sub_2^I(A, B) = \prod_i (\overline{A(x_i)} + A(x_i)B(x_i))^{I(x_i)} \tag{15}$$

However, the interpretation of the importance weights from the perspective of a conjunction operator and our particular case of documents classification is different. Namely, in the former case, (14) properly limits the influence of the less important elements on the result of the conjunction as the arguments x_i with the lowest values most strongly influence the result of the product. If some of them are of a low importance $w_i < 1$, then taking the power as in $x_i^{w_i}$ effectively raises their values and reduces their influence, cf. particularly the case of $w_i = 0$ which leads to $x_i^{w_i} = 1$ which is a neutral element of the product (and of any t-norm). On the other hand, for our purposes, as explained in Sect. 3.2, the computation of a subsethood index—with its final step being the weighted conjunction represented by the product—is carried out for a given d^* and each category and case considered, and then their combined results should be compared. Thus, we would wish less important keywords essentially neither decrease nor increase the result of the weighted conjunction. Formula (14) secures only the former and, again, in the extreme case when $w_i = 0$, $x_i^{w_i}$ is lifted to 1 even for a very small x_i which may negatively influence the comparison mentioned above.

Thus, we propose to use of a number of possible formulas to account for the varying importance of the keywords:

(1) (15), i.e., increasing the value of the index for less important keywords;
(2) its modification based on a revised version of (14):

$$T((w_1, x_1), \ldots, (w_n, x_n)) = \prod_i x_i^{2-w_i} \tag{16}$$

reducing the value of the index w.r.t. less important keywords. Based on the above we employ the following subsethood index with importances, where $I(x_i) = w_i$:

$$sub_2^{I_2}(A, B) = \prod_i (\overline{A(x_i)} + A(x_i)B(x_i))^{2-I(x_i)} \tag{17}$$

3) a combination of the above based on the following formula for the product based weighted conjunction with importances:

$$T((w_1, x_1), \ldots, (w_n, x_n)) =$$
$$= \prod_i x_i^{x_i(2-w_i)+(1-x_i)w_i} = \prod_i x_i^{2x_i+w_i} \tag{18}$$

which leads to:

$$sub_2^{I_3}(A, B) =$$
$$= \prod_i (\overline{A(x_i)} + A(x_i)B(x_i))^{x_i(2-I(x_i))+(1-x_i)I(x_i)} \tag{19}$$

where, as above, $I(x_i) = w_i$. For x_i's of high values and low importance their value is reduced while for x_i's of low values and low importance their values are increased, all that to reduce the influence of the low importance keywords on the final result of the matching.

Notice that we expect only a few keywords to be identified as shared by a group of sibling category leaves in H or by the cases from the same category, using one of the methods from Sect. 3.1, and they will get reduced importance weights. All the remaining keywords will get the importance weight equal 1 so that their influence on the result of the matching will be left intact.

3.2.3 Łukasiewicz' Implication and the *t*-norm Based Subsethood Index sub_3

Here, similarly as for sub_1^I, we include the importance in the calculus of linguistically quantified propositions and interpret the mean operator vi the unitary linguistic quantifier Q_{avg} given by $\mu_{Q_{avg}}(y) = y$ for $y \in [0, 1]$. Then, with the importance of keywords as a fuzzy set I in the space of all keywords and using (9), we obtain:

$$sub_3^I(A, B) = \frac{\sum_i \min(\overline{A(x_i)} + B(x_i), 1, I(x_i))}{\sum_i I(x_i)} \tag{20}$$

In a rare case when the denominator of (20) becomes 0 the value of the index is assumed to be 0.5 is meant to represent a lack of information as to the subsethood of A in B when all keywords are not important.

4 Computational Experiments

The dataset used is much larger than in our previous paper [5] though derived from the same ACL Anthology Reference Corpus (ACL ARC) [22] of selected scientific papers on computational linguistics. Each paper is explicitly divided into sections and we treat a whole paper as a case (sequence of documents) σ and its sections as the documents. We use 664 papers, i.e. we have 664 cases which comprise in total 6884 documents. First, however, the papers are grouped via a hierarchical clustering algorithm using the standard `hclust` function of the R platform with the default settings, into 6 clusters then treated as categories C of the documents. The hierarchy H has three levels: the 6 leaf categories form the first one, the pairs of categories 1 and 4, 2 and 3, 5 and 6 form three nodes of the second level and, finally, there is a root node of H on the third level. The hierarchy H is derived by an analysis of the overlap of the leaf categories in terms of the weights of keywords in their profile.

In each experiment we randomly select 40 cases as ongoing cases (cf. Sect. 3), then randomly select a cut-off position in each of the ongoing cases and all the documents located at that position and beyond are removed and may be later used as the testing documents (in the currently reported experiments we use as test documents only the documents located at the cut-off points). All the remaining documents from the ongoing cases and all closed cases are used as the training documents. The R platform [23] was used, specifically the `tm` package [24] and some specific scripts were used.

We carried out a series of 100 experiments. We tested all combinations of 15 subsethood measures including: (a) 3 original measures sub_1, sub_2 and sub_3 presented in Table 1, (b) two variants of the first measure listed in Table 1, sub_1^I and $sub_1^{I_2}$ for each form of the importance I as defined by (7) and (8), thus giving rise to 4 Kosko's subsethood measures with importances, (c) three variants of the second measure listed in Table 1, $sub_2^{I_1}$, $sub_2^{I_2}$ and $sub_2^{I_3}$ for each form of the importance I as defined by (7) and (8), thus giving rise to 6 Reichenbach's implication with product t-norm based subsethood measures with importances, and (d) a variant of the third measure listed in Table 1, denoted sub_3^{I}, for each form of the importance I as defined by (7) and (8), thus giving rise to 2 Łukasiewicz' implication and t-norm based subsethood measures with importances.

Now, we denote the subsethood measures referring to the importance defined w.r.t. the sibling cases, i.e., according to (8), with the superscript 1 at I, e.g., $sub_1^{I_2^1}$ and those referring to the importance defined w.r.t. the sibling categories, i.e., according to (7), with the superscript 2 at I, e.g., $sub_1^{I_2^2}$. The weights w_{seq} and w_{cat} (5) are assumed equal which turned out to be a good choice, cf. [5].

The experiments were meant to identify the best combinations of the subsethood measures taking into account (cf. our previous work [5]): (a) the number of documents properly classified to their cases, (b) the number of documents for which their actual case is the second best according to our algorithm, (c) the number of documents for which their actual case is the third best according to our algorithm, and (d) the number of documents classified to a proper category; with the first as the most important as it directly corresponds to the classical measure of the accuracy of classification.

The best mean accuracy, equal 59%, over all 100 runs have been obtained for the pairs of subsethood measures (sub_3, $sub_1^{I^1}$) and ($sub_3^{I^1}$, $sub_1^{I^1}$). For the same pairs the best assignment of the category took place—the best mean accuracy here amounts to 82%. The best single experiments, with the accuracy of the assignment of cases equal 80%, has been obtained for a number of the combination of the subsethood measures but only the measures from the first and the third families were involved for the matching of documents against cases. Concerning the assignment of the categories, the best individual run has been characterized by the accuracy of 98% and the use of only the subsethood measures of the first two families for matching documents against cases and mostly the measures of the first family for matching documents against categories.

Table 2 is meant to support the analysis of the advantages brought by the use of the extra information based on hierarchical relations between categories and cases, as advocated in we present the analysis . There is a row for each of 9 combinations of the original subsethood measures listed in Table 1. The columns contain the following information: (a) columns 1 and 2 present the names of the original subsethood measures used to compute $sim(d^*, \sigma)$ and $sim(d^*, c)$, respectively; (b) column 3 shows the mean accuracy of the combination identified in columns 1–2; (c) columns 4 and 5 show the best, importance based, replacements for the measures shown in columns 1 and 2, i.e., a measure named in columns 4 is an extension of the measure named in column 1 obtained with an account for the importance of keywords, and similarly a measure in column 5; there are usually several replacements of the equal quality and they are all displayed in subsequent subrows of the table; and (d) column 6 shows the mean accuracy of the combination identified in columns 4–5.

Table 2 The performance of the extended versions of the subsethood measures compared to their original forms

Sub4Seq	Sub4Cat	Mean accuracy	Sub4Seq w/imp	Sub4Cat w/imp	Mean accuracy
sub_1	sub_1	23.33	sub_1	$sub_1^{I_1}$	**23.54**
sub_1	sub_2	22.54	$sub_1^{I_1}$ $sub_1^{I_1}$ $sub_1^{I_1}$	sub_2 $sub_2^{I_1}$ $sub_2^{I_2^2}$	22.64
sub_1	sub_3	22.78	$sub_1^{I_1}$	$sub_3^{I_2}$	22.85
sub_2	sub_1	23.25	sub_2 $sub_2^{I_1}$ $sub_2^{I_2^2}$	$sub_1^{I_1}$ $sub_1^{I_1}$ $sub_1^{I_1}$	23.26
sub_2	sub_2	23.15	$sub_3^{I_2^2}$	$sub_2^{I_2}$	23.19
sub_2	sub_3	23.29	sub_2 sub_2 sub_2 $sub_2^{I_1}$ $sub_2^{I_1}$ $sub_2^{I_1}$ $sub_2^{I_2^2}$ $sub_2^{I_2^1}$ $sub_2^{I_2^1}$	sub_3 $sub_3^{I_1}$ $sub_3^{I_2}$ sub_3 $sub_3^{I_1}$ $sub_3^{I_2}$ sub_3 $sub_3^{I_1}$ $sub_3^{I_2}$	23.29
sub_3	sub_1	23.60	sub_3 $sub_3^{I_1}$	$sub_1^{I_1}$ $sub_1^{I_1}$	23.65
sub_3	sub_2	23.19	sub_3 sub_3 sub_3 $sub_3^{I_1}$ $sub_3^{I_1}$ $sub_3^{I_1}$	sub_2 $sub_2^{I_1}$ $sub_2^{I_2^2}$ sub_2 $sub_2^{I_1}$ $sub_2^{I_2^2}$	23.19
sub_3	sub_3	23.33	sub_3 sub_3 $sub_3^{I_1}$ $sub_3^{I_1}$	sub_3 $sub_3^{I_1}$ sub_3 $sub_3^{I_1}$	23.33

The results shown in Table 2 indicate that in many cases the inclusion of the extra information has helped to obtain slightly better results. However, only in one case, illustrated by the first row of Table 2, the improvement is statistically significant using the Wilcoxon test. On the other hand, as mentioned earlier, the extended subsethood measures were always a part of the best solutions in the experiments we report.

5 Conclusion

In our previous works we proposed the concept of the multiaspect text categorization (MTC), and here we considerably extended it by using subsethood measures of fuzzy sets enhanced with information pertaining to the hierarchical relations between the categories and cases. The proposed extension proves to be promising.

Acknowledgements This work is supported by the National Science Centre under contracts no. UMO-2011/01/B/ST6/06908 and UMO-2012/05/B/ST6/03068.

References

1. F. Sebastiani, Machine learning in automated text categorization. ACM Comput. Surv. **34**(1), 1–47 (2002)
2. M. Ceci, D. Malerba, Classifying web documents in a hierarchy of categories: a comprehensive study. J. Intell. Inf. Syst. **28**(1), 37–78 (2007)
3. S. Zadrożny, J. Kacprzyk, M. Gajewski, M. Wysocki, A novel text classification problem and two approaches to its solution, in *Proceedings of the International Congress on Control and Information Processing 2013* (Cracow University of Technology, 2013)
4. S. Zadrożny, J. Kacprzyk, M. Gajewski, M. Wysocki, A novel text classification problem and its solution, in *Technical Transactions*, vol. 4-AC (2013), pp. 7–16
5. S. Zadrożny, J. Kacprzyk, M. Gajewski, A novel approach to sequence-of-documents focused text categorization using the concept of a degree of fuzzy set subsethood, in *Proceedings of the Annual Conference of the North American Fuzzy Information Processing Society NAFIPS'2015 and 5th World Conference on Soft Computing 2015* (Redmond, WA, USA, 17–19 Aug 2015)
6. S. Zadrożny, J. Kacprzyk, M. Gajewski, A new two-stage approach to the multiaspect text categorization, in *IEEE Symposium on Computational Intelligence for Human-like Intelligence, CIHLI 2015* (IEEE, Cape Town, South Africa, 8–10 Dec 2015). pp. 1484–1490
7. M. Gajewski, J. Kacprzyk, S. Zadrożny, Topic detection and tracking: a focused survey and a new variant. Informatyka Stosowana **2014**(1), 133–147 (2014)
8. S. Zadrożny, J. Kacprzyk, M. Gajewski, A new approach to the multiaspect text categorization by using the support vector machines, in *Challenging Problems and Solutions in Intelligent Systems*, ed. by G. De Tré et al.(Springer) (to appear)
9. S. Zadrożny, J. Kacprzyk, M. Gajewski, Multiaspect text categorization problem solving: a nearest neighbours classifier based approaches and beyond. J. Autom. Mob. Robot. Intell. Syst. **9**, 58–70 (2015)
10. S. Zadrożny, J. Kacprzyk, M. Gajewski, On the detection of new cases in multiaspect text categorization: a comparison of approaches, in *Proceedings of the Congress on Information Technology, Computational and Experimental Physics* (AGH University of Science and Technology, 2015), pp. 213–218
11. D. Koller, M. Sahami, Hierarchically classifying documents using very few words, in *Proceedings of the Fourteenth International Conference on Machine Learning (ICML 1997)*, ed. by D. H. Fisher (Nashville, Tennessee, USA, 8–12 July 1997), pp. 170–178
12. A.S. Weigend, E.D. Wiener, J.O. Pedersen, Exploiting hierarchy in text categorization. Inf. Retr. **1**(3), 193–216 (1999)
13. J.W.T. Wong, W. Kan, and G.H. Young, ACTION: automatic classification for full-text documents. SIGIR Forum **30**(1), 26–41 (1996). https://doi.org/10.1145/381984.381987

14. S. D'Alessio, K.A. Murray, R. Schiaffino, A. Kershenbaum, The effect of using hierarchical classifiers in text categorization, in *Computer-Assisted Information Retrieval (Recherche d'Information et ses Applications)—RIAO, 6th International Conference, Proceedings*, ed. by J. Mariani and D. Harman, vol. 2000 (College de France, France, 12–14 Apr 2000) CID, pp. 302–313

15. S. Zadrożny, J. Kacprzyk, M. Gajewski, M. Wysocki, A novel text classification problem and its solution, in *Technical Transactions. Automatic Control*, vol. 4-AC (2013), pp. 7–16

16. J. Allan (ed.), *Topic Detection and Tracking: Event-Based Information* (Kluwer Academic Publishers, 2002)

17. L. Zadeh, A computational approach to fuzzy quantifiers in natural languages. Comput. Math. Appl. **9**, 149–184 (1983)

18. B. Kosko, Fuzzy entropy and conditioning. Inf. Sci. **40**(2), 165–174 (1986)

19. V.R. Young, Fuzzy subsethood, in *Fuzzy Sets and Systems*, vol. 77 (1996), pp. 371–384

20. L.A. Zadeh, Probability measures of fuzzy events. J. Math. Anal. Appl. **23**, 421–427 (1968)

21. R.R. Yager, Weighted triangular norms using generating functions. Int. J. Intell. Syst. **19**(3), 217–231 (2004). https://doi.org/10.1002/int.10162

22. S. Bird et al., The ACL anthology reference corpus: a reference dataset for bibliographic research in computational linguistics, in *Proceedings of Language Resources and Evaluation Conference (LREC 08)* (Marrakesh, Morocco), pp. 1755–1759

23. R Core Team, R: A Language and Environment for Statistical Computing, in *R Foundation for Statistical Computing* (Vienna, Austria, 2014), http://www.R-project.org

24. I. Feinerer, K. Hornik, D. Meyer, Text mining infrastructure. R. J. Stat. Softw. **25**(5), 1–54 (2008)

Adaptive Neuro-Fuzzy Inference System for Classification of Texts

Aida-zade Kamil, Samir Rustamov, Mark A. Clements and Elshan Mustafayev

Abstract In this work, we applied Adaptive Neuro-Fuzzy Inference System to three different classification problems: (1) sentence-level subjectivity detection, (2) sentiment analysis of texts, and (3) detecting user intention in natural language call routing system. We used English dataset for the first and second problems, but Azerbaijani dataset for the third problem based on same features. Our feature extraction algorithm calculates a feature vector based on the statistical occurrences of words in a corpus without any lexical knowledge.

1 Introduction

Sentiment analysis, which attempts to identify and analyze opinions and emotions, has become a popular research topic in recent years. The main goal of sentiment analysis is to understand subjective information such as opinions, attitudes, and feelings. Basical research has focused on two subproblems for sentimental analysis: detecting whether a segment of text, either a whole document or a sentence, is subjective or objective, and detecting the overall polarity of the text, i.e., positive or negative.

Call routing is the task of directing callers to the right place in the call center, which could be either the appropriate live agent or an automated service.

A. Kamil (✉) · E. Mustafayev
Institute of Control Systems, Baku, Azerbaijan
e-mail: kamil_aydazade@rambler.ru

E. Mustafayev
e-mail: elshan.mustafayev@gmail.om

S. Rustamov
Institute of Control Systems, ADA University, Baku, Azerbaijan
e-mail: srustamov@ada.edu.az

M. A. Clements
Georgia Institute of Technology, Atlanta, GA, USA
e-mail: clements@ece.gatech.edu

© Springer International Publishing AG, part of Springer Nature 2018
L. A. Zadeh et al. (eds.), *Recent Developments and the New Direction in Soft-Computing Foundations and Applications*, Studies in Fuzziness and Soft Computing 361, https://doi.org/10.1007/978-3-319-75408-6_6

Call centers typically employ a complex hierarchy of touch-tone or speech-enabled menus to provide self-service using Interactive Voice Response (IVR) which enables skills-based routing. A touch-tone menu implements call routing by having a caller select from a list of options using a touch-tone keypad. If there are more than a few routing destinations, several touch-tone menus are arranged in hierarchical layers. Natural language call routing (NLCR) lets callers describe the reason for their calls in their own words, instead of presenting them with a closed list of menu options. NLCR directs more callers to the right place, thus saving both caller and agent time.

In recent years, several different supervised and unsupervised learning algorithms were investigated for classification of texts. There are machine learning algorithms, such as Naive Bayes (NB), Maximum Entropy (ME), Support Vector Machines (SVM), and unsupervised learning.

Pang, Lee and Vaithyanathan Pang (2002) employed SVM, NB and ME classifiers using a diverse set of features, such as unigrams, bigrams, binary and with or without part-of-speech labels. They concluded that the SVM classifier with binary unigram based features produces the best results [1].

Prabowo and Thelwall (2009) proposed a hybrid classification process by combining, in sequence, several rule-based classifiers with a SVM classifier. Their experiments showed that combining multiple classifiers can result in better effectiveness than any individual classifier, especially when sufficient training data is not available [2].

Yulan He (2010) proposed subjLDA for sentence-level subjectivity detection by modifying the latent Dirichlet allocation (LDA) model through adding an additional layer to model sentence-level subjectivity labels [3].

Rustamov et al. (2013) applied hybrid Neuro-Fuzzy and HMMs to document level sentiment analysis and sentence level subjectivity analysis of movie reviews [4–8].

The vector-based information retrieval technique was applied for determination of user intention in NLCR. In the vector-based technique, for every topic in the training corpus, queries are represented as vectors of features representing the frequencies of terms that occur within them. Then a distance is computed between the query vector and each topic vector. The topic that is the closest to the query is chosen by the classifier [9].

A Markov Decision Process (MPD) framework is applied for the design of a dialogue agent. The basic assumption in MPDs is that the current state and action of the system determine the next state of the system (Markov property). Partially Observable MPDs have been demonstrated that are proper candidates for modeling dialogue agents [10].

Boosting is an iterative method for improving the accuracy of any given learning algorithm. Its basic idea is to build a highly accurate classifier by combining many "weak" or "simple" base classifiers. The algorithm operates by learning a weak rule at each iteration so as to minimize the training error rate [11].

Aidazade et al., applied a Hybrid Fuzzy Control and HMM system for understanding user intention in the NLCR problem [12, 13].

In this paper we investigated three type text classification problems by using Adaptive Neuro-Fuzzy Inference System (ANFIS): sentiment analysis of texts; sentence-level subjectivity detection, and detecting user intention in natural language call routing system. This model can be applied to any language and call routing domain, i.e. there is no lexical, grammatical, syntactic analysis used in the understanding process. Our feature extraction algorithm calculates a feature vector based on statistical occurrences of words in the corpus without any lexical knowledge.

2 Data Preparation and Feature Extraction

Feature extraction algorithms are a major part of any machine learning method. We describe a feature extraction algorithm which intuitive, computationally efficient and does not require additional human annotation or lexical knowledge. This algorithm consists of two part: pre-processing (data preparation) and calculation of feature vectors.

2.1 Pre-processing

As mentioned above, we use 3 DataSets: a sentiment polarity dataset 2v.0: 1000 positive and 1000 negative processed movie reviews [Pang/Lee ACL 2004], subjectivity database from the "Rotten Tomatoes" movie reviews (see http://www.cs. cornell.edu/people/pabo/movie-review-data) and Azerbaijani DataSet for NLCR.

In a machine learning based classification, two sets of documents are required: a training set and a test set. A training set is used by an automatic classifier to learn the differentiating characteristics of documents, and a test set is used to validate the performance of the automatic classifier. We now introduce the data distribution in the corpus (DataSet). Preliminary steps were taken to remove rating information from the text files. Thus, if the original review contains several instances of rating information, potentially given in different forms, those not recognized as valid ratings remain part of the review text. To build the term list, the following operations are carried out:

- Combine all files from the corpus and make one text file;
- Convert the text to an array of words;
- Sort the array of words: $\begin{pmatrix} A \\ Z \downarrow \end{pmatrix}$;
- Code: $V = \{v_1, \ldots, v_M\}$, where M—is the number of different words (terms) in the corpus.

As our target does not use lexical knowledge, we consider every word as one code word. In our algorithm we do not combine verbs in different tenses, such as present and past ("remind" vs "reminded") nor nouns as singular or plural ("reminder" vs "reminders"). We consider them as the different code words. We divided datasets randomly into 2 parts (training and testing) and made 10 folds.

2.2 Calculation of Feature Vectors

Below, we describe some of the parameters:

- N is the number of classes;
- M is the number of different words (terms) in the corpus;
- R is the number of observed sequences in the training process;
- $\mu_{i,j}$ describes the association between i-th term (word) and the j-th class $i = 1, \ldots, M; j = 1, 2, \ldots, N$;
- $c_{i,j}$ is the number of times i-th term occurred in the j-th class;
- $t_i = \sum_j c_{i,j}$ denotes the occurrence times of the i-th term in the corpus;
- Frequency of the i-th term in the j-th class

$$\bar{c}_{i,j} = \frac{c_{i,j}}{t_i};$$

- Pruned ICF (Inverse-Class Frequency) [5]

$$ICF_i = \log_2\left(\frac{N}{dN_i}\right),$$

where i is a term, dN_i is the number of classes containing the term i, which $\bar{c}_{i,j} > q$, where

$$q = \frac{1}{\delta \cdot N},$$

the value of δ is found empirically for the corpus investigated.

Fig. 1 The structure of ANFIS

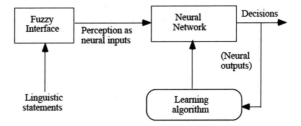

3 Application ANFIS for Classification of Texts

The general structure of ANFIS is illustrated in Fig. 1. In response to linguistic statements, the fuzzy interface block provides an input vector to a multi-layer neural network. The neural network can be adapted (trained) to yield desired command outputs or decisions [14].

The membership degree of the terms $(\mu_{i,j})$ for appropriate classes can be estimated by experts or can be calculated by analytical formulas. Since a main goal is to avoid using human annotation or lexical knowledge, we calculated the membership degree of each term by an analytical formula as follows $(i = 1, \ldots, M; j = 1, 2, \ldots, N)$:

$$\mu_{i,j} = \bar{c}_{i,j} \cdot ICF_i. \tag{1}$$

3.1 Fuzzification Operations

Maximum membership degree is found with respect to the classes for every term of the r-th request

$$\bar{\mu}^r_{s,j} = \mu^r_{s,js}, \\ j_s = \arg \max_{1 \leq v \leq N} \mu^r_{i,v}, s = 1, \ldots, T_r. \tag{2}$$

Means of maxima are calculated for all classes:

$$\bar{\bar{\mu}}^r_j = \frac{\sum_{k \in Z^r_j} \bar{\mu}^r_{k,j}}{T_r}, \\ Z^r_j = \left\{ i : \bar{\mu}^r_{i,j} = \max_{1 \leq v \leq N} \mu^r_{i,v} \right\}, \\ j = 1, \ldots, N. \tag{3}$$

Fig. 2 The structure of
MANN in ANFIS

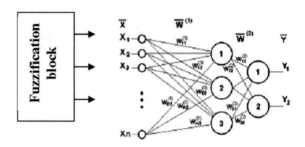

We use the Center of Gravity Defuzzification (CoGD) method for the defuzzification operation. The CoGD method avoids the defuzzification ambiguities which may arise when an output degree of membership comes from more than one crisp output value.

We used statistical estimation of membership degree of terms by (1) instead of linguistic statements at the first stage. Then we applied fuzzy operations (2) and (3). MANN was applied to the output of the fuzzification operation. Outputs of MANN are taken as indexes of classes appropriate to the requests (Fig. 2). MANN is trained by the back-propagation algorithm.

For application of ANFIS to the current test, the membership degree of the words of request are calculated by a fuzzification model in the testing process. By using of trained neural networks parameter estimated index of class.

We set two boundary conditions for an acceptance decision:

(1) $$\bar{y}_k \geq \Delta_1$$

(2) $$\bar{y}_k - \tilde{y}_p \geq \Delta_2$$

where y_i is the output vector of MANN and

$$\bar{y}_k = \max_{1 \leq i \leq N} y_i, k = \arg \max_{1 \leq i \leq N} y_i$$

$$\tilde{y}_p = \max_{1 \leq i \leq k-1;\, k+1 \leq i \leq N} y_i.$$

There is shown results of sentimental analysis of movie reviews by ANFIS with different values of Δ_2 and Δ_3 in Table 1.

There is shown results of subjectivity detection in movie reviews by ANFIS with different values of Δ_2 and Δ_3 in Table 2.

In Table 3 is shown the results of classification of user requests by ANFIS with different values of Δ_2 and Δ_3.

Table 1 Results of ANFIS for classification of movie reviews

	Correct (%)	Rejection (%)	Error (%)
$\Delta_2 = 0.5; \Delta_3 = 0.5$	67	24.6	8.4
$\Delta_2 = 0.1; \Delta_3 = 0.5$	76.35	10.85	12.8
No restriction	83	0	7

Table 2 Average results of 10 folds cross validation accuracy ANFIS based on TF · ICF for subjectivity detection in movie reviews

	Correct (%)	Rejection (%)	Error (%)
$\Delta_2 = 0.8; \Delta_3 = 0.5$	78.66	18.84	2.5
$\Delta_2 = 0.5; \Delta_3 = 0.5$	85.77	8.62	5.61
No restriction	92.16	0.01	7.83

Table 3 Results of ANFIS for NLCR

Boundary conditions	Correct (%)	Rejection (%)	Error (%)
No restriction	92	0	8
$\Delta_1 = 0.1; \Delta_2 = 0.5$	88	8	4
$\Delta_1 = 0.3; \Delta_2 = 0.5$	86	10	4
$\Delta_1 = 0.5; \Delta_2 = 0.5$	84	14	2

4 Conclusions

We have described ANFIS classification system structures and applied to three different classification problems: (1) sentence-level subjectivity detection, (2) sentiment analysis of texts, and (3) detecting user intention in natural language call routing system. A goal of the research was to formulate methods that did not depend on linguistic knowledge and therefore would be applicable to any language. An important component of these methods is the feature extraction process. We focused on analysis of informative features that improve the accuracy of the systems with no language-specific constraints.

When comparing the current system with others, it is necessary to emphasize that the use of linguistic knowledge does improve accuracy. Since we do not use such knowledge, our results should only be compared with other methods having similar constraints, such as those which use features based on bags of words that are tested on the same data set. Examples include studies by Pang and Lee and Martineau and Finin [15, 16]. Pang and Lee report 92% accuracy on sentence-level subjectivity classification using Naïve Bayes classifiers and 90% accuracy using SVMs on the same data set. Martineau and Finin (2009) reported 91.26% accuracy using SVM Difference of TFIDFs. The currently reported results: ANFIS (92.16%) are similar. It is anticipated that when IF-THEN rules and expert knowledge are inserted into ANFIS, accuracy will improve to a level commensurate with human judgment.

Acknowledgements This work was supported by 5th Mobility Grant of the Science Development Foundation under the President of the Republic of Azerbaijan.

References

1. B. Pang, L. Lee, S. Vaithyanathan, Thumbs up? Sentiment classification using machine learning techniques, in *Proceedings of CoRR* (2002)
2. R. Prabowo, M. Thelwall, Sentiment analysis: a combined approach. J. Info. **3**(2), 143–157 (2009)
3. Y. He, Bayesian models for sentence-level subjectivity detection, in *Technical Report KMI-10-02, June 2010* (2010)
4. S.S. Rustamoy, E.E. Mustafayev. M.A. Clements, Sentiment analysis using neuro-fuzzy and hidden Markov models of text, in *IEEE Southeast Conference 2013* (Jacksonville, USA, 2013) (in press)
5. S.S. Rustamov, M.A. Clements, Sentence-level subjectivity detection using neuro-fuzzy and hidden Markov models, in *Proceedings of the 4th Workshop on Computational Approaches to Subjectivity, Sentiment and Social Media Analysis in NAACL-HLT 2013* (Atlanta, USA, 2013) pp. 108–114
6. S.S. Rustamov, E.E. Mustafayev, M.A. Clements, An application of hidden Markov models in subjectivity analysis, in *IEEE 7th International Conference on Application of Information and Communication Technologies AICT2013* (Baku, Azerbaijan, 2013), pp. 64–67
7. S.S. Rustamov, An application of neuro-fuzzy model for text and speech understanding systems/PCI'2012, in *The IV International Conference "Problems of Cybernetics and Informatics"*, vol. I (Baku, Azerbaijan, 2012), pp. 213–217
8. S.S. Rustamov, On an understanding system that supports human-computer dialogue/ PCI'2012, in *The IV International Conference "Problems of Cybernetics and Informatics"*, vol. I (Baku, Azerbaijan, 2012), pp. 217–221
9. J. Chu-Carroll, B. Carpenter, Vector-based natural language call routing. Comput. Linguist. **25**(3), 361–388 (1999)
10. F. Doshi, N. Roy, Efficient model learning for dialogue management, in *Proceedings of the ACM/IEEE international conference on Human-Robot Interaction (HRI'07)* (2007) pp. 65–72
11. I. Zitouni, H.-K.J. Kuo, C.-H. Lee, Combination of boosting and discriminative training for natural language call steering systems, in *Proceedings of the International Conference on Acoustics, Speech and Signal Processing* (Orlando, USA, 2002)
12. K.R. Aida-zade, S.S. Rustamov, E.A. Ismayilov, N.T. Aliyeva, Using fuzzy set theory for understanding user's intention in human-computer dialogue systems, in *Translation of ANAS, Series of Physical-Mathematical and Technical Sciences*, vol. XXXI, no. 6 (Baku, 2011), pp. 80–90 (in Azerbaijani)
13. K.R. Aida-zade, S.S. Rustamov, E.E. Mustafayev, N.T. Aliyeva, Human-computer dialogue understanding hybrid system, in *International Symposium on Innovations in Intelligent Systems and Applications (INISTA 2012)* (Trabzon, Turkey, 2012)
14. R. Fuller, Neural Fuzzy Syst. (1995)
15. J. Martineau, T. Finin, Delta TFIDF: An improved feature space for sentiment analysis, in *Proceedings of the 3rd AAAI International Conference on Weblogs and Social Media* (2009)
16. B. Pang, L. Lee, A sentimental education: Sentiment analysis using subjectivity summarization based on minimum cuts. *In Proceedings of the 42nd Annual Meeting on Association for Computational Linguistics (ACL)* (2004), pp. 271–278

Part II
Fundamentals of Fuzzy Sets

Game Approach to Fuzzy Measurement

V. L. Stefanuk

Abstract The game approaches are rather popular in many applications, where a collective of automata is used. In the present paper such a collective consists in a group of learning automata characterized with simple number parameters. The fuzzy measuring implemented as a game which is played sequentially with one automaton at a time, the result of the game defines next automaton to be played with. This game provides some measuring system that is very close to the procedure of collecting statistics in Probability Theory. For measuring of an unknown membership function a new concept has been introduced called Cognitive Generator which transforms a fuzzy singleton to ordinary crisp logic value. Considerations on various types of axiomatic approaches shows that the Cognitive Generator, as well as the Evidence Combination Axiomatic, belongs to one class of axiomatic theories, which may be used in application directly. The present paper contains also some programming examples aimed to illustrate our general approach.

1 Introduction

After deep theoretical study of Fuzzy Sets Theory, proposed by Prof. L. A. Zadeh [1], came the epoch of practical applications of this theory, which turned out to be quite successful, especially in the area of various Control Systems. Yet, some other practical questions still exist. The most intriguing question is related to the problem of understanding what actual decision is taken by a person or a machine when she/he/it being asked about the value of a phenomenon given with some membership value for it. It is a question of practical interpretation of fuzzy values. In another words, how to go from fuzzy membership functions to ordinary values used in the real

V. L. Stefanuk (✉)
Institute for Information Transmission Problems, Peoples' Friendship University of Russia, Moscow, Russian Federation
e-mail: stefanuk@iitp.ru

© Springer International Publishing AG, part of Springer Nature 2018
L. A. Zadeh et al. (eds.), *Recent Developments and the New Direction in Soft-Computing Foundations and Applications*, Studies in Fuzziness and Soft Computing 361, https://doi.org/10.1007/978-3-319-75408-6_7

physical world, where one deals with such crisp values as voltage, pressure, mass, and etc., and where the only uncertainty might be related to the precision of measurements of those values by technical tools and sometimes may be related to unknown probability of events.

Actually, in application of the Fuzzy Set Theory clever engineers had to work somehow with the problem of practical interpretation of fuzzy values. One of the first solution was the use of procedures of fuzzification and defuzzification [2], that provide the bridge between Fuzzy World and ordinary Physical Phenomena.

In [3] one reads: "The basic structure of Fuzzy approach consists of the following components: (1) Knowledge Base. (2) Fuzzification. (3) Inference Engine. (4) De-fuzzification". The authors then propose to use neuro-fuzzy technique to retrieve the data in faster manner with high speed and better usage of the above components.

Nevertheless in the paper [4] it is mentioned: "Mamdani type fuzzy controller with constant center for output classes may not perform satisfactory in all operating conditions". The authors also proposed to use neural network technology, which due to the neuron training will present an adaptive fuzzy controller. In our opinion this work is going in a right direction, yet the problem of learning was tackled in [4] outside of the fuzzy part of the system. In some sense their approach as well as approaches in [5, 6], may be considered as a modification of Mamdani approach.

Also we would like to stress that successful results in publications [4–6] were achieved for concrete installations, such as Power Control, and there is no general approach proposed that may be used in different fuzzy systems.

In present paper we continue our research [7–9], which aimed to find a way to measure fuzzy values using for this purpose learning automata. This line of research was first proposed in paper [7]. There our goal was formulated, namely to construct something that is reminiscent of statistics in Probability Theory.

In papers [8, 9] we obtained formulas for finite learning automaton in a fuzzy environment and explained the idea of measuring fuzzy singletons using this automaton. The idea was based on some important properties discovered and proved for such automata.

In order to reach above mentioned general goal the present paper provides five sections. In the next Sect. 2 we will give a brief survey of some appropriate results obtained in our previous publications.

In the Sect. 3 considerations will be provided concerning the concept of *Cognitive Generator* that helps to relate fuzzy values with the ordinary crisp ones. We will argue that the transformation must leads from a single membership value to one of two possible crisp values, that roughly speaking, are *False* and *True*. In a certain sense it is some new interpretation of a fuzzy value this time based on use of *PRN*.

To demonstrate the consistency and realistic of our approach in Sect. 4 there are some computer programs. These programs are in LISP symbolic language to keep the programming ideas obvious for reader.

In the Sect. 5 we put some obviously important considerations for various theoretical and practical schemes that are being built using various axiomatic means. We will provide a number of examples of different type of axiomatic tools, which may be classified either as *formal* or as *constructive* versions of axiomatic.

In Sect. 6 (conclusion) we try to visualize our plans for the future research.

2 Some Previous Results

In papers [8, 9] there were obtained formulas for finite learning automata being put in a fuzzy environment and it was explained how we expect to measure fuzzy values using those automata after some important properties have been proved for the automata.

The finite automaton under consideration was proposed by Tsetlin [10]. It is able to perform two actions, which will be denoted with the symbols **1** and **2**. It operates in discrete time $t = 0, 1, \ldots$, when it obtains penalties (or rewards) for its actions performed in each moment t. In the publication [10] the study of this automaton was performed for the case of *random environment*, which issues rewards or penalties for automaton actions with certain probabilities.

In [8, 9] and below it is assumed that each moment of time the finite automaton obtains either penalty with the membership value $\lambda^{(i)}$, or non-penalty with the complimentary fuzzy membership value $(1 - \lambda^{(i)})$, $i = 1, 2$. The defined environment is a stationary one, i.e. $\lambda^{(1)}$ and $\lambda^{(2)}$ do not change with time. The state transitions of the finite-state automaton are shown in the following figure.

In all the states belonging to the left side in Fig. 1 the automaton performs the first action (**1**) and changes its inner states when it obtains the penalty as shown in the upper graph, or changes its inner states when it obtains the reward as shown in the bottom graph in Fig. 1. The right side in Fig. 1 shows the inner state transitions when the automaton performs the second action (**2**).

Thus the situation here is different from that studied in [10], as presently this automaton is being put into a fuzzy environment and performs its actions in correspondence with the fuzzy membership values.

It may be shown that the performance of the linear automaton of Fig. 2 is controlled with a Generalized Markov Chain defined in [11], which might be referred to as the Markov-Stefanuk chain. It was possible to obtain the final

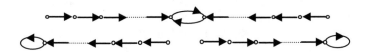

Fig. 1 Finite-state automaton with linear tactic [10]

expressions, describing the behavior of the linear automaton, taking into account also some results from [7].

Indeed, let the *memory depth* of the automaton shown in Fig. 1 be equal to n. It means that the automaton has exactly n inner states for each of its actions (**1** and **2**). Also it may be seen that the Markov-Stefanuk chain is ergodic one and one may speak about *final* membership values.

If $M^{(1)}$ is the final membership value for the first action and $M^{(2)}$ is the final membership value for the second action, then it was shown in [7] that the following expressions are valid (in the time limit, when $t \to \infty$):

$$M^{(1)} = 1 - \prod_{k=1}^{n} \left(1 - \mu_1^{(1)} \left(\frac{1 - \lambda^{(1)}}{\lambda^{(1)}} \right)^{k-1} \right) \tag{1}$$

$$M^{(2)} = 1 - \prod_{k=1}^{n} \left(1 - \mu_1^{(1)} \frac{\lambda^{(1)}}{\lambda^{(2)}} \left(\frac{1 - \lambda^{(2)}}{\lambda^{(2)}} \right)^{k-1} \right) \tag{2}$$

In [7] it was also shown that under additional condition

$$\frac{1 - \lambda^{(i)}}{\lambda^{(i)}} > 1, \; i = 1, 2, \tag{3}$$

which will not be discussed in current paper, the finite automata of Fig. 1 is able to learn to perform the optimal action, obtaining eventually the lesser penalty.

When $n \to \infty$ one may see from (1), (2) that $M^{(1)} \to 1$, $M^{(2)} \to 0$. It means that this finite automaton in our fuzzy environment has the important property of *asymptotic optimality* [10]. In other words, this automaton lets one to establish for sure, which of the following relations are true, either $\lambda^{(1)} > \lambda^{(2)}$, or $\lambda^{(1)} < \lambda^{(2)}$.

3 Cognitive Generator

The membership value is a theoretical concept showing up to what degree corresponding fuzzy value belongs to a set [1]. Our purpose in this chapter is to provide a means allowing interpret membership function as a crisp value, when it is needed.

For example, consider a point x on the abscises axis and corresponding membership function value $\mu(x)$ for some fuzzy value x, say graphically represented. Being requested the Cognitive Generator produces one of two results, corresponding to *True* or *False* in the crisp world. In another words, each moment when the Generator is requested it may produce one of two values depending on $\mu(x)$. It might be 1 (True) or 0 (False) in case the *Feedback Value* [8, 9] for the learning

Fig. 2 Learning machine built in 1957–1961 [17]

automaton in a fuzzy environment, or *Emptiness Value,* if one is estimating how empty is the glass of water.

The graphical representation is the most popular in Fuzzy set text books [12]. The scholars usually demonstrate a membership curve, describing the membership function for each point *x*, showing the membership value for *Feedback* or *Emptiness* in this point. Yet, having graphical representation, a person still have to take *a decision* on the phenomenon under consideration and say False or True, or, for mathematical convenience 0 or 1.

Thus the graphical description is not a total description of some fuzzy value, as it does not show what *final decision* would be taken with respect to *Emptiness* or *Feedback.* For the final decision one needs some binary *Cognitive Generator*, which takes decision formulated somehow in the brain of a person, or which was produced in some technical device, like our finite learning machine shown in Fig. 2 above.

Trying to built such a generator and make its decisions as close as possible to the human brain decisions, please note that in practice the result of Cognitive Generator is not always the same even for the same point *x*. This fact was observed during teaching our learning machine Fig. 2 originally intended to work in a probabilistic environment.

In our research [8, 9] we specially stressed that the probability is not observed value by itself. One might observe only some statistically collected numbers. Those old days (1957–1961) people did not know such a science as Fuzzy Set Theory discovered by Professor Lotfi A. Zadeh in 1965 [1]. As it was mentioned in [7], myself, being the teacher of the machine in Fig. 2, in fact used some fuzzy considerations for punishments or rewards.

It is important that, though the fuzzy value is not observed either, yet the parameter $\mu(x)$ is a number that maybe easily stored in the brain as it is. And if the brain holds some Generator that we are looking for in this chapter, it would not be a problem, what button should be pushed in the machine depicted in Fig. 2.

Now we have to acknowledge that this Cognitive Generator in the brain acts in parallel with many other processes and hence it works in a very noisy environment. Yet, it is not the noise as it is considered in Information Theory and it is not a collection of errors of some kind either. This noise has a logical origin.

We have considered this type of the interferences in our invention of Collective of Radio Stations [13] that many decades later became known throughout the world as Mobile or Cellular Communication and found various engineering implementations. It was shown in [13] that the regular transmission between stations A and B appears as a noise for any other parties of speakers.

Now continuing our imitation of brain functioning one may conclude that the logical noise might be produced using some logical device, which is known as *Pseudo Random Number Generator (PRNG)*. It is important that unlike the probability p the numbers generated with PRNG are actual ones, i.e. they are observed number.[1]

All the above considerations were made to convince the reader that for interpretation of certain fuzzy value one needs not only membership function $\mu(x)$ but also a *PRN* number r produced with *PRN* or something similar (see below). It is important to stress that each moment t, when the Generator obtains a request, it may produce 1 or 0, and the value of $r(t)$ is unpredictable.

Probably this explains why having the same value $\mu(x)$ our Brain returns decisions that is related to this value but not in a straightforward manner.

One may propose several implementations for such a generator.

The value produced with the binary Cognitive Generator is related to the fuzzy value $\mu(x)$ as we mentioned above.

One may say for sure that the Cognitive Generator, obtaining $G(x(t), \mu(t))$, may produce the *same quantity* for a fuzzy value $\mu(x)$ in the moments $t = 1, 2, \ldots$ only in two extreme cases, namely when $\mu(x(t)) = 0$ or when $\mu(x(t)) = 1$. In all other cases the produced quantity might be either 0 or 1, as it depends on the performance of Cognitive Generator.

For technical implementation of such a Generator the following versions may be considered:

1. Simple Cognitive Generator: provides 1 if $\mu(x(t)) \geq 0.5$ and provides 0 otherwise.
2. Type 2 Cognitive Generator, when a binary function is provided for each $\mu(x(t))$. This is closely related to the Type 2 fuzzy theory introduced by L. A. Zadeh. Probably, this function is different for different fuzzy phenomena.
3. Frequency generator. In this case the frequency of generation 1 will be more the greater is the value of $\mu(x(t))$. Again, exact construction of the frequency generator depends on application.

---

[1]Such Pseudo Random Numbers Generators are used in most of computers to imitate probability.

4. In the case of our learning automata [8, 9] and in some other technical devices the Cognitive Generator may be described somehow else. In automata mentioned in Chapter 2 an input to the technical device is sent as the result of some initial generations of λ_i, $i = 1.2$. Then the membership functions are calculated following some previous values. It is quite suitable for theoretical studies using appropriate Markov-Stefanuk chains.

5. In general case the Cognitive Generator might be an arbitrary function $f(\mu, z)$ that produces 0 or 1, provided that $f(1, z) = 1$ and $f(0, z) = 0$. This approach reminds the axioms that were used in [14] to get an expression for the formula of combining evidences that turned out to be a T-norm.

In present paper we propose the following expression for the $f(\mu, z)$ that we exhibit as a function in *LISP* language. This function uses a *LISP* library function RANDOM, the latter produces PRN, of course.

```
(defun generator ( μ )
        (prog(z)
            (setq z (random 1.0))
            (print z)
            (if (>= (-1  μ ) z) (return 0)(return 1))))
```

Examples:

```
> (generator 0.8)
0.9087889365840982
1

> (generator 0.8)

0.3109750013996597
1
> (generator 0.8)
```

We expect that such Cognitive Generator may be better for usage, than the procedures of fuzzification and de-fuzzification mentioned above.

Besides, the concept of Cognitive Generator promises to help to create *Perfectly Fuzzy Systems (PFS)*, which avoids above mentioned heuristics, and hence may improve the performance of various technical systems.

4 Demo Programs

The next program *tsetlin* describes the behavior of Linear Automaton by M. L. Tsetlin [10], performing two actions with two membership functions, certain initial state, and having n inner states for each action. The result of its learning is obtained after certain number of steps (see examples below).

```
(defun tsetlin (depth lambd1 lambd2 steps)
       (prog (state action n m1 m2)
         (setq state 1) ; initial state=1, action=1
         (setq action 1)
         (setq m1 1)    ; m1 is fuzzy-value for action 1
         (setq m2 0)    ; m2 is fuzzy-value for action 2
         (setq n (+ 1 steps))
       tag1
         (setq n (- n 1))
         (cond ((<= n 0)(return (list m1 m2))))
         (cond ((eq action 1)(if (eq (generator lambd1) 0)
                    (progn
                     (setq action 2)
                     (setq m2 (+ m2 1))
                     (go tag1))
                    (progn
                     (setq m1 (+ m1 1))
                     (go tag1)))))
         (cond ((eq action 2)
                  (if (eq (generator lambd2) 0)
                     (progn
                      (setq action 1)
                      (setq m1 (+ m1 1))
                      (go tag1))
                     (progn
                      (setq m2 (+ m2 1))
                      (go tag1)))))
       (t t)))
```

Examples:

(tsetlin 1 0.4 0.3 100) → (60 41)

(tsetlin 1 0.4 0.4 100) → (50 51)

(tsetlin 1 0.4 0.5 100) → (45 56)

An auxiliary function *direct* uses above function *tsetlin*:

```
(defun direct (depth lambd1 alpha steps)
       (prog (list temp1 temp2 temp3setq)
         (setq list (tsetlin depth lambd1 alpha steps))
         (if (> (car list) (cadr list)) (return ">")
             (return "<"))))
```

and the final function *reply*, used to get the value of unknown *lambd1*, is the following one:

```
(defun reply (depth lambd1 steps)
       (prog (numbers temp1 temp2 temp3 temp4)
         (setq numbers '(1 0.9 0.8 0.7 0.6 0.5 0.4 0.3 0.2 0.1 0))
       tag1
         (setq temp1 (car numbers))
         (setq numbers (cddr numbers))
           (when (null numbers) (return temp1))
           (if (string-equal(direct depth lambd1 temp1 steps) "<")
              (go tag1) (return temp1))))
```

Examples:

(reply 4 0.4 100) → 0.4

(reply 4 0.65 100) → 0.6

Thus, *reply* shows that unknown membership (0.4) is 0.4.

Also function *reply* shows that unknown membership (0.65) is equal to 0.6 with decimal precision.

The linear tactic automaton in above examples was taken with the depth $n = 4$ and we have been waiting 100 steps for the final decision. (Thus, unknown membership values are represented within the programs as some sequences of ones and zeroes during 100 steps, using *Cognitive Generator*. The learning automaton decides with the *reply* function what is the value under measuring.)

5 Axiomatic Systems

Sometimes in axiomatic theories the relation between theory and reality, which is the area of its application, is broken. Indeed, it is not always clear how to transfer from the theoretical concepts to their practical counterparts.

Let us consider some examples.

1. *Non-deterministic automata*, having excellent logical and theoretical consistency does not provide a direct possibility of its use in applied systems and tools. (Note that in the present paper we used deterministic automatf having deterministic number of actions, states, and state transactions.)
2. *Category Theory* allows to establish a connection among various CT schemas. It allows to find some optimal decisions within the frames of the theory, and etc. However CT creates only conceptual connections among elements, the latter may not be used directly in practice, giving only some intuitive intellectual hints for practical solutions, and providing theoretical explanation for decisions using an analogy with those theoretical schemas that have being built within the theory [15]. That is why for the theoretical constructions in CT one needs a process of their interpretation [15].
3. *Probability Theory* in many cases may be directed applied to practical problems, which is the result of existence of some limit theorems, which create a bridge between theory and practice.
4. *Fuzzy Sets Theory* created by the outstanding research by Prof. Zadeh, has some properties making it difficult sometimes to apply it directly. Hence, for an interpretation of fuzzy values there were proposed some procedures of fuzzification and de-fuzzification.

Another possibility is our proposal of Cognitive Generator described above in this paper.

Again remaining strictly within the fuzzy theory one may obtain strict logical relations connecting one set of membership functions to another [2]. In the present paper we introduced other ways of interpretation of the concept of fuzziness trying to solve the problem of measuring fuzzy values [3–7].

For this purpose we made a research of behavior of deterministic learning automata in fuzzy environment and proved that the automaton is having an important property of asymptotic optimality that was previously discovered by

Tsetlin for probabilistic environment [10], and having some other important behavioral properties that may be used in practice.

Returning to Fuzzy Sets in the frames of axiomatic approach introduced in [1], while working with the membership functions, it is not easy to "make a decision". Yet, it became possible due to the important results on evidence accumulation obtained in [14], which were in a good agreement with practical axioms for the evidences, and hence may be directly transferred from theory to practice (and back). Also a theory of Markov-Stefanuk chains was helpful for the purpose, being generalization of classic Markovian Chains [11].

The behavior of the learning automata was studied strictly within the Fuzzy Sets Theory, however the many properties of the results look quite natural and applying our special way of fuzzy interpretation we obtained a practically valuable result.

5. *Quantum Mechanics* also should be mentioned among the axiomatic theories where physicists found interpretations for essential theoretical concepts to find their correspondence with concrete physical observations. However some new for the practice concepts were introduced that were absent before the axiomatic has been created (the uncertainty relation, Planck constant and etc.).

After some time practical physicists got used to the new concepts and accepted them as an inherent physical properties and observations. In this respect it is worth to point to the recent publication [16] where Prof. Vladik Kreinovich has pointed to the deep correspondence between Quantum Theory and another axiomatic theory—Fuzzy Set Theory.

Mamdani [2] produced one of the possible ways to connect theory with reality at least in the area of Control Systems.

Another approach related to direct interpretation of membership functions is proposed in present paper. Within the axiomatic of Fuzzy Sets we have one value – membership function $\mu(x)$, showing the grade of belonging x to a Fuzzy Set A.

Speaking on practical use of the value $\mu(x)$ it should be understood that the actual practical decision, that a person must take, consists in either one of two crisp values.

Indeed given graphical representation for phenomenon A at each point x the person has to produce only one of two values *True* or *False*: "Whether the person of age x is young or not". In this paper we offer a concept of *Cognitive Generator* to generate *True* or *False*. It gives the corresponding interpretation of $\mu(x)$ applying the pseudorandom numbers. In Sect. 3 we use a concrete formula for such an interpretation, understanding that this is not the only possibility.

There are two considerations to support our approach. One is the practical axiomatic that our formula must satisfy (see above). In addition the results obtained are quite satisfactory as they open a possibility to measure membership function with a given precision, finally resolving the problem formulated in sequence of publications [7–9].

Yet, we keep for the future some further considerations of the problem of such generation, realizing that it is quite possible that there is a set of such interpretations, which may depend upon the Subject Domain in question.

At this point we would like to stress that both the probability of $\Pr(A)$ and the fuzzy membership $\mu(x)$ are unobserved values, and we feel that this is the main point that relates Fuzzy Set Theory to Probability Theory. In this respect please see the careful study of relation of Fuzzy and Probability in the corresponding publications by Prof. L. A. Zadeh.

In present paper we showed that Cognitive Generator allows us to implement our central idea of measuring fuzzy value with the help of learning automation, and to resolve our problem of building an analog for statistics in probability area.

Our approach was called Game Approach as a collection of automata are used for measuring $\mu(x)$. The power of this set depends on the precision of measuring. Note that the Western scientists prefer to call the *collective of automata* [17] as *Multiagent systems*, which is not completely fair as the term *collective behavior* was introduced in 1965 in [17], i.e. some two dozens years before the term multiagent systems was ever mentioned in AI literature.

6 Conclusion

The main point under consideration is that within the frames of axiomatic theory, allowing to work with membership functions, it is impossible to reach a definite decision, if a subject comes across with such a necessity. Yet it is natural to propose that depending on the value $\mu(x)$ human being or machine are taking a decision *True* or *False*. The Cognitive Generator or an interpreter solves this problem using *GPRN*. In the paper we proposed a concrete scheme of such a generator, realizing that it is not a unique possibly.

In future we plan to establish experimentally, which interoperations are used by human. Maybe many versions of such interpretation do exist, possibly depending even on the subject domain. Presently we do not know it.

As it was demonstrated in this paper, having such a Cognitive Generator it was not difficult to finalize our idea of measuring fuzzy value with the help of learning automaton [10] and resolve our central idea, formulated in [7–9], of designing some Fuzzy Set Theory analog of statistics being used in Probability Theory.

Acknowledgement The research is partially supported with Russian Fond for Basic Research, Grant 15-07-07486, and with Program 1.5Π of Presidium of Russian Academy of Sciences.

References

1. L.A. Zadeh, Fuzzy sets. Inf. Control (USA) (8), 338–348
2. E.H. Mamdani, Application of fuzzy logic to approximate reasoning using linguistic synthesis. IEEE Trans. Comput. **26**(12), 1182–1191 (1977)
3. S. Katke, M. Pawar, T. Kadam, S. Gonge, Combination of neuro fuzzy techniques give better performance. Int. J. Comput. Sci. Inf. Technol. **6**(1), 550–553 (2015)
4. S. Mishra, P.K. Hota, P. Mohanty, *A Neuro-fuzzy based Unified Power Flow Controller for Improvement of Transient Stability Performance*. CiteSeer
5. S. Mishra, A.K. Pradhan. P.K. Hota, Development and implementation of a fuzzy logic based constant speed DC drive. J. Inst. Eng. (India) (pt EL) **79**, 146 (1998)
6. M. Senthil Kumar, P. Renuga, K. Saravanan, Adaptive neuro-fuzzy based transient stability improvement using UPFC. Int. J. Recent Trends Eng. **2**(7), 127 (2009)
7. V.L. Stefanuk, Behavior of Tsetlin's learning automata in a fuzzy environment, in *Second World Conference on Soft Computing (WConSC)* (Letterpress, Baku, Azerbaijan, 2012), pp. 511–513
8. V.L. Stefanuk, Interaction using qualitative data, in *4th World Conference on Soft Computing. Program of Conference*, Plenary talk (abstracts) (2014), pp. 43–44
9. V. L. Stefanuk, How to measure qualitative data, in *Proceedings of American Fuzzy Information Processing Society NAFIPS'2015 and 5th World Conference on Soft Computing, Redmond, USA* (2015), pp. 37–40
10. M.L. Tsetlin, Some problems of finite automata behaviour. Doklady USSR Acad. Sci. (Moscow) **139**(4), (1961)
11. V.L. Stefanuk, Deterministic Markovian chains. Inf. Process. (Moscow: IITP RAS (in Russian)) **11**(4), 702–709 (2011)
12. T. Munakata, *Fundamentals of the New Artificial Intelligence* (Springer, New York, 1998), pp. 231
13. V.L. Stefanuk, M.L. Tzetlin, On power control in the collective of radio stations. Inf. Trans. Prob. **3**(4), 59–67 (1967)
14. V.L. Stefanuk, Should one trust evidences? in *Proceedings of the* All-country *AI Conference*, vol. 1. (Moscow, 1988), pp. 406–410
15. A. Zhozhikashvili, V.L. Stefanuk, Theory of category approach to knowledge based programming, in Knowledge-Based Software Engineering, Communication in Computer and Information Science, vol. 466, *Proceedings of the 11th Joint Conference, JCKBSE 2014* (Springer International Publishing, Switzerland, 2014), pp. 735–746
16. V. Kreinovich, Formalizing the informal, precisiating the imprecise: how fuzzy logic can help mathematicians and physicist by formalizing their intuitive ideas, in *NAFIPS 2015*, Keynote Speech
17. V.L. Stefanuk, An example of collective behaviour of two automata. Autom. Remote Control (Moscow (in Russian)) **24**(6), 781–784 (1963)

Towards Real-Time Łukasiewicz Fuzzy Systems

Barnabas Bede and Imre J. Rudas

Abstract Łukasiewicz fuzzy systems are fuzzy systems based on Łukasiewicz implication and Łukasiewicz t-norm and t-conorm as fuzzy operations. They are deeply rooted in classical logic while being fuzzy systems, so they establish a connection between classical logic and fuzzy logic. Łukasiewicz fuzzy systems with Center of Gravity defuzzification have been shown to have good approximation properties, however Center of Gravity defuzzification makes them to be computationally not very efficient. In the present paper we develop a real-time Łukasiewicz fuzzy system, using the Mean of Maxima defuzzification. This defuzzification will be directly computable for Łukasiewicz systems with certain properties. We investigate approximation properties of such systems and we obtain a generalization of a previous universal approximation result.

1 Introduction

Currently, fuzzy systems of Mamdani [1] and Takagi-Sugeno types [2] are wastly dominating the fuzzy logic literature and they are both very successful in applications. Mamdani systems have a full linguistic interpretation, which makes their design easy and their applicability immediate. Takagi-Sugeno systems, have piecewise linear, or possibly higher order output, and they are performing much better from the point of view of computational complexity, learning and adaptability. This makes Takagi-Sugeno systems perform much better in real-time applications, however their interpretability on the output domain is lost as compared to Mamdani systems. Łukasiewicz logic [3, 4] is a fuzzy logic based on Łukasiewicz operators.

B. Bede (✉)
Department of Mathematics, DigiPen Institute of Technology, 9931 Willows Rd. NE,
Redmond, WA 98052, USA
e-mail: bbede@digipen.edu

I. J. Rudas
Óbuda University, Bécsi út 96/b, Budapest 1034, Hungary
e-mail: rudas@uni-obuda.hu

© Springer International Publishing AG, part of Springer Nature 2018   85
L. A. Zadeh et al. (eds.), *Recent Developments and the New Direction
in Soft-Computing Foundations and Applications*, Studies in Fuzziness
and Soft Computing 361, https://doi.org/10.1007/978-3-319-75408-6_8

Łukasiewicz fuzzy systems have been shown to be a viable alternative to the Mamdani approach from the point of view of their approximation properties, and also, from the point of view of their theoretical background. Ł ukasiewicz logic has deep roots in classical logic, making it possible to make connections between fuzzy logic and various other areas of classical logic. Approximation properties of fuzzy systems were investigated by several authors [5–10] but mainly in the above mentioned Mamdani and Takasi-Sugeno approaches. Recently in [11, 12], Łukasiewicz fuzzy systems were investigated from the point of view of their approximation properties. We continue the line of research in [11] but also, we would like to construct a new type of Łukasiewicz fuzzy system with smaller computational complexity, as compared to the approach proposed in [11]. To achieve an improvement in this direction we replace the Center of Gravity defuzzification procedure with the Mean of Maxima defuzzification, which for a Łukasiewicz fuzzy system can be directly computed without needing the evaluation or approximation of an integral. This way, we get a real-time Ł ukasiewicz fuzzy system. Also, in the present paper the requirements of [11] are weakened, so the antecedents and consequences are no longer required to add up to 1 at every value in the corresponding domains. Let us mention here that the present paper does not want to replace Mamdani and Takagi-Sugeno fuzzy system, instead we want to enlarge the set of possible choices with a new type of fuzzy system that is fully linguistic, computationally powerful and with strong connections to classical logic.

The paper begins with introducing and showing the graphical interpretation of a Łukasiewicz fuzzy system, followed by the discussion of the Łukasiewicz system with Mean of maxima defuzzification, and is concluded by examples with one or more antecedents.

2 Łukasiewicz Fuzzy System

Let us consider a fuzzy rule base

$$\text{If } x \text{ is } A_i \text{ then } y \text{ is } B_i, i = 1, \dots, n.$$

Mamdani systems model the implication via the min operation $x \wedge y = \min\{x, y\}$, which is a t-norm, and combine the fuzzy rules via the max t-conorm $x \vee y = \max\{x, y\}$ as follows:

$$R(x, y) = \bigvee_{i=1}^{n} A_i(x) \wedge B_i(y).$$

Interpretation of a rule base via the Mamdani approach is possible using a disjunction. In Mamdani's approach some of the rules are active to certain degree (firing level) while others are inactive. The most active rules have a higher weight in the composition. Mamdani system in fact can be more precisely modeled by fuzzy conjunctive rules

$$x \text{ is } A_i \text{ and } y \text{ is } B_i, i = 1, \ldots, n,$$

with a disjunctive composition for a rule base

$$\text{Or}_{i=1}^{n}(x \text{ is } A_i \text{ and } y \text{ is } B_i).$$

Let us assume that the fuzzy system described above receives a fuzzy input A'. Then the fuzzy output is

$$B'(y) = \bigvee_{i=1}^{n} \left(\bigvee_{x \in X} A'(x) \wedge A_i(x) \right) \wedge B_i(y).$$

The loss of interpolation property of Mamdani fuzzy systems is a well known property and it states that if $A' = A_i$ then in general we have $B' \neq B_i$. It shows that the implicative formulation of fuzzy rules is not necessarily precise in view of the modus ponens rule. The property which we consider is that if the fuzzy system of Mamdani type receives a fuzzy output that coincides with an antecedent, the Mamdani rule base would not return the corresponding consequence. If the input is A_i then the fuzzy output will not be B_i. The disjunctive-conjunctive formulation of the Mamdani rule base does not contradict the interpolation property. The properties discussed here surely do not limit the applicability of the Mamdani systems, which are extremely successful. It is sure that they will continue to be the number one choice of designers who wish to use a fuzzy system in a certain application, but we need to provide a more precise formulation of their properties, to avoid some criticism.

For the case of a singleton input $x = x_0$ the fuzzy output can be calculated in a simple way [13]

$$B'(y) = \bigvee_{i=1}^{n} A_i(x_0) \wedge B_i(y).$$

The other drawback of the Mamdani fuzzy system is the computational complexity, that comes from using the Center of Gravity defuzzification. If the support of the ouptut $B'(y)$ of a fuzzy system is W then we have

$$COG(B') = \frac{\int_W y \cdot B'(y) dy}{\int_W B'(y) dy}.$$

Surely, numerical approximation of the values of the integrals is a time consuming operation.

Despite enumerating some deficiencies of the Mamdani fuzzy systems, the goal of the present paper is not to diminish their importance or to advise researchers against them. In contrary, we recognize the huge importance of the Mamdani approach and we are convinced that Mamdani fuzzy systems should be the first choice when implementing a fuzzy controller. The goal of the present paper is to open up new directions, increasing the overlap between classical and fuzzy logic, as we are convinced of the fact that these are not competing but complimentary research directions, with a large non-empty intersection.

In what follows we will construct a Łukasiewicz fuzzy system. Let us first recall that the Łukasiewicz t-norm and t-conorm are defined respectively as

$$x \wedge_L y = \max\{x + y - 1, 0\},$$

$$x \to_L y = \min\{1 - x + y, 1\}$$

Łukasiewicz fuzzy systems are modeling a cause-effect relationship. The fuzzy rule base that we consider is

$$\text{if } x \text{ is } A_i \text{ then } y \text{ is } B_i, i = 1, \ldots, n.$$

This fuzzy rule base can be interpreted through a conjunctive-implicative model

$$\text{And }_{i=1}^{n} (\text{if } x \text{ is } A_i \text{ then } y \text{ is } B_i).$$

This fuzzy rule base can be now naturally modeled by the fuzzy relation

$$R(x, y) = (\wedge_L)_{i=1}^{n} (A_i(x) \to_L B_i(y)).$$

If an input $A'(x)$ is considered then the fuzzy output is

$$B'(y) = (\wedge_L)_{x \in X} \left[A'(x) \to_L \left((\wedge_L)_{i=1}^{n} (A_i(x) \to_L B_i(y)) \right) \right].$$

It is known [4] that Łukasiewicz fuzzy systems have the fuzzy interpolation property, i.e., if the input is one of the antecedents $A' = A_i$ then the output of the fuzzy system will be the corresponding consequence B_i. For a singleton input $x = x_0$, similar to the Mamdani system, we have

$$B'(y) = (\wedge_L)_{i=1}^{n} (A_i(x_0) \to_L B_i(y)).$$

For the construction of a single input single output system we need to add a defuzzification procedure, which can be the center of gravity defuzzification, which is used in the Mamdani systems as well. In [11] this approach was investigated. The resulting fuzzy systems can be written as

$$F(f,x) = \frac{\int\limits_{c}^{d} \left[(\wedge_L)_{i=1}^{n} A_i(x) \rightarrow_L B_i(y) \right] \cdot y \cdot dy}{\int\limits_{c}^{d} \left[(\wedge_L)_{i=1}^{n} A_i(x) \rightarrow_L B_i(y) \right] \cdot dy}.$$

These systems have good approximation properties, however, they have the same drawback as the Mamdani fuzzy systems, i.e., the complexity of the defuzzification algorithm.

Takagi-Sugeno fuzzy systems, have fuzzy input and crisp output, and they are computationally much more convenient as compared with the Mamdani approach, but their interpretability is reduced, as the output is not a fuzzy set any more, but a function with possibly learned parameters.

In the present paper we propose a novel fuzzy system, based on the Łukasiewicz fuzzy system, with a Mean of Maxima defuzzification. The proposed defuzzification will have advantages both with respect to Mamdani and the Łukasiewicz systems proposed in [11]. The first advantage will be theoretical. It helps us to release some hard requirements in the results of [11] where both antecedents and consequences need to form a Ruspini partition, otherwise the resulting fuzzy output will not have a compact support, so it does not allow calculation of a COG. The second advantage is computational. We will show that in a large class of Ł ukasiewicz fuzzy systems, i.e. those with L-R fuzzy sets modeling antecedents and consequences the defuzzification step is directly computable.

In [11] an interesting graphical interpretation was given. Let us recall that interpretation as it will be important for the following discussion. To show this interpretation we consider a fuzzy rule base with 3 fuzzy rules

$$\text{If } x \text{ is } A_i \text{ then } y \text{ is } B_i, i = 1, 2, 3,$$

with Gaussian membership functions for both antecedents and consequences. In [11] a similar example was shown with triangular functions. Here we illustrate the Łukasiewicz fuzzy system with Gaussian functions and we observe that the support of the fuzzy input is in general unbounded. Figure 1 illustrates the antecedents and the crisp input considered.

In Fig. 2 the output of the fuzzy system is shown together with the COG defuzzification. Łukasiewicz t-norm is used to model the conjunction of the outputs of rule 2 and rule 3. The output of the first rule is not shown on the figure, as it is constant 1, so it is neutral with respect to Łukasiewicz t-norm.

We observe if we denote by B' the fuzzy output and if we assume an unbounded domain we obtain that $\lim_{y \to -\infty} B'(y) \neq 0$ and $\lim_{y \to \infty} B'(y) \neq 0$, unless the antecedents and consequences form a Ruspini partition, i.e.,

$$\sum_{i=1}^{n} A_i(x) = \sum_{i=1}^{n} B_i(y) = 1.$$

Fig. 1 Antecedents of a
Łukasiewicz fuzzy system

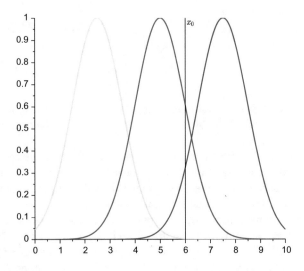

Fig. 2 Consequences,
individual output of the rules
(cyan = rule 1, blue = rule 2,
green = rule 3) the fuzzy
output (dashed line) and the
COG defuzzification of the
fuzzy output

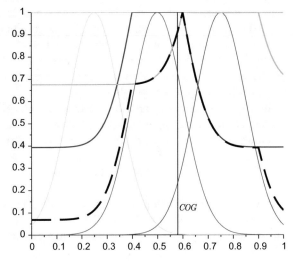

So if the Ruspini conditions do not hold then the considered fuzzy systems on
unbounded domain will not have a COG defuzzification, as the integrals diverge.
Practical situation of course will avoid that as the domains are bounded, but we would
like to bring correct solution in any situation.

3 Łukasiewicz Fuzzy Systems with Mean of Maxima Defuzzification

To get a more meaningful and computationally less expensive defuzzification for the Łukasiewicz fuzzy systems we consider Mean of Maxima defuzzification, instead of the Mean of Maxima. First we will prove a lemma that calculates the individual rule outputs.

Lemma 1 *The output of the ith fuzzy rule in a Łukasiewicz fuzzy system is given by*

$$A_i(x) \to_L B_i(y) = \begin{cases} 1 - A_i(x) & \text{if} \quad B_i(y) = 0 \\ 1 - A_i(x) + B_i(y) & \text{if } 0 < B_i(y) \leq A_i(x) \\ 1 & \text{if} \quad A_i(x) < B_i(y) \end{cases}.$$

We will prove an approximation theorem for a fuzzy system of Łukasiewicz type with MOM defuzzification. To construct an approximation operator we can consider any continuous function $f : [a, b] \to [c, d]$. Let us consider the data $x_0, x_1, \ldots, x_{n+1} \in [a, b]$ and $f(x_i) = y_i$, $i = 1, \ldots, n + 1$. We consider the fuzzy rules

If x is about x_i then y is about y_i, $i = 1, \ldots, n$.

The following theorem will give a formula for the Mean of Maxima defuzzification for a large class of Łukasiewicz fuzzy systems.

The error estimate that we obtain is using the modulus of continuity

$$\omega(f, \delta) = \sup\{|f(x) - f(y)| : |x - y| \leq \delta\},$$

where $f : [a, b] \to \mathbb{R}$ is a continuous function.

We denote by $(A)_0 = cl\{x \in X : A(x) > 0\}$ the closure of the support of a fuzzy set, and in general, by $(A)_\alpha = \{x \in X : A(x) \geq \alpha\}$, its α-level set, for $\alpha \in (0, 1]$ (please note that $(A)_1$ stands for the core of the fuzzy set).

The following theorem shows that any continuous function can be approximated by a Łukasiewicz fuzzy system under more relaxed conditions as compared with the results in [11].

Theorem 1 *Let us consider a continuous function $f : [a, b] \to \mathbb{R}$ and $y_i = f(x_i)$, $i = 0, \ldots, n + 1$. Then f can be uniformly approximated by a Łukasiewicz fuzzy system*

$$F(f, x) = MOM \left[(\wedge_L)_{i=1}^n (A_i(x) \to_L B_i(y)) \right]$$

with continuous membership functions for the antecedents and consequences $A_i, B_i, i = 1, \ldots, n$ satisfying

(i) $(A_i)_0 \subseteq [x_{i-1}, x_i, x_{i+1}]$
(ii) $\sum_{i=1}^n A_i(x) \geq 1$.

(iii) $(B_i)_0 \subseteq [\min\{y_{i-1}, y_i, y_{i+1}\}, \max\{y_{i-1}, y_i, y_{i+1}\}]$
(iv) $(B_i)_1 \neq \emptyset$.
Moreover the following error estimate holds true

$$\|F(f, x) - f(x)\| \leq 3\omega (f, \delta),$$

with $\delta = \max_{i=1,\dots,n}\{x_i - x_{i-1}\}$.

Proof First let us observe that if $A_i(x) = 0$ then the corresponding individual rule output is $A_i(x) \to_L B_i(y) = 1$, which is the neutral element with respect to the Łukasiewicz conjunction $\wedge_L$ so they do not impact the result of the fuzzy system. So we observe that under the given conditions there are two rules that are active at a time. If $x \in [x_j, x_{j+1}]$ these are rules j and $j + 1$.

$$|F(f, x) - f(x)| = |MOM((\wedge_L)_{i=j}^{j+1} A_i(x) \to_L B_i(y)) - f(x)|.$$

Let us assume that y is outside the supports of both B_j and B_{j+1}. Then we have $B_j(y) = B_{j+1}(y) = 0$ and we get

$$(A_j(x) \to_L B_j(y)) \wedge_L (A_{j+1}(x) \to_L B_{j+1}(y)) = (1 - A_j(x)) \wedge_L (1 - A_{j+1}(x))$$
$$= \max\{1 - A_j(x) - A_{j+1}(x), 0\}.$$

The condition (ii) of the theorem ensures that $A_j(x) + A_{j+1}(x) \geq 1$ which gives

$$(A_j(x) \to_L B_j(y)) \wedge_L (A_{j+1}(x) \to_L B_{j+1}(y)) = 0,$$

for any $y \in Y$ outside of the support of B_j and B_{j+1}.
    Taking now into account that

$$(B_i)_0 = [\min\{y_{i-1}, y_i, y_{i+1}\}, \max\{y_{i-1}, y_i, y_{i+1}\}]$$

we can restrict the output to the union of the supports of B_j and B_{j+1}.
    Based on Lemma 1, and since B_i is normal (iv) and continuous, for any fixed $x \in X$ there exist $y \in Y$ such that $B_i(y) \geq A_i(x)$ MOM is the average of the elements y such that $B_j(y) \geq A_j(x)$ in which case the output of the rule number j is 1, so it is neutral in the fuzzy system on the given interval. If we repeat the reasoning for the output of rule $j + 1$ then we have $B_{j+1}(y) \geq A_{j+1}(y)$. Then if the intersection

$$\{y : |B_j(y) \geq A_j(x)\} \cap \{y|B_{j+1}(y) \geq A_{j+1}(x)\}$$

is non-empty then the MOM is in the intersection, which is a value

$$y \in [\min\{y_{j-1}, y_j, y_{j+1}, y_{j+2}\}, \max\{y_{j-1}, y_j, y_{j+1}, y_{j+2}\}].$$

If the intersection is empty then the Łukasiewicz t-norm will reduce the joint output of rules j and $j+1$ and then the output of $\wedge_L$ is within the two intervals, as output of other rules is at 1. As Łukasiewicz t-norm, implication, A_i and B_i are all continuous, the output of the Łukasiewicz fuzzy system is continuous as a function of y so, it is bounded and it attains its bounds, based on the Bolzano-Weierstarass property. Then as a conclusion we have

$$|F(f,x) - f(x)| = |y - f(x)|$$

for some

$$y \in [\min\{y_{j-1}, y_j, y_{j+1}, y_{j+2}\}, \max\{y_{j-1}, y_j, y_{j+1}, y_{j+2}\}].$$

Further, it is easy to see that there exists a $z \in [x_{j-1}, x_{j+2}]$ such that $f(z) = y$.
Finally we get

$$|F(f,x) - f(x)| = |f(z) - f(x)| \leq \omega(f, |z - x|) \leq\leq \omega(f, 3\delta) \leq 3\omega(f, \delta).$$

∎

Let us remark here that the conditions in the universal approximation result in [11] were stronger if we compare them to the conditions in the previous theorem, as in [11] both antecedents and consequences were adding up to one at every point within the respective domain, while in condition (iii) we only require that the antecedents sum up to at least 1. So the previous theorem is a generalization of the approximation result in [11].

An immediate generalization is obtained by using the same construction and allowing $r \geq 2$, overlapping antecedents.

Theorem 2 *Any continuous function $f : [a, b] \to [c, d]$ can be approximated by the Łukasiewicz fuzzy system with MOM defuzzification*

$$F(f,x) = MOM\left[(\wedge_L)_{i=1}^n (A_i(x) \to_L B_i(y))\right]$$

with continuous antecedents and consequences $A_i, B_i, i = 1, \ldots, n$ such that there exist $0 < \varepsilon, r \in \mathbb{N}, r < n$, such that

(i) $(A_i)_0 \subseteq [x_i, x_{i+r}], i = 1, \ldots, n;$
(ii) $\sum_{i=1}^n A_i(x) \geq 1.$
(iii) $(B_i)_0 \subseteq [\min_{j=i,\ldots,i+r}\{y_j\}, \max_{j=i,\ldots,i+r}\{y_j\}].$
(iv) $(B_i)_1 \neq \emptyset.$

Moreover the following error estimate holds true

$$\|F(f,x) - f(x)\| \leq (r + 1)\omega(f, \delta)$$

with

$$\delta = \max_{i=1,\ldots,n} \{x_i - x_{i-1}\}$$

Proof The proof is based on the fact that $x \in [a, b]$ belongs to the 0-level set of at most r antecedents. Suppose that $x \in (A_j)_0 \cup \ldots \cup (A_{j+r})_0$. If $i \notin \{i, \ldots, i + r\}$ then $A_i(x) = 0$ and by the definition of the Łukasiewicz implication we have $A_i(x) \rightarrow B_i(y) = 1$ and similar to the reasoning in the previous theorem we have exactly r active consequences and we get

$$|F(f, x) - f(x)| \leq |y - f(x)|,$$

where

$$y \in [\min\{y_i, \ldots, y_{i+r}\}, \max\{y_i, \ldots, y_{i+r}\}].$$

Taking into account that $y = f(z)$ with $z \in [x_i, x_{i+r}]$ we have $|y - f(x)| < (r + 1)\omega(f, \delta)$ and finally we obtain

$$|F(f, x) - f(x)| \leq (r + 1)\omega(f, \delta).$$

∎

In some cases, when the antecedents and consequences are LR fuzzy numbers then we can obtain a very simple expression for the output of a Łukasiewicz fuzzy system. In fact if the LR fuzzy numbers have the same left and right shape functions, the Łukasiewicz fuzzy system becomes a piecewise linear interpolation. In general this is not the case as it will be shown in the examples in Sect. 4.

4 Applications

Łukasiewicz fuzzy systems that satisfy the properties of the theorems in the previous section can be implemented very efficiently in the case of consequences begin L-R fuzzy sets with invertible left and right shape functions L and R. Let us recall that such fuzzy sets can be denoted $B_j(y) = (a, b, c, d)_{L,R}$, and can be written as

$$B_j(y) = \begin{cases} L\left(\frac{y-a}{b-a}\right) & y \in [a, b) \\ 1 & y \in [b, c] \\ R\left(\frac{d-y}{d-c}\right) & y \in (c, d] \\ 0 & otherwise \end{cases}.$$

As it is shown in the proof of the theorems of the previous section a maximum point satisfies $B_j(y) \geq A_j(x)$ and $B_{j+1}(y) \geq A_{j+1}(x)$. Then if we solve the equations $B_j(y) = A_j(x)$ and $B_{j+1}(y) = A_{j+1}(x)$ we obtain the left and right endpoints y_l and y_r of the interval j of maxima as

$$(y_l)_i = y_{i-1} + L^{-1}(A_i(x))(y_i - y_{i-1}),$$
$$(y_r)_i = y_{i+1} - R^{-1}(A_{i+1}(x))(y_{i+1} - y_i),$$

for $i \in \{j, j+1\}$. If the two maxima are at the same value, e.g., at 1 then the intersection of the two intervals will give the intervals of maxima, in which case the MOM can be calculated as the average of the two endpoints of the interval of maxima. If one of the maximum values is higher then we obtain the MOM as the average of the endpoints of the given interval. In any case, using the endpoints calculated above we can directly calculate the defuzzification without the time consuming step of the COG defuzzification.

We observe that if the antecedents form a Ruspini partition and if $B_j(y) + B_{j+1}(y) = 1$ for any $y \in [y_j, y_{j+1}]$ then there is a unique intersection point of the intervals of maxima $(y_r)_j = (y_l)_{j+1}$. Also, if the antecedents and consequences are both linear, then the Łukasiewicz fuzzy system becomes a piecewise linear interpolation. Indeed, if both L, R shape functions for antecedents and consequences are piecewise linear then the output of the fuzzy system becomes piecewise linear.

As an example we consider $f : [0, 1] \rightarrow [0, 1]$, $f(x) = \sin(6x)$ with $x \in [0, 1]$ and we construct the Łukasiewicz fuzzy system that approximates this function using 10 fuzzy rules. The antecedents represented in Fig. 3 are LR-fuzzy numbers (see [13]), defined based on monotonic splines as considered in [14]. For the consequences we have triangular membership functions Fig. 4. The proposed approximation is presented in Fig. 5.

$$p(t; \beta_0, \beta_1) = \frac{t^2 + \beta_0 t(1 - t)}{1 + (\beta_0 + \beta_1 - 2)t(1 - t)}.$$

Fig. 3 Antecedents for a Łukasiewicz fuzzy systems for function approximation

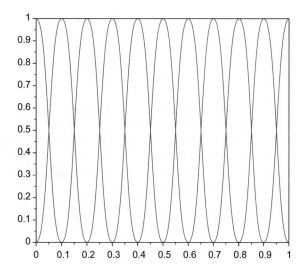

Fig. 4 Consequences for
our fuzzy rule base

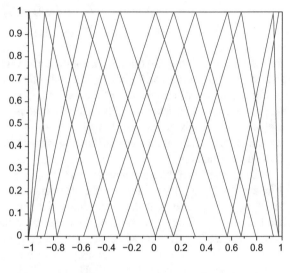

Fig. 5 Function
$f(x) = \sin(6x)$ (red) is
approximated by a
Łukasiewicz (blue) fuzzy
system

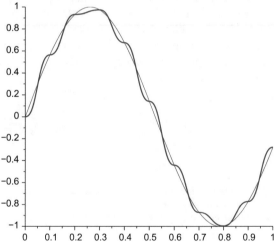

In what follows we compare the proposed Łukasiewicz fuzzy system with MOM
defuzzification with the Łukasiewicz fuzzy system with COG defuzzification and
with the Mamdani fuzzy system with COG defuzzification (see Fig. 6)

We observe that Łukasiewicz fuzzy system with MOM defuzzification better follows the peaks of the function to be approximated. It rescues this way the smoothing
effect of the COG defuzzification.

Fig. 6 Function
$f(x) = x^2 + \sin(10x)$ (black)
is approximated by a
Łukasiewicz (red) fuzzy
system with MOM
defuzzification, Łukasiewicz
(blue) fuzzy system with
COG defuzzification and
Mamdani (green) fuzzy
system with COG
defuzzification

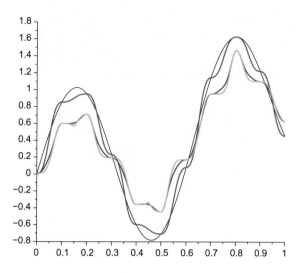

5 Conclusion and Further Research

A Łukasiewicz fuzzy system with MOM defuzzification can be calculated very efficiently and the resulting fuzzy system has approximation properties similar to Mamdani and Łukasiewicz systems with COG defuzzification, often presenting an improvement with respect to these.

References

1. E.H. Mamdani, S. Assilian, An experiment in linguistic synthesis with a fuzzy logic controller. J. Man Mach. Stud. **7**, 1–13 (1975)
2. M. Sugeno, An introductory survey of fuzzy control. Inf. Sci. **36**, 59–83 (1985)
3. P. Hájek, *Metamathematics of Fuzzy Logic* (Kluwer, Dordrecht, 1998)
4. V. Novák, Which logic is the real fuzzy logic? Fuzzy Sets Syst. **157**, 635–641 (2006)
5. L.T. Kóczy, A. Zorat, Fuzzy systems and approximation. Fuzzy Sets Syst. **85**, 203–222 (1997)
6. B. Kosko, Fuzzy systems as universal approximators. IEEE Trans. Comput. **43**, 1329–1333 (1994)
7. Yong-Ming Li, Zhong-Ke Shi, Zhi-Hui Li, Approximation theory of fuzzy systems based upon genuine many-valued implications-SISO cases. Fuzzy Sets Syst. **130**, 147–157 (2002)
8. Yong-Ming Li, Zhong-Ke Shi, Zhi-Hui Li, Approximation theory of fuzzy systems based upon genuine many-valued implications-MIMO cases. Fuzzy Sets Syst. **130**, 159–174 (2002)
9. A.H. Sonbol, M.S. Fadali, TSK fuzzy function approximators: design and accuracy analysis. IEEE Trans. Syst. Man Cybern. **42**, 702–712 (2012)
10. L.-X. Wang, J.M. Mendel, Fuzzy basis functions, universal approximation, and orthogonal least-squares learning. IEEE Transactions on Neural Networks **3**, 807–814 (1992)
11. B. Bede, I.J. Rudas, E.D. Schwab, G. Schwab, Approximation properties of Łukasiewicz fuzzy systems, in Fuzzy information processing society (NAFIPS) held jointly with 2015 5th world conference on soft computing (WConSC), 2015 annual conference of the north american. (IEEE, Aug 2015), pp. 1–5

12. I. Perfilieva, V. Kreinovich, A new universal approximation result for fuzzy systems which reflects CNF-DNF duality. Int. J. Intel. Syst. **17**, 1121–1130 (2002)
13. B. Bede, *Mathematics of Fuzzy Sets and Fuzzy Logic*. (Springer, 2013)
14. L. Stefanini, L. Sorini, M.L. Guerra, Parametric representation of fuzzy numbers and application to fuzzy calculus. Fuzzy Sets Syst. **157**, 2423–2455 (2006)

Rankings and Total Orderings on Sets of Generalized Fuzzy Numbers

Li Zhang and Zhenyuan Wang

Abstract Ranking and ordering generalized fuzzy numbers are hot topics in decision making under uncertainty. By extension principle, addition and multiplication of generalized fuzzy numbers are presented for establishing goodness criteria. This paper initially proposes four criteria for judging the goodness of a given ranking or total ordering defined on a set of generalized fuzzy numbers and then discusses the methods of rankings and total orderings which satisfy these goodness criteria. Besides, the cardinality of the set of all generalized fuzzy numbers is, for the first time, determined.

Keywords Rankings · Total orderings · Generalized fuzzy numbers
Criteria for goodness · Decision making

1 Introduction

Ranking and total ordering fuzzy numbers have been widely applied to many areas such as optimization, data mining, decision making, and artificial intelligence since fuzzy set theory was introduced by L.A. Zadeh. Up to date, with a variety of ranking and ordering methods for fuzzy numbers proposed and applied in decision making and artificial intelligence [1–4, 7, 8, 13], a useful set of criteria for ranking and total ordering of fuzzy numbers has been formed [12]. Further, there has been much discussion regarding ranking and total ordering of generalized fuzzy numbers. However, there is no existing criterion for judge the goodness of the various methods of ranking and total ordering for generalized fuzzy numbers.

L. Zhang (✉)
Department of Epidemiology, University of Nebraska, Omaha, NE, USA
e-mail: lwestman@unomaha.edu

Z. Wang
Department of Mathematics, University of Nebraska, Omaha, NE, USA
e-mail: zhenyuanwang@unomaha.edu

© Springer International Publishing AG, part of Springer Nature 2018
L. A. Zadeh et al. (eds.), *Recent Developments and the New Direction in Soft-Computing Foundations and Applications*, Studies in Fuzziness and Soft Computing 361, https://doi.org/10.1007/978-3-319-75408-6_9

It is necessary to give a general discussion for ranking or totally ordering generalized fuzzy numbers. Then we may introduce some goodness criteria for judging various ranking and totally ordering methods.

After the introduction, the paper is arranged as follows. The necessary preliminary knowledge is presented in Sect. 2. In order to give a general discussion of ranking and totally for generalized fuzzy numbers, some concepts on relation are shown in Sect. 3. Before introducing goodness criteria in Sect. 5, we defined the addition and multiplication for generalized fuzzy numbers in Sect. 4. In Sect. 6, a general discussion on rankings and total orderings defined on sets of fuzzy numbers and on sets of generalized fuzzy numbers is given. In addition, the cardinality of the set of all generalized fuzzy is discussed in Sect. 7, followed by some conclusion in Sect. 8.

2 Preliminary Knowledge on Fuzzy Numbers and Generalized Fuzzy Numbers

Fuzzy numbers are used to describe fuzzy concepts on real numbers such as "between 10 and 10.6", "around 1.5", etc.

Definition 1 Let $R = (-\infty, \infty)$. A fuzzy subset of R, denoted by $\tilde{e}$, is called a fuzzy number when its membership function $m_e : R \to [0, 1]$ satisfies the following conditions:

(FN1) Set $\{x | m_e(x) > 0\}$, the support set of $\tilde{e}$ (denoted by supp e), is bounded.
(FN2) Set $\{x | m_e(x) \geq \alpha\}$, the α-cut of $\tilde{e}$ (denoted by e_α), is a closed interval, denoted as $\left[e_\alpha^l, e_\alpha^r\right]$, for every $\alpha \in (0, 1]$

$\alpha = 1$ means that membership degree of a fuzzy number is 1 for at least one point. However, a number of many papers discuss the concept of generalized fuzzy number for which there may be no point at which the membership degree is 1.

Definition 2 Let $R = (-\infty, \infty)$. A fuzzy subset of R, denoted by $\tilde{e}$, is called a generalized fuzzy number when its membership function $m_e : R \to [0, 1]$ satisfies the following conditions:

(GFN1) Set $\{x | m_e(x) > 0\}$, the support set of $\tilde{e}$ (denoted by supp e), is bounded.
(GFN2) There exists h belongs to $(0, 1]$ such that set $\{x | m_e(x) \geq \alpha\}$, the α-cut of $\tilde{e}$ (denoted by e_α), is a closed interval, denoted as $\left[e_\alpha^l, e_\alpha^r\right]$, for every $\alpha \in (0, h]$ and is the empty set for every $\alpha \in (h, 1]$.

From above definitions, we know that any fuzzy number is a special case of generalized fuzzy number, where $h = 1$. A ranking on a set of generalized fuzzy numbers is a transitive and totally comparable relation on this set, while a total ordering is an anti-symmetric, transitive, and totally comparable relation on a set of generalized fuzzy number. As is well known, total comparability implies the reflexivity.

In literature, there are many ways to define a ranking or a totally ordering on sets of generalized fuzzy numbers, mostly, restricted on sets of generalized triangle or trapezoidal fuzzy numbers, based on some geometric properties of the membership function such as area, centroid, and the moment [5, 6], which are usually called an ranking index. For generalized fuzzy number a, its ranking index is denoted by $r(a)$. Generally speaking, for any given set, A, of generalized fuzzy numbers, if we can establish a mapping from A to the real line, $R = (-\infty, \infty)$ regarded as a reference system with the natural ordering [12], then a ranking on A can be obtained. For any generalized fuzzy number $a \in A$, its mapping image $r(a)$ is just the ranking index. Thus, a ranking, denoted by R, on A can be defined as follows. For any pair of generalized fuzzy numbers $a, b \in A$, we say aRb iff $r(a) \leq r(b)$. When the mapping is one-to-one, the established ranking is a total ordering.

When we have more than one mappings with respective ranking indexes, a lexicography can be used for establishing a ranking. If all of these mapping are one-to-one, the obtained ranking is a total ordering.

3 Equivalence Relation Among Generalized Fuzzy Numbers Based on a Given Ranking

Let A be a nonempty set of generalized fuzzy numbers, a and b belong to A, and R be a ranking on A. According to ranking R, a preceding b is denoted by aRb.

Definition 3 For any given ranking R on A, we say that a is equivalent to b, denoted as $a \approx_R b$, iff aRb and bRa, that is, a and b have the same rank.

$\approx_R$ is an equivalence relation on A. If there is no confusion, $a \approx_R b$ can be simply written as $a \approx b$.

Definition 4 Let m_a and m_b be the membership functions of a and b respectively. If there exists real number $k \in [1, \infty)$ such that $km_a = m_b$, then we say that a is a contraction of b and b is an expansion of a. In case b is a fuzzy number, it is called the maximal expansion of a, denoted by $\bar{a}$.

Example 1 Let a and b be two generalized fuzzy numbers with membership functions shown in Fig. 1.

$$m_a(x) = \begin{cases} \frac{1}{2}x & \text{if } x \in [0, 1] \\ 0 & \text{otherwise} \end{cases}$$

and

Fig. 1 Membership
functions of generalized fuzzy
numbers a and b

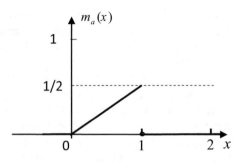

(a).Membership function of generalized fuzzy number a

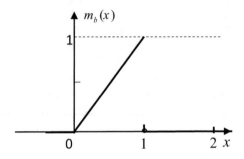

(b). Membership function of generalized fuzzy number b

$$m_b(x) = \begin{cases} x & \text{if } x \in [0, 1] \\ 0 & \text{otherwise} \end{cases}.$$

We have $2\, m_a(x) = m_b(x)$. So, a is a contraction of b and b is an expansion of a. Since b is a fuzzy number, it is a maximal expansion of a.

Definition 5 a and b are similar to each other iff a is a contraction of b or b is a contraction of a.

It is easy to know that generalized fuzzy numbers a and b are similar iff they have the same maximal expansion.

The similarity shown in Definition 5 above is an equivalence relation on A. Considering A being the set of all generalized fuzzy numbers, According to this equivalence relation, the set of all fuzzy numbers is just the quotient of A.

4   Addition and Multiplication of Generalized Fuzzy Numbers

By using Zadeh's extension principle, we may define the sum and the product of two generalized fuzzy numbers as follows.

Let m_a and m_b be the membership functions of generalized fuzzy numbers a and b respectively.

Definition 6 The membership function of the sum of a and b is

$$m_{a+b}(z) = \sup_{z=x+y}[m_a(x) \wedge m_b(y)] \text{ for } z \in (-\infty, \infty).$$

Definition 7 The membership function of the product of a and b is

$$m_{ab}(z) = \sup_{z=x \cdot y}[m_a(x) \wedge m_b(y)] \text{ for } z \in (-\infty, \infty).$$

The sum and the product of two generalized fuzzy numbers are still generalized fuzzy numbers. As for the scalar multiplication of generalized fuzzy numbers, it can be regarded as a special case of the multiplication given in Definition 7 since any real number can be regarded as a special generalized fuzzy number.

5 Goodness Criteria for Ranking

In literature, there are various methods for ranking and totally ordering generalized fuzzy numbers based on areas, the centroid, and moments [5, 6]. Beyond the complexity of calculation, there is no criterion proposed for comparing the goodness of ranking methods. Therefore, it is necessary to introduce some goodness criteria shown below for ranking and totally ordering generalized fuzzy numbers.

Let R be a ranking on a nonempty set, A, of generalized fuzzy numbers and $r(a)$ be corresponding ranking index, if any, of a in A.

5.1 Geometric Intuition

Let a and b be two generalized fuzzy numbers in A, and let $\bar{a}$ and $\bar{b}$ be their maximal expansions respectively. Both $\bar{a}_\alpha$ and $\bar{b}_\alpha$ are intervals. If $\bar{a}_\alpha \leq \bar{b}_\alpha$ for every $\alpha \in (0, 1]$, then aRb, that is, $r(a) \leq r(b)$. Here, the inequality for closed intervals is defined as follows: $[c_1, d_1] \leq [c_2, d_2]$ iff $c_1 \leq c_2$ and $d_1 \leq d_2$.

5.2 Continuity

If, for any given $a \in A$ with the maximal membership degree h, and any sequence $\{b_i | i = 1, 2, 3, \ldots\} \subseteq A$, $(b_i)_\alpha$ converges to $(a)_\alpha$ for every $\alpha \in (0, h]$ and to the empty set when $\alpha \in (h, 1]$ uniformly with respect to α, then we say that sequence $\{b_i | i = 1, 2, 3, \ldots\}$ is strongly convergent to a. Here $\{(b_i)_\alpha | i = 1, 2, 3, \ldots\}$ is a sequence of intervals or the empty set. Its convergence to interval $(a)_\alpha$ means that right and left endpoints, if any, of the sequence of intervals converge to the left and right endpoints of interval $(a)_\alpha$ respectively Ranking R is continuous iff for any sequence $\{b_i | i = 1, 2, 3, \ldots\}$ that is strongly convergent to a, we have $\lim_i r(b_i) = r(a)$.

5.3   Hereditability for Addition

$a_1 R b_1$ and $a_2 R b_2$ imply $(a_1 + a_2) R(b_1 + b_2)$ for any $a_1, b_1, a_2, b_2 \in A$.

5.4   Hereditability for Nonnegative Scalar Multiplication

$a R b$ implies $(ca) R(cb)$ for any $a, b \in A$ and any nonnegative real number c.

These criteria can also be used to judge totally ordering on sets of general fuzzy numbers since any totally ordering is a special case of ranking.

In literature, most ranking methods based on the area or the centroid violate some or even all above-defined criteria. The ranking method that we try to propose in the next section satisfies all of these criteria.

6   Rankings and Total Orderings

For any given ranking R_0, which may be a total ordering, on a set, F, of fuzzy numbers, we may introduce a ranking, R, on set G of generalized fuzzy numbers, where

$$G = \{a | a \text{ is a contraction of some fuzzy number in } F\},$$

by defining $a R b$ iff $\bar{a} R_0 \bar{b}$. Ranking R on G is called the generated ranking by ranking R_0 on F.

In our previous works, we have defined rankings and total orderings that satisfy all above-mentioned four criteria on the set of all fuzzy numbers [9, 12], however, what's new here is that we redefined those criteria on the set of all generalized fuzzy numbers based on maximal expansion, which is different from our previous paper [12]. Now for each of these ranking and total orderings, we can produce a ranking on a set of generalized fuzzy numbers satisfying all these four criteria.

As one of the important results for totally order generalized fuzzy numbers, let's try to introduce total orderings defined on the set of all generalized fuzzy numbers from a given total ordering on the set of all fuzzy numbers by lexicography.

Let R_0 be any given total ordering on the set of all fuzzy numbers and a and b be generalized fuzzy numbers. A total ordering, R, on the set of all generalized fuzzy numbers is defined as follows. We say aRb iff $\bar{a}R\bar{b}$ when $\bar{a} \neq \bar{b}$ or $h_a \leq h_b$ when $\bar{a} = \bar{b}$, where h_a and h_b are the maximal membership degree of a and b, respectively.

We have learned that there are infinitely many different ways to define total orderings on the sets of all fuzzy numbers [9, 11] as well as on any infinite set of generalized fuzzy numbers. For any given total ordering on a nonempty set of generalized fuzzy numbers, based on an equivalence relation on the same set, we can produce a ranking on this set by selecting a representative in each equivalence class.

Example 2 Let A be the set of all generalized triangular fuzzy numbers. Each generalized triangular fuzzy number can be identified by 4-tupe (h, c, f, g) illustrated in Fig. 2.

A total ordering, R, on A is defined by lexicography as follows. For any two generalized triangular fuzzy number a_1 and a_2, identified by 4-tuples (h_1, c_1, f_1, g_1) and (h_2, c_2, f_2, g_2) respectively, $a_1 R a_2$ iff $c_1 < c_2$, or $f_1 < f_2$ when $c_1 = c_2$, or $g_1 < g_2$ when $c_1 = c_2$ and $f_1 = f_2$, or $h_1 < h_2$ when $c_1 = c_2$ and $f_1 = f_2$ and $g_1 = g_2$ as well, We take an equivalence relation, R_e, on A by setting any two generalized triangular fuzzy number a_1 and a_2 being equivalent iff $c_1 = c_2, f_1 = f_2$, and $g_1 = g_2$. In each equivalent class, there is a unique generalized triangularfuzzy number that is a triangular fuzzy number, that is, whose height is $h = 1$. We select it as the representative of this equivalence class. Thus, a rank, R_e, defined on A is obtained. This rank can be described as follows. For any two generalized triangular fuzzy numbers, identified by (h_1, c_1, f_1, g_1) and (h_2, c_2, f_2, g_2) respectively, $a_1 R_e a_2$ iff $c_1 < c_2$, or $f_1 < f_2$ when $c_1 = c_2$, or $g_1 < g_2$ when $c_1 = c_2$ and $f_1 = f_2$.

From Example 2, we can see that the procedure of constructing a total ordering from an existing ranking is just the inverse procedure of introducing a ranking from a total ordering.

Fig. 2 The membership function of a generalized triangular fuzzy number

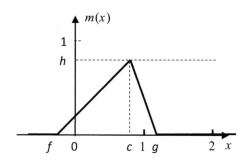

7 The Cardinality of the Set of All Generalized Fuzzy Numbers

We have proved that the cardinality of the set of all fuzzy numbers is $\aleph_1$ [10]. The set of all generalized fuzzy numbers can be regarded as the product set of the set including all fuzzy numbers and set consisting of all real numbers in interval (0, 1]. As is well known, the cardinality of interval $(0, 1]$ is $\aleph_1$. Therefore, the cardinality of the set of all generalized fuzzy numbers is $\aleph_1 \times \aleph_1$, which is equal to $\aleph_1$.

8 Conclusion

Ranking and ordering generalized fuzzy numbers are indispensable tools used for decision making and data analysis in fuzzy environment. By extension principle, addition and multiplication of generalized fuzzy numbers are reviewed. This paper initially proposes four criteria for judging the goodness of a given ranking or total ordering defined on a set of generalized fuzzy numbers, followed by the methods of rankings and total orderings which satisfy the goodness criteria. Besides, the cardinality of the set of all generalized fuzzy numbers is concluded for the first time.

References

1. S. Abbasbandy, B. Asady, Ranking of fuzzy numbers by sign distance. Inf. Sci. **176**(16), 2405–2416 (2006)
2. T.C. Chu, C.T. Tsao, Ranking fuzzy numbers with an area between the centroid point and original point. Comput. Math Appl. **43**, 111–117 (2002)
3. B. Farhadinia, Ranking fuzzy numbers on lexicographical ordering. Int. J. Appl. Math. Comput. Sci. **5**(4), 248–251 (2009)
4. N. Hassasi, R. Saneifard, On the central value of fuzzy numbers. J. Appl. Sci. Res. **7**, 1146–1152 (2011)
5. P.P. Rao, N.R. Shankar, Ranking generalized fuzzy numbers using area, mode, spreads and weights. Int. J. Appl. Sci. Eng. **10**(1), 41–57 (2012)
6. S. Rezvani, Ranking generalized trapezoidal fuzzy numbers with Euclidean distance by the in centre of centroids. Math. Aeterna **3**(2), 103–114 (2013)
7. Y.J. Wang, H.S. Lee, The revised method of ranking fuzzy numbers with an area between the centroid point and original point. Comput. Math Appl. **55**, 2033–2042 (2008)
8. Y.-M. Wang, J.-B. Yang, D.-L. Xu, K.S. Chin, On the centroid of fuzzy numbers. Fuzzy Sets Syst. **157**, 919–926 (2006)
9. W. Wang, Z. Wang, Total ordering defined on the set of all fuzzy numbers. Fuzzy Sets Syst. **234**, 31–41 (2014)
10. Z. Wang, L. Zhang-Westman, The cardinality of the set of all fuzzy numbers, in *Proceedings of the IFSA-NAFIPS*, pp. 1045–1049 (2013)

11. H.-C. Wu, Decomposition and construction of fuzzy sets and their applications to the arithmetic operations on fuzzy quantities. Fuzzy Sets Syst. **233**, 1–25 (2013)
12. L. Zhang, Z. Wang, Ranking fuzzy numbers with goodness criteria and its applications in stock performance evaluation. J. Res. Appl. Econ. **3**, 1–8 (2016). https://doi.org/10.14355/rae.2016.03.001
13 L. Zhang-Westman, Z. Wang, Ranking Fuzzy numbers by their left and right wingspans, in *Proceedings of the IFSA-NAFIPS*, pp. 1039–1044 (2013)

Part III
Novel Population-Based Optimization Algorithms

Bio-inspired Optimization Metaheuristic Algorithm Based on the Self-defense of the Plants

Camilo Caraveo, Fevrier Valdez and Oscar Castillo

Abstract In this work a new method of bio-inspired optimization based on the self-defense mechanism of plants applied to mathematical functions is presented. Habitats on the planet have gone through changes, so plants have had to adapt to these changes and adopt new techniques to defend from natural predators (herbivores). There are many works in the literature have shown that plants have mechanisms of self-defense to protect themselves from predators. When the plants detect the presence of invading organisms this triggers a series of chemical reactions that are released to air and attract natural predators of the invading organism (Bennett and Wallsgrove New Phytol 127(4):617–63, 1994 [1]; Melin et al Expert Syst Appl 40(8):3196–3206, 2013 [10]; Neyoy et al Recent Advances on Hybrid Intelligent Systems, 2013 [11]). For the development of this algorithm we consider as a main idea the predator prey model of Lotka and Volterra.

Keywords Predator prey model · Plants · Self-defense · Mechanism

1 Introduction

Throughout history there have been tested and developed multiple methods of search and optimization inspired by natural processes. This with the goal of solving particular problems in the area of computer science has tried different bio-inspired methods such as PSO, ACO, GA, BCO etc. [8, 9, 12, 14] Trying to get the resolution of a specific problem with a smaller error. In this paper a new optimization algorithm inspired in the self-defense mechanisms of plants is presented.

C. Caraveo (✉) · F. Valdez · O. Castillo
Division of Graduate Studies, Tijuana Institute of Technology, Tijuana, Mexico
e-mail: Camilo.caraveo@gmail.com

F. Valdez
e-mail: fevrier@tectijuana.mx

O. Castillo
e-mail: Ocastillo@tectijuana.mx

© Springer International Publishing AG, part of Springer Nature 2018
L. A. Zadeh et al. (eds.), *Recent Developments and the New Direction in Soft-Computing Foundations and Applications*, Studies in Fuzziness and Soft Computing 361, https://doi.org/10.1007/978-3-319-75408-6_10

This in order to compete against existing optimization methods. In nature, plants are exposed to many different pathogens in the environment. However, only a few can affect them. If a particular pathogen is unable to successfully attack a plant, it is said that it is resistant to it, in other words, cannot be affected by the pathogen [3, 4]. The proposed approach takes as its main basis the Lotka and Volterra predator-prey model, which is a system formed by a pair of first order differential equations, nonlinear for moderating the growth of two populations that interact with each other (predator and prey) [7], and maintain a balance between the two populations.

2 Self-defense Mechanisms of the Plants

In the habitat, plants as well as animals are exposed to a large number of invading organisms, such as insects, fungi, bacteria and viruses that can cause various types of diseases, and even death [3, 4, 11, 16].

Defense mechanisms also named by others as copying strategies are automatic processes that protect the individual against external or internal threats. The plant is able to react to external stimuli. When it detects the presence or attack of an organism triggers a series of chemical reactions that are released into the air that attracts natural predator of the assailant also chemical reactions liberated by plant attracts insect pollinators such as birds, bees and other insects, in order to reproduce and let before dying offspring or cause internal damage to the aggressor [10] sometimes even dead. In Fig. 1 a general scheme is shown to illustrate the behavior of the plant when it detects the presence or attack by a predator [3, 4, 19].

The leaves of the plants normally release into the air small amounts of volatile chemicals, but when a plant is damaged by herbivorous insects, the amount of

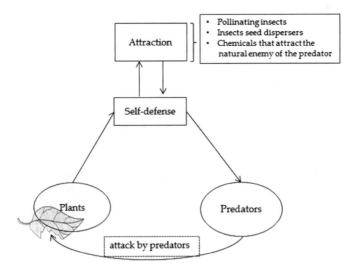

Fig. 1 Representation of the process of self-defense of the plant

chemicals tends to grow depending on the size damage caused by the predator. Volatile chemicals vary depending on the plant species and species of herbivorous insects [13]. These chemicals attract both predators, pollinators and parasitic insects that are natural enemies of herbivores see Fig. 1. Such chemicals, which work in the communication between two or more species, as well as those who serve as messengers between members of the same species are called semi-chemicals [1, 2, 6, 11].

3   Predator-Prey Model

In nature organisms live in communities, forming intricate relationships of inter-action, where each species directly or indirectly dependent on the presence of the other. One of the tasks of Ecology is to develop a theory of community organization for under-standing the causes of diversity and mechanisms of interaction. In this work, we consider the interaction of two whose population size at time t is x(t) and y(t) species [2, 3, 5]. Furthermore, we assume that the change in population size can be written as:

$$\frac{dy}{dt} = I(x, y) \tag{1}$$

$$\frac{dx}{dt} = P(x, y) \tag{2}$$

There are different kinds of biological interaction that can be represented mathematically with this system of equations [3]. As $P(x, y)$ and $I(x, y)$ determining the growth rate of each of the populations; there is the case where one of these species is fed from the other, then the system of survival is given by (Eq. 3):

$$P_y(x, y) < 0$$
$$I_x(x, y) > 0 \tag{3}$$

That is, the change of the prey population relative to the predator decreases and the change of the predator population relative to the prey increases. These are some of the conditions that must meet a set of predator prey equations [3, 15].

This model of Lotka and Volterra is based on the following assumptions.

1. The population grows proportionally to its size, and has enough space and food. If this happens and x(t) represents the prey population (in the absence of predators), then the population growth is given by:

$$\frac{dx}{dt} = ax, a > 0,$$

$$x(t) = x_0 e^{at}. \tag{4}$$

The population of prey in the absence of the predator grows exponentially.

2. The predator y(t) only feeds on the prey x(t). Thus, if there is no prey, their size decreases with a rate proportional to its population is represented by (Eq. 5).

$$\frac{dy}{dt} = -dy, d > 0,$$

$$y(t) = y_0 e^{-dt}. \tag{5}$$

The population of predators in the absence of prey decreases exponentially to extinction.

3. The number of encounters between predator and prey is proportional to the product of their populations. Each of the number of encounters favor predators and reduces the number of prey.

The presence of prey helps the growth of the predator and is represented by (Eq. 6).

$$cxy, c > 0. \tag{6}$$

While the interaction between them, reduces the growth of prey is represented by (Eq. 7).

$$-bxy, b > 0. \tag{7}$$

Under the above hypothesis, we have a model of interaction between x(t) and y(t) is given by the following system (Eqs. 8 and 9):

$$\frac{dx}{dt} = Ax - Bxy \tag{8}$$

$$\frac{dy}{dt} = -Cxy + Dy \tag{9}$$

X Is the number of prey.
Y Is the number of predators.

$\frac{dx}{dt}$ Is the growth of the population of prey time t.
$\frac{dy}{dt}$ Is the growth of the population of predator at time t.

A It represents the birth rate of prey in the absence of predator.
B It represents the death rate of predators in the absence of prey.
C Measures the susceptibility of prey.
D Measures the ability of predation.

4 Proposed Optimization Algorithm Based on the Self-defense Mechanisms of Plants with Lévy Flights

The proposed approach takes as its main basis the Lotka and Volterra predator-prey model, which is a system formed by a pair of differential equations of first order nonlinear moderating the growth of two populations that interact with each other (predator and prey). In Fig. 2 a general scheme of our proposal is shown and takes as a basis the traditional model of predator prey, using the principle of the dynamics of both populations the evolutionary process of plants is generated to develop the techniques of self-defense [3, 4].

In nature, plants have different methods of reproduction, in our approach we consider only the most common: **clone**, **graft** and **pollen**. **Clone**: the offspring identical to the parent plant. **Graft**: it takes a stem of a plant and is encrusted on

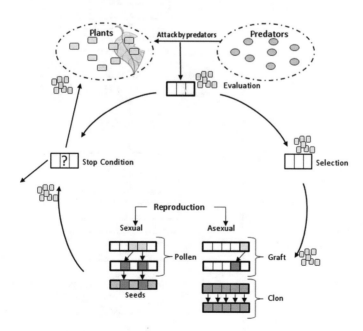

Fig. 2 General illustration of our proposal

another to generate an alteration in the structure of the plant. **Pollen**: one plant pollinates other flowers and generates a seed and the descent is a plant with characteristics of both plants, in [17] the authors develop a new optimization algorithm called pollination of flowers algorithm, they use other method of pollination, such as water, air and animals [3, 4].

To generate the initial population of the algorithm we use the equations of the model of Lotka and Volterra, the mathematical representation is shown in Sect. 3.1, Eqs. (8) and (9). Equation 8 is used to generate the population of prey (plants), and Eq. 9 is used to generate the population of predators (herbivores), as mentioned above functions predator prey model is used to model our proposed model variables adapted to the proposal [3, 4]. In Fig. 3 describes the steps of the optimization algorithm proposed [3, 4].

The initial sizes of both populations (prey, predators) are defined by the user, the parameters (a, b, c, d) are also defined by the user, the model of Lotka and Volterra recommended the following parameter values $a = 0.4$, $b = 0.37$, $c = 0.3$, $d = 0.05$. Both populations that initiated these populations interact with each other prey and preda-tor, use this method to generate new offspring of plants, these plant

Fig. 3 Flowchart of the proposed algorithm

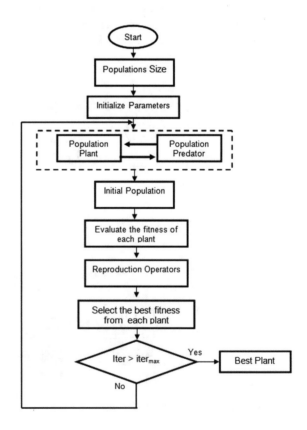

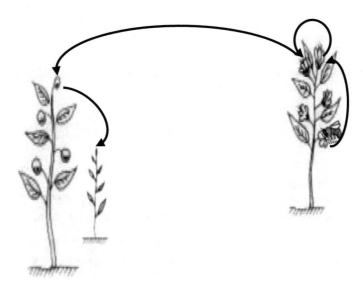

Fig. 4 The process of pollination by insect pollinators

reproduction in biological processes are applied. The population is re-evaluated and if the stop criterion is not satisfied, return the iterative cycle of the algorithm [3, 4].

In [3, 4] we can find published a work using the same optimization algorithm, but with a variation in the method of biological reproduction, in [3, 4] the method of biological reproduction of plants used is the cloning process. In this paper the method of reproduction type used is pollinated by insect pollinators. in Fig. 4 the process of pollination is represented.

To maintain better performance of the proposed algorithm, we propose to use the method of reproduction pollination by insects, to maintain a better balance between exploitation and exploration of the algorithm, we use the method flights levy to control flying insects pollinators. In [18] the authors proposed a new optimization algorithm based on the pollination of flowers, and we are using their model of Levy flights to integrate it as a method of reproduction in plants. The insect pollinators located a group of plants and select the one with better fitness, this plant was used as a base to pollinate other neighboring plants, and its next plant visit is determined using a probability calculated using flight levy [18], the probability may be, apply self-pollination or fly to one of the neighboring plants.

5   Results

For test the performance of the proposed algorithm, we used a set of benchmark functions, where the goal is to approximate the value of the function to zero. Thirty experiments were performed for the following mathematical functions the evaluation is for 10 Variable, the mathematical definition of the functions described below.

Powell Function

The function is usually evaluated on the hypercube $x_i \in [-4, 5]$, for all $i = 1 \dots$ d.

$$f(x) = \sum_{i=1}^{d/4} [(x_{4i-3} + 10x_{4i-2})^2 + 5(x_{4i-1} - x_{4i})^2 + (x_{4i-2} - 2x_{4i-1})^4 + 10(x_{4i-3} - x_{4i})^2]$$

(10)

Ackley Function

The function is usually evaluated on the hypercube $x_i \in [-32.768, 32.768]$, for all $i = 1 \dots$ d, although it may also be restricted to a smaller domain.

$$f(x) = -a \cdot \exp\left(-b \cdot \sqrt{\frac{1}{n} \sum_{i=1}^{n} x_i^2}\right) - \exp\left(\frac{1}{n} \sum_{i=1}^{n} \cos(cx_i)\right) + a + \exp(1)$$

(11)

Griewank Function

The function is usually evaluated on the hypercube $x_i \in [-600, 600]$, for all $i = 1 \dots$ d.

$$f(x) = \sum_{i=1}^{d} \frac{x_i^2}{4000} - \prod_{i=1}^{d} \cos\left(\frac{x_i}{\sqrt{i}}\right) + 1$$

(12)

Rastrigin Function

The function is usually evaluated on the hypercube $x_i \in [-5.12, 5.12]$, for all $i = 1 \dots$ d.

$$f(x) = 10d + \sum_{i}^{b} [x_i^2 - 10\cos(2\pi x_i)]$$

(13)

Schwefel Function

The function is usually evaluated on the hypercube $x_i \in [-500, 500]$, for all $i = 1 \dots$ d.

$$f(x) = \sum_{i=1}^{d} \frac{x_i^2}{4000} - \prod_{i=1}^{d} \cos\left(\frac{x_i}{\sqrt{i}}\right) + 1$$

(14)

Sphere Function

The function is usually evaluated on the hypercube $x_i \in [-5.12, 5.12]$, for all $i = 1 \dots$ d.

Table 1 Experimental results

Function	Experimental results with 6 variables						
	Parameters					Values	
	A	B	C	D	Best	σ	Average
Ackley	0.34	0.04	0.23	0.36	7.99E−15	0.4814	3.32E−01
Griewank	0.29	0.06	0.27	0.320	2.89E−15	0.1412	4.38E−02
Rastrigin	0.29	0.02	0.10	0.170	3.13E−13	4.3820	1.39
Schwefel	0.50	0.08	0.51	0.740	1.88E−04	0.0115	3.57E−03
Sphere	0.37	0.02	0.10	0.330	2.13E−49	0.5061	2.132E−01
Powell	0.84	0.20	0.41	0.444	1.11E−04	0.4581	1.29E−01
Rosenbrock	0.33	0.19	0.22	0.36	2.5573	7.145	5.2213

$$f(x) = \sum_{i=1}^{n} x_i^2 \tag{15}$$

Rosenbrock Function

The function is usually evaluated on the hypercube $x_i \in [-5, 10]$, for all $i = 1\ldots d$, although it may be restricted to the hypercube $x_i \in [-2.048, 2.048]$, for all $i = 1\ldots d$.

$$f(x) = \sum_{i=1}^{d} \frac{x_i^2}{4000} - \prod_{i=1}^{d} \cos\left(\frac{x_i}{\sqrt{i}}\right) + 1 \tag{16}$$

In Table 1 the results of the mathematical functions used in this work for 6 variables is presented, in the table the most important parameters of the proposed algorithm is shown, the values shown are as follows, A, B, C, D, these parameters explained in Eqs. 8 and 9, the size, the populations of plants and herbivores were used in the following range, plants [300–350], herbivores [200–250].

The values shown in Table 1 are the results of 30 experiments for each mathematical function, the values of A, B, C, D were changed manually to analyze the behavior of the algorithm with these values, the algorithm shown some weaknesses in some mathematical functions, but remember that the algorithm is in the early stages therefore in need to make some adaptation and modifications.

6 Conclusions

The proposal is to create, develop and test a new optimization algorithm bio-inspired by the self-defense mechanisms of the plants, the first challenge is to adapt the predator-prey model and test the algorithm in an optimization problem, in this case, we decided to test the performance in mathematical functions and we have found acceptable results. When we move the parameters manually observe that the algorithm has better performance when the values are in a range of values, these observations are only for this problem, we need to apply it to other optimization problems to analyze the behavior. throughout this research we have been making some modifications and adding new methods of biological reproduction of plants, to date we have successfully adapted flights levy to control the pollination of the plants, we achieved acceptable results but we consider that the results can be improved.

References

1. R.N. Bennett, R.M. Wallsgrove, Secondary metabolites in plant defense mechanisms. New Phytol. **127**(4), 617–633 (1994)
2. A.A. Berryman, The origins and evolution of predator-prey theory. Ecology, 1530–1535 (1992)
3. C. Caraveo, F. Valdez, O. Castillo, A new bio-inspired optimization algorithm based on the self-defense mechanisms of plants, in *Design of Intelligent Systems Based on Fuzzy Logic, Neural Networks and Nature-Inspired Optimization* (Springer International Publishing, 2015), pp. 211–218
4. C. Caraveo, F. Valdez, O. Castillo, Bio-inspired optimization algorithm based on the self-defense mechanism in plants, in *Advances in Artificial Intelligence and Soft Computing* (Springer International Publishing, 2015), pp. 227–237
5. J.M.L. Cruz, G.B. González, Modelo depredador-presa. Revista de Ciencias Básicas UJAT **7** (2), 25–34 (2008)
6. J.M. García-Garrido, J.A. Ocampo, Regulation of the plant defense response in arbuscular mycorrhizal symbiosis. J. Exp. Bot. **53**(373), 1377–1386 (2002)
7. D. Karaboga, B. Basturk, A powerful and efficient algorithm for numerical function optimization: artificial bee colony (ABC) algorithm. J. Global Optim. **39**(3), 459–471 (2007)
8. J.H. Law, F.E. Regnier, Pheromones. Annu. Rev. Bio-chem. **40**(1), 533–548 (1971)
9. G.G. Lez-Parra, A.J. Arenas, M.R. Cogollo, Numerical-analytical solutions of predator-prey models. WSEAS Trans. Biol. Biomed. **10**(2) (2013)
10. P. Melin, F. Olivas, O. Castillo, F. Valdez, J. Soria, M. Valdez, Optimal design of fuzzy classification systems using PSO with dynamic parameter adaptation through fuzzy logic. Expert Syst. Appl. **40**(8), 3196–3206 (2013)
11. H. Neyoy, O. Castillo, J. Soria, Dynamic fuzzy logic parameter tuning for ACO and its application in TSP problems, in *Recent Advances on Hybrid Intelligent Systems* (Springer, Berlin, Heidelberg, 2013), pp. 259–271
12. K.M. Ordeñana, Mecanismos de defensa en las interacciones planta-patógeno. Revista Manejo Integrado de Plagas. Costa Rica **63**, 22–32 (2002)
13. P.W. Paré, J.H. Tumlinson, Plant volatiles as a defense against insect herbivores. Plant Physiol. **121**(2), 325–332 (1999)
14. Teodorovic, Bee colony optimization (BCO), in *Innovations in Swarm Intelligence*, ed. by C. P. Lim, L.C. Jain, S. Dehuri, 65, 215 (Springer, 2009), pp. 39–60
15. L. Tollsten, P.M. Müller, Volatile organic compounds emitted from beech leaves. Phytochemistry **43**, 759–762 (1996)
16. J.M. Vivanco, E. Cosio, V.M. Loyola-Vargas, H.E. Flores, Mecanismos químicos de defensa en las plantas. Investigación y ciencia **341**(2), 68–75 (2005)
17. Y. Xiao, L. Chen, Modeling and analysis of a predator–prey model with disease in the prey. Math. Biosci. **171**(1), 59–82 (2001)

18. X.S. Yang, Flower pollination algorithm for global optimization, in *Unconventional Computation and Natural Computation* (Springer, Berlin, Heidelberg, 2012), pp. 240–249
19. T. Yoshida, L.E. Jones, S.P. Ellner, G.F. Fussmann, N.G. Hairston, Rapid evolution drives ecological dynamics in a predator–prey system. Nature **424**(6946), 303–306 (2003)

Experimenting with a New Population-Based Optimization Technique: FUNgal Growth Inspired (FUNGI) Optimizer

A. Tormási and L. T. Kóczy

Abstract In this paper the experimental results of a new evolutionary algorithm are presented. The proposed method was inspired by the growth and reproduction of fungi. Experiments were executed and evaluated on discretized versions of common functions, which are used in benchmark tests of optimization techniques. The results were compared with other optimization algorithms and the directions of future research with many possible modifications/extension of the presented method are discussed.

1 Introduction

One of the main principles of computational intelligence [1, 2] besides the fuzzy logic [3] and artificial neural networks [4, 5] are the various nature-inspired methods used in optimization for example genetic algorithms, swarm intelligence and evolutionary algorithms, which are widely used in complex problems.

The main contribution of this paper is a new nature-inspired optimization technique, called FUNgal Growth Inspired (FUNGI) optimizer. The proposed method is tested with a set of benchmark functions and the results are compared to two other methods, namely genetic algorithm (GA) [6] and bacterial evolutionary algorithm (BEA) [7].

After the introduction in Sect. 2 the background of optimization techniques, the usual evaluation of those and the basic properties of fungi are outlined. In Sect. 3 the concept of the proposed FUNGI optimization technique is detailed. The used

A. Tormási (✉) · L. T. Kóczy
Department of Information Technology, Széchenyi István University, Győr, Hungary
e-mail: tormasi@sze.hu

L. T. Kóczy
e-mail: koczy@sze.hu; koczy@tmit.bme.hu

L. T. Kóczy
Department of Telecommunications and Media Informatics, Budapest University of Technology and Economics, Budapest, Hungary

© Springer International Publishing AG, part of Springer Nature 2018
L. A. Zadeh et al. (eds.), *Recent Developments and the New Direction in Soft-Computing Foundations and Applications*, Studies in Fuzziness and Soft Computing 361, https://doi.org/10.1007/978-3-319-75408-6_11

benchmark functions, test parameters and results of the executed experiment are detailed in Sect. 4. In Sect. 5 conclusions from the experiment results are discussed and directions of future research including possible extensions of the proposed method are pointed out.

2   Background

2.1   Metaheuristic Optimization

The conventional type of optimization algorithms (like hill climbing [8]) are deterministic; in the other type are the stochastic optimization methods including heuristics and metaheuristics, which have some sort of randomness in the algorithm. Metaheuristics means higher level heuristic algorithms [9]. There is no unified definition of metaheuristic optimization techniques, but in general it could be summarized as strategies to guide an iterative process in which the search space is efficiently explored to find (near)optimum solution. There are several ways to classify metaheuristic methods which could be briefly summarized as follows according to [10]:

(1) *Nature- or non-nature inspired*: based on the origins of the algorithm's concept.
(2) *Population-based or single point search*: population-based if many solution exists at any time.
(3) *Dynamic or static objective function*: dynamic if the objective function changes during the search, static otherwise.
(4) *One or various neighborhood structures*: one neighborhood structure if the fitness landscape topology does not change in the course of the search algorithm.
(5) *Memory usage or memory-less methods*: the algorithm is memory-less if it does not use the search history.

2.2   Evaluation of Optimization Techniques

The most common way to validate and compare optimization techniques are through tests with benchmark functions. Several benchmark functions with various properties could be found in literature [11–13].

The main difference between the characteristics of benchmark functions are the number of global and local minima and maxima, the distribution of these in the search space. Some examples of 2-Dimensional benchmark functions are shown in Fig. 1.

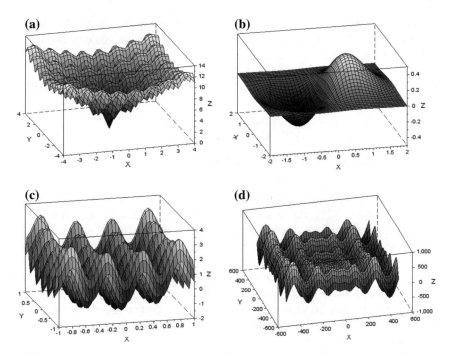

Fig. 1 Examples of benchmark functions with various properties: **a** Ackley's function, **b** Double Dip function, **c** Rastrigin's function and **d** Schwefel's function

The most common properties of benchmark functions, which could be used as a basis for the categorization of these functions, are continuity, differentiability, separability, scalability and modality.

2.3 Basic Parts and Properties of Fungi

Despite the common belief fungi are not plants, neither animal; they are an independent life kingdom in the eukaryote empire. They evolved about 1.5 billion years ago from a common ancestor of fungi and animals. The plants diverged earlier from the common plant-animal-fungi ancestor [14, 15]. The diversity of fungi is about ten times bigger than the plants'.

The most common part of the fungi are the hyphae, which are thread-like food and water seeking entities consisting of one or more cells with cell walls of chitin.

The mycelium is a common element of fungi, a network-like structure of hyphae usually grown belowground or inside a host and they are responsible for the feeding. The mycelium helps to produce the energy to grow the fruiting bodies or mushrooms (which grows from and consists of hyphae) in some types of fungi and could be considered as the roots.

Fungi can reproduce by both sexual and asexual means. In both cases spores are formed, which are usually spread by the fruiting bodies. Spores could be considered as the seeds of fungi from which hyphae are growing.

Fungi are able to sense their environment [16, 17], one common form of this ability is the quorum sensing [17], which can be observed also in many types of bacteria.

3 The Concept of FUNGI Optimizer

FUNGI is an evolutionary algorithm, which uses a simplified model or simulation of fungal growth [18] and reproduction. According to the categorization of metaheuristic algorithms summarized above, FUNGI is a nature-inspired, population-based metaheuristic optimization algorithm, which uses memory, static objective function and one neighborhood structure.

The first step of the proposed algorithm is to randomly generate some candidate solutions, called spores in the search space. Each spore grows a particular number of (first-level) hyphae in random directions and distances in a ring shaped area from the spores. The hyphae are representing candidate solutions as well. Each spore with the connected hyphae represents a mycelia.

In the second step each spore's hyphae grows further, second- or higher-level hyphae in the direction with a random deviance (between 0 and a defined maximum value of deviance) of the hypha with the lowest fitness value. The growth of hyphae is also ring shaped. This step is repeated until the maximum length of hyphae is reached.

When mycelium are full grown (the length of the hyphae reaches the defined maximum), then a particular number of hyphae with the best fitness are growing fruiting bodies and start to spread a predefined number of new spores. A particular number of spores with the best fitness values are selected to survive, while the others are eliminated. The algorithm continues with repetition of the above steps with the new spores until it reaches the maximum number of generations.

The guided (second- or higher-level) hyphae growth in the algorithm could be considered as a global search in the problem space. It uses information from other solutions in the population to control the search in the space. The spore dispersal operator could be considered as a local search in the sense of searching for new candidate solutions in a small limited area near a particular candidate solution.

The proposed algorithm uses 13 parameters, which are the following:

(1) *Maximum number of generations*
(2) *Number of initial spores*
(3) *Number of first-level hyphae per spore*
(4) *Number of higher-level hyphae per first-level hyphae*
(5) *Number of hyphae levels*

(6) *Inner radius of hyphae growing ring*
(7) *Outer radius of hyphae growing ring*
(8) *Maximum angle of growth deviation*
(9) *Number of hyphae growing fruits*
(10) *Number of dispersed spores per best fruits*
(11) *Inner radius of spore dispersal ring*
(12) *Outer radius of spore dispersal ring*
(13) *Number of surviving spores for the next generation*

The most similar algorithm compared to the FUNGI optimizer is the Cuckoo Optimization Algorithm (COA) [19], but it also differs at many points, which could be observed in population handling, in reproduction patterns, in memory usage and so on.

4 Experiments and Results

4.1 Benchmark Functions

10 well known benchmark functions were selected with 2-Dimensional search space for the experiment, namely:

(1) *Ackley's function*, evaluated on $x, y \in [-4, 4]$;
(2) *Booth function*, evaluated on $x, y \in [-10, 10]$;
(3) *Double Dip function*, evaluated on $x, y \in [-2, 2]$;
(4) *Drop-Wave function*, evaluated on $x, y \in [-5.12, 5.12]$;
(5) *Easom function*, evaluated on $x, y \in [-100, 100]$;
(6) *Matyas function*, evaluated on $x, y \in [-10, 10]$;
(7) *Rastrigin's function*, evaluated on x and $y = [-5.12, 5.12]$;
(8) *Schaffer's function N. 2*, evaluated on $x, y = [-100, 100]$;
(9) *Schwefel's function*, evaluated on $x, y = [-500, 500]$;
(10) *Six-Hump Camel function*, evaluated on $x \in [-3, 3], y \in [-2, 2]$.

A decision was made to discretize each benchmark function in order to make the experiment easier to follow and observe. This modification had a significant effect on the benchmark functions; the search space, the values and the locations of the local/global minimum/maximum were changed according to the sampling as part of the discretization. The properties of the search spaces, the new global minimum values and their locations for each discretized benchmark functions are summarized in Table 1.

The experiment's goal was to find global minimum of these discretized functions using genetic algorithm (GA), bacterial evolutionary algorithm (BEA) and FUNGI optimizer.

Table 1 Properties of discretized benchmark functions

Benchmark function	x	y	Global minimum value	Global minimum coordinate
Ackley's function	1...1053	1...1053	0.00990673572320988	(527,527)
Booth function	1...1053	1...1053	0.0003480926	(685,579)
Double Dip function	1...1035	1...1035	−0.428877520506912	(336,518)
Drop-Wave function	1...1053	1...1053	−0.998287090164812	(527,527)
Easom function	1...1053	1...1053	−0.9998225	(543,543)
Matyas function	1...1053	1...1053	0.000003607344	(527,527)
Rastrigin's function	1...1053	1...1053	−1.99991504067884	(527,527)
Schaffer's function N. 2	1...1053	1...1053	0.00001803623	(527,527)
Schwefel's function	1...1053	1...1053	−837.921529876121	(970,970)
Six-Hump Camel function	1...702	1...1053	−1.031618711244	(476,511)

4.2 Parameters of Algorithms Used in the Experiment

The (x, y) coordinates were selected as the parameters of candidate solutions (e.g. alleles of chromosomes or bacteria) in the experiment. The values of fitness functions of the 3 used algorithms were the values of the benchmark functions at the given coordinates represented by the candidate solutions of the algorithms. The methods' goal was to minimize the value of the fitness function. The parameter values of the used methods were selected in order to have a similar number of fitness evaluations in a generation for each algorithm. The reason of this choice was that in most cases the computational cost of optimization algorithms mostly depends on the number of the fitness evaluations.

The implemented GA had 4 parameters: population size, maximum number of generations, the probability of crossovers and the probability of mutations, the values were 285, 50, 0.7 and 0.001 respectively. The number of fitness evaluations in the initial population was 285, but after that only about 70% of the population was changed in crossovers and 0.1% in mutations, which means about 199.785 fitness evaluations/generation (depending on the exact number of crossovers and mutations in the particular generation), since the fitness is the same for the unchanged chromosomes. Roulette wheel method was used for selecting the 2 parents in the crossover step; the new chromosomes were generated from the randomly selected coordinate (x or y) of the first parent and the other coordinate from the second parent. The mutated coordinate was selected by a random (normally distributed) function and the new value was an integer generated between the ranges of the given coordinate for the function. The initial generation was consisted of randomly generated chromosomes.

The BEA had 5 parameters: population size, maximum numbers of generations, number of clones in bacterial mutation, number of infections in a generation and the point of division in infection to distinguish good and bad bacteria, the values were 20, 50, 5, 5 and 50% respectively. 20 fitness evaluations were required for the randomly generated initial population. After that 205 fitness evaluations/generation were required, this consists of calculations of 5 clones for 20 times 5 for each 2 mutated alleles and 5 evaluations for the infected bacterium. The order of allele mutation, the good and the bad bacteria and the allele to transfer were randomly selected.

The FUNGI optimizer had 13 parameters. The number of randomly selected initial spores and the number of maximum generations were both set to 10. The level of hyphae was set to 2, the number of first-level hyphae (grown from the spores randomly) was set to 2, while the number of second-level hyphae (hyphae grown from first-level hyphae in the direction of the hypha with the best value) was set to 4. The inner radius of hyphae growing ring was set to 20, while the outer radius was set to 200 and the angle of growing was set to $10°$. The number of fruiting hyphae (the number of mushrooms) was 10 and each fruit spread 10 spores. The spores were spread in a ring with an inner radius of 1 and outer radios of 30. The best 10 spores were selected from the 100 generated spores. The 10 spores and the two first-level hyphae per spore require 30, while the 4 second-level hyphae per first-level hyphae requires 80 fitness evaluations. The 10 spores spread by the 10 best hyphae require 100 evaluations of the fitness function, which added to the previous number gives 200 fitness evaluations/generation plus 10 in the first generation for the randomly generated initial spores.

4.3 Results

The experiment was executed on a desktop PC with a 3.6 GHz 8-core AMD FX–8150 processor, 4 GB RAM running the 64 bit version Microsoft Windows 8 Pro. The algorithms were implemented as single threaded windowed applications in C# (.NET 4.5). In this environment FUNGI evaluated a generation in 7.104 s in average based on the 10 executions on the 10 benchmark functions, while GA and BEA required 7.835 and 28.129 s/generation respectively. These runtimes are not able to express the efficiency of the methods in general, because these highly depend on their implementation and the parameters of the environment, but could be used as a basis to compare the algorithms to each other in this given test.

The following results were compared for GA, BEA and FUNGI during the benchmark test for each function:

(1) *Best run (BR)*: the 1 run out of 10 in which the population reached the best result (a candidate with the lowest fitness) for a particular benchmark function.

If two or more different runs reached the same results, then the one is selected in which it appeared in the earliest generation.

(2) *Worst run (WR)*: the 1 run out of 10 in which the population's best candidate reached the worst result (the greatest fitness value) for a particular benchmark function. If two or more different runs had the same results, then the one is selected in which the value appeared in a latter generation.

(3) *Average of populations (AP)*: the average value of the best candidates' fitness values of populations in 10 runs for a particular benchmark function.

The experiments run on Ackley's function showed that BEA could find the global minimum in each run; FUNGI reached it 7 times out of 10 runs, while GA couldn't find it in any runs. It is important to highlight the fact that the FUNGI could evaluate the 10 runs (for each test case) in the third time compared to BEA. In BR BEA found the best solution 1 generation earlier than FUNGI. In AP BEA reached slightly better (a lower fitness with 0.00215) results, but FUNGI converged faster to the optimum value.

In BR for Booth function FUNGI reached the global minimum 7 generations earlier than BEA, while GA could not find it in any of the runs. In WR none of the algorithms could find the optimum, but FUNGI reached a solution with lower (better) fitness by 0.004726. In AP FUNGI reached better fitness values and had better convergence than BEA in each generation.

The results of benchmark test for Double Dip function showed that in BR FUNGI could reach the global optimum 5 generations earlier than BEA, while in WR and AP BEA reached slightly lower fitness (by 0.0000003 and 0.0000001 respectively).

In BR for Drop-Wave function FUNGI reached the global minimum 15 generations earlier than BEA. In WR FUNGI reached better results by 0.0620418 than BEA. In AP FUNGI found solutions lower than BEA by 0.016461 shown in Fig. 2.

In BR for Easom function FUNGI reached the global minimum 15 generations earlier, than BEA which is more impressive if the runtime is also considered. In WR FUNGI reached better result by 0.008335 compared to BEA. In AP BEA found solutions with lower fitness by 0.004026 as shown in Fig. 3.

In BR and WR for Matyas function FUNGI reached better results than BEA and GA by $2.6 * 10^{-10}$ and 0.004 respectively. In AP FUNGI found solution with lower fitness by 0.00016 compared to BEA.

In BR for Rastrigin's function BEA reached the global optimum in 20 generations earlier than FUNGI, while in WR and AP (shown in Fig. 4) BEA found better solution than FUNGI by 0.270433 and 0.027106 respectively.

Fig. 2 APs for Drop-Wave function

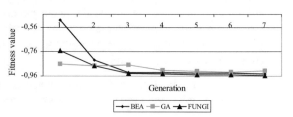

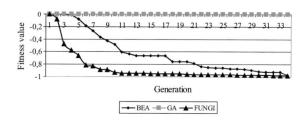

Fig. 3 APs for Easom function

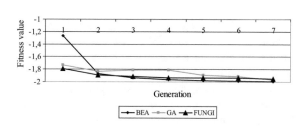

Fig. 4 APs for Rastrigin's function

In BR, WR and AP for Schaffer's N. 2 function FUNGI reached better results compared to BEA by $6.5 * 10^{-10}$, 0.0021 and 0.00045 respectively. The AP results are shown in Fig. 5.

The benchmark results for the Schwefel's function showed that in BR FUNGI reached the same result as BEA 5 generations earlier and both achieved better fitness, than GA by 0.02699. In WR FUNGI performed definitely worse fitness values than GA (by 222.4196) and BEA (by 236.9071). In AP both BEA and GA reached better results compared to FUNGI by 82.90251 and 78.40624 respectively as shown in Fig. 6.

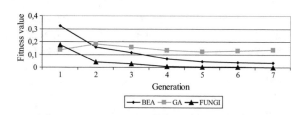

Fig. 5 APs for Schaffer's N. 2 function

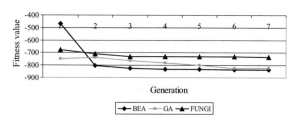

Fig. 6 APs for Schwefel's function

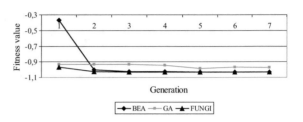

Fig. 7 APs for Six-Hump Camel function

In BR for Six-Hump Camel function FUNGI reached the global minimum 8 generations earlier than BEA. In WR both FUNGI and BEA reached the same result, but FUNGI found that value 5 generations earlier. In AP BEA could find solutions with lower fitness than FUNGI by $6.6341 * 10^{-9}$ as shown in Fig. 7.

The benchmark results clearly showed that in APs GA performed definitely worse than BEA in all cases and 9 times out of 10 than FUNGI. In most cases BEA and FUNGI reached quite similar results, except for Rastrigin's and for Schwefel's functions in which FUNGI performed worse. If we also consider the runtime of the algorithms, then the performance of FUNGI are more impressive, because it required about 25–33% of BEA's run time to reach the results.

5 Conclusions and Future Work

A new fungal growth inspired (FUNGI) optimization method was presented. The method was implemented and an experiment was executed on 10 different 2-Dimensional discretized search spaces. The benchmark results and the properties of the proposed algorithm were compared with GA and BEA. The introduced method reached promising results in the benchmark test compared to the other algorithms. The results of the experiments showed that the FUNGI optimizer had better convergence or reached same or lower fitness values in BRs for 8 cases out of 10 benchmark functions, while the average evaluation time of a generation was similarly low as GA's (about the third of BEA's).

There are several plans for further research, which could be grouped into two main directions: (1) the examination of parameters and basic properties of the method (e.g. random functions) and (2) the investigation of various extensions of the method (e.g. used operators).

The main goal of the following research will be an extensive examination of the FUNGI optimizer's parameters to determine the effects of those on the speed and success of the algorithm.

The effects of the random functions' distribution and shape will be also considered as a main direction of the following work. For example a square shaped space of hyphae growth and spore spreading could be as good as the ring shaped used in the algorithm, but it would be easier to implement especially for problems with higher number of dimensions in the search space.

The most trivial and important extension of the method is enabling it to handle N-Dimensional spaces. In most cases of optimization problems there are more than 2 dimensions. A similarly significant modification of the algorithm would be to handle continuous search spaces, which is also regular in optimization problems. These two extensions of the proposed algorithm would significantly increase its possible applications.

The current model uses the hypha with best fitness to determine the direction for growing second- and further level hyphae. This may result in the system as convergence to a local minimum, which could be avoided by the use of aggregated fitness of mycelia. In this case the hyphae would grow to the direction of the mycelia, which has the highest aggregated fitness. A similar approach is used in imperialist competitive algorithm, where an aggregated fitness of an empire and its colonies is also used.

The firefly algorithm (FA) [20] uses a distance-based weighted fitness of the fireflies to attract each other. A similar approach could be used in FUNGI optimizer, where the hyphae would grow in the direction of the hypha with best distance- or age-based fitness. With the higher value of distance or the age the attractiveness of hyphae would decrease. Hypha with higher distance would force even the most distant hypha to grow in the direction of a (possibly) local minimum ignoring a possible better solution in a near position. This way the population might be forced to migrate big distances without guaranteeing better solution and resulting in the avoidance of the global minimum near a given hypha. This could be handled with the distance-based weighted fitness. The age-based fitness weights could have an advantage in the case if the highest fitness value is a local minimum point in the search space and it does not change even in several generations. If the fitness of the hypha decreases with its age (the difference between the current number of generation and in which it was generated), then the system might be asserted to converge in a new direction, where a hypha has lower fitness, but could be closer to a global minimum.

The current system uses a fix sized ring as the possible area of hyphae growth even if the best hypha has a close position compared to the hypha which grows the new hyphae. A possible modification of the system would be the use of a growing ring, where the distance between the new hyphae and the hypha in which direction it grows would be used to determine the inner and outer radius of the ring. This way the new hyphae would have a small growth if the best hypha is close; and it would grow bigger otherwise. Other possible modifications of the proposed algorithm could be a dynamically changing or unlimited number of hyphae branches and/or hyphae length based on the fitness or the current number of the generation.

The proposed method in this paper keeps all the generated mycelium in the population even for later generations; it does not require further computations in the algorithm, but slightly increases the used memory. A possible extension of the presented method could be the death of mycelium and/or hyphae with low fitness or a greater age. Beside the possible decrease in the systems memory usage it would be reasonable to eliminate them in order to reach a better convergence in the system.

The presented algorithm uses the asexual (without the exchange of genetic information) concept, which is common in fungi reproduction. However some types of fungi use sexual reproduction, in which two fungi may exchange genetic information through the hyphae. This concept could be also applied in a modified version of the proposed method, which way there would be a global information exchange between the entities in the population. This could be modeled and integrated in the method as the gene transfer operator in BEA.

Finally some parts of the algorithm could be easily modified to use multiple threads. The extension of the algorithm with parallel computing capacity could dramatically decrease the computational time required by the method.

Acknowledgements This paper was partially supported by the National Research, Development and Innovation Office (NKFIH) K105529, K108405. The implementations of the used benchmark functions are based on the work of J. D. McCaffrey [21], S. Surjanovic and D. Bingham [22].

References

1. J. Bezdek, On the relationship between neural networks, pattern recognition and intelligence. Int. J. Approx. Reason. **6**(2), 85–107 (1992)
2. R.J. Marks, Intelligence: computational versus artificial. IEEE Trans. Neural Netw. **4**(5), 737–739 (1993)
3. L.A. Zadeh, Fuzzy sets. Inf. Control **8**(3), 338–353 (1965)
4. W.S. McCulloch, W. Pitts, A logical calculus of the ideas immanent in nervous activity. Bull. Math. Biophys. **5**(4), 115–133 (1943)
5. F. Rosenblatt, The perceptron: a probabilistic model for information storage and organization in the brain. Psychol. Rev. **65**(6), 386–408 (1958)
6. J.H. Holland, *Adaption in Natural and Artificial Systems* (The MIT Press, Cambridge, Massachusetts, 1992)
7. N.E. Nawa, T. Furuhashi, Fuzzy system parameters discovery by bacterial evolutionary algorithm. IEEE Trans. Fuzzy Syst. **7**(5), 608–616 (1999)
8. S. Forrest, M. Mitchell, Relative building-block fitness and the building-block hypothesis, in *Foundations of Genetic Algorithms 2*, ed. by L.D. Whitley (Morgen Kauffman, San Mateo, CA, 1993)
9. X.-S. Yang, *Nature-Inspired Metaheuristic Algorithms* (Luniver Press, Cambridge, UK, 2010)
10. C. Blum, A. Roli, Metaheuristics in combinatorial optimization: overview and conceptual comparison. ACM Comput. Surv. **35**(3), 268–308 (2003)
11. N. Chase, M. Rademacher, E. Goodman, R. Averill, R. Sidhu, *A Benchmark Study of Optimization Search Algorithms* (Red Cedar Technology, MI, USA, 2010), pp. 1–15
12. J. Dieterich, B. Hartke, Empirical review of standard benchmark functions using evolutionary global optimization. Appl. Math. **3**(10A), 1552–1564 (2012)
13. M. Jamil, X.S. Yang, A literature survey of benchmark functions for global optimisation problems. Int. J. Math. Model. Numer. Optim. **4**(2), 150–194 (2013)
14. B.H. Bowman, J.W. Taylor, A.G. Brownlee, J. Lee, S.D. Lu, T.J. White, Molecular evolution of the fungi: relationship of the Basidiomycetes, Ascomycetes, and Chytridiomycetes. Mol. Biol. Evol. **9**(2), 285–296 (1992)

15. D.S. Heckman, D.M. Geiser, B.R. Eidell, R.L. Stauffer, N.L. Kardos, S.B. Hedges, Molecular evidence for the early colonization of land by fungi and plants. Science **293**(5532), 1129–1133 (2001)
16. M. Johnston, Feasting, fasting and fermenting: glucose sensing in yeast and other cells. Trends Genet. **15**(1), 29–33 (1999)
17. P. Albuquerque, A. Casadevall, Quorum sensing in fungi—a review. Med. Mycol. **50**(4), 337–345 (2012)
18. A. Meškauskas, M.D. Fricker, D. Moore, Simulating colonial growth of fungi with the neighbour-sensing model of hyphal growth. Mycol. Res. **108**(11), 1241–1256 (2004)
19. R. Rajabioun, Cuckoo optimization algorithm. Appl. Soft Comput. **11**(8), 5508–5518 (2011)
20. X.-S. Yang, Firefly algorithms for multimodal optimization, in *SAGA 2009, LNCS 5792*, ed. by O. Watanabe, T. Zeugmann (Springer, Berlin, Heidelberg, 2009), pp. 169–178
21. J.D. McCaffrey, Software research, development, testing, and education, https://jamesmccaffrey.wordpress.com/. Accessed 12 Feb 2016
22. S. Surjanovic, D. Bingham, Virtual library of simulation experiments: test functions and datasets, http://www.sfu.ca/~ssurjano. Accessed 12 Feb 2016

A Hybrid Genetic Algorithm for Minimum Weight Dominating Set Problem

O. Ugurlu and D. Tanir

Abstract Minimum Weight Dominating Set (MWDS) belongs to the class of NP-hard graph problem which has several real life applications especially in wireless networks. In this paper, we present a new hybrid genetic algorithm. Also, we propose a new heuristic algorithm for MWDS to create initial population. We test our hybrid genetic algorithm on (Jovanovic et al., Proceedings of the 12th WSEAS international conference on automatic control, modeling and simulation, 2010) [3] data set. Then the results are compared with existing algorithms in the literature. The experimental results show that our hybrid genetic algorithm can yield better solutions than these algorithms and faster than these algorithms.

1 Introduction

A dominating set for $G = (V, E)$ is a subset of vertices $S \in V$; such that every vertex in $V \setminus S$ is adjacent to at least one vertex in S. The dominating set problem consists of finding smallest dominating set in a graph. The cardinality of the smallest dominating set is called the domination number of the graph. The problem is also known to be NP-hard [1]. In fact, the minimum dominating set problem for general graphs is polynomially equivalent to the Set Cover problem [2]. Many real life problems can be formulated as instances of the minimum dominating set. Examples of such areas where the minimum dominating set problem occurs positioning retail, sensor networks and Mobile Ad hoc Networks (MANETs) [3].

O. Ugurlu (✉) · D. Tanir
Faculty of Science, Department of Mathematics, Ege University, Izmir, Turkey
e-mail: onurugurlu@mail.ege.edu.tr

D. Tanir
e-mail: tanirdeniz35@gmail.com

© Springer International Publishing AG, part of Springer Nature 2018
L. A. Zadeh et al. (eds.), *Recent Developments and the New Direction in Soft-Computing Foundations and Applications*, Studies in Fuzziness and Soft Computing 361, https://doi.org/10.1007/978-3-319-75408-6_12

One of the crucial extensions of the minimum dominating set is Minimum Weighted Dominating Set MWDS problem. Given a graph $G = (V, E)$ with vertex weight function $C: V \rightarrow R^+$, the MWDS problem is to find a dominating set of graph G such that its total weight is minimum. The problem of finding a MWDS has been proven to be NP-hard [1]. MWDS has numerous real life applications as routing and clustering in wireless networks, reducing the number of full wavelength converters and so on [4].

Due to the significant real-life applications, a great numbers of algorithms have been proposed for MWDS in last decade. [5–8] are some of the approximations algorithms which have been proposed for MWDS. Jovanovic et al. [3] have proposed an ant colony optimization algorithm and a data set for MWDS. Moreover, Potluri and Singh [4] have been proposed three hybrid metaheuristic algorithms for MWDS and have tested their algorithms performance on [3] data set.

In this paper, a new hybrid genetic algorithm has been proposed for MWDS. We have also proposed a heuristic algorithm for MSDS. The proposed hybrid genetic algorithm has been compared with other metaheuristics [3, 4] algorithms for MWDS. The experimental results show that our hybrid genetic algorithm is far better than the other existing metaheuristic algorithms on Type I instances [3, 4] and can give quality solutions on Type II.

The paper is organized as follows: Sect. 2 describes the proposed hybrid genetic algorithms and heuristic algorithm for initial population. In Sect. 3, graph models used in the experiments are briefly described, and the proposed algorithm is compared with other metaheuristic algorithms on [3] data set. Section 4 summarizes and concludes the paper.

2   Hybrid Genetic Algorithm

Hybrid genetic algorithms have received significant interest in recent years and are being increasingly used to solve for optimization problems. A genetic algorithm can incorporate other techniques within its framework to produce a hybrid that reaps the best from the combination [9].

In the proposed algorithm, we have combined a new local search with a genetic algorithm to solve MWDS problem.

2.1   Initial Heuristic

To create an initial population for the genetic algorithm, we have proposed a new heuristic algorithm. The pseudo-code of the proposed algorithm as follows ($N(v)$: neighbors of vertex v):

```
1.  S = ∅
2.  While ( S is not a dominating set) do
3.      For (all v which are not in S )
4.          Find the min  d(v_i)/w(v_i)  value and mark the  v_i
5.      end
6.      For (all v in N(v_i) )
7.          Find the max  d(v*)/w(v*) value and mark the  v*
8.      end
9.      S = S ∪ v*
10. end
```

We have also used random function in step 4 (0.5 chance for random v_i, and 0.5 chance for *step 4*). We have tried to ensure diversity in population by selecting v_i randomly.

2.2   Genetic Algorithm

We have set population size to $\mu = n$ for genetic algorithm. For fitness function, we have used the sum of weights of all vertices in the dominating set. The elitism rate has been set to 0.2, and the remaining 0.8 of the population have been reproduced in every generation. To generate new solutions, we have used the tournament selection ($k = 3$) method and single-point crossover technique. Since we have a powerful local search, we only add vertices to the solutions in mutation operator. 0.02 of population size have selected randomly and have been applied mutation. And 0.02 of n have selected randomly and have been added to the selected solutions. The main purpose of the mutation operator is that ensure the diversity the population to prevent the genetic algorithm from being stuck in a local minimum. After the mutation operator, all solution have been checked by repair procedure. The pseudo-code of the repair procedure as follows:

```
1.  While ( S is not a dominating set) do
2.      rep[] ← 0
3.      For (all v_i which are not in S )
4.          For (all v_j in N(v_i) )
5.              rep[v_j]=rep[v_j]+1
6.          end
7.      end
8.      For (all v which are not in S )
9.          Find the max rep[v*]
10.     end
11.     S = S ∪ v*
12. do
```

After checking all solution via repair procedure, local search has carried out to all solutions. If the best solution in population is not improved over 100 iterations, the algorithm will stop.

2.3 New Local Search Technique

If all neighbors of vertex v, are covered by other vertices of the dominating set and at least, one of the neighbors of vertex v belongs to the dominating set then we can call vertex v redundant vertex, and we can remove v from the dominating set. We have used this technique to minimize the cardinality of the dominating set. However, when the solutions converge towards the local optima in the genetic algorithm, it is almost impossible to find a redundant vertex. Therefore, we have used the following Procedure2 and Procedure3. These procedures provide the population diversity by making small moves around the solution neighbors and create more redundant vertex. Since in our problem vertices have weight, changing in dominating set, provide much more quality solutions even if the cardinality of the dominating set does not change. Main steps of the proposed local search as follows:

- Procedure1: If there is a redundant v vertex in solution, remove v from solution. Repeat this procedure till the solution has no redundant vertices.
- Procedure2: If all neighbors of vertex v, are covered by other vertices of the solution and none of them is in solution, then define v as can be shift vertex and remove vertex v from solution and add its one neighbor to the solution which has minimum weight. Perform the Procedure1.
- Procedure3: If all neighbors of vertex v, are covered by other vertices of the solution except for only one vertex w, then define v as can be shift vertex and remove the v from solution and add the w to the solution. Perform the Procedure1.

We have applied local search to the each solution until the solutions have no redundant or shift vertices.

3 Experimental Results

The proposed HGA was implemented by the authors in C++ and compiler with gcc version 4.9.2 with –o3 optimization. The experiments were carried out on Intel Core i7 4700HQ 2.4 GHz CPU with 32 GB of RAM running Windows 8.1 64-bit Edition. Since reading the graph and building the adjacency lists are common to all algorithms, they are not included in running times.

The proposed HGA was tested on [3] data set. This data set consists of two types of instances. Type I instances are connected undirected graphs with vertices weights randomly distributed in [20, 70]. In Type II instances, the weights of the vertices are determined by a function of the degree of the vertex and randomly distributed in $[1, d(v)^2]$. The number of vertices in the graphs varies from 50 to 1000. For each vertex and edge number, there are 10 instances. For each of these instances, a total of 10 runs of the algorithm are performed, and best solution is selected. In Tables 1 and 2, the average value of 10 instances for each vertex and edge number are given.

Table 1 Results for Type I instances

n	m	ACO1	HGA1	ACO2	ACO3	HGA
50	50	539.8	**531.3**	**531.3**	532.6	**531.3**
50	100	391.9	371.2	371.2	371.5	**362.2**
50	250	195.3	**175.7**	176	**175.7**	175.8
50	500	112.8	94.9	94.9	95.2	**94.5**
50	750	69	**63.1**	**63.1**	63.2	**63.1**
50	1000	44.7	**41.5**	**41.5**	**41.5**	41.8
100	100	1087.2	1081.3	1066.9	1065.4	**1061.4**
100	250	698.7	626.2	627.2	627.4	**618.9**
100	500	442.8	358.3	362.5	363.2	**356.2**
100	750	313.7	261.2	263.5	265	**256.5**
100	1000	247.8	205.6	209.2	208.8	**204.2**
100	2000	125.9	108.2	108.1	108.4	**107.8**
150	150	1630.1	1607	1582.8	1585.2	**1580.7**
150	250	1317.7	1238.6	1237.2	1238.3	**1220**
150	500	899.9	763	767.7	768.6	**748.4**
150	750	674.4	558.5	565	562.8	**548.6**
150	1000	540.7	438.7	446.8	448.3	**433.6**
150	2000	293.1	245.7	259.4	255.6	**245,9**
150	3000	204.7	169.2	173.4	175.2	**168.8**
200	250	2039.2	1962.1	1934.3	1927	**1912.9**
200	500	1389.4	1266.3	1259.7	1260.8	**1236.2**
200	750	1096.2	939.8	938.7	940.1	**913**
200	1000	869.9	747.8	751.2	753.7	**726.3**
200	2000	524.1	432.9	440.2	444.7	**415.9**
200	3000	385.7	308.5	309.9	315.2	**300.4**
250	250	NA	2703.4	2655.4	2655.4	**2638.8**
250	500	NA	1878.8	1850.3	1847.9	**1809.2**
250	750	NA	1421.1	1405.2	1405.5	**1369.3**
250	1000	NA	1143.4	1127.1	1122.9	**1096.9**
250	2000	NA	656.6	672.8	676.4	**626.7**
250	3000	NA	469.3	474.1	478.3	**453.8**
250	5000	NA	300.5	310.4	308.7	**293.1**
300	300	NA	3255.2	3198.5	3205.9	**3186.9**
300	500	NA	2509.8	2479.2	2473.3	**2437.7**
300	750	NA	1933.9	1903.3	1913.9	**1867.2**
300	1000	NA	1560.1	1552.5	1555.8	**1499**
300	2000	NA	909.6	916.8	916.5	**875.5**
300	3000	NA	654.9	667.8	670.7	**634.9**
300	5000	NA	428.3	437.4	435.9	**415.5**
500	500	5476.3	5498.3	5398.3	5387.7	**5338.5**

(continued)

Table 1 (continued)

n	m	ACO1	HGA1	ACO2	ACO3	HGA
500	1000	4069.8	3798.6	3714.8	3698.3	**3654.9**
500	2000	2627.5	2338.2	2277.6	2275.9	**2215.8**
500	5000	1398.5	1122.7	1115.3	1110.2	**1069.2**
500	10,000	825.7	641.1	652.8	650.9	**631.7**
800	1000	8098.9	8017.7	8117.6	8068	**7785.1**
800	2000	5739.9	5318.7	5389.9	5389.6	**5096.8**
800	5000	3116.5	2633.4	2616	2607.9	**2531.4**
800	10,000	1923	1547.7	**1525.7**	1535.3	1537.3
1000	1000	10924.4	11095.2	11035.5	11022.9	**10706.3**
1000	5000	4662.7	3996.6	4012	4029.8	**3796.5**
1000	10000	2890.3	2334.7	2314.9	**2306.6**	2326.9
1000	15,000	2164.3	1687.5	**1656.3**	1657.4	1672.1
1000	20,000	1734.3	1337.2	**1312.8**	1315.8	1372.3

The proposed hybrid genetic algorithm has been compared with other meta-heuristics algorithm for MWDS. Jovanovic et al. [3] have proposed an ant colony optimization algorithm for MWDS and Potluri and Singh [4] have proposed three metaheuristics (one hybrid genetic algorithm and two ant colony optimization algorithms) algorithms for MWDS. For a detail description of these algorithms, the reader is referred to Potluri and Singh [4]. Note that we have set population size to $\mu = 100$ for $n > 500$ to find solutions in more reasonable times.

To compare results, we use following notations in tables:

ACO1: The ant-colony optimization algorithm proposed by Raka et al. [3].
HGA1: The hybrid genetic algorithm proposed by Potluri and Singh [4].
ACO2: The ant-colony optimization algorithm with local search (ACO-LS) proposed by Potluri and Singh [4].
ACO3: The ant-colony optimization algorithm with local search and pre-processing (ACO-PP-LS) proposed by Potluri and Singh [4].
HGA: Our hybrid genetic algorithm.

The results of these algorithms are given in Tables 1 and 2 and the running time of these algorithms are given in Tables 3 and 4.

4 Conclusions

A new hybrid genetic algorithm has been proposed for minimum weight dominating set. The algorithm has been compared with other metaheuristic algorithms [3, 4] for this problem. The experimental results show that proposed algorithm superior to [3] and far better than [4] on Type I instances. On Type II instances,

Table 2 Results for Type II instances

n	m	ACO1	HGA1	ACO2	ACO3	HGA
50	50	62.3	**60.8**	**60.8**	**60.8**	**60.8**
50	100	98.4	90.3	90.3	90.3	**90.1**
50	250	202.4	**146.7**	**146.7**	**146.7**	147.4
50	500	312.9	**179.9**	**179.9**	**179.9**	195.8
50	750	386.3	**171.1**	**171.1**	**171.1**	182.9
50	1000	NA	**146.5**	**146.5**	**146.5**	170.4
100	100	126.5	124.5	123.6	**123.5**	**123.5**
100	250	236.6	211.4	**210.2**	210.4	215.5
100	500	404.8	**306**	307.8	308.4	331.1
100	750	615.1	**385.3**	385.7	386.3	417.5
100	1000	697.3	**429.1**	430.3	430.3	445.5
100	2000	1193.7	**550.6**	558.8	559.8	609.7
150	150	190.1	186	**184.7**	184.9	185.2
150	250	253.9	234.9	**233.2**	233.4	235.3
150	500	443.2	**350**	351.9	351.9	378.8
150	750	623.3	455.8	456.9	**454.7**	518
150	1000	825.3	**547.5**	551.4	549	651.8
150	2000	1436.4	**720.1**	725.7	725.7	1023.4
150	3000	1751.9	**792.6**	794	806.2	1065.6
200	250	293.2	275.1	**272.6**	**272.6**	275.2
200	500	456.5	390.7	388.6	**388.4**	408.8
200	750	657.9	507	501.7	**501.4**	558
200	1000	829.2	**601.1**	605.9	605.8	732.8
200	2000	1626	893.5	**891**	892.9	1251.2
200	3000	2210.3	**1021.3**	1027	1034.4	1592.6
250	250	NA	310.1	**306.5**	306.7	307.5
250	500	NA	444	443.8	**443.2**	461.4
250	750	NA	578.2	**573.1**	575.9	639.2
250	1000	NA	672.8	**671.8**	675.1	791.4
250	2000	NA	**1030.8**	1033.9	1031.5	1404.8
250	3000	NA	**1262**	1288.5	1277	1937.1
250	5000	NA	**1480.9**	1493.6	1520.1	2501.3
300	300	NA	375.6	**371.1**	371.2	373.2
300	500	NA	484.2	**480.8**	481.2	500.8
300	750	NA	623.8	621.6	**618.3**	670
300	1000	NA	751.1	744.9	**743.5**	844.6
300	2000	NA	**1106.7**	1111.6	1107.5	1516.3
300	3000	NA	**1382.1**	1422.8	1415.3	2103.8
300	5000	NA	**1686.3**	1712.1	1698.6	3014.6
500	500	651.2	632.9	627.5	**627.3**	632.7

(continued)

Table 2 (continued)

n	m	ACO1	HGA1	ACO2	ACO3	HGA
500	1000	1018.1	919.2	913	**912.6**	1002.8
500	2000	1871.8	1398.2	1384.9	**1383.9**	1664.3
500	5000	4299.8	**2393.2**	2459.1	2468.8	4148
500	10,000	8543.5	**3264.9**	3377.9	3369.4	8234.4
800	1000	1171.2	1128.2	1126.4	**1125.1**	7785.1
800	2000	1938.7	**1679.2**	1693.7	1697.9	5116..4
800	5000	4439.0	**3003.6**	3121.9	3120.9	4170.2
800	10,000	8951.1	**4268.1**	4404.1	4447.9	8573.6
1000	1000	1289.3	1265.2	1259.3	**1258.6**	1272.7
1000	5000	4720.1	**3220.1**	3411.6	3415.1	4414.8
1000	10,000	9407.7	**4947.5**	5129.1	5101.9	7982.8
1000	15,000	14433.5	**6267.6**	6454.6	6470.6	13470.4
1000	20,000	19172.6	**7088.5**	7297.4	7340.8	19346.6

Table 3 Running times for Type I instances in sec.

n	m	HGA1	ACO2	ACO3	HGA
50	50	2.7	1.2	1.1	0.02
50	100	2.6	1	0.9	0.03
50	250	2.3	0.6	0.6	0.1
50	500	2	0.5	0.5	0.2
50	750	2.1	0.3	0.3	0.6
50	1000	2.1	0.3	0.3	0.7
100	100	9.6	4.3	3.9	0.2
100	250	8.4	3.1	2.8	0.3
100	500	5.8	2.3	2	0.7
100	750	4.8	2	1.9	0.9
100	1000	4.9	1.7	1.7	1.7
100	2000	4.9	1.2	1.2	3.7
150	150	23.4	9.9	8.8	0.6
150	250	20.8	8.6	7.5	0.9
150	500	15.3	6.1	5.5	1.3
150	750	12.2	5	4.5	2
150	1000	11.8	4.5	4	2.6
150	2000	9.2	3.3	3.3	5.5
150	3000	8.3	3	2.8	10.3
200	250	41.7	17.7	15.3	2
200	500	33.4	13.2	11.5	3.5
200	750	28.1	10.6	9.2	5
200	1000	24.9	9	8	5.8

(continued)

Table 3 (continued)

n	m	HGA1	ACO2	ACO3	HGA
200	2000	15.2	6.5	6	10.3
200	3000	14.4	5.7	5.4	15
250	250	72.7	32.5	28.4	3.8
250	500	59.9	25.1	21.9	7
250	750	55	20	17.4	9.5
250	1000	47.3	17.3	15.2	31.2
250	2000	26.1	11.6	10.7	20.8
250	3000	23	9.5	9	52.8
250	5000	21.8	8.4	8.1	86.5
300	300	116.3	49.2	42.2	8.5
300	500	109	41	35.7	13.6
300	750	93.2	34	30	14.1
300	1000	80.1	28.2	24.5	17.5
300	2000	47.6	18.8	17	40.4
300	3000	37.3	15.4	14.3	50.8
300	5000	27.4	12.5	12	76.3
500	500	412.3	180.3	156	50.5
500	1000	359.8	143.7	116.4	69.2
500	2000	219.2	90.3	81.7	89.2
500	5000	114.4	51.3	45.7	148.2
500	10,000	64	36.9	35.8	314.9
800	1000	1459.5	769.8	709.6	9.8
800	2000	1094.3	572.6	554.8	16
800	5000	551.5	263.5	237.3	35.6
800	10,000	246.1	147.5	145.4	83.3
1000	1000	2829.6	1320.7	1189.9	11.6
1000	5000	1152.1	627.7	600.8	53.8
1000	10,000	566.2	308.8	289.9	133
1000	15,000	356.3	227.7	211	219.4
1000	20,000	251.1	204.9	200.3	343.7

Table 4 Running times for Type II instances in sec.

n	m	HGA1	ACO2	ACO3	HGA
50	50	3.8	1.2	1.2	0.01
50	100	3.8	1.1	1.1	0.03
50	250	3.4	0.8	0.7	0.03
50	500	2.8	0.6	0.6	0.2
50	750	2.5	0.5	0.4	0.4
50	1000	1.9	0.3	0.2	0.7
100	100	11	4.2	4	0.1
100	250	11	3.5	3.2	0.3
100	500	10.2	2.7	2.5	0.7
100	750	8.6	2.4	2.3	1
100	1000	8.5	2	1.9	1.5
100	2000	7.8	1.6	1.5	3.8
150	150	29.5	9.1	8.6	0.5
150	250	29.1	8.6	7.8	1.4
150	500	24.5	7	6.2	1.9
150	750	21.8	5.9	5.6	2.5
150	1000	19.7	5.5	5.1	2.9
150	2000	17.5	4.4	4.1	6.4
150	3000	15.4	3.6	3.6	10.3
200	250	53.9	17.1	15.6	2
200	500	49.9	14.6	12.8	6.7
200	750	45.9	12.5	11.1	7.5
200	1000	37.3	11.2	10.2	9.3
200	2000	31.9	8.9	8.1	14.5
200	3000	29.9	7.2	7	24.6
250	250	88	28	25.5	5
250	500	89.1	25.4	22.8	12.8
250	750	77.1	23	20.6	18.2
250	1000	74.2	20.8	19	16.3
250	2000	56.8	15.9	15.1	16
250	3000	49	14	13.5	35.9
250	5000	46.9	11.2	10.9	34.3
300	300	142.3	43	39.2	12.3
300	500	143.4	40	35.8	15.6
300	750	124.5	36.9	32.9	16.3
300	1000	113	33.7	30.5	19.7
300	2000	87.7	24.3	22.7	23.9
300	3000	70.9	23.9	20.9	33.6
300	5000	66.7	18.2	17.4	41
500	500	522	149.1	135.4	32.7

(continued)

Table 4 (continued)

n	m	HGA1	ACO2	ACO3	HGA
500	1000	428.8	137.8	122.1	93.8
500	2000	354.3	109.9	98.7	134.4
500	5000	217.5	91.3	90.7	213.8
500	10,000	191.1	60.1	60.4	429.1
800	1000	1810.5	498.8	444.9	10
800	2000	1560.5	516.6	474.7	18.7
800	5000	955.7	413.7	407.5	50.9
800	10,000	653.6	249.9	249.7	95
1000	1000	3481.4	931.2	832.8	18.8
1000	5000	1995.4	832.6	827.4	76.6
1000	10,000	1250.5	546.5	548.6	192.7
1000	15,000	977.7	398.7	398.9	247.3
1000	20,000	817.3	322.8	325.5	347.4

HGA1 which is proposed by [4] gives best results in general. Also, experimental results show that our algorithm is extremely fast on sparse graphs.

The main contribution of this paper is the idea of local search technique. In future works, the authors aim to improve solutions quality on Type II instances by improving the local search technique and test the algorithm performance on disk graphs.

Acknowledgements The authors were supported in part by TUBITAK (The Scientific and Technological Research Council of Turkey) fellowship. The authors also would like to thank anonymous referees for helpful comments and suggestions.

References

1. M.R. Garey, D.S. Johnson, *Computers and Intractability: A Guide to the Theory of NP-Completeness* (W. H. Freeman, New York, 1979)
2. R. Bar-Yehuda, S. Moran, On approximation problems related to the independent set and vertex cover problem. Discrete Appl. Math. **9**, 1–10 (1984)
3. R. Jovanovic, M. Tuba, D. Simian, Ant colony optimization applied to minimum weight dominating set problem, in *Proceedings of the 12th WSEAS International Conference on Automatic Control, Modeling and Simulation* (World Scientific and Engineering Academy and Society (WSEAS), 2010), pp. 322–326
4. A. Potluri, A. Singh, Hybrid Metaheuristic algorithms for minimum weight dominating set. Appl. Soft Comput. **13**(1), 76–88 (2013)
5. D. Dai, C. Yu, $5+\in$-approximation algorithm for minimum weighted dominating set in unit disk graph. Theor. Comput. Sci. **410**, 756–765 (2009)
6. F. Zou, Y. Wang, X.-H. Xu, X. Li, H. Du, P. Wan, W. Wu, New approximation for minimum-weighted dominating sets and minimum-weighted connected dominating sets on unit disk graphs. Theor. Comput. Sci. **412**(3), 198–208 (2011)

7. Y. Wang, W. Wang, X.-Y. Li, Efficient distributed low-cost backbone formation for wireless networks. IEEE Trans. Parallel Distrib. Syst. **17**(7), 681–693 (2006)
8. X. Zhu, W. Wang, S. Shan, Z. Wang, W. Wu, A PTAS for the minimum weighted dominating set problem with smooth weights on unit disk graphs. J. Comb. Optim. **23**(4), 443–450 (2012)
9. T.A. El-Mihoub, A.A. Hopgood, L. Nolle, A. Battersby, Hybrid genetic algorithms: a review. Eng. Lett. **13**(2), 124–137 (2006)

Part IV
Ensemble Neural Networks

Optimization of Ensemble Neural Networks with Type-1 and Type-2 Fuzzy Integration for Prediction of the Taiwan Stock Exchange

Martha Pulido and Patricia Melin

Abstract This paper describes an optimization method based on genetic algorithms and particle swarm optimization for ensemble neural networks with type-1 and type-2 fuzzy aggregation for forecasting complex time series. The time series that was considered in this paper to compare the hybrid approach with traditional methods is the Taiwan Stock Exchange (TAIEX), and the results shown are for the optimization of the structure of the ensemble neural network with type-1 and type-2 fuzzy integration. Simulation results show that the ensemble approach produces good prediction of the Taiwan Stock Exchange.

Keywords Ensemble neural networks · Optimization · Genetic algorithms Time series prediction · Particle swarm

1 Introduction

Time series are usually analyzed to understand the past and to predict the future, enabling managers or policy makers to make properly informed decisions. Time series analysis quantifies the main features in data, like the random variation. These facts, combined with improved computing power, have made time series methods widely applicable in government, industry, and commerce. In most branches of science, engineering, and commerce, there are variables measured sequentially in time. Reserve banks record interest rates and exchange rates each day. The government statistics department will compute the country's gross domestic product on a yearly basis. Newspapers publish yesterday's noon temperatures for capital cities from around the world. Meteorological offices record rainfall at many different sites with differing resolutions. When a variable is measured sequentially in time over or

M. Pulido (✉) · P. Melin
Division of Graduate Studies, Tijuana Institute of Technology, Tijuana, BC, Mexico
e-mail: marthapulido_84@hotmail.com

P. Melin
e-mail: epmelin@hafsamx.org

© Springer International Publishing AG, part of Springer Nature 2018   151
L. A. Zadeh et al. (eds.), *Recent Developments and the New Direction in Soft-Computing Foundations and Applications*, Studies in Fuzziness and Soft Computing 361, https://doi.org/10.1007/978-3-319-75408-6_13

at a fixed interval, known as the sampling interval, the resulting data form a time series [1].

Time series predictions are very important because based on them we can analyze past events to know the possible behavior of futures events and thus can take preventive or corrective decisions to help avoid unwanted circumstances.

Time series predictions are very important because based on them we can analyze past events to know the possible behavior of futures events and thus we can take preventive or corrective decisions to help avoid unwanted circumstances.

The choice and implementation of an appropriate method for prediction has always been a major issue for enterprises that seek to ensure the profitability and survival of business. The predictions give the company the ability to make decisions in the medium and long term, and due to the accuracy or inaccuracy of data this could mean predicted growth or profits and financial losses. It is very important for companies to know the behavior that will be the future development of their business, and thus be able to make decisions that improve the company's activities, and avoid unwanted situations, which in some cases can lead to the company's failure. In this paper we propose a hybrid approach for time series prediction by using an ensemble neural network and its optimization with genetic algorithms and particle swarm optimization. In the literature there have been recent produced work of time series [2–10].

This method consists of a hybrid approach to optimization ensemble neural networks [11–27] with genetic algorithms [28–32] and particle swarm optimization [32, 33] and integration of the responses of the neural network with type-1 and type-2 fuzzy systems [34, 35].

2 Problem Statement and Proposed Method

The goal of this work was to implement a Genetic Algorithm and particle swarm optimization to optimize the ensemble neural network architectures. In this case the optimization is for each of the modules, and thus to find a neural network architecture that yields optimum results in each of the Time Series to be considered. In Fig. 1 we have the historical data of each time series prediction, then the data is provided to the modules that will be optimized with the genetic algorithm for the ensemble network and then these modules are integrated with type-2 fuzzy integration method.

Historical data of the Taiwan Stock Exchange time series was used for the ensemble neural network trainings, where each module was fed with the same information, unlike the modular networks, where each module is fed with different data, which leads to architectures that are not uniform.

The Taiwan Stock Exchange (Taiwan Stock Exchange Corporation) is a financial institution that was founded in 1961 in Taipei and began to operate as stock exchange on 9 February 1962. It is regulated by the Financial Supervisory Commission. The index of the Taiwan Stock Exchange is the TWSE [36].

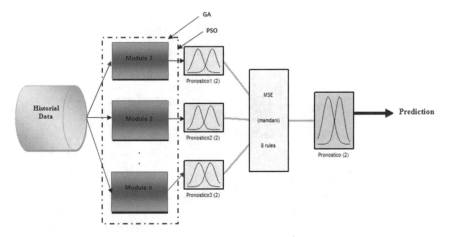

Fig. 1 General architecture of the proposed ensemble model

Data of the Taiwan Stock Exchange time series: We are using 800 points that correspond to a period from 03/04/2011 to 05/07/2014 (as shown in Fig. 2). We used 70% of the data for the ensemble neural network trainings and 30% to test the network [36].

The objective function is defined to minimize the prediction error as follows:

$$EM = \left(\sum_{t=1}^{D} |a_i - x_i| \right) / D \qquad (1)$$

where a, corresponds to the predicted data depending on the output of the network modules, X represents real data, D the Number of Data points and EM is the total prediction error.

The corresponding particle structure is shown in Fig. 3.

The corresponding particle structure is shown in Fig. 4.

Fig. 2 Taiwan stock exchange

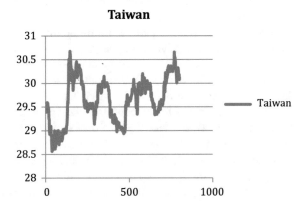

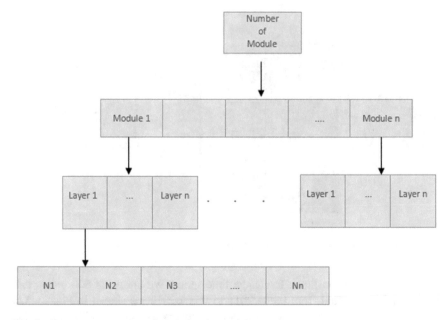

Fig. 3 Chromosome structure to optimize the neural network

Number of Modules	Number of Layers	Neurons 1	...	Neurons n

Fig. 4 Particle structure to optimize the ensemble neural network

Figure 4 represents the Particle Structure to optimize the ensemble neural network, where the parameters that are optimized are the number of modules, number of layers, and number of neurons of the ensemble neural network. PSO determines the number of modules, number of layers and number of neurons per layer that the neural network ensemble should have, to meet the objective of achieving the better Prediction error.

The parameters for the particle swarm optimization algorithm are: 100 Particles, 100 iterations, Cognitive Component (C1) = 2, Social Component (C2) = 2, Constriction coefficient of linear increase (C) = (0–0.9) and Inertia weight with linear decrease (W) = (0.9–0). We consider a number of 1–5 modules, number of layers of 1–3 and neurons number from 1 to 30.

The aggregation of the responses of the optimized ensemble neural network is performed with type-1 and type-2 fuzzy systems. In this work the fuzzy system consists of 5 inputs depending on the number of modules of the neural network ensemble and one output is used. Each input and output linguistic variable of the fuzzy system uses 2 Gaussian membership functions. The performance of the type-2 fuzzy aggregators is analyzed under different levels of uncertainty to find out

Fig. 5 Fuzzy inference system for integration of the ensemble neural network

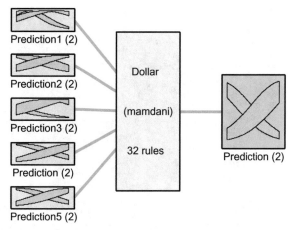

Prediction1 (2)

Prediction2 (2)

Prediction3 (2)

Prediction (2)

Prediction5 (2)

Dollar

(mamdani)

32 rules

Prediction (2)

System Dollar: 5 inputs, 1 outputs, 32 rules

the best design of the membership functions for the 32 rules of the fuzzy system. Previous tests have been performed only with a three input fuzzy system and the fuzzy system changes according to the responses of the neural network to give us better prediction error. In the type-2 fuzzy system we also change the levels of uncertainty to obtain the best prediction error.

Figure 5 shows a fuzzy system consisting of 5 inputs depending on the number of modules of the neural network ensemble and one output. Each input and output linguistic variable of the fuzzy system uses 2 Gaussian membership functions. The performance of the type-2 fuzzy aggregators is analyzed under different levels of uncertainty to find out the best design of the membership functions for the 32 rules of the fuzzy system. Previous experiments were performed with triangular, and Gaussian and the Gaussian produced the best results of the prediction.

Figure 6 represents the 32 possible rules of the fuzzy system; we have 5 inputs in the fuzzy system with 2 membership functions, and the outputs with 2 membership functions. These fuzzy rules are used for both the type-1 and type-2 fuzzy systems. In previous work several tests were performed with 3 inputs, and the prediction error obtained was significant and the number of rules was greater, and this is why we changed to 2 inputs.

3 Simulation Results

In this section we present the simulation results obtained with the genetic algorithm and particle swarm optimization for the Taiwan Stock Exchange.

We consider working with a genetic algorithm to optimize the structure of an ensemble neural network and the best architecture obtained was the following (shown in Fig. 7).

1. If (Prediction1 is Pron1Bajo) and (Prediction2 is Pron2Bajo) and (Prediction3 is Pron3Bajo) and (Prediction4 is Pron4Bajo) and (Prediction5 is Pron5Bajo) then (Prediction is Bajo)

2. If (Prediction1 is Pron1Alto) and (Prediction2 is Pron2Alto) and (Prediction3 is Pron3Alto) and (Prediction4 is Pron4Alto) and (Prediction5 is Pron5Alto) then (Prediction is Alto)

3. If (Prediction1 is Pron1Bajo) and (Prediction2 is Pron2Bajo) and (Prediction3 is Pron3Bajo) and (Prediction4 is Pron4Bajo) and (Prediction5 is Pron5Alto) then (Prediction is Bajo)

4. If (Prediction1 is Pron1Alto) and (Prediction2 is Pron2Alto) and (Prediction3 is Pron3Alto) and (Prediction4 is Pron4Alto) and (Prediction5 is Pron5Alto) then (Prediction is Alto)

5. If (Prediction1 is Pron1Bajo) and (Prediction2 is Pron2Bajo) and (Prediction3 is Pron3Bajo) and (Prediction4 is Pron4Alto) and (Prediction5 is Pron5Alto) then (Prediction is Bajo)

6. If (Prediction1 is Pron1Alto) and (Prediction2 is Pron2Alto) and (Prediction3 is Pron3Alto) and (Prediction4 is Pron4Bajo) and (Prediction5 is Pron5Bajo) then (Prediction is Alto)

7. If (Prediction1 is Pron1Bajo) and (Prediction2 is Pron2Bajo) and (Prediction3 is Pron3Alto) and (Prediction4 is Pron4Alto) and (Prediction5 is Pron5Alto) then (Prediction is Alto)

8. If (Prediction1 is Pron1Alto) and (Prediction2 is Pron2Alto) and (Prediction3 is Pron3Bajo) and (Prediction4 is Pron4Bajo) and (Prediction5 is Pron5Bajo) then (Prediction is Bajo)

9. If (Prediction1 is Pron1Bajo) and (Prediction2 is Pron2Alto) and (Prediction3 is Pron3Alto) and (Prediction4 is Pron4Alto) and (Prediction5 is Pron5Alto) then (Prediction is Alto)

10. If (Prediction1 is Pron1Alto) and (Prediction2 is Pron2Bajo) and (Prediction3 is Pron3Bajo) and (Prediction4 is Pron4Bajo) and (Prediction5 is Pron5Bajo) then (Prediction is Bajo)

11. If (Prediction1 is Pron1Bajo) and (Prediction2 is Pron2Alto) and (Prediction3 is Pron3Bajo) and (Prediction4 is Pron4Alto) and (Prediction5 is Pron5Bajo) then (Prediction is Bajo)

12. If (Prediction1 is Pron1Alto) and (Prediction2 is Pron2Bajo) and (Prediction3 is Pron3Alto) and (Prediction4 is Pron4Bajo) and (Prediction5 is Pron5Alto) then (Prediction is Alto)

13. If (Prediction1 is Pron1Bajo) and (Prediction2 is Pron2Alto) and (Prediction3 is Pron3Bajo) and (Prediction4 is Pron4Bajo) and (Prediction5 is Pron5Bajo) then (Prediction is Bajo)

14. If (Prediction1 is Pron1Alto) and (Prediction2 is Pron2Bajo) and (Prediction3 is Pron3Alto) and (Prediction4 is Pron4Alto) and (Prediction5 is Pron5Alto) then (Prediction is Alto)

15. If (Prediction1 is Pron1Bajo) and (Prediction2 is Pron2Bajo) and (Prediction3 is Pron3Alto) and (Prediction4 is Pron4Bajo) and (Prediction5 is Pron5Bajo) then (Prediction is Bajo)

16. If (Prediction1 is Pron1Alto) and (Prediction2 is Pron2Alto) and (Prediction3 is Pron3Bajo) and (Prediction4 is Pron4Alto) and (Prediction5 is Pron5Alto) then (Prediction is Alto)

17. If (Prediction1 is Pron1Bajo) and (Prediction2 is Pron2Bajo) and (Prediction3 is Pron3Bajo) and (Prediction4 is Pron4Alto) and (Prediction5 is Pron5Bajo) then (Prediction is Bajo)

18. If (Prediction1 is Pron1Alto) and (Prediction2 is Pron2Alto) and (Prediction3 is Pron3Alto) and (Prediction4 is Pron4Bajo) and (Prediction5 is Pron5Alto) then (Prediction is Alto)

19. If (Prediction1 is Pron1Bajo) and (Prediction2 is Pron2Bajo) and (Prediction3 is Pron3Alto) and (Prediction4 is Pron4Alto) and (Prediction5 is Pron5Bajo) then (Prediction is Bajo)

20. If (Prediction1 is Pron1Alto) and (Prediction2 is Pron2Alto) and (Prediction3 is Pron3Bajo) and (Prediction4 is Pron4Bajo) and (Prediction5 is Pron5Alto) then (Prediction is Alto)

21. If (Prediction1 is Pron1Bajo) and (Prediction2 is Pron2Alto) and (Prediction3 is Pron3Alto) and (Prediction4 is Pron4Bajo) and (Prediction5 is Pron5Bajo) then (Prediction is Bajo)

22. If (Prediction1 is Pron1Alto) and (Prediction2 is Pron2Bajo) and (Prediction3 is Pron3Bajo) and (Prediction4 is Pron4Alto) and (Prediction5 is Pron5Alto) then (Prediction is Alto)

23. If (Prediction1 is Pron1Bajo) and (Prediction2 is Pron2Bajo) and (Prediction3 is Pron3Alto) and (Prediction4 is Pron4Alto) and (Prediction5 is Pron5Bajo) then (Prediction is Bajo)

24. If (Prediction1 is Pron1Alto) and (Prediction2 is Pron2Alto) and (Prediction3 is Pron3Bajo) and (Prediction4 is Pron4Bajo) and (Prediction5 is Pron5Alto) then (Prediction is Alto)

25. If (Prediction1 is Pron1Bajo) and (Prediction2 is Pron2Alto) and (Prediction3 is Pron3Alto) and (Prediction4 is Pron4Bajo) and (Prediction5 is Pron5Alto) then (Prediction is Alto)

26. If (Prediction1 is Pron1Alto) and (Prediction2 is Pron2Bajo) and (Prediction3 is Pron3Bajo) and (Prediction4 is Pron4Alto) and (Prediction5 is Pron5Alto) then (Prediction is Bajo)

27. If (Prediction1 is Pron1Bajo) and (Prediction2 is Pron2Alto) and (Prediction3 is Pron3Bajo) and (Prediction4 is Pron4Alto) and (Prediction5 is Pron5Bajo) then (Prediction is Bajo)

28. If (Prediction1 is Pron1Alto) and (Prediction2 is Pron2Bajo) and (Prediction3 is Pron3Alto) and (Prediction4 is Pron4Bajo) and (Prediction5 is Pron5Alto) then (Prediction is Alto)

29. If (Prediction1 is Pron1Bajo) and (Prediction2 is Pron2Alto) and (Prediction3 is Pron3Alto) and (Prediction4 is Pron4Alto) and (Prediction5 is Pron5Alto) then (Prediction is Bajo)

30. If (Prediction1 is Pron1Alto) and (Prediction2 is Pron2Bajo) and (Prediction3 is Pron3Bajo) and (Prediction4 is Pron4Alto) and (Prediction5 is Pron5Alto) then (Prediction is Bajo)

31. If (Prediction1 is Pron1Bajo) and (Prediction2 is Pron2Alto) and (Prediction3 is Pron3Bajo) and (Prediction4 is Pron4Alto) and (Prediction5 is Pron5Alto) then (Prediction is Alto)

32. If (Prediction1 is Pron1Alto) and (Prediction2 is Pron2Bajo) and (Prediction3 is Pron3Alto) and (Prediction4 is Pron4Bajo) and (Prediction5 is Pron5Bajoo) then (Prediction is Bajo)

Fig. 6 Rules of the type-2 fuzzy system

In this architecture we have three layers in each module. In module 1, in the first layer we have 12 neurons, In module 2 we used 12 neurons in the first layer, in module 3 we have 1 neurons in the first layer, in module 4 we have 22 neurons in the first layer and in module 5 we have 30 neurons in the first layer,

Fig. 7 The best architecture
GA for ensemble network

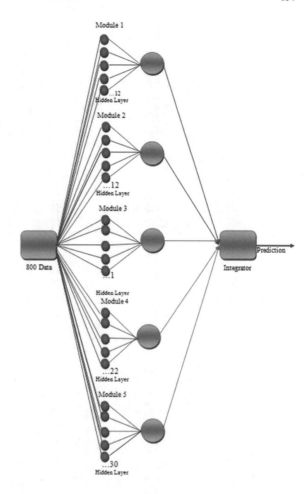

the Levenberg-Marquardt (LM) training method was used; 3 delays for the network were considered.

Table 1 shows the genetic algorithm results (as shown in Fig. 7) where the prediction error is of 0.0011421.

Fuzzy integration is performed initially by implementing a type-1 fuzzy system in which the best result is in experiment of row number 7 of Table 2 with an error of: 0.0129.

As a second phase, to integrate the results of the optimized ensemble neural network a type-2 fuzzy system is implemented, where the best results that are obtained are as follows: with a degree uncertainty of 0.3 a forecast error of 0.0124 is obtained, with a degree of uncertainty of 0.4 the error is of 0.0136 and with a degree of uncertainty of 0.5 the error is of 0.001828, as shown in Table 3.

Figure 8 shows the plot of real data against the predicted data generated by the monolithic neural network optimized with the genetic algorithm.

Table 1 Genetic algorithm results for the ensemble neural network

No.	Gen.	Ind.	GGP	Selection	Mutation	Pm	Crossover	Pc	Num. modules	Num. layers	Num. neurons	Duration	Prediction error
1	100	100	0.85	rws	mutbga	0.09	xovsp	0.9	5	1	18 1 16 10 23	02:16:10	0.0012029
2	100	100	0.85	rws	mutbga	0.05	xovsp	0.5	5	1	13 21 25 23 20	01:31:17	0.001264
3	100	100	0.85	rws	mutbga	0.09	xovsp	1	5	1	30 28 8 14 4	01:21:41	0.0012303
4	100	100	0.85	rws	mutbga	0.08	xovsp	0.6	5	1	16 19 25 23 23	01:37:16	0.0012434
5	100	100	0.85	rws	mutbga	0.09	xovsp	0.55	5	1	1 18 5 20 25	01:16:11	0.0012743
6	100	100	0.85	rws	mutbga	0.05	xovsp	1	5	1	21 28 28 9 3	01:08:44	0.0012785
7	**100**	**100**	**0.85**	**rws**	**mutbga**	**0.06**	**xovsp**	**1**	**5**	**1**	**12 16 1 22 30**	**01:05:45**	**0.0011421**
8	100	100	0.85	rws	mutbga	0.05	xovsp	0.05	5	1	21 24 4 1 3	01:14:17	0.0012777
9	100	100	0.85	rws	mutbga	0.02	xovsp	0.04	5	1	1 13 14 20 2	01:24:07	0.0011632
10	100	100	0.85	rws	mutbga	0.06	xovsp	0.3	5	1	5 23 11 1 4	01:43:18	0.0012637

Table 2 Results GA for type-1 fuzzy integration of the TAIEX

Experiments	Prediction error with fuzzy integration Type-1
Experiment 1	0.0165
Experiment 2	0.0169
Experiment 3	0.0153
Experiment 4	0.0147
Experiment 5	0.0140
Experiment 6	0.0160
Experiment 7	**0.0129**
Experiment 8	0.0139
Experiment 9	0.0149
Experiment 10	0.0157

Table 3 Results of type 2 fuzzy integration of TAIEX

Experiment	0.3 Uncertainty	0.4 Uncertainty	0.5 Uncertainty
Experiment 1	0.0235	0.033	0.0281
Experiment 2	0.0289	0.0494	0.01968
Experiment 3	0.0189	0.0182	0.0247
Experiment 4	0.0193	0.0228	0.0263
Experiment 5	0.0348	0.0276	0.0277
Experiment 6	0.0131	0.0120	0.01953
Experiment 7	**0.0124**	**0.0136**	**0.01828**
Experiment 8	0.0387	0.0277	0.0267
Experiment 9	0.0325	0.0599	0.0575
Experiment 10	0.0247	0.0363	0.0382

Fig. 8 Prediction with the optimized ensemble neural network with GA of the TAIEX

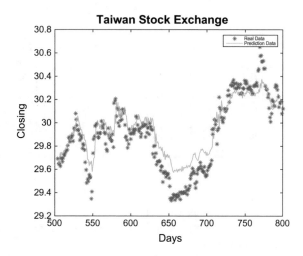

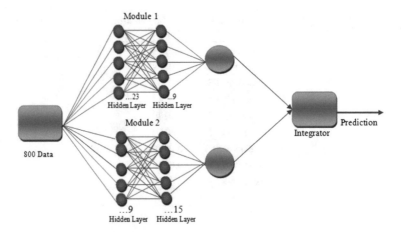

Fig. 9 Prediction with the optimized ensemble neural network with GA of the TAIEX

Table 4 Particle swarm optimization results for the ensemble neural network

No.	Iterations	Particles	Number of modules	Number of layers	Number of neurons	Duration	Prediction error
1	100	100	2	3	13, 16, 2 18, 20, 18	01:48:30	0.002147
2	100	100	2	2	3, 9 14, 19	01:03:09	0.0021653
3	100	100	2	2	20, 4 10, 7	01:21:02	0.0024006
4	100	100	2	2	16, 19 3, 12	01:29:02	0.0019454
5	100	100	2	2	19, 19 24, 17	02:20:22	0.0024575
6	100	100	2	3	21, 14, 23 14, 24, 20	01:21:07	0.0018404
7	**100**	**100**	**2**	**2**	**23, 9 9, 15**	**01:19:08**	**0.0013066**
8	*100*	*100*	2	2	15, 17 9, 22	01:13:20	0.0018956
9	*100*	*100*	2	2	22, 15 20, 16	01:13:35	0.0023377
10	*100*	*100*	2	2	23, 8 10, 17	01:04:23	0.0023204

We also consider working with a particle swarm optimization to the structure of an ensemble neural network and the best architecture obtained was the following (shown in Fig. 9).

Table 5 Results PSO for type-1 fuzzy integration of the TAIEX

Experiments	Prediction error with fuzzy integration Type-1
Experiment 1	0.0473
Experiment 2	0.0422
Experiment 3	0.0442
Experiment 4	0.0981
Experiment 5	0.0253
Experiment 6	0.0253
Experiment 7	0.0253
Experiment 8	**0.0235**
Experiment 9	0.0253
Experiment 10	0.0253

Table 6 Results PSO of type 2 fuzzy integration of TAIEX

Experiment	0.3 Uncertainty	0.4 Uncertainty	0.5 Uncertainty
Experiment 1	0.0335	0.033	0.0372
Experiment 2	0.0299	0.5494	0.01968
Experiment 3	0.0382	0.0382	0.0387
Experiment 4	0.0197	0.0222	0.0243
Experiment 5	0.0433	0.0435	0.0488
Experiment 6	0.0121	0.0119	0.0131
Experiment 7	**0.01098**	**0.01122**	**0.01244**
Experiment 8	0.0387	0.0277	0.0368
Experiment 9	0.0435	0.0499	0.0485
Experiment 10	0.0227	0.0229	0.0239

Figure 9 shows the plot of real data against the predicted data generated by the monolithic neural network optimized with the genetic algorithm.

Table 4 shows the particle swarm optimization (as shown in Fig. 9) where the prediction error is of 0.0013066.

Fuzzy integration is performed initially by implementing a type-1 fuzzy system in which the best result is in experiment of row number 8 of Table 5 with an error of: 0.0235.

As a second phase, to integrate the results of the optimized ensemble neural network a type-2 fuzzy system is implemented, where the best results that are obtained are as follows: with a degree uncertainty of 0.3 a forecast error of 0.01098 is obtained, with a degree of uncertainty of 0.4 the error is of 0.01122 and with a degree of uncertainty of 0.5 the error is of 0.001244, as shown in Table 6.

Figure 10 shows the plot of real data against the predicted data generated by the monolithic neural network optimized with the genetic algorithm.

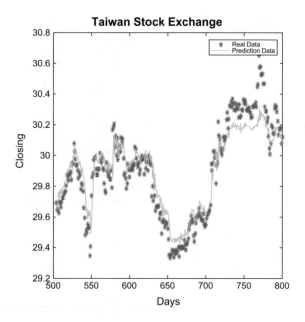

Fig. 10 Prediction with the optimized ensemble neural network with PSO of the TAIEX

4 Conclusion

The best result when applying the genetic algorithm to optimize the ensemble neural network was: 0.0011421 (as shown in Fig. 7 and Table 2). We also Implemented a type 2 fuzzy system for ensemble neural network, in which the results where for the best evolution as obtained a degree of uncertainty of 0.3 yielded a forecast error of 0.0124, with a 0.4 uncertainty error: 0.0136, and 0.5 uncertainty error of 0.01828, as shown in Table 3. The best result when applying the particle swarm to optimize the ensemble neural network was: 0.0013066 (as shown in Fig. 8 and Table 4). Implemented a type 2 fuzzy system for ensemble neural network, in which the results where for the best evolution as obtained a degree of uncertainty of 0.3 yielded a forecast error of 0.01098, with an 0.4 uncertainty error: 0.01122, and 0.5 uncertainty error of 0.01244, as shown in Table 6. After achieving these results, we have verified efficiency of the algorithms applied to optimize the neural network ensemble architecture. In this case, the method was efficient but it also has certain disadvantages, sometimes the results are not as good, but genetic algorithms can be considered as good technique a for solving search and optimization problems.

References

1. P. Cowpertwait, A. Metcalfe, Time series, in *Introductory Time Series with R* (Springer, Dordrecht, Heidelberg, London, New York, 2009), pp. 2–5
2. O. Castillo, P. Melin, Hybrid intelligent systems for time series prediction using neural networks, fuzzy logic, and fractal theory. IEEE Trans Neural Netw **13**(6), 1395–1408 (2002)
3. O. Castillo, P. Melin, Simulation and forecasting complex economic time series using neural networks and fuzzy logic, in *Proceeding of the International Neural Networks Conference 3* (2001), pp. 1805–1810
4. O. Castillo, P. Melin, Simulation and forecasting complex financial time series using neural networks and fuzzy logic, in *Proceedings the IEEE the International Conference on Systems, Man and Cybernetics 4* (2001), pp. 2664–2669
5. N. Karnik, M. Mendel, Applications of type-2 fuzzy logic systems to forecasting of time-series. Inf. Sci. **120**(1–4), 89–111 (1999)
6. A. Kehagias, V. Petridis, Predictive modular neural networks for time series classification. Neural Netw. **10**(1), 31–49, 245–250 (2000)
7. L.P. Maguire, B. Roche, T.M. McGinnity, L.J. McDaid, Predicting a chaotic time series using a fuzzy neural network. Inf. Sci. **112**(1–4), 125–136 (1998)
8. P. Melin, O. Castillo, S. Gonzalez, J Cota, W. Trujillo, P. Osuna, *Design of Modular Neural Networks with Fuzzy Integration Applied to Time Series Prediction*, vol. 41 (Springer, Berlin, Heidelberg, 2007), pp. 265–273
9. R.N. Yadav, P.K. Kalra, J. John, Time series prediction with single multiplicative neuron model. Soft computing for time series prediction. Appl. Soft Comput. **7**(4), 1157–1163 (2007)
10. L. Zhao, Y. Yang, PSO-based single multiplicative neuron model for time series prediction. Expert Syst. Appl. **36**(2 (Part 2)), 2805–2812 (2009)
11. P.T. Brockwell, R.A. Davis, *Introduction to Time Series and Forecasting* (Springer, New York, 2002), pp. 1–219
12. N. Davey, S. Hunt, R. Frank, *Time Series Prediction and Neural Networks* (University of Hertfordshire, Hatfield, UK, 1999)
13. J.S.R. Jang, C.T. Sun, E. Mizutani, *Neuro-Fuzzy and Soft Computing* (Prentice Hall, 1996)
14. I.M. Multaba, M.A. Hussain, *Application of Neural Networks and Other Learning Technologies in Process Engineering* (Imperial Collage Press, 2001)
15. A. Sharkey, *Combining Artificial Neural Nets: Ensemble and Modular Multi-net Systems* (Springer, London, 1999)
16. P. Sollich, A. Krogh, Learning with ensembles: how over-fitting can be useful, in *Advances in Neural Information Processing Systems 8*, ed. by D.S. Touretzky, M.C. Mozer, M.E. Hasselmo (Denver, CO, MIT Press, Cambridge, MA, 1996), pp. 190–196
17. L.K. Hansen, P. Salomon, Neural network ensembles. IEEE Trans. Pattern Anal. Mach. Intell. **12**(10), 993–1001 (1990)
18. A. Sharkey, *One Combining Artificial of Neural Nets* (Department of Computer Science University of Sheffield, U.K., 1996)
19. S. Gutta, H. Wechsler, Face recognition using hybrid classifier systems, in *Proceedings of ICNN-96*, Washington, DC (IEEE Computer Society Press, Los Alamitos, CA, 1996), pp. 1017–1022
20. F.J. Huang, Z. Huang, H-J. Zhang, T.H. Chen, Pose invariant face recognition, in *Proceedings of 4th IEEE International Conference on Automatic Face and Gesture Recognition*, Grenoble, France (IEEE Computer Society Press, Los Alamitos, CA)
21. H.. Drucker, R. Schapire, P. Simard, Improving performance in neural networks using a boosting algorithm, in *Advances in Neural Information Processing Systems 5*, Denver, CO, ed. by S.J. Hanson, J.D. Cowan, C.L. Giles (Morgan Kaufmann, San Mateo, CA, 1993), pp. 42–49
22. J. Hampshire, A. Waibel, A novel objective function for improved phoneme recognition using time-delay neural networks. IEEE Trans. Neural Netw. **1**(2), 216–228 (1990)

23. J. Mao, A case study on bagging, boosting and basic ensembles of neural networks for OCR, in *Proceedings of IJCNN-98*, vol. 3, Anchorage, AK (IEEE Computer Society Press, Los Alamitos, CA, 1998), pp. 1828–1833
24. K.J. Cherkauer, Human expert level performance on a scientific image analysis task by a system using combined artificial neural networks, in *Proceedings of AAAI-96 Workshop on Integrating Multiple Learned Models for Improving and Scaling Machine Learning Algorithms*, Portland, OR, ed. by P. Chan, S. Stolfo, D. Wolpert (AAAI Press, Menlo Park, CA, 1996), pp. 15–21
25. P. Cunningham, J. Carney, S. Jacob, Stability problems with artificial neural networks and the ensemble solution. Artif. Intell. Med. **20**(3), 217–225 (2000)
26. Z.-H. Zhou, Y. Jiang, Y.-B. Yang, S.-F. Chen, Lung cancer cell identification based on artificial neural network ensembles. Artif. Intell. Med. **24**(1), 25–36 (2002)
27. Y.N. Shimshon, Intrator Classification of seismic signal by integrating ensemble of neural networks. IEEE Trans. Signal Process. **461**(5), 1194–1201 (1998)
28. A. Antoniou, W. Sheng (eds.), *Practical Optimization Algorithms and Engineering Applications "Introduction Optimization"* (Springer, 2007), pp. 1–4
29. J. Holland, *Adaptation in Natural and Artificial Systems* (University of Michigan Press, 1975)
30. D. Goldberg, *Genetic Algorithms in Search, Optimization and Machine Learning* (Addison Wesley, 1989)
31. K. Man, K. Tang, S. Kwong, Genetic algorithms and designs, in *Introduction, Background and Biological Background* (Springer, London Limited, 1998), pp. 1–62
32. R. Eberhart, J. Kennedy, A new optimizer using swarm theory, in *Proceedings of 6th International Symposium on Micro Machine and Human Science (MHS)*, Oct 1995, pp. 39–43
33. J. Kennedy, R. Eberhart, Particle swarm optimization, in *Proceedings of IEEE International Conference on Neural Network (ICNN)*, Nov 1995, vol. 4, pp. 1942–1948
34. L.A. Zadeh, *Fuzzy Sets Information and Control*, vol. 8 (1965), pp. 338–353
35. J. Mendel, *Uncertain Rule-Based Fuzzy Logic Systems "Introduction an New Directions"* (Prentice-Hall Inc., 2001), pp. 213–231
36. Taiwan Bank Database. www.twse.com.tw/en. Accessed 03 Apr 2011

Optimization of Modular Neural Network Architectures with an Improved Particle Swarm Optimization Algorithm

Alfonso Uriarte, Patricia Melin and Fevrier Valdez

Abstract According to the literature of Particle Swarm Optimization (PSO), there are problems of getting stuck at local minima and premature convergence with this algorithm. A new algorithm is presented in this paper called the Improved Particle Swarm Optimization using the gradient descent method as an operator incorporated into the Algorithm, as a function to achieve a significant improvement. The gradient descent method (BP Algorithm) helps not only to increase the global optimization ability, but also to avoid the premature convergence problem. The Improved PSO Algorithm (IPSO) is applied to the design of Neural Networks to optimize their architecture. The results show that there is an improvement with respect to using the conventional PSO Algorithm.

Keywords Modular Neural Network · Particle Swarm Optimization
Pattern recognition

1 Introduction

The Particle Swarm Optimization (PSO) algorithm is inspired by the movements of bird flocks and the mutual collaboration among them-selves in seeking food within the shortest period of time [1]. PSO is one of the most popular optimization algorithms due to its extremely simple procedure, easy implementation and very fast convergence rate. Apart from all the advantages, the algorithm has its draw-backs too. The PSO algorithm faces problems with premature convergence as it is easily trapped into local optima. It is known that it is almost impossible for the PSO

A. Uriarte · P. Melin (✉) · F. Valdez
Division of Graduate Studies and Research, Tijuana Institute of Technology,
Tijuana, Mexico
e-mail: pmelin@tectijuana.mx

© Springer International Publishing AG, part of Springer Nature 2018   165
L. A. Zadeh et al. (eds.), *Recent Developments and the New Direction
in Soft-Computing Foundations and Applications*, Studies in Fuzziness
and Soft Computing 361, https://doi.org/10.1007/978-3-319-75408-6_14

algorithm to escape from the local optima once it has been trapped, causing the algorithm to fail in achieving the global optimum result. Many methods have been proposed throughout the years to counter this drawback of the PSO algorithm [2, 3].

Although some improvement measures have been proposed, such as increasing the population scale, the dynamic adjustment of the coefficient of inertia, this to a certain extent improving the optimization algorithm performance, but the algorithm itself has some of the nature problem that is not solved.

In order to apply the particle swarm algorithm in training neural network avoiding slow convergence speed and so on, this paper combines the particle swarm algorithm with the gradient descending method, the article proposes an improved particle swarm optimization algorithm, this method has the purpose of combining the advantages of particle swarm algorithm global parallel search and the gradient descending method local certainty search, improve of convergence speed and avoid the local minima.

The remainder of the paper is organized as follows: Sect. 2 defines Modular Neural Networks; Sect. 3 presents the Gradient Descent Method, Sect. 4 the Particle Swarm Optimization, Sect. 5 the Improved Particle Swarm Optimization and Sect. 6 the Experimental Results.

2 Modular Neural Networks

The modular neural network consists on a network of interconnected simple neural networks (also called modules) each of which solves a sub problem from a complete problem. The greatest benefit of modular neural networks is the inherent separation of the overall problem into sub problems.

A modular neural network is a set of simple networks and an integrating unit that determines how the outputs of the networks are combined to find the final output. The simple network is also called an expert network. The expert networks and the integrating unit work together to learn how to divide a task into subtasks that are functionally independent. The integration unit measures the competence of each expert network and assigns different networks to learn each task. This modular architecture avoids learning a global control policy set that cannot be good for the entire control of each task under different operating points.

Neural networks are particularly adaptable to scientific research of mathematical nonlinear variables models. In these models, the effect of the variables interrelationship attempt to solve them by using algorithmic methods, but it has been proven that these are very difficult to solve. Figure 1 shows the architecture of a Modular Neural Network.

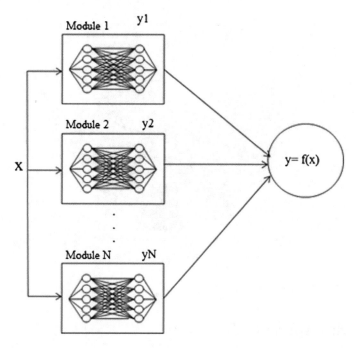

Fig. 1 Modular neural network architecture

3   Gradient Descent Method

The gradient descent algorithm is a method used to find the local minimum of a function. It works by starting with an initial guess of the solution and it takes the gradient of the function at that point. It moves the solution in the negative direction of the gradient and it repeat the process. The algorithm will eventually converge where the gradient is zero (which corresponds to a local minimum).

A similar algorithm, the gradient ascent, finds the local maximum nearer the current solution by stepping it towards the positive direction of the gradient. They are both first-order algorithms because they take only the first derivative of the function.

The BP algorithm is a kind of simple deterministic local search method, it uses local adjustments aspects and shows a strong performance, along the direction of gradient descent and can quickly find local optimal solution, but the BP algorithm is very sensitive to the choice of the initial position, and does not ensure that the optimal solution is the global optimal. Figure 2 shows the Gradient Descent Method Visualization.

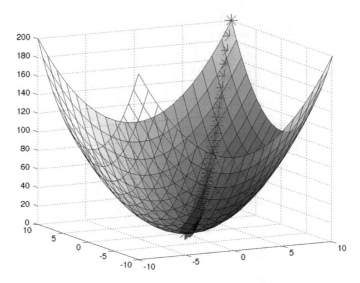

Fig. 2 Gradient descent method visualization

4 Particle Swarm Optimization

PSO was originally formulated by Eberhart and Kennedy in 1995. The thought process behind the algorithm was inspired by the social behavior of animals, such as bird flocking or fish schooling. PSO is similar to the continuous Genetic Algorithm (GA) in that it begins with a random population matrix. Unlike the GA, PSO has no evolutionary operators such as crossover and mutation. The rows in the matrix are called particles (similar to the GA chromosome). They contain the variable values and are not binary encoded. Each particle moves about the cost surface with certain velocity. The particles update their velocities and positions based on the local and global best solutions:

The equation to update the velocity is as follows:

$$V_{ij}(t+1) = V_{ij}(t) + c_1 r_{1j}(t)[y_{ij}(t) - x_{ij}(t)] + c_2 r_2(t)[\hat{y}_j(t) - x_{ij}(t)] \qquad (1)$$

The equation to update the position is expressed as:

$$x_i(t+1) = x_i(t) + v_i(t+1) \qquad (2)$$

In Eqs. (1) and (2), x_i is the position particle i, v_i is the velocity particle i, y_i is best position of particle i, $\hat{y}$ is the best global position, c_1 is cognitive coefficient, c_2 social coefficient, r_{1j} and r_{2j} are random values.

The PSO algorithm updates the velocity vector for each particle and then adds that velocity to the particle position or values. Velocity updates are influenced by both the best global solution associated with the lowest cost ever found by a particle and the best local solution associated with the lowest cost in the present population. If the best local solution has a cost lower than the cost of the current global solution, then the best local solution replaces the best global solution.

The particle velocity is reminiscent of local minimizers that use derivative information, because velocity is the derivative of position. The constant C_1 is called the cognitive parameter.

The constant C_2 is called the social parameter. The advantages of PSO are that it is easy to implement and there are few parameters to adjust.

5 Improved Particle Swarm Optimization

The process is as follows: all particles first are improved by PSO in the group of every generation algorithm, according to the Eqs. (1) and (2) update each particle speed and position, and calculate for each particle the fitness function value.

According to the particle fitness value choose one or several of the adaptable particles. These particles are called elite individuals. The elite individual is not directly into the next generation of algorithm in the group, but rather through the BP operator to improve the performance.

That is the BP operator in the area around to develop individual elite more excellent performance of particle position, and they lead a rapid evolution to the particles in the next generation.

From the overall perspective, the improved particle swarm optimization algorithm is in the algorithm of the next generation groups merge with BP Algorithm (Gradient Descent Method). Figure 3 shows a flow diagram of the Improved Particle Swarm Optimization.

IPSO Pseudocode

```
-Begin
    -Start population
    -While (no stop condition is met) do
        -Evaluate first population
        -Update velocity and position of particle
        -Evaluate new population
        -If the best local particle is better than best
                global particle
          - Move to the new population
        -Else
          -Evaluate with Gradient Descent
                        Method
    -End while
-End
```

Fig. 3 Improved particle
swarm optimization diagram

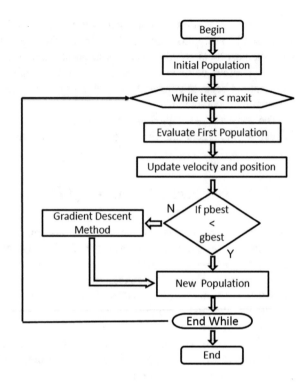

6 Experimental Results

In this case 6-dimensional particles are generated, the first dimension represents the
layers and the other 5 dimensions are the neurons per layer. Figure 4 shows the
architecture of a particle of the PSO algorithm with 3 layers.

Figure 5 shows the architecture of the particles of the PSO algorithm with 4
layers. Figure 6 shows the architecture of a particle of the PSO algorithm with 5
layers.

Fig. 4 Architecture of a
particle of the PSO algorithm
with 3 layers

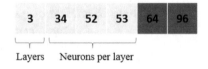

Fig. 5 Architecture of a
particle of the PSO algorithm
with 4 layers

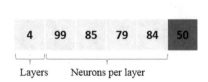

Fig. 6 Architecture of a
particle of the PSO algorithm
with 5 layers

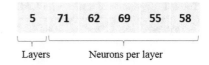

| 5 | 71 | 62 | 69 | 55 | 58 |

Layers   Neurons per layer

Fig. 7 Sound 1 of database
without noise

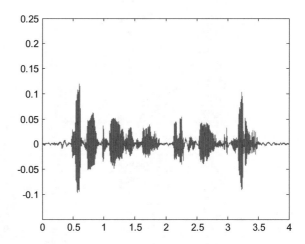

To train the modular neural network, we used The CMU_ARCTIC databases, which were recorded at the Language Technologies Institute at Carnegie Mellon University as phonetically balanced, US English single speaker databases designed for unit selection speech synthesis research.

The IPSO algorithm was executed with an initial population of 30 individuals, with which the neural network is evaluated. A modular neural network with the database is trained and recognition tests are made.

The database of American men was selected, which includes a subset of 500 speeches of the BD CMU Artic. Figure 7 shows one of the sounds of the database without noise.

Table 1 shows the parameters of training with IPSO. Table 2 shows the experimental results with IPSO. Table 3 shows the experimental results with simple PSO.

Table 1 Parameters of
training with IPSO

Parameter	Value
Training method	Trainscg
Epochs	300
Goal error	0.02
Transfer function hidden layer	Sigmoidal tangent
Transfer function output layer	Sigmoidal tangent

Table 2 Experimental results with IPSO

# Training	Goal error	Training method	Epochs	Error	% Recognition
1	0.02	Trainscg	300	0.27530284	72.47
2	0.02	Trainscg	300	0.16310281	83.69
3	0.02	Trainscg	300	0.12044402	87.96
4	0.02	Trainscg	300	0.29992361	70.01
5	0.02	Trainscg	300	0.11952153	88.05
6	0.02	Trainscg	300	0.2379497	76.21
7	0.02	Trainscg	300	0.22739126	77.26
8	0.02	Trainscg	300	0.18417452	81.58
9	0.02	Trainscg	300	0.10459964	89.54
10	0.02	Trainscg	300	0.2662329	73.38
11	0.02	Trainscg	300	0.11546172	88.45
12	0.02	Trainscg	300	0.14153354	85.85
13	0.02	Trainscg	300	0.15189437	84.81
14	0.02	Trainscg	300	0.27950404	72.05
15	0.02	Trainscg	300	0.29540453	70.46
16	0.02	Trainscg	300	0.23227976	76.77
17	0.02	Trainscg	300	0.13733892	86.27
18	0.02	Trainscg	300	0.16698889	83.30
19	0.02	Trainscg	300	0.10573184	89.43
20	0.02	Trainscg	300	0.25490314	74.51
21	0.02	Trainscg	300	0.19766658	80.23
22	0.02	Trainscg	300	0.19286712	80.71
23	0.02	Trainscg	300	0.10235316	89.76
24	0.02	Trainscg	300	0.18789587	81.21
25	0.02	Trainscg	300	0.26156607	73.84
26	0.02	Trainscg	300	0.12784975	87.22
27	**0.02**	**Trainscg**	**300**	**0.09717687**	**90.28**
28	0.02	Trainscg	300	0.12208608	87.79
29	0.02	Trainscg	300	0.12491552	87.51
30	0.02	Trainscg	300	0.14531977	85.47

7 Conclusion

In order to reduce the search process for the particle swarm algorithm and the existing early-convergence problem, a kind of "variation" into the idea of the particle swarm algorithm is proposed, the gradient descent method (BP algorithm) as a particle swarm operator is embedded in the particle swarm algorithm and helps

Table 3 Experimental results with simple PSO

# Training	Goal error	Training method	Epochs	Error	% Recognition
1	0.02	Trainscg	300	0.25925926	74.07
2	0.02	Trainscg	300	0.25925926	74.07
3	0.02	Trainscg	300	0.14814815	85.19
4	0.02	Trainscg	300	0.22222222	77.78
5	0.02	Trainscg	300	0.25925926	74.07
6	0.02	Trainscg	300	0.25925926	74.07
7	0.02	Trainscg	300	0.18518519	81.48
8	0.02	Trainscg	300	0.18518519	81.48
9	0.02	Trainscg	300	0.25925926	74.07
10	0.02	Trainscg	300	0.22222222	77.78
11	0.02	Trainscg	300	0.25925926	74.07
12	0.02	Trainscg	300	0.2962963	70.37
13	0.02	Trainscg	300	0.25925926	74.07
14	**0.02**	**Trainscg**	**300**	**0.14814815**	**85.19**
15	0.02	Trainscg	300	0.18518519	81.48
16	0.02	Trainscg	300	0.25925926	74.07
17	0.02	Trainscg	300	0.2962963	70.37
18	0.02	Trainscg	300	0.25925926	74.07
19	0.02	Trainscg	300	0.25925926	74.07
20	0.02	Trainscg	300	0.25925926	74.07
21	0.02	Trainscg	300	0.22222222	77.78
22	0.02	Trainscg	300	0.18518519	81.48
23	0.02	Trainscg	300	0.2962963	70.37
24	0.02	Trainscg	300	0.2962963	70.37
25	0.02	Trainscg	300	0.2962963	70.37
26	0.02	Trainscg	300	0.33333333	66.67
27	0.02	Trainscg	300	0.18518519	81.48
28	0.02	Trainscg	300	0.22222222	77.78
29	0.02	Trainscg	300	0.2962963	70.37
30	0.02	Trainscg	300	0.22222222	77.78

to solve problems of local minimum and premature convergence with faster convergence velocity in this way approaching the global optimal solution for the best design of modular neural network.

The results of experiments show that the average of recognition increases using the proposed method when compared to the simple PSO algorithm.

References

1. J. Kennedy, R. Eberhart, Particle swarm optimization, in *Proceedings of IEEE International Conference on Neural Networks, IV* (IEEE Service Center, Piscataway, NJ, 1995), pp. 1942–1948
2. R.D. Palupi, M.S. Siti, Particle swarm optimization: technique, system and challenges. Int. J. Appl. Inf. Syst. **1**, 19–27 (2011)
3. Q.H. Bai, Analysis of particle swarm optimization algorithm. Comput. Inf. Sci. **3**, 180–184 (2010)

Ensemble Neural Network with Type-2 Fuzzy Weights Using Response Integration for Time Series Prediction

Fernando Gaxiola, Patricia Melin, Fevrier Valdez and Juan R. Castro

Abstract In this paper an ensemble of three neural networks with type-2 fuzzy weights is proposed. One neural network uses type-2 fuzzy inference systems with Gaussian membership functions for obtain the fuzzy weights; the second neural network uses type-2 fuzzy inference systems with triangular membership functions; and the third neural network uses type-2 fuzzy inference systems with triangular membership functions with uncertainty in the standard deviation. Average integration and type-2 fuzzy integrator are used for the results of the ensemble neural network. The proposed approach is applied to a case of time series prediction, specifically in the Mackey-Glass time series.

1 Introduction

We are presenting an ensemble with three neural networks for the experiments. The final result for the ensemble was obtained with average integration and type-2 fuzzy integration. The time series prediction area is the study case for this paper, and particularly the Mackey-Glass time series is used to test the proposed approach.

This research uses the managing of the weights of a neural networks using type-2 fuzzy inference systems and due to the fact that these affect the performance of the learning process of the neural network, the used of type-2 fuzzy weights are an important part in the training phase for managing uncertainty.

F. Gaxiola (✉) · P. Melin · F. Valdez
Tijuana Institute of Technology, Tijuana, Mexico
e-mail: fergaor_29@hotmail.com

P. Melin
e-mail: pmelin@tectijuana.mx

F. Valdez
e-mail: fevrier@tectijuana.mx

J. R. Castro
UABC University, Tijuana, Mexico
e-mail: jrcastror@uabc.edu.mx

© Springer International Publishing AG, part of Springer Nature 2018
L. A. Zadeh et al. (eds.), *Recent Developments and the New Direction
in Soft-Computing Foundations and Applications*, Studies in Fuzziness
and Soft Computing 361, https://doi.org/10.1007/978-3-319-75408-6_15

One type of supervised neural network and its variations is the one that would be of most interest in our study, which is the backpropagation network. This type of network is the most commonly used in the above mentioned areas.

The weights of a neural network are an important part in the training phase, because these affect the performance of the learning process of the neural network.

This conclusion is based on the practice of neural networks of this type, where some research works have shown that the training of neural networks for the same problem initialized with different weights or its adjustment in a different way but at the end is possible to reach a similar result.

The next section presents the basic concepts of neural networks and type-2 fuzzy logic. Section 3 presents a review of research about modifications of the back-propagation algorithm, different management strategies of weights in neural networks and time series prediction. Section 4 explains the proposed ensemble neural network. Section 5 describes the simulation results for the ensemble neural network with average integration and the type-2 fuzzy integrator proposed in this paper. Finally, in Sect. 6, some conclusions are presented.

2 Basic Concepts

2.1 Neural Network

An artificial neural network (ANN) is a distributed computing scheme based on the structure of the nervous system of humans. The architecture of a neural network is formed by connecting multiple elementary processors, this being an adaptive system that has an algorithm to adjust their weights (free parameters) to achieve the performance requirements of the problem based on representative samples [1, 2]. The most important property of artificial neural networks is their ability to learn from a training set of patterns, i.e. they are able to find a model that fits the data [3, 4].

The artificial neuron consists of several parts (see Fig. 1). On one side are the inputs, weights, the summation, and finally the adapter function. The input values are multiplied by the weights and added: $\sum x_i w_{ij}$. This function is completed with the addition of a threshold amount i. This threshold has the same effect as an entry with value -1. It serves so that the sum can be shifted left or right of the origin.

Fig. 1 Scheme of an artificial neuron

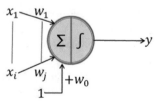

After addition, we have the function f applied to the sum, resulting the final value of the output, also called y_i [5], obtaining the following equation.

$$y_i = f\left(\sum_{i=1}^{n} x_i w_{ij}\right). \tag{1}$$

where f may be a nonlinear function with binary output $+ -1$, a linear function f $(z) = z$, or as sigmoidal logistic function:

$$f(z) = \frac{1}{1 + e^{-z}}. \tag{2}$$

2.2 Type-2 Fuzzy Logic

The concept of a type-2 fuzzy set, was introduced by Zadeh (1975) as an extension of the concept of an ordinary fuzzy set (henceforth called a "type-1 fuzzy set"). A type-2 fuzzy set is characterized by a fuzzy membership function, i.e., the membership grade for each element of this set is a fuzzy set in [0, 1], unlike a type-1 set where the membership grade is a crisp number in [0, 1] [6, 7].

Such sets can be used in situations where there is uncertainty about the membership grades themselves, e.g., uncertainty in the shape of the membership function or in some of its parameters [8]. Consider the transition from ordinary sets to fuzzy sets. When we cannot determine the membership of an element in a set as 0 or 1, we use fuzzy sets of type-1 [9–11]. Similarly, when the situation is so fuzzy that we have trouble determining the membership grade even as a crisp number in [0, 1], we use fuzzy sets of type-2 [12–17].

3 Historical Development

The backpropagation algorithm and its variations are the most useful basic training methods in the area of research of neural networks. When applying the basic backpropagation algorithm to practical problems, the training time can be very high. In the literature we can find that several methods have been proposed to accelerate the convergence of the algorithm [18–21].

There exist many works about adjustment or managing of weights but only the most important and relevant for this research will be considered here [22–25].

Ishibuchi et al. [26], proposed a fuzzy network where the weights are given as trapezoidal fuzzy numbers, denoted as four trapezoidal fuzzy numbers for the four parameters of trapezoidal membership functions.

Ishibuchi et al. [27], proposed a fuzzy neural network architecture with symmetrical fuzzy triangular numbers for the fuzzy weights and biases, denoted by the lower, middle and upper limit of the fuzzy triangular numbers.

Momentum method—Rumelhart, Hinton and Williams suggested adding in the increased weights expression a momentum term β, to filter the oscillations that can be formed a higher learning rate that lead to great change in the weights [5, 28].

Adaptive learning rate—focuses on improving the performance of the algorithm by allowing the learning rate changes during the training process (increase or decrease) [28].

Castro et al. [29], proposed interval type-2 fuzzy neurons for the antecedents and interval of type-1 fuzzy neurons for the consequents of the rules.

Kamarthi and Pittner [30], focused in obtaining a weight prediction of the network at a future epoch using extrapolation. Feuring [31], developed a learning algorithm in which the backpropagation algorithm is used to compute the new lower and upper limits media weights. The modal value of the new fuzzy weight is calculated as the average of the new computed limits.

Recent works on type-2 fuzzy logic have been developed in time series prediction, like that of Castro et al. [32], and other researchers [33, 34].

4   Proposed Ensemble Neural Network

The focus of this work is to use ensemble neural networks with three neural networks with type-2 fuzzy weights to allow the neural network to handle data with uncertainty; we used an average integration approach and type-2 fuzzy integrator for the final result of the ensemble. The approach is applied in time series prediction for the Mackey Glass time series (for $\tau = 17$).

The three neural network works with type-2 fuzzy weights [35], one network works with two-sided Gaussian interval type-2 membership functions with uncertain mean and standard deviation in the two type-2 fuzzy inference systems (FIST2) used to obtain the weights (one in the connections between the input and hidden layer and the other between the hidden and output layer); the other two networks work with triangular interval type-2 membership function with uncertain and triangular interval type-2 membership function with uncertain standard deviation, respectively (see Fig. 2).

We considered a three neural network architecture, and each network works with 30 neurons in the hidden layer and 1 neuron in the output layer. These neural networks handle type-2 fuzzy weights in the hidden layer and output layer. In the hidden layer and output layer of the networks we are working with a type-2 fuzzy inference system obtaining new weights in each epoch of the networks [36–39].

We used two similar type-2 fuzzy inference systems to obtain the type-2 fuzzy weights in the hidden and output layer for the neural network.

The weight managing in the three neural networks will be done differently to the traditional management of weights performed with the backpropagation algorithm

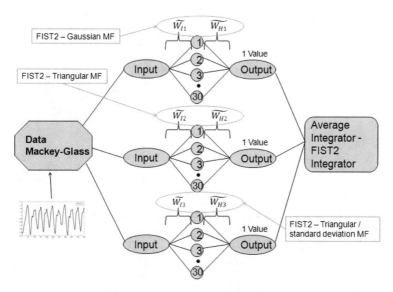

Fig. 2 Proposed ensemble neural network architecture with interval type-2 fuzzy weights using average integration or type-2 fuzzy integrator

(see Fig. 3); the method works with interval type-2 fuzzy weights, taking into account the change in the way we work internally in the neuron (see Fig. 4) [40].

The activation function f (-) used in this research was the sigmoid function in the neurons of the hidden layer and the linear function in the neurons of the output for the three neural networks.

The three neural networks used two type-2 fuzzy inference systems with the same structure (see Fig. 5), which have two inputs (the current weight in the actual epoch and the change of the weight for the next epoch) and one output (the new weight for the next epoch).

In the first neural network, the inputs and the output for the type-2 fuzzy inference systems used between the input and hidden layer are delimited with two Gaussian membership functions with their corresponding range (see Fig. 6); and the inputs and output for the type-2 fuzzy inference systems used between the hidden

Fig. 3 Schematic of the management of numerical weights for input of each neuron

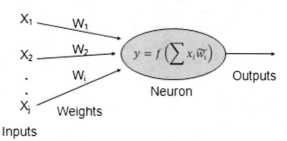

Fig. 4 Schematic of the
management of interval type 2
fuzzy weights for input of
each neuron

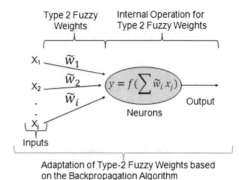

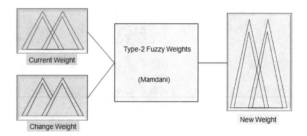

Fig. 5 Structure of the six
type-2 fuzzy inference
systems used in the three
neural networks

and output layer are delimited with two Gaussian membership functions with their corresponding range (see Fig. 7).

In the second neural network, the inputs and the output for the type-2 fuzzy inference systems used between the input and hidden layer are delimited with two triangular membership functions with their corresponding ranges (see Fig. 8); and the inputs and output for the type-2 fuzzy inference systems used between the hidden and output layer are delimited with two triangular membership functions with their corresponding ranges (see Fig. 9).

In the third neural network, the inputs and the output for the type-2 fuzzy inference systems used between the input and hidden layer are delimited with two triangular membership functions with standard deviation with their corresponding range (see Fig. 10); and the inputs and output for the type-2 fuzzy inference systems used between the hidden and output layer are delimited with two triangular membership functions with uncertainty in the standard deviation with their corresponding ranges (see Fig. 11).

The rules for the six type-2 fuzzy inference systems are the same, we used six rules for the type-2 fuzzy inference systems, corresponding to the four combinations of two membership functions and we added two rules for the case when the change of weight is null (see Fig. 12).

We obtain the prediction result for the ensemble neural network using the average integration and type-2 fuzzy integrator.

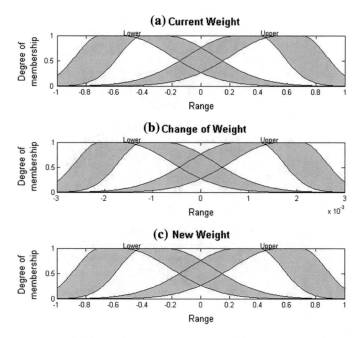

Fig. 6 Inputs (**a** and **b**) and output (**c**) of the type-2 fuzzy inference system used between the input and hidden layer for the first neural network

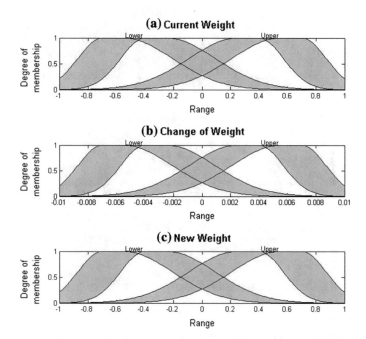

Fig. 7 Inputs (**a** and **b**) and output (**c**) of the type-2 fuzzy inference system used between the hidden and output layer for the first neural network

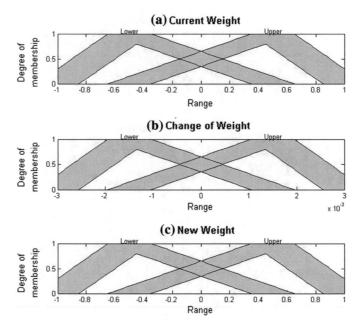

Fig. 8 Inputs (**a** and **b**) and output (**c**) of the type-2 fuzzy inference system used between the input and hidden layer for the second neural network

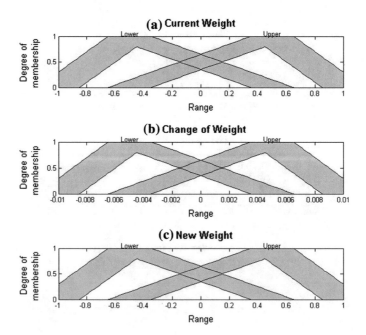

Fig. 9 Inputs (**a** and **b**) and output (**c**) of the type-2 fuzzy inference system used between the hidden and output layer for the second neural network

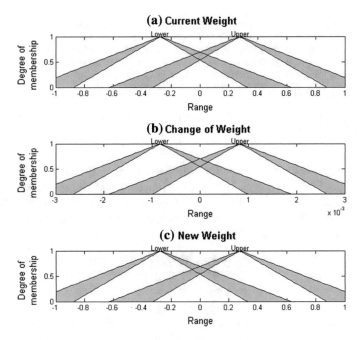

Fig. 10 Inputs (**a** and **b**) and output (**c**) of the type-2 fuzzy inference system used between the input and hidden layer for the third neural network

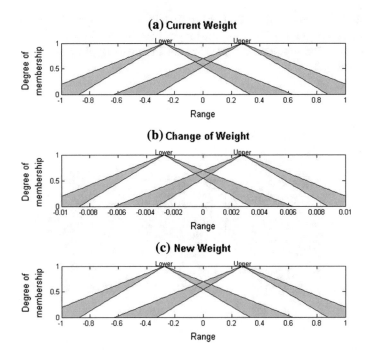

Fig. 11 Inputs (**a** and **b**) and output (**c**) of the type-2 fuzzy inference system used between the hidden and output layer for the third neural network

1. (Current_Weight is lower) and (Change_Weight is lower) then (New_Weight is lower)
2. (Current_Weight is lower) and (Change_Weight is upper) then (New_Weight is lower)
3. (Current_Weight is upper) and (Change_Weight is lower) then (New_Weight is upper)
4. (Current_Weight is upper) and (Change_Weight is upper) then (New_Weight is upper)
5. (Current_Weight is lower) then (New_Weight is lower)
6. (Current_Weight is upper) then (New_Weight is upper)

Fig. 12 Rules of the type-2 fuzzy inference system used in the six FIST2 for the neural networks with type-2 fuzzy weights

The average integration is performed with the Eq. 3 (prediction of the neural network with FIST2 Gaussian MF: NNGMF, prediction of the neural network with FIST2 triangular MF: NNTMF, prediction of the neural network with FIST2 triangular SD MF: NNTsdMF, number of neural networks in the ensemble: #NN, and prediction of the ensemble: PE).

$$PE = \frac{\text{NNGMF} + NNTMF + NNTsdMF}{\#NN} \tag{3}$$

The structure of the type-2 fuzzy integrator consists of three inputs: the prediction for the neural network with type-2 fuzzy weights using Gaussian membership functions (MF), triangular MF and triangular MF with uncertainty in the standard deviation; and one output: the final prediction of the integration (see Fig. 13)

We used three triangular membership functions in the inputs and output for the type-2 fuzzy integrator (T2FI) and the range is established in the interval for 0–1.5 (see Fig. 14). The footprint and positions of the membership functions are established empirically.

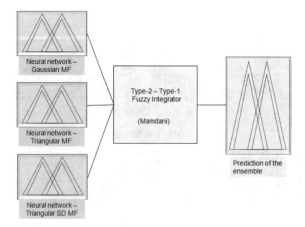

Fig. 13 Structure of the type-2 fuzzy integrator

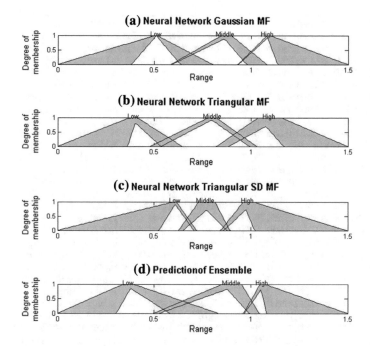

Fig. 14 Structure of the type-2 fuzzy integrator

In the type-2 fuzzy integrator we utilized 30 rules, 27 for the combination of the three inputs with "and" operator and there are also 3 rules using the "or" operator (see Fig. 15).

5 Simulation Results

The results for the experiments for the ensemble neural network with average integration (ENNAI) are shown on Table 1 and Fig. 16. The best prediction error is of 0.0346, and the average error is of 0.0485.

We presented 10 experiments of simulations for the ensemble neural network with the average integration and the type-2 fuzzy integrator, but the average error was calculated considering 30 experiments with the same parameters and conditions. The results for the experiments for the ensemble neural network with type-2 fuzzy integrator (ENNT2FI) are shown on Table 2. The best prediction error is of 0.0265, and the average error is of 0.0561.

We show in Table 3 a comparison for the prediction for the Mackey-Glass time series between the results for the monolithic neural network (MNN), the neural

1.	(NNGMF is Low) and (NNTMF is Low) and (NNTsdMF is Low) then (PE is Low)
2.	(NNGMF is Low) and (NNTMF is Low) and (NNTsdMF is Middle) then (PE is Low)
3.	(NNGMF is Low) and (NNTMF is Low) and (NNTsdMF is High) then (PE is Low)
4.	(NNGMF is Low) and (NNTMF is Middle) and (NNTsdMF is Low) then (PE is Low)
5.	(NNGMF is Low) and (NNTMF is Middle) and (NNTsdMF is Middle) then (PE is Middle)
6.	(NNGMF is Low) and (NNTMF is Middle) and (NNTsdMF is High) then (PE is Middle)
7.	(NNGMF is Low) and (NNTMF is High) and (NNTsdMF is Low) then (PE is Low)
8.	(NNGMF is Low) and (NNTMF is High) and (NNTsdMF is Middle) then (PE is Middle)
9.	(NNGMF is Low) and (NNTMF is High) and (NNTsdMF is High) then (PE is Middle)
10.	(NNGMF is Middle) and (NNTMF is Low) and (NNTsdMF is Low) then (PE is Middle)
11.	(NNGMF is Middle) and (NNTMF is Low) and (NNTsdMF is Middle) then (PE is Middle)
12.	(NNGMF is Middle) and (NNTMF is Low) and (NNTsdMF is High) then (PE is Middle)
13.	(NNGMF is Middle) and (NNTMF is Middle) and (NNTsdMF is Low) then (PE is Middle)
14.	(NNGMF is Middle) and (NNTMF is Middle) and (NNTsdMF is Middle) then (PE is Middle)
15.	(NNGMF is Middle) and (NNTMF is Middle) and (NNTsdMF is High) then (PE is Middle)
16.	(NNGMF is Middle) and (NNTMF is High) and (NNTsdMF is Low) then (PE is Middle)
17.	(NNGMF is Middle) and (NNTMF is High) and (NNTsdMF is Middle) then (PE is Middle)
18.	(NNGMF is Middle) and (NNTMF is High) and (NNTsdMF is High) then (PE is High)
19.	(NNGMF is High) and (NNTMF is Low) and (NNTsdMF is Low) then (PE is High)
20.	(NNGMF is High) and (NNTMF is Low) and (NNTsdMF is Middle) then (PE is High)
21.	(NNGMF is High) and (NNTMF is Low) and (NNTsdMF is High) then (PE is High)
22.	(NNGMF is High) and (NNTMF is Middle) and (NNTsdMF is Low) then (PE is High)
23.	(NNGMF is High) and (NNTMF is Middle) and (NNTsdMF is Middle) then (PE is Middle)
24.	(NNGMF is High) and (NNTMF is Middle) and (NNTsdMF is High) then (PE is Middle)
25.	(NNGMF is High) and (NNTMF is High) and (NNTsdMF is Low) then (PE is High)
26.	(NNGMF is High) and (NNTMF is High) and (NNTsdMF is Middle) then (PE is High)
27.	(NNGMF is High) and (NNTMF is High) and (NNTsdMF is High) then (PE is High)
28.	(NNGMF is Low) or (NNTMF is Low) or (NNTsdMF is Low) then (PE is Low)
29.	(NNGMF is Middle) or (NNTMF is Middle) or (NNTsdMF is Middle) then (PE is Middle)
30.	(NNGMF is High) or (NNTMF is High) or (NNTsdMF is High) then (PE is High)

Fig. 15 Rules for the type-2 fuzzy integrator

Table 1 Results for the ensemble neural network with average integration for Mackey-Glass time series

No.	Prediction error
E1	0.0350
E2	0.0496
E3	0.0553
E4	0.0375
E5	0.0428
E6	0.0523
E7	0.0623
E8	**0.0346**
E9	0.0437
E10	0.0372
Average	0.0485

network with type-2 fuzzy weights (NNT2FW), the ensemble neural network with average integration (ENNAI) and the ensemble neural network with type-2 fuzzy integrator (ENNT2FI).

Fig. 16 Graphic of real data again prediction data of ENNAI for the Mackey-Glass time series

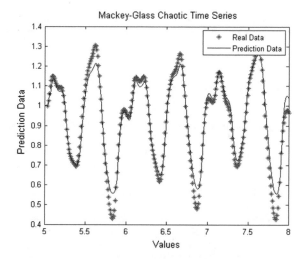

Table 2 Results for the ensemble neural network with the type-2 fuzzy integrator for time series Mackey-Glass

No.	Prediction error
E1	0.0290
E2	0.0429
E3	0.0539
E4	**0.0265**
E5	0.0428
E6	0.0488
E7	0.0340
E8	0.0555
E9	0.0306
E10	0.0666
Average	0.0561

Table 3 Comparison results for the Mackey-Glass time series

Method	Prediction error
MNN [35]	0.0530
NNT2FW [35]	0.0390
ENNAI	0.0346
ENNT2FI	0.0265

6 Conclusions

In the experiments, we observe that using an ensemble neural network with average integration and type-2 fuzzy integrator, we can achieve better results than the monolithic neural network and the neural network with type-2 fuzzy weights for the

Mackey-Glass time series. The ensemble with type-2 fuzzy integrator presents better results in almost all the experiments than the optimization with PSO.

REFERENCES

1. M. Cazorla, F. Escolano, Two Bayesian methods for junction detection. IEEE Trans. Image Process. **12**(3), 317–327 (2003)
2. G. Martinez, P. Melin, D. Bravo, F. Gonzalez, M. Gonzalez, Modular neural networks and fuzzy Sugeno integral for face and fingerprint recognition. Adv. Soft Comput. **34**, 603–618 (2006)
3. O. De Wilde, The magnitude of the diagonal elements in neural networks. Neural Netw. **10** (3), 499–504 (1997)
4. P.A. Salazar, P. Melin, O. Castillo, A new biometric recognition technique based on hand geometry and voice using neural networks and fuzzy logic, in *Soft Computing for Hybrid Intelligent Systems* (2008), pp. 171–186
5. V.V. Phansalkar, P.S. Sastry, Analysis of the back-propagation algorithm with momentum. IEEE Trans. Neural Networks **5**(3), 505–506 (1994)
6. O. Castillo, P. Melin, *Soft Computing for Control of Non-linear Dynamical Systems* (Springer, Heidelberg, Germany, 2001)
7. P. Melin, O. Castillo, *Hybrid Intelligent Systems for Pattern Recognition Using Soft Computing* (Springer, Heidelberg, 2005), pp. 2–3
8. L.A. Zadeh, Fuzzy sets. J. Inf. Control **8**, 338–353 (1965)
9. M. Okamura, H. Kikuch, R. Yager, S. Nakanishi, Character diagnosis of fuzzy systems by genetic algorithm and fuzzy inference, in *Proceedings of the Vietnam-Japan Bilateral Symposium on Fuzzy Systems and Applications*, Halong Bay, Vietnam (1998), pp. 468–473
10. W. Wang, S. Bridges, *Genetic Algorithm Optimization of Membership Functions for Mining Fuzzy Association Rules* (Department of Computer Science Mississippi State University, 2000)
11. J.S.R. Jang, C.T. Sun, E. Mizutani, *Neuro-fuzzy and Soft Computing: A Computational Approach to Learning and Machine Intelligence* (Prentice Hall, 1997)
12. O. Castillo, P. Melin, Type-2 Fuzzy Logic Theory and Applications (Springer, Berlin, 2008), pp. 29–43
13. J. Castro, O. Castillo, P. Melin, An interval type-2 fuzzy logic toolbox for control applications, in *FUZZ-IEEE* (2007), pp. 1–6
14. J. Castro, O. Castillo, P. Melin, A. Rodriguez-Diaz, Building fuzzy inference systems with a new interval type-2 fuzzy logic toolbox. Trans. Comput. Sci. **1**, 104–114 (2008)
15. D. Hidalgo, O. Castillo, P. Melin, Type-1 and type-2 fuzzy inference systems as integration methods in modular neural networks for multimodal biometry and its optimization with genetic algorithms, in *Soft Computing for Hybrid Intelligent Systems* (2008), pp. 89–114
16. D. Sanchez, P. Melin, Optimization of modular neural networks and type-2 fuzzy integrators using hierarchical genetic algorithms for human recognition, in *IFSA World Congress*, Surabaya, Indonesia, OS-414 (2011)
17. R. Sepúlveda, O. Castillo, P. Melin, A. Rodriguez, O. Montiel, Experimental study of intelligent controllers under uncertainty using type-1 and type-2 fuzzy logic. Inf. Sci. **177**(11), 2023–2048 (2007)
18. T.G. Barbounis, J.B. Theocharis, Locally recurrent neural networks for wind speed prediction using spatial correlation. Inf. Sci. **177**(24), 5775–5797 (2007)
19. T. Gedeon, Additive neural networks and periodic patterns. Neural Netw. **12**(4–5), 617–626 (1999)

20. M. Meltser, M. Shoham, L. Manevitz, Approximating functions by neural networks: a constructive solution in the uniform norm. Neural Netw. **9**(6), 965–978 (1996)
21. D. Yeung, P. Chan, W. Ng, Radial basis function network learning using localized generalization error bound. Inf. Sci. **179**(19), 3199–3217 (2009)
22. D. Casasent, S. Natarajan, A classifier neural net with complex-valued weights and square-law nonlinearities. Neural Netw. **8**(6), 989–998 (1995)
23. S. Draghici, On the capabilities of neural networks using limited precision weights. Neural Netw. **15**(3), 395–414 (2002)
24. R.S. Neville, S. Eldridge, Transformations of Sigma–Pi Nets: obtaining reflected functions by reflecting weight matrices. Neural Netw. **15**(3), 375–393 (2002)
25. J. Yam, T. Chow, A weight initialization method for improving training speed in feedforward neural network. Neurocomputing **30**(1–4), 219–232 (2000)
26. H. Ishibuchi, K. Morioka, H. Tanaka, A fuzzy neural network with trapezoid fuzzy weights, fuzzy systems, in *IEEE World Congress on Computational Intelligence*, vol. 1 (1994), pp. 228–233
27. H. Ishibuchi, H. Tanaka, H. Okada, Fuzzy neural networks with fuzzy weights and fuzzy biases, in *IEEE International Conference on Neural Networks*, vol. 3 (1993), pp. 160–165
28. M.T. Hagan, H.B. Demuth, M.H. Beale, *Neural Network Design* (PWS Publishing, Boston, 1996), p. 736
29. J. Castro, O. Castillo, P. Melin, A. Rodríguez-Díaz, A hybrid learning algorithm for a class of interval type-2 fuzzy neural networks. Inf. Sci. **179**(13), 2175–2193 (2009)
30. S. Kamarthi, S. Pittner, Accelerating neural network training using weight extrapolations. Neural Netw. **12**(9), 1285–1299 (1999)
31. T. Feuring, Learning in fuzzy neural networks, in *IEEE International Conference on Neural Networks*, vol. 2 (1996), pp. 1061–1066
32. J. Castro, O. Castillo, P. Melin, O. Mendoza, A. Rodríguez-Díaz, An interval type-2 fuzzy neural network for chaotic time series prediction with cross-validation and Akaike test, in *Soft Computing for Intelligent Control and Mobile Robotics* (2011), pp. 269–285
33. N. Karnik, J. Mendel, Applications of type-2 fuzzy logic systems to forecasting of time-series. Inf. Sci. **120**(1–4), 89–111 (1999)
34. R. Abiyev, A type-2 fuzzy wavelet neural network for time series prediction. Lect. Notes Comput. Sci. **6098**, 518–527 (2010)
35. F. Gaxiola, P. Melin, F. Valdez, O. Castillo, Interval type-2 fuzzy weight adjustment for backpropagation neural networks with application in time series prediction. Inf. Sci. **260**, 1–14 (2014)
36. O. Castillo, P. Melin, A review on the design and optimization of interval type-2 fuzzy controllers. Appl. Soft Comput. **12**(4), 1267–1278 (2012)
37. P. Melin, *Modular Neural Networks and Type-2 Fuzzy Systems for Pattern Recognition* (Springer, 2012), pp. 1–204
38. H. Hagras, Type-2 fuzzy logic controllers: a way forward for fuzzy systems in real world environments, in *IEEE World Congress on Computational Intelligence* (2008), pp. 181–200
39. R. Sepúlveda, O. Castillo, P. Melin, O. Montiel, An efficient computational method to implement type-2 fuzzy logic in control applications, in *Analysis and Design of Intelligent Systems using Soft Computing Techniques* (2007), pp. 45–52
40. M.D. Monirul Islam, K. Murase, A new algorithm to design compact two-hidden-layer artificial neural networks. Neural Netw. **14**(9), 1265–1278 (2001)

Part V
Fuzziness in Human and Resource Management

Resource Selection with Soft Set Attribute Reduction Based on Improved Genetic Algorithm

Absalom E. Ezugwu, Shahnaz N. Shahbazova, Aderemi O. Adewumi and Sahalu B. Junaidu

Abstract In principle, distributed heterogeneous commodity clusters can be deployed as a computing platform for parallel execution of user application, however, in practice, the tasks of first discovering and then configuring resources to meet application requirements are difficult problems. This paper presents a general-purpose resource selection framework that addresses the problems of resources discovery and configuration by defining a resource selection scheme for locating distributed resources that match application requirements. The proposed resource selection method is based on the frequencies of weighted condition attribute values of resources and the outstanding overall searching ability of genetic algorithm. The concept of soft set condition attributes reducts, which is dependent on the weighted conditions' attribute value of resource parameters is used to achieve the required goals. Empirical results are reported to demonstrate the potential of soft set condition attribute reducts in the implementation of resource selection decision models with relatively higher level of accuracy.

A. E. Ezugwu (✉) · A. O. Adewumi
School of Mathematics, Statistics and Computer Science, University of KwaZulu-Natal, Private Bag X54001, Durban 4001, South Africa
e-mail: ezugwu.absalom@fulafia.edu.ng

A. O. Adewumi
e-mail: adewumia@ukzn.ac.za

S. N. Shahbazova
Department of IT and Programming, Azerbaijan Technical University, Baku, Azerbaijan
e-mail: shahnazova@gmail.com

A. E. Ezugwu
Department of Computer Science, Federal University Lafia, Lafia, Nasarawa State, Nigeria

S. B. Junaidu
Department of Mathematics, Ahmadu Bello University, Zaria, Kaduna State, Nigeria
e-mail: sahalu@abu.edu.ng

© Springer International Publishing AG, part of Springer Nature 2018
L. A. Zadeh et al. (eds.), *Recent Developments and the New Direction in Soft-Computing Foundations and Applications*, Studies in Fuzziness and Soft Computing 361, https://doi.org/10.1007/978-3-319-75408-6_16

193

1 Introduction

The selection of suitable resource for user applications amidst proliferation of large sets of computing resources inside distributed computing environments poses unprecedented challenges. These environments are remarkably dynamic in nature, resources are often heterogeneous, and their availability marked with several inconsistences and uncertainties. Therefore, managing these resources raises a lot of challenges, including determining the right resource subset for a specific application and scheduling a job on it [1]. Researchers realize that in order to achieve high resource utilization and optimal application makespan, feature resource selection procedures is an indispensable component of the entire distributed system scheduling process [2, 3]. In distributed computing generally, most of the existing resource selection approach takes enormous amount of time to find minimal subset of qualified candidate resources, capable of executing specific user applications.

There are several important existing mathematical theories or tools, used for dealing with uncertainties in relation to some complex problems that involve data which are not always crisp, as pointed by Molodtsov [4]. This can also be extended to similar uncertainties associated with the distribution and availability of computing resources in a dynamically distributed computing environment. Some examples of these theories include: theory of probability, theory of fuzzy sets, theory of intuitionistic fuzzy sets, theory of vague sets, theory of interval mathematics, and theory of rough sets. However, all these theories have their own inherent limitations as pointed out in [4]. It is based on this trend that Molodtsov initiated the concept of soft theory as a new mathematical tool for dealing with uncertainties which is free from the inadequacy of the parameterization tool of the existing theories [5].

Soft set is a parameterized mathematical tool that deals with a collection of approximate descriptions of objects. Each approximate description has two parts, a predicate and an approximate value set [4–6]. Soft set theory has a rich potential for applications in several directions, few of which had been shown by Molodtsov in his pioneer work [4]. Other application areas of soft set theory can also be found in [7–11].

The new researches in the area of distributed resource selection focus on reducing resources search times or schedule lengths for the purpose of achieving optimal resource throughput and utilization. Following the concept presented in the work of Zhao [12], a new resource selection technique, which is based on the utilization of the unique parameterization properties of soft set theory is proposed. The concept is based on the calculation of weighted-average of all possible decision values of parameterized object (or machine) and the weight of each possible condition attribute value which is decided by the distribution of other objects (machines). Hence, this work proposes an effective and efficient resource selection method, fine-tuned by the concept of condition attribute reduct-soft-set theoretic model.

The concepts of attribute reduction with condition attributes in soft set theoretic and genetic algorithm (GA) are employed to obtain minimal reduction in resource discovery and selection time. This is achieved from a decision table under existing conditions by coupling the outstanding overall searching ability of GAs with soft set condition attributes reduct technique. A fitness function and selection algorithm are proposed and applied to the GAs, which accelerated the speed of convergence. Therefore, by coupling soft set decision rule with genetic algorithms proficiency problem search ability, the authors envisaged that an induction engine, which is able to induce probable decision rules from inconsistent resource information, can possibly be developed.

The rest of the paper is organized as follows. In Sect. 2, a review of related literature is presented; In Sect. 3, related soft set concept is briefly recalled; the principle of Genetic algorithm is explained in Sect. 4; while the empirical result is reported in Sect. 5 and Sect. 6 concludes the paper.

2 Related Work

Due to the robustness of soft set theory, in dealing with uncertainty and vagueness, many researchers tend to combine it with inductive learning techniques so as to achieve better results. In [13] soft set theory was applied to solve a decision making problem. The author developed a specialized algorithm that selects objects based on its optimal choice. This algorithm uses fewer parameters to select the optimal object for a decision problem. Conversely, in decision making problem, there is a straightforward relationship between the condition attribute values of objects and the conditional parameters. That is, the condition attribute values are computed with respect to the conditional parameters. This basic fact is also true for resource selection. For example, when selecting list of resources from a cluster, the attribute values of the resources are dependent on the priority placed on each resource parameters'.

Genetic algorithm is known to be conceptually simple but computationally powerful, which has made it a promising research area since its inception. Genetic algorithm and its variants have been used to handle a wide variety of applications, especially in the areas of numerical optimization, search optimization and machine learning [14–17]. Genetic algorithm-based learning techniques take advantage of the unique search engine of GAs to glean probable decision rules from its search space. Genetic algorithm have been combined with soft set based induction rule to solve the problems of GA poor convergence speed and plunging into local minimums [18]. Many other related work have addressed specific issues discussed in this paper, specifically, object selection (e.g., [5, 13, 19, 20]). However, for the current work, the authors do not claim innovation in this area, but instead emphasize the merits of the techniques presented in this paper and, in particular, the soft set techniques used to integrate adaptive mechanisms into the GA architecture. In the next section, the proposed approach employed to solve the resource selection problem is discussed.

3   Cluster Information System Based on Soft Set Theory

Let C denote a universal set of clusters of machines, and E a set of parameters (each parameter could be a word or a sentence, such as, "processing speed", "memory size", "band width", and so on) with certain attributes $A \subseteq E$. Also let $P(C)$ denote the power set of C.

Definition 1 A pair (F, E) is called a soft set over C, where F is a mapping given by $F: E \to P(C)$, (See [4, 13]).

For $\varepsilon \in E$, $F(\varepsilon)$ may be considered as the set of ε-elements of the soft set (F, E) or as the set of ε-approximate elements of the soft set. It is noteworthy that the sets $F(\varepsilon)$ may be empty for some $\varepsilon \in E$.

In this case, let us treat a cluster as an information system defined as a 4-tuple $C = (M, A, V_{\{0,1\}}, F$, where $M = \{m_1, m_2, \ldots, m_{|M|}\}$ is a non-empty finite set of machines (where $|M|$ denotes the cardinality of M), $A = \{a_1, a_2, \ldots, a_{|A|}\}$ is a non-empty finite set of attributes (where $|A|$ denotes the cardinality of A), $V = \bigcup_{a \in A} V_a, V_a$ is the domain (value set) of attribute a, $F: M \times A \to V$ is the cluster information function such that $F(m, a) \in V_a$ for every $(m, a) \in M \times A$, called cluster information or knowledge function.

In essence, a heterogeneous cluster can be seen as a knowledge representation system or an attribute-valued system. Thus, information about a cluster can be intuitively expressed in terms of an information table (see Table 1). Adopted from [19, 21].

Also, in a cluster knowledge representation system given by $C = (M, A, V, F)$, if $V_a = \{0, 1\}$ for every $a \in A$, then C is called a Boolean-valued information system. A simple example of a Boolean-valued information system is shown in Table 2. To illustrate the idea presented so far, let us consider the following example.

Table 1 A cluster information system

M	a_1	a_2	$\cdots$	a_k	$\cdots$	$a_{|A|}$
m_1	$F(m_1, a_1)$	$F(m_2, a_2)$	$\cdots$	$F(m_1, a_k)$	$\cdots$	$F(m_1, a_{|A|})$
m_2	$F(m_2, a_1)$	$F(m_2, a_2)$	$\cdots$	$F(m_2, a_k)$	$\cdots$	$F(m_2, a_{|A|})$
$\vdots$	$\vdots$	$\vdots$	$\ddots$	$\vdots$	$\ddots$	$\vdots$
$m_{|M|}$	$F(m_{|M|}, a_1)$	$F(m_{|M|}, a_2)$	$\cdots$	$F(m_{|M|}, a_k)$	$\cdots$	$F(m_{|M|}, a_{|A|})$

Table 2 Tabular representation of a soft set in the above example

M	e_1	e_2	e_3	e_4	e_5
m_1	1	1	1	1	1
m_2	1	1	0	1	0
m_3	0	1	1	1	1
m_4	1	0	0	0	0
m_5	1	0	1	1	1

Example 1 For instance, consider a soft set (F, E) which describes the "capabilities of a cluster" that the application J requires for its application.

Suppose that there are five machines in the universal set C under consideration, then

$$M = \{m_1, m_2, m_3, m_4, m_5\} \text{ and } E = \{e_1, e_2, e_3, e_4, e_5\}$$

is a set of conditional parameters, where e_1 stands for the parameter "processing speed", e_2 stands for the parameter "band width", e_3 stands for the parameter "memory size", e_4 stands for the parameter "cache size" and e_5 stands for the parameter "number of processors". Considering the mapping:

$F: E \to P(C)$, given by "machines $(\cdot)$" where $(\cdot)$ is to be filled in by one of parameters $e \in E$

$$F(e_1) = \{m_1, m_2, m_4, m_5\}, F(e_2) = \{m_1, m_2, m_3\}, F(e_3) = \{m_1, m_3, m_5\},$$
$$F(e_4) = \{m_1, m_2, m_3\}, F(e_5) = \{m_1, m_4, m_5\}$$

Therefore, $F(e_1)$ means machines with "high processing speeds" whose functional value is the set $\{m_1, m_2, m_4, m_5\}$.

Thus, we can view the soft set (F, E) as a collection of approximations as below:

$$(F, E) = \left\{ \begin{array}{l} \text{machines with high processing speeds} = \{m_1, m_2, m_4, m_5\} \\ \text{machines with high network bandwidths} = \{m_1, m_2, m_3\} \\ \text{machines with high memory sizes} = \{m_1, m_3, m_5\} \\ \text{machines with high cache sizes} = \{m_1, m_2, m_3\} \\ \text{machines with high number of processors} = \{m_1, m_4, m_5\} \end{array} \right\}$$

Each approximation has two parts, a predicate p and an approximate value set v. For example, for the approximation "high processing speed machines $= \{m_1, m_2, m_4, m_5\}$", we have the predicate name of high processing speed machines and the approximate value set or value set of m_1, m_2, m_4, m_5. Thus, a soft set (F, E) can be viewed as a collection of approximations, as shown below:

$$(F, E) = \{p_1 = v_1, p_2 = v_2, \ldots, p_n = v_n\}$$

Thus, a soft set now can be viewed as a knowledge representation system where the set of attributes is to be replaced by a set of parameters describing resource or machine specification characteristics values.

3.1 Reduct Soft Set

Consider the soft set (F, E). Clearly, for any $P \subset E$, (F, P) is a soft subset of (F, E). A reduct soft set of the soft set (F, P) is defined as follows:

Definition 2 Intuitively, a reduct-soft-set (F, Q) of the soft set (F, P) is that essential part of (F, P), which is sufficient enough to describe all basic approximate descriptions of the soft set [13].

3.2 Condition Attribute Value of a Machine m_i

A generalization selection technique of the classical soft set framework is made by constructing discernibility function from the available cluster information system. The condition attribute value of a machine $m_i \in M$ is r_i, given by

$$r_i = \sum_j m_{i,j} \tag{1}$$

where $m_{i,j}$ are the entries in the table of the reduct-soft-set.

It may happen that in selecting a machine, all the parameters belonging to P are not of equal importance to the task. In that case, the task would impose weights on its condition parameters, i.e., corresponding to each element $p_i \in P$, there is a weight $w_i \in (0, 1]$.

3.3 Weighted Condition Attribute Value of a Cluster m_i

The weighted condition attributes value of a machine $m_i \in M$ given by

$$r_i = \sum_j d_{i,j} \tag{2}$$

where $d_{i,j} = w_i \times m_{i,j}$. Imposing weights on its condition parameters, the task or application can use algorithm 1 to arrive at its final decision. With this improvement, the cluster information table presented in Table 1 can be re-adjusted and presented as shown in Table 3.

Table 3 A cluster information system with weighted parameters

M	a_1w_1	a_2w_2	$\cdots$	a_2w_k	$\cdots$	$a_{|a|}w_{|a|}$	Condition attributes
m_1	$F(m_1, a_1w_1)$	$F(m_2, a_2w_2)$	$\cdots$	$F(m_1, a_kw_k)$	$\cdots$	$F(m_1, a_{|A|}w_{|A|})$	$\sum_j F(m_1, a_{|A|}w_{|A|})$
m_2	$F(m_2, a_1w_1)$	$F(m_2, a_2w_2)$	$\cdots$	$F(m_2, a_kw_k)$	$\cdots$	$F(m_2, a_{|A|}w_{|A|})$	$\sum_j F(m_2, a_{|A|}w_{|A|})$
$\cdots$	$\cdots$	$\cdots$		$\cdots$		$\cdots$	$\cdots$
$m_{|m|}$	$F(m_{|M|}, a_1w_1)$	$F(m_{|M|}, a_2w_2)$	$\cdots$	$F(m_{|M|}, a_kw_k)$	$\cdots$	$F(m_{|M|}, a_{|A|}w_{|A|})$	$\sum_j F(m_{|M|}, a_{|A|}w_{|A|})$

4   Genetic Algorithm

Genetic algorithms are stochastic heuristic search methods, which use principles inspired by natural genetics to evolve solutions to problems [14]. Specifically, GA usually operates on multiple solutions simultaneously. This characteristic makes GA a good solution method that efficiently exploit both exploration and exploitation of the problem search space. The GA starts its search process by first initializing a population of individuals. Individual solutions are selected from the population, and then mate to form new solutions. The mating process which is usually achieved by combining genetic material from both parents to form new genetic material for a new solution confers the data from one generation of solution to the next one. Subsequently, diversity is promoted through the periodic application of random mutation. If the new solution is better than those in the population, the individuals in the population are replaced by the new solution [15]. The process is illustrated in Fig. 1.

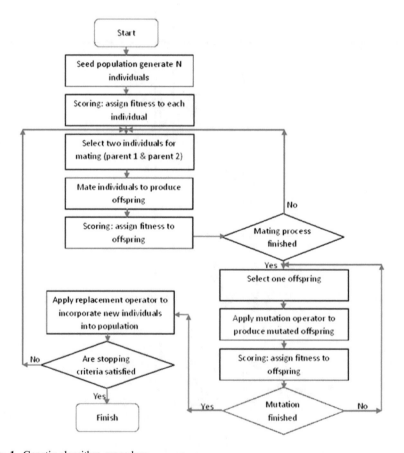

Fig. 1 Genetic algorithm procedure

4.1 Genetic Algorithm Operators

The traditional GA generally includes the following genetic operators: selection operator, crossover operator, mutation operator and optimal preservation strategy. The one-point crossover operator is used in this paper. The concrete implementation process is described as follows:

Chromosomes formulation: The optimization process of genetic algorithm does not directly act on the data in the solution space, but act on the corresponding code, which is expressed as a genotype string in the genetic space through encoding [14, 16]. Fixed-length binary array of one-dimension is used here, the length of chromosome equal to the number of condition attributes. Each gene bit does correspond to a condition attribute of either 1 or 0. The condition is such that, if the bit is 1, then the corresponding condition attribute is selected, and if otherwise, then the corresponding condition attribute is not selected. For example, the code 0101 represents condition attributes of 4 bits and the reduction is done by selecting the corresponding first and fourth condition attributes. This coding characteristic of the genetic algorithm makes it very easy to operate.

Initialization: The initial population is generated in such a way that, the gene bits corresponding to the resource parameters with attributes in the core are set as 1, while others are set as 0. The notion is that, machines with high attribute rating, in terms of speed, strength and capacities of their parameters are assigned attribute value of 1 and those with low attribute rating are assigned attribute value of 0.

Fitness function: In the GA, the fine level of the optimal solution can be achieved by making use of the fitness function which measures each individual's optimization calculation in the groups. In this paper, the fitness of each machine r is defined according to the expression given in Eq. 2 and is presented as follows:

$$fitness(r_i) = \sum_{i=1}^{n} r_i \qquad (3)$$

Selection process: To select best machine, the optimal preservation strategy is used [14, 17]. The process is such that, the number of population (that is the number of machines in a cluster) is set to n, and the fitness function of the individual machine r is given by $fitness(r)$, then the probability of machine r_i being selected (which is denoted by $P(r_{si})$) is expressed as follows:

$$P(r_{si}) = \frac{fitness(r_i)}{\sum_i^n fitness(r_i)} \qquad (4)$$

Summarily, the general selection process is based on comparing all machine fitness function and selecting the machine with the maximum fitness function, also referred to as the best machine or $best(r_i)$. This is illustrated in Table 4 and expressed as shown in Eq. 5.

$$best(r_i) = \max_{i \in \{1, 2, \ldots, |n|\}} fitness(r_i) \tag{5}$$

Crossover: The crossover operator is one of the most important genetic functions in evolutionary algorithms. This is a process where chromosomes exchange segments via recombination. An investigation of the standard one point crossover method and a "eugenic" best chromosome crossover method was carried out, in which a certain percentage of offspring was forced to be generated from the chromosomes with the highest fitness values. The new individual that emerges is defined as the crossover of them. Instead of adopting a random crossover, a uniform crossover is rather chosen, where the probability that the *i*th variable of the new individual is equal to the *i*th variable of the first or second parent is proportional to the fitness of the first or second parent. See [14, 22] for further details.

Mutation: mutation is a genetic operator used to maintain genetic diversity from one generation of a population of chromosomes to the next. Mutation is an important part of the genetic search, which help to prevent the population from stagnating at any local optima. Two mutation operators (swap and random) are used to introduce some extra variability into the population. A swap mutation swaps two selected genes from randomly selected genes in the chromosome; this is the primary method used to introduce variability. Random mutation changes a randomly selected gene value to another possible value; this method generally has an adverse effect on fitness due to poor 'distribution', but it can introduce 'lost' gene values back into the chromosome, thereby preventing premature convergence. A combination of these two methods show better results than using them individually. See [14, 22] for further details.

4.2 Algorithm Description

The following algorithm is used by the user application to select the resources that meets its execution requirements:

Algorithm: Improved reduct-soft-set genetic algorithm

Input: a decision table $C = (M, A, V, F), A = C \cup D$, where C is the condition attribute, D is the condition attribute values

Output: one or more advantage weak reduction

Step 1: Compute: $\sum_j F(m_{|M|}, a_{|A|} w_{|A|})$.

Step 2: Initializing the population: randomly generate m numbers initial population that length is n

Step 3: According to the type (1) compute: $fitness(r_i) = \sum_{i=1}^{n} r_i$

Step 4: Select the improved selection operator.

Step 5: Apply the one-point crossover operator

Step 6: Apply the basic bit mutation operator

Step 7: Compute the fitness values of new individuals; and then taking the optimal preservation strategy

Step 8: If all individuals are 1, turn to step 9, otherwise, turn to step 3

Step 9: Find i, for which $r_i = best\{r_i\}$

Step 10: Judging whether $best(r_i) = max\{r_i\}$, if true, then select r_i as candidate resources, otherwise, return to step 2.

Example 2 Let us consider the same soft set given in Sect. 3 above, which is expressed as a collection of approximations as shown below:

$$(F, E) = \left\{ \begin{array}{l} \text{machines with high processing speeds} = \{m_2, m_4, m_5, m_7, m_8, m_{11}\} \\ \text{machines with high network bandwidths} = \{m_2, m_3, m_5, m_6, m_{10}\} \\ \text{machines with high memory sizes} = \{m_1, m_4, m_5, m_7, m_8, m_9, m_{11}\} \\ \text{machines with high number of processors} = \{m_1, m_2, m_8, m_{10}, m_{12}\} \end{array} \right\}$$

Suppose that an application requires a resource $m_i \in M$ with the following weights for the parameters of m_i: for the parameter "*PS*", $w_1 = 0.4$, for the parameter "*BW*", $w_2 = 0.4$, for the parameter "*MS*" $w_3 = 0.2$, and for the parameter "*NP*", $w_4 = 0.1$.

Incorporating the condition attribute values, the reduct-soft-set is thus represented as shown in Table 4.

Here $max\{r_i\} = m_5 = 0.9$.

Decision: The application can opt for the machine m_5.

5 Empirical Analysis

In this section, an empirical evaluation of the resource selection technique was implemented using C++ Program and Microsoft Excel. The Excel software contains the resources tables, which consist of twelve clusters, with their parameters descriptions. The Improved Genetic Algorithm with Reduct-Softset proposed in this paper is demonstrated using the C++ program we wrote to test the proposed

Table 4 Resource selection decision table for cluster with 12 machines

M	Condition attributes				Decision attribute
	$e_1, w_1 = 0.4$	$e_2, w_1 = 0.3$	$e_3, w_1 = 0.2$	$e_4, w_1 = 0.1$	Decision value
m_1	0	0	1	1	$m_1 = 0.3$
m_2	1	1	0	1	$m_2 = 0.8$
m_3	0	1	0	0	$m_3 = 0.3$
m_4	1	0	1	0	$m_4 = 0.6$
m_5	1	1	1	0	$m_5 = 0.9$
m_6	0	1	0	0	$m_6 = 0.3$
m_7	1	0	1	0	$m_7 = 0.6$
m_8	1	0	1	1	$m_8 = 0.7$
m_9	0	0	1	0	$m_9 = 0.2$
m_{10}	0	1	0	1	$m_{10} = 0.4$
m_{11}	1	0	1	0	$m_{11} = 0.6$
m_{12}	0	0	0	1	$m_{12} = 0.1$

algorithm. There are three stages in the selection process, which involves the following.

1. Resource filtering stage
2. Resource classification according to assigned weights and fitness
3. Selection of resources with best fitness characteristics that meets the application resource requirement description of the user

The simulation starts by supplying input in the format represented in Table 5. This is usually presented in the form of application or job resource requirement. All the resources values were randomly generated for the purpose of the experiment.

Table 5 Jobs Resource Requirement

Jobs resource requirement				
Job ID	# processors	Processor speed	Memory size	Bandwidth
Job 1	1	2608	3244	111
Job 2	1	3062	2849	70
Job 3	4	2608	3244	100

The second phase of the simulation starts with resource classification according to assigned weights and computation of fitness values for each specific machines per cluster. There are twelve sets of clusters whose resource parameters are generated randomly. Weights are assigned to the parameters according their priorities. Since, four parameters are considered for the computing resources, $\beta_1, \beta_2, \beta_3$ and β_4 are the weights assigned to the four parameters respectively. The four weights are selected in such a way that they sum up to one. e.g. $\beta_1 = 0.4$, $\beta_2 = 0.3$, $\beta_3 = 0.2$ and $\beta_4 = 0.1$. The selection condition is set in such a way that, job resource requirement parameters must be less or equal to the available resource parameters, which the resource provider can offer. The result of the simulation process is illustrated in Table 6 (See appendix).

After the second phase of the simulation process, what follows is the selection of resource cluster having the best fitness characteristics that fulfills the job resource requirement description of the user. This is the third simulation phase. In this case, all the machines in cluster 2 fulfill the resource requirements of the user job and thus, it is selected for the job execution as shown in Table 7 (See appendix).

6 Conclusion and Future Work

In this paper we proposed an efficient resource selection heuristic that is based on the concept of soft set attribute reduction enhanced with improved genetic algorithm framework. This algorithm makes use of the frequency information of individual resource parameterized attribute in the cluster information system or knowledge base, and develops a weighting mechanism to rank resources based on their attributes values. Even though, the proposed method does not guarantee finding optimal solution, but empirical results show that in most situations it does. The ideas presented in this paper on theoretical application of soft sets, provides an intuitive solution to the problem of resource configuration and selection problem. The resource selection method presented in this paper is generic and can be applied to other resource discovery and selection problems. Further research direction includes enhancing the proposed resource selection method with other machine learning algorithms and developing a more efficient weighting mechanism.

Appendix: Empirical Results

Table 6 Resource filtering and classification

Cluster 1

C1Node ID	# processors	Processor speed	Memory size	Bandwidth	Job 1	Job2	Job3	Decision value			Qualified Machines for Job 1 in Cluster1				
machine_1	4	8853	8187	200	machine_1	machine_1	machine_1	1	1	1	machine_1	4	8853	8187	200
machine_2	10	6832	4052	300	machine_2	machine_2	machine_2	1	1	1	machine_2	10	6832	4052	300
machine_3	4	4341	3694	170	machine_3	machine_3	machine_3	1	1	1	machine_3	4	4341	3694	170
machine_4	9	4341	6326	140	machine_4	machine_4	machine_4	1	1	1	machine_4	9	4341	6326	140
machine_5	5	3661	5335	114	machine_5	machine_5	machine_5	1	1	1	machine_5	5	3661	5335	114
machine_6	4	4728	5627	105	No Match	machine_6	machine_6	0	1	1					
machine_7	5	5021	7349	169	machine_7	machine_7	machine_7	1	1	1	machine_7	5	5021	7349	169
machine_8	10	2488	2352	161	No Match	No Match	No Match	0	0	0					
machine_9	9	2843	5612	128	machine_9	No Match	machine_9	1	0	1	machine_9	9	2843	5612	128
machine_10	1	3466	3268	156	machine_10	No Match	No Match	1	0	0	machine_10	1	3466	3268	156
TOTAL	61	46574	51802	1643		Decision values for Cluster 1 =		0.80	0.80	0.80					

Cluster 2

C2Node ID	# processors	Processor speed	Memory size	Bandwidth	Job 1	Job2	Job3	Decision value			Qualified Machines for Job 1 in Cluster2				
machine_1	4	5030	9851	200	machine_1	machine_1	machine_1	1	1	1	machine_1	4	5030	9851	200
machine_2	6	6031	7481	250	machine_2	machine_2	machine_2	1	1	1	machine_2	6	6031	7481	250
machine_3	10	5806	15400	151	machine_3	machine_3	machine_3	1	1	1	machine_3	10	5806	15400	151
machine_4	10	5909	15282	137	machine_4	machine_4	machine_4	1	1	1	machine_4	10	5909	15282	137
machine_5	4	5411	6722	176	machine_5	machine_5	machine_5	1	1	1	machine_5	4	5411	6722	176
machine_6	4	5662	6537	171	machine_6	machine_6	machine_6	1	1	1	machine_6	4	5662	6537	171
machine_7	6	5496	4609	300	machine_7	machine_7	machine_7	1	1	1	machine_7	6	5496	4609	300
machine_8	8	5108	13368	189	machine_8	machine_8	machine_8	1	1	1	machine_8	8	5108	13368	189
machine_9	8	3838	3423	120	machine_9	machine_9	machine_9	1	1	1	machine_9	8	3838	3423	120
machine_10	9	3803	11185	131	machine_10	machine_10	machine_10	1	1	1	machine_10	9	3803	11185	131
TOTAL	69	52094	93858	1825		Decision values for Cluster 2 =		1.00	1.00	1.00					

Cluster 3

C3Node ID	# processors	Processor speed	Memory size	Bandwidth	Job 1	Job2	Job3	Decision value			Qualified Machines for Job 1 in Cluster3				
machine_1	10	5081	12895	211	machine_1	machine_1	machine_1	1	1	1	machine_1	10	5081	12895	211
machine_2	24	7661	10093	170	machine_2	machine_2	machine_2	1	1	1	machine_2	24	7661	10093	170
machine_3	20	3027	1545	211	No Match	No Match	No Match	0	0	0					
machine_4	14	2461	6441	163	No Match	No Match	No Match	0	0	0					
machine_5	10	5886	7715	223	machine_5	machine_5	machine_5	1	1	1	machine_5	10	5886	7715	223
machine_6	6	2751	9280	203	machine_6	No Match	machine_6	1	0	1	machine_6	6	2751	9280	203
machine_7	6	5478	6896	185	machine_7	machine_7	machine_7	1	1	1	machine_7	6	5478	6896	185
machine_8	6	5752	10048	195	machine_8	machine_8	machine_8	1	1	1	machine_8	6	5752	10048	195
machine_9	14	5871	10969	226	machine_9	machine_9	machine_9	1	1	1	machine_9	14	5871	10969	226
machine_10	6	4398	12064	172	machine_10	machine_10	machine_10	1	1	1	machine_10	6	4398	12064	172
TOTAL	116	48366	87946	1959		Decision values for Cluster 3 =		0.80	0.70	0.80					

Cluster 4

C4Node ID	# processors	Processor speed	Memory size	Bandwidth	Job 1	Job2	Job3	Decision value			Qualified Machines for Job 1 in Cluster4				
machine_1	9	5127	1619	236	No Match	No Match	No Match	0	0	0					
machine_2	4	6067	7270	267	machine_2	machine_2	machine_2	1	1	1	machine_2	4	6067	7270	267
machine_3	7	4842	5085	283	machine_3	machine_3	machine_3	1	1	1	machine_3	7	4842	5085	283
machine_4	10	3770	2500	207	No Match	No Match	No Match	0	0	0					
machine_5	8	5596	2955	227	No Match	machine_5	No Match	0	1	0					
machine_6	10	5748	6258	255	machine_6	machine_6	machine_6	1	1	1	machine_6	10	5748	6258	255
machine_7	6	2794	2125	256	No Match	No Match	No Match	0	0	0					
TOTAL	54	33944	27812	1731		Decision values for Cluster 4 =		0.43	0.57	0.30					

Cluster 5

C5Node ID	# processors	Processor speed	Memory size	Bandwidth	Job 1	Job2	Job3	Decision value			Qualified Machines for Job 1 in Cluster5				
machine_1	10	3699	5194	203	machine_1	machine_1	machine_1	1	1	1	machine_1	10	3699	5194	203
machine_2	20	2857	6681	235	machine_2	No Match	machine_2	1	0	1	machine_2	20	2857	6681	235
machine_3	12	3927	5014	208	machine_3	machine_3	machine_3	1	1	1	machine_3	12	3927	5014	208
machine_4	24	3608	2555	252	No Match	No Match	No Match	0	0	0					
machine_5	10	3074	7246	257	machine_5	machine_5	machine_5	1	1	1	machine_5	10	3074	7246	257
TOTAL	76	17165	26690	1155		Decision values for Cluster 5 =		0.80	0.60	0.80					

Cluster 6

C6Node ID	# processors	Processor speed	Memory size	Bandwidth	Job 1	Job2	Job3	Decision value			Qualified Machines for Job 1 in Cluster6				
machine_1	20	4341	5531	196	machine_1	machine_1	machine_1	1	1	1	machine_1	20	4341	5531	196
machine_2	10	5235	3052	173	No Match	machine_2	No Match	0	1	0					
machine_3	14	8737	3694	170	machine_3	machine_3	machine_3	1	1	1	machine_3	14	8737	3694	170
machine_4	9	8965	6326	280	machine_4	machine_4	machine_4	1	1	1	machine_4	9	8965	6326	280
machine_5	45	3661	5335	200	machine_5	machine_5	machine_5	1	1	1	machine_5	45	3661	5335	200
machine_6	7	4728	5627	200	machine_6	machine_6	machine_6	1	1	1	machine_6	7	4728	5627	200
machine_7	20	5021	7349	260	machine_7	machine_7	machine_7	1	1	1	machine_7	20	5021	7349	260
machine_8	10	3488	4352	161	machine_8	machine_8	machine_8	1	1	1	machine_8	10	3488	4352	161
machine_9	9	6843	5612	128	machine_9	machine_9	machine_9	1	1	1	machine_9	9	6843	5612	128
machine_10	22	3466	3268	156	machine_10	machine_10	machine_10	1	1	1	machine_10	22	3466	3268	156
TOTAL	166	54485	50146	1924		Decision values for Cluster 6 =		0.90	1.00	0.90					

Table 6 (continued)

Cluster 7 — Condition Attributes

C7Node ID	# processors	Processor speed	Memory size	Bandwidth
machine_1	22	5030	9851	109
machine_2	3	6031	7481	108
machine_3	10	5806	15400	151
machine_4	7	5909	15282	137
machine_5	4	5411	6722	176
machine_6	2	5662	6537	171
machine_7	30	5496	4609	159
machine_8	8	5108	13368	189
machine_9	14	3838	3423	113
machine_10	9	3803	11185	131
TOTAL	109	52094	93858	1444

Job 1	Job2	Job3	Decision value		
No Match	machine_1	machine_1	0	1	1
No Match	machine_2	No Match	0	1	0
machine_3	machine_3	machine_3	1	1	1
machine_4	machine_4	machine_4	1	1	1
machine_5	machine_5	machine_5	1	1	1
machine_6	machine_6	No Match	1	1	0
machine_7	machine_7	machine_7	1	1	1
machine_8	machine_8	machine_8	1	1	1
machine_9	machine_9	machine_9	1	1	1
machine_10	machine_10	machine_10	1	1	1
Total Matching Score for Cluster 7 =			0.80	1.00	0.80

Qualified Machines for Job 1 in Cluster7

machine_3	10	5806	15400	151
machine_4	7	5909	15282	137
machine_5	4	5411	6722	176
machine_6	2	5662	6537	171
machine_7	30	5496	4609	159
machine_8	8	5108	13368	189
machine_9	14	3838	3423	113
machine_10	9	3803	11185	131

Cluster 8 — Condition Attributes

C8Node ID	# processors	Processor speed	Memory size	Bandwidth
machine_1	4	5081	12895	211
machine_2	4	7661	10093	170
machine_3	6	3027	1545	211
machine_4	3	2461	6441	163
machine_5	10	5886	7715	223
machine_6	2	2751	9280	203
machine_7	6	5478	6896	185
machine_8	6	5752	10048	140
machine_9	5	5871	10969	226
machine_10	6	4398	12064	200
TOTAL	52	48366	87946	1932

Job 1	Job2	Job3	Decision value		
machine_1	machine_1	machine_1	1	1	1
machine_2	machine_2	machine_2	1	1	1
No Match	No Match	No Match	0	0	0
No Match	No Match	No Match	0	0	0
machine_5	machine_5	machine_5	1	1	1
machine_6	No Match	No Match	1	0	0
machine_7	machine_7	machine_7	1	1	1
machine_8	machine_8	machine_8	1	1	1
machine_9	machine_9	machine_9	1	1	1
machine_10	machine_10	machine_10	1	1	1
Decision values for Cluster 8 =			0.80	0.70	0.70

Qualified Machines for Job 1 in Cluster8

machine_1	4	5081	12895	211
machine_2	4	7661	10093	170
machine_5	10	5886	7715	223
machine_6	2	2751	9280	203
machine_7	6	5478	6896	185
machine_8	6	5752	10048	140
machine_9	5	5871	10969	226
machine_10	6	4398	12064	200

Cluster 9 — Condition Attributes

C9Node ID	# processors	Processor speed	Memory size	Bandwidth
machine_1	9	5127	1619	236
machine_2	4	6067	7270	267
machine_3	7	4842	5085	283
machine_4	10	3770	2500	207
machine_5	8	5596	2955	227
machine_6	10	5748	6258	255
machine_7	6	2794	2125	256
TOTAL	54	33944	27812	1731

Job 1	Job2	Job3	Decision value		
No Match	No Match	No Match	0	0	0
machine_2	machine_2	machine_2	1	1	1
machine_3	machine_3	machine_3	1	1	1
No Match	No Match	No Match	0	0	0
No Match	machine_5	No Match	0	1	0
machine_6	machine_6	machine_6	1	1	1
No Match	No Match	No Match	0	0	0
Decision values for Cluster 9 =			0.43	0.57	0.43

Qualified Machines for Job 1 in Cluster9

machine_2	4	6067	7270	267
machine_3	7	4842	5085	283
machine_6	10	5748	6258	255

Cluster 10 — Condition Attributes

C10Node ID	# processors	Processor speed	Memory size	Bandwidth
machine_1	10	3699	5194	203
machine_2	28	2857	6681	235
machine_3	14	3927	5014	208
machine_4	8	3608	2555	252
machine_5	10	3074	7246	257
TOTAL	70	17165	26690	1155

Job 1	Job2	Job3	Decision value		
machine_1	machine_1	machine_1	1	1	1
machine_2	No Match	machine_2	1	0	1
machine_3	machine_3	machine_3	1	1	1
No Match	No Match	No Match	0	0	0
machine_5	machine_5	machine_5	1	1	1
Decision values for Cluster 10 =			0.80	0.60	0.80

Qualified Machines for Job 1 in Cluster10

machine_1	10	3699	5194	203
machine_2	28	2857	6681	235
machine_3	14	3927	5014	208
machine_5	10	3074	7246	257

Cluster 11 — Condition Attributes

C11Node ID	# processors	Processor speed	Memory size	Bandwidth
machine_1	7	4341	5531	196
machine_2	8	5235	3052	173
machine_3	22	2737	3694	170
machine_4	9	3965	6326	140
machine_5	16	3661	5335	114
machine_6	7	4728	5627	105
machine_7	30	5021	7349	169
machine_8	10	2488	2352	161
machine_9	9	2843	5612	128
machine_10	12	3466	3268	156
TOTAL	130	38485	48146	1512

Job 1	Job2	Job3	Job 1	Job 2	Job 3
machine_1	machine_1	machine_1	1	1	1
No Match	machine_2	No Match	0	1	0
machine_3	No Match	machine_3	1	0	1
machine_4	machine_4	machine_4	1	1	1
machine_5	machine_5	machine_5	1	1	1
No Match	machine_6	machine_6	0	1	1
machine_7	machine_7	machine_7	1	1	1
machine_8	No Match	No Match	1	0	0
machine_9	machine_9	machine_9	1	0	1
machine_10	machine_10	machine_10	1	1	1
Decision values for Cluster 11 =			0.70	0.70	0.80

Qualified Machines for Job 1 in Cluster11

machine_1	7	4341	5531	196
machine_3	22	2737	3694	170
machine_4	9	3965	6326	140
machine_5	16	3661	5335	114
machine_7	30	5021	7349	169
machine_9	9	2843	5612	128
machine_10	12	3466	3268	156

Cluster 12 — Condition Attributes

C12Node ID	# processors	Processor speed	Memory size	Bandwidth
machine_1	8	5030	9851	109
machine_2	3	6031	7481	108
machine_3	10	5806	15400	180
machine_4	7	5909	15282	137
machine_5	4	5411	6722	176
machine_6	6	5662	6537	200
machine_7	18	5496	4609	159
machine_8	8	5108	13368	189
machine_9	22	3838	3423	113
machine_10	9	3803	11185	200
TOTAL	95	52094	93858	1571

Job 1	Job 2	Job 3	Job 1	Job 2	Job 3
No Match	machine_1	machine_1	0	1	1
No Match	machine_2	No Match	0	1	0
machine_3	machine_3	machine_3	1	1	1
machine_4	machine_4	machine_4	1	1	1
machine_5	machine_5	machine_5	1	1	1
machine_6	machine_6	machine_6	1	1	1
machine_7	machine_7	machine_7	1	1	1
machine_8	machine_8	machine_8	1	1	1
machine_9	machine_9	machine_9	1	1	1
machine_10	machine_10	machine_10	1	1	1
Decision values for Cluster 12 =			0.80	1.00	0.90

Qualified Machines for Job 1 in Cluster12

machine_3	10	5806	15400	180
machine_4	7	5909	15282	137
machine_5	4	5411	6722	176
machine_6	6	5662	6537	200
machine_7	18	5496	4609	159
machine_8	8	5108	13368	189
machine_9	22	3838	3423	113
machine_10	9	3803	11185	200

Table 7 Selection of cluster with the best sets of machines fitness

Cluster 2

C2Node ID	# processors	Processor speed	Memory size	Bandwidth
machine_1	4	5030	9851	200
machine_2	6	6031	7481	250
machine_3	10	5806	15,400	151
machine_4	10	5909	15,282	137
machine_5	4	5411	6722	176
machine_6	4	5662	6537	171
machine_7	6	5496	4609	300
machine_8	8	5108	13,368	189
machine_9	8	3838	3423	120
machine_10	9	3803	11,185	131
Total	69	52,094	93,858	1825

References

1. A.E. Ezugwu, M.E. Frincu, S.B. Junaidu, Performance characterization of heterogeneous distributed commodity cluster resources, in *2014 IEEE 6th International Conference on Adaptive Science and Technology (ICAST)* (IEEE, 2014), pp. 1–8
2. G.F. Coulouris, J. Dollimore, T. Kindberg, *Distributed Systems: Concepts and Design* (Pearson Education, 2005)
3. R. Buyya, D. Abramson, J. Giddy, Nimrod/G: An architecture for a resource management and scheduling system in a global computational grid, in *The Fourth International Conference/ Exhibition on High Performance Computing in the Asia-Pacific Region, 2000. Proceedings*, vol. 1 (IEEE, 2000), pp. 283–289
4. D. Molodtsov, Soft set theory—first results. Comput. Math Appl. **37**(4), 19–31 (1999)
5. P.K. Maji, R. Biswas, A. Roy, Soft set theory. Comput. Math Appl. **45**(4), 555–562 (2003)
6. D.A. Kumar, R. Rengasamy, Parameterization reduction using soft set theory for better decision making, in *2013 International Conference on Pattern Recognition, Informatics and Mobile Engineering (PRIME)* (IEEE, 2013), pp. 365–367
7. D.K. Sut, An application of fuzzy soft relation in decision making problems. Int. J. Math. Trends Technol. **3**(2), 51–54 (2012)
8. K. Gong, Z. Xiao, X. Zhang, The bijective soft set with its operations. Comput. Math Appl. **60**(8), 2270–2278 (2010)
9. N. Çağman, I. Deli, Products of FP-soft sets and their applications. Hacettepe J. Math. Stat. **41** (3) (2012)
10. N. Çağman, F. Çıtak, S. Enginoğlu, Fuzzy parameterized fuzzy soft set theory and its applications. Turk. J. Fuzzy Syst. **1**(1), 21–35 (2010)
11. S. Alkhazaleh, A.R. Salleh, N. Hassan, Soft multisets theory. Appl. Math. Sci. **5**(72), 3561–3573 (2011)
12. Y. Zhao, F. Luo, S.K.M. Wong, Y.Y. Yao, A general definition of an attribute reduct. Lect. Notes Artif. Intell. **4481**, 101–108 (2007)
13. P.K. Maji, A.R. Roy, R. Biswas, An application of soft sets in a decision making problem. Comput. Math. Appl. **44**(8), 1077–1083 (2002)
14. K. Sastry, D.E. Goldberg, G. Kendall, Genetic algorithms. In Search methodologies. (Springer, US, 2014), pp. 93–117
15. K.F. Man, K.S. Tang, S. Kwong, Genetic algorithms: concepts and applications. IEEE Trans. Ind. Electron. **43**(5), 519–534 (1996)
16. A. Ezugwu, N. Okoroafor, S. Buhari et al., Grid resource allocation with genetic algorithm using population based on multisets. J. Intell. Syst. (2016). https://doi.org/10.1515/jisys-2015-0089. Accessed 12 Feb 2016
17. D.E. Goldberg, K. Deb, A comparative analysis of selection schemes used in genetic algorithms. Found. Genet. Algorithms **1**, 69–93 (1991)
18. Z. Wu, J. Zhang, Y. Gao, An attribute reduction algorithm based on genetic algorithm and discernibility matrix. J. Softw. **7**(11), 2640–2648 (2012)
19. S. Vijayabalaji, A. Ramesh, A new decision making theory in soft matrices. Int. J. Pure Appl. Math. **86**(6), 927–939 (2013)
20. L. Chen, H. Liu, Z. Wan, An attribute reduction algorithm based on rough set theory and an improved genetic algorithm. J. Softw. **9**(9), 2276–2282 (2014)
21. T. Herawan, A.N.M. Rose, M.M. Deris, Soft set theoretic approach for dimensionality reduction, in *Database Theory and Application* (Springer, Berlin, Heidelberg, 2009), pp. 171–178
22. J. Zhang, H.S.H. Chung, W.L. Lo, Clustering-based adaptive crossover and mutation probabilities for genetic algorithms. IEEE Trans. Evol. Comput. **11**(3), 326–335 (2007)

Fuzzy Multi-criteria Method to Support Group Decision Making in Human Resource Management

M. H. Mammadova and Z. G. Jabrayilova

Abstract The objective of this research is to develop a methodological approach to the making managerial decisions in HRM tasks, which have such specific features as multi-objectivity and heterogeneity of data, the hierarchal, quantitative, and qualitative nature of criteria, their ambiguity, the need for considering the expert evaluation of their weight, and the influence of the experts' competence on the made decision. To ensure the adaptability of multi-criteria decision-making in HRM a modified TOPSIS method is proposed. Introducing additional components into the decision-making algorithm, this modification excludes the hierarchal structure of criteria and takes into account the competence of experts. The method is tested on the employment case study.

1 Introduction

Globalization and rapid change of technologies precondition changes in the labor market, which, in its turn, causes considerable transformations in personnel relations, and requires the development of new conceptual approaches and scientifically substantiated methods in the policy that regulates these relations, depending on a specific HRM task. According to this concept, HRM is a special type of managerial activity. In this case, the main managed object is the human and his competencies, including knowledge, skills, and professional abilities, personal and behavioral qualities, motivational principles, intellectual and qualification potential, while HRM is aimed at supporting the organization's activity strategy under the growing role and importance of the human factor [1–3]. Therefore, in order to make decisions that are adequate to the new conditions with regard to personnel planning, selection, recruitment, adaptation to the changing market environment, retention, dismissal, promotion, development, training, and motivation of personnel, the

M. H. Mammadova (✉) · Z. G. Jabrayilova
Institute of Information Technology of National Academy Science of Azerbaijan,
B.Vahabzadeh str., 9, AZ1141 Baku c., Republic of Azerbaijan
e-mail: depart15@iit.ab.az

© Springer International Publishing AG, part of Springer Nature 2018
L. A. Zadeh et al. (eds.), *Recent Developments and the New Direction in Soft-Computing Foundations and Applications*, Studies in Fuzziness and Soft Computing 361, https://doi.org/10.1007/978-3-319-75408-6_17

decision-maker should evaluate and consider a wide range of information regarding the competencies of employees, be able to compare applicants, based on a multiple heterogeneous attributes (criteria), select the optimal solution (candidate) with the consideration of multiple influences, preferences, interests, and possible consequences. All these peculiarities of HRM tasks determine their multi-objectivity. At that, one should consider the volume, quantitative and qualitative nature, complexity and inconsistency of the flow of information that the decision-maker receives, which allow identifying HRM tasks as semi-structured tasks, for which the construction of objective models is either impossible or extremely difficult. Along with the abovementioned problems that arise during the generation and selection of managerial decisions, one should consider the decision-maker's preferences, and experts' competence (knowledge, intuition, experience, etc.).

Therefore, in human resource management tasks, the handling of such data requires the application of models and methods, based on the fuzzy set and fuzzy logic theories [4–6]. In order to overcome the abovementioned difficulties one needs to select, create, and develop methodological approaches to multiple-criteria analysis and decision-making in human resource management, based on intelligent technologies, methods, and computer systems of decision-making support [7, 8].

2 Multi-criteria Methods of Decision-Making in HRM Tasks: Literature Survey

The literature survey shows among the HRM tasks' decision-making processes that require intelligent support, including employment management, assessment, and organization of personnel remuneration system, career planning, formation of the reserve, authors pay most attention to the selection and recruitment of personnel resources, which is caused by the greatest practical applicability of the latter.

At present, while solving personnel selection tasks, developers mostly prefer multi-criteria methods of decision-making, including methods of decision tree analysis [9], analytic hierarchy process (AHP) [10–12], the Technique for Order of Preference by Similarity to Ideal Solution (TOPSIS) [13–21], expert systems [4, 22], etc.

According to the authors of works [10–12], the FAHP-based computer system of decision-making support eliminates restrictions to information on candidates, and aids managers in making optimal decisions (selecting the best candidate) under fuzzy conditions. However, although these approaches allow making the best decision among the possible ones, they do not allow selecting an alternative that is preferable by all criteria, i.e. is the most similar to the ideal (optimal) solution. This possibility is provided by the TOPSIS method, which was first suggested by work [13]. With a view to selecting personnel, [14] suggests group decision-making, based on the TOPSIS fuzzy method, where membership functions of alternatives on criteria are described by values of fuzzy linguistic variables, presented as fuzzy triangular numbers. With a view to selecting personnel, [15] introduces an additional stage, which allows aggregating heterogeneous information by using ordered weighted averaging (OWA). [16] also suggests a modified TOPSIS method by the example of selecting university lecturers. In this case, the modification consists in suggesting a new measure for calculating the distance from the positive ideal and negative ideal solution during multi-criteria decision-making. In [17], the TOPSIS-based support of decision-making during the selection of the most skilled human resources includes a new concept of ranking (arranging in order) alternatives, which consists in that the final decision is determined by a veto set, established by the decision-maker, rather than by the proximity of alternatives to the ideal solution. The authors of [18], suggest another TOPSIS modification, which consists in the introduction into the final decision making algorithm, based on a veto set that is established by decision-makers, the relative importance of the latter. [18] uses a modified TOPSIS method to solve the task of selecting middle managers for a Greek IT-company; [19] demonstrates the application of the TOPSIS-based multi-criteria decision-making method with a view to selecting human resources for a Greek bank; [20] implements a TOPSIS-based fuzzy model of the decision-making support system with a view to improving the selection and recruitment of personnel for the Iran Khodro Industrial Group. Paper [21] offers a modified TOPSIS for decision-making support in recruitment tasks. This modification accounts for the hierarchal structure of criteria that are typical for HRM tasks.

However, it should be noted that all abovementioned modifications of TOPSIS only deal with certain aspects of specific features of HRM tasks. In the present research, the authors attempted to account for all components of decision-making in HRM tasks.

For example, the suggested TOPSIS modification consists in the integration into the decision-making algorithm of additional components, which, at the first stage, provide a calculation, based on the analysis of weight coefficients of partial criteria, which allows disposing of the hierarchal structure of criteria, and, at the second stage, allow introducing into the algorithm pre-estimated competence coefficients of expert, who participate in the evaluation of alternatives. With a view to ensuring the robustness of criteria to the confidence interval boundaries, this paper, when using group decision-making, describes the membership functions of alternatives on criteria by values of fuzzy linguistic variables, presented as fuzzy trapezoidal numbers.

3 Generalized Conceptual Model of Human Resource Management Tasks

The authors of works [21, 23–25] studied the main HRM tasks, emphasized those that require intelligent support, and determined their specific features. For example, the tasks, the solution whereof comes down to making efficient decisions, include the task of managing employment processes (selection, evaluation, and recruitment), the task of assessment (determining the conformity of personnel to the held position), the task of organizing a personnel remuneration system, the task of employees' career planning (promotion), the task of forming a reserve, etc. The analysis of mentioned HRM tasks allowed distinguishing the following peculiar features of the latter: multi-objectivity and heterogeneity of data that describe HRM tasks; the multilevel hierarchal structure of criteria, expressed in the fact that each individual upper-level criterion is based on the aggregation of partial criteria of the next lower level; quantitative and qualitative nature of criteria; the impossibility of unambiguous determination of criteria and the variableness of their value range; different extent of the influence of criteria and indicators on the considered variants (objects, alternatives), and the need to consider the difference in their weights. This determines the need for involving experts (information carriers) in decision-making, and the consideration of their opinions; the influence of the experts' competence on the quality of the made decision; the presence of a vast number of heterogeneous partial criteria in real situations, which complicate the formal comparison of alternatives.

The abovementioned peculiarities of HRM tasks allow identifying them as tasks of multiple-criteria analysis and decision-making in a fuzzy environment [26].

Based on the comprehensive approach to the consideration of the specificity of human resource management, the generalized conceptual model of decision-making in HRM tasks can be presented by the following set of information: $M_{HRM} = (X, K, Y, E, V, P, L, W)$, where:

- $X = \{x_1, x_2, \ldots, x_n\} = \{x_i, i = \overline{1, n}\}$ is the set of admissible alternatives; $K = \{K_1, K_2, \ldots, K_m\} = \{K_j, j = \overline{1, m}\}$ is the set of choice criteria that characterize alternatives; $K_j = \{k_{j1}, k_{j2}, \ldots, k_{jT}\} = \{k_{jt}, t = \overline{1, T}\}$ is the set of sub-criteria that characterize each criterion; Y is the range of definition of each partial criterion's value; E is the group of experts, participating in decision-making; V is the set of relation between experts in accordance with the decision-maker's preferences; P is the relations between the X, K and E sets; L is the linguistic expressions that reflect the level of partial criteria's satisfaction by alternatives (membership level); W is the relations between criteria and partial criteria.

According to the conceptual model, the essence of decision making in HRM tasks is: (1) finding a systematized list of alternatives ($X \rightarrow X^*$), ranked from the best (optimal) one to the worst one (or vice versa); (2) choosing the best (optimal) variant from among the alternatives.

4   General Formulation

With the specific peculiarities of decision-making tasks in HRM and the suggested conceptual model in mind, it is necessary to formulate in general the task of multi-objective ranking/choice of alternatives. A multi-objective optimization task is generally understood as finding the maximum and minimum vector-valued criterion in a feasible set of alternatives. Given the following components of HRM tasks in organization:

1. $X = \{x_i, i = \overline{1, n}\}$ is the set of alternatives;
2. $K = \{K_j, j = \overline{1, m}\}$ is the set of criteria;
3. $K_j = \{k_{jt}, t = \overline{1, s_j}\}$ is the set of partial criteria;
4. $E = \{e_l, l = \overline{1, g}\}$ is the set of experts;
5. $w_j, j = \overline{1, m}$ is   the   coefficients   of   criteria's   relative   importance $(K = \{k_j, j = \overline{1, m}\})$;
6. $w_{jt}, t = \overline{1, s_j}, j = \overline{1, m}$ is the coefficients of partial criteria's relative importance $(k_j = \{k_{jt}, t = \overline{1, s_j}\})$;
7. $v_l, l = \overline{1, g}$ is the experts' competence coefficients.

Assume $f(x)$ is an objective function that guarantees the choice of the best alternatives:

(1) $f(x) = \max(f(x_1), f(x_2), \ldots, f(x_n))$ и $f(x) \rightarrow [0, 1]$,
   where $f(x_i)$ is the resultant vector of the evaluation of alternative $x_i \in X$ in accordance with integral criterion K, i.e. $f(x_i) \rightarrow K(x_i)$.
(2) $K(x_i) = (p(x_i), w, v)$ is the integral evaluation of alternative x_i, according to the set of evaluation criteria, the weight of partial criteria in the integral criterion K and the coefficient of the relative importance of experts' competence, where —$p(x_i)$ is the integral evaluation of alternative x_i, $i = \overline{1, n}$ in accordance with the values of the linguistic variable by the experts' preference; $w = (w_1, \ldots, w_Z)$ are the weights of partial criteria in the integral criterion K, $z = \overline{1, Z}$. Z is the total number of partial criteria; $v = (v_1, \ldots, v_g)$ is the coefficient of the relative importance of experts' competence, according to the decision-maker's preferences.
(3) $f(x_i) > 0$, provided that $p(x_i) \geq 0$.
(4) $g(K(x), w, v) \in G, x \in X$,

$$w_j > 0, j = \overline{1, m}, \ \sum_{j=1}^{m} w_j = 1, \quad w_{jt} > 0, t = \overline{1, s_j}, \ \sum_{t=1}^{s_j} w_{jt} = 1,$$

$$w_z > 0, z = \overline{1, Z}., \quad v_l > 0, l = \overline{1, g}, \ \sum_{l=1}^{g} v_l = 1.$$

It is necessary to find the alternatives that best correspond with the objective functions and restrictions. According to the task's formulation, the set of feasible solutions is formed by eliminating from the initial set of alternatives the ones that do not satisfy the set objective and restrictions. As shown above, multi-objective HRM tasks are semi-structured problems that contain many criteria (both qualitative and quantitative) of decision quality evaluation. The decision-maker is guided by his or her subjective preferences with regard to the efficiency of possible alternatives and the importance of different criteria. Constructing a decision-maker preference model produces a large volume of information. It is difficult and, sometimes, impossible to estimate efficiency indicators and choose a single best decision by analytical methods. Therefore, the existing concepts of evaluating preferences are based on heuristic methods and the inclusion of the decision-maker (experts) as a main component of the decision-making task.

Effective instruments are needed to build complex decision-making procedures and to evaluate a wide range of alternatives. The present paper preferred a modern multi-objective choice method—TOPSIS, modified to suit the conditions of the solved problem.

5 TOPSIS Method

The main idea of the TOPSIS method is that the most preferable alternative should have the shortest distance from the ideal solution and the longest geometric among all alternatives from the inadmissible solution [13]. Based on the essence of the TOPSIS method, the use of the latter is efficient in solving tasks of fuzzy multi-objective optimization, which constitute the mathematical basis of decision-making support in human resource management tasks. In the decision-making theory, multi-objective optimization is understood as the selection of the best solution among the possible alternatives [4]. The solution of optimization tasks with the use of TOPSIS assumes the need for translating the values of qualitative linguistic variables that express the level of satisfaction of the criteria by this or than alternative into fuzzy numbers. A fuzzy number is a fuzzy subset of a universal set of real numbers, which has a normal or convex membership function, for which there exists such a carrier value, where the membership function is

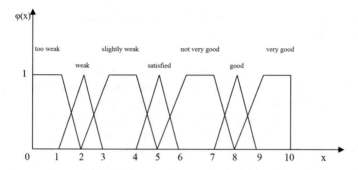

Fig. 1 Transformation of linguistic values into fuzzy trapezoidal numbers

Table 1 Linguistic values and their respective fuzzy trapezoidal numbers

Linguistic values	Fuzzy trapezoidal numbers
Too weak	(0, 0, 1, 2)
Weak	(1, 2, 2, 3)
Slightly weak	(2, 3, 4, 5)
Satisfactory	(4, 5, 5, 6)
Not very good	(5, 6, 7,8)
Good	(7, 8, 8, 9)
Very good	(8, 9, 10, 10)

1, while the membership function decreases during leftwards or rightwards deviation [27]. According to [28], fuzzy expert judgments that were formulated in the natural language can be described by fuzzy triangular and fuzzy trapezoidal numbers. This paper, taking into consideration the need for ensuring the robustness of criteria to the confidence interval boundaries, uses a fuzzy trapezoidal number. Operations on fuzzy numbers are introduced by means of operations on membership functions, based on the segmental principle [29]. In order to implement this method, one should handle linguistic variables and their values that express verbal ranking scales for measuring attributes. Here, the levels are arranged in the order of ascension of these attributes' intensity. In this case, the number of linguistic variables' values (ranks) is seven. Figure 1 shows a graphical representation of the transformation of linguistic values into numeric equivalents.

Table 1. shows the 7-level values of the linguistic variable and respective fuzzy trapezoidal numbers.

According to Table 1, a numeric equivalent can be found for each linguistic variable value.

6   TOPSIS-Based Algorithm of Multi-objective Optimization in HRM Tasks

The goal of the task is to rank alternatives, based on the evaluations of experts, taking into consideration the competence of the latter. The solution of the task assumes the performance of the following sequence of actions:

Step 1. In order to perform TOPSIS-based multi-objective optimization of HRM tasks, one should first dispose of the hierarchal structure of criteria. For this purpose, based on Saaty's AHP, by relative importance coefficients of criteria $\{K_j, j = \overline{1, m}\}$ and partial criteria $\{k_{jt}, t = \overline{1, s_j}\}$, weights are determined [26], with which the latter will enter the calculation of the K integral criterion. In a formalized form, w_{jt}^K—the weight of the k_{jt} partial criterion in the calculation of the integral criterion $K = \{k_j, j = \overline{1, m}\}$, i.e. $w_{jt}^K = w_{jt} \cdot w_j$, is determined by the multiplication w_j, where $\sum_{j=1}^m w_j = 1$, and w_{jt}, where $\sum_{t=1}^{s_j} w_{jt} = 1$. As a result, the two-level hierarchal structure of choice criteria $K = \{K_j, j = \overline{1, m}\}$ that characterize alternatives comes down to the calculation of the integral criterion that includes the weights of partial criteria $\{k_{jt}, t = \overline{1, s_j}\}$, which allows disposing of the hierarchal structure.

During subsequent steps, all partial criteria are united into a single G set, with a view to simplifying indexes.

$$G = \{k_{jt}, j = \overline{1, m}, t = \overline{1, s_j}\} = \{k_z, z = \overline{1, Z}\}, z = s_{j-1} + t, j = \overline{1, m}, t = \overline{1, s_j}, s_0 = 0. \quad (1)$$

Here, Z is the overall number of partial criteria that characterize alternatives, i.e.

$$Z = \sum_{j=1}^m s_j.$$

In this case, $w_z = w_{jt}^K$.

Step 2. The level of membership (relation) of alternatives on partial criteria is evaluated by linguistic values (see Table 1) and expressed by trapezoidal numbers $R^l = (r_{iz}^l) = (a_{iz}^l, b_{iz}^l, c_{iz}^l, d_{iz}^l)$. The expert evaluation of alternatives' membership on partial criteria results in the following matrix:

$$R^l = [r_{iz}^l], l = \overline{1, g} \Leftrightarrow \{a_{iz}^l, b_{iz}^l, c_{iz}^l, d_{iz}^l\}, l = \overline{1, g}. \quad (2)$$

Step 3. This step assumes the pre-estimation of experts' competence coefficient $v_l, l = \overline{1, g}$. For this purpose, the authors applied the modified method that integrates into the algorithm pre-estimated coefficients of competence of experts, who participate   in   the   evaluation   of   alternatives.   The   matrix $R^{v_l} = [r_{iz}^{v_l}], l = \overline{1, g} \Leftrightarrow \{a_{iz}^{v_l}, b_{iz}^{v_l}, c_{iz}^{v_l}, d_{iz}^{v_l}\}, l = \overline{1, g}$ is formed, taking into account the experts' competence coefficient $v_l, l = \overline{1, g}$. The elements of this matrix are trapezoidal numbers that express the level of satisfaction by alternative x_i of partial criteria k_z, taking into account the experts' competence. The elements are calculated as follows:

$$a_{iz}^{v_l} = a_{iz}^l \cdot v_l;$$
$$b_{iz}^{v_l} = b_{iz}^l \cdot v_l;$$
$$c_{iz}^{v_l} = c_{iz}^l \cdot v_l;$$
$$d_{iz}^{v_l} = d_{iz}^l \cdot v_l. \tag{3}$$

Step 4. This step determines the single aggregated matrix:

$$R^{v_l} = \left[r_{iz}^{v_l}\right], l = \overline{1, g} \Leftrightarrow \left\{a_{iz}^{v_l}, b_{iz}^{v_l}, c_{iz}^{v_l}, d_{iz}^{v_l}\right\}, l = \overline{1, g} \Rightarrow R_{iz}$$
$$= [r_{iz}] \Leftrightarrow \{a_{iz}, b_{iz}, c_{iz}, d_{iz}\}.$$

The elements of this matrix are determined as follows:

$$a_{iz} = \left\{\min a_{iz}^{v_l}, l = \overline{1, g}\right\};$$
$$b_{iz} = \frac{1}{g}\sum_{l=1}^{g} b_{iz}^{v_l};$$
$$c_{iz} = \frac{1}{g}\sum_{l=1}^{g} c_{iz}^{v_l};$$
$$d_{iz} = \left\{\max d_{iz}^{v_l}, l = \overline{1, g}\right\}. \tag{4}$$

Step 5. The elements of matrix $R_{iz} = [r_{iz}] \Leftrightarrow \{a_{iz}, b_{iz}, c_{iz}, d_{iz}\}$ are multiplied by the weights of partial criteria. This operation builds the weighed fuzzy matrix $R_{iz}^w = \left[r_{iz}^w\right] \Leftrightarrow \left\{a_{iz}^w, b_{iz}^w, c_{iz}^w, d_{iz}^w\right\}$. Here:

$$a_{iz}^w = a_{iz} \cdot w_z;$$
$$b_{iz}^w = b_{iz} \cdot w_z;$$
$$c_{iz}^w = c_{iz} \cdot w_z;$$
$$d_{iz}^w = d_{iz} \cdot w_z. \tag{5}$$

Step 6. The obtained matrix is normalized. For this purpose, the *Hsu* and *Cehn* method [30] is used, which determines $d_z^+ = \max d_{iz}^w, i = \overline{1, n}$. Then, based on the expression

$$R_{iz}^N = \left[r_{iz}^N\right] \Leftrightarrow \left\{a_{iz}^N, b_{iz}^N, c_{iz}^N, d_{iz}^N\right\} \Leftrightarrow \left\{\frac{a_{iz}^w}{d_z^+}, \frac{b_{iz}^w}{d_z^+}, \frac{c_{iz}^w}{d_z^+}, \frac{d_{iz}^w}{d_z^+}\right\}. \tag{6}$$

the elements of the normalized decision-making matrix are determined.

Step 7. Based on the weighed values, the positive ideal (optimal) solution (PIS) X^* is determined. For this purpose, for each $k_z, z = \overline{1, Z}$,

$$d_z^* = \{\max d_{iz}^N, i = \overline{1, n}\} \tag{7}$$

is selected, and the

$$X^* = [d_z^*] = [(d_1^*, d_1^*, d_1^*, d_1^*), \ldots, (d_Z^*, d_Z^*, d_Z^*, d_Z^*)]. \tag{8}$$

matrix is formed. In accordance with expression (8), $d_z^* = 1$ for $\forall z$, i.e. all elements of matrix X^* are 1.

Step 8. The negative (worst) ideal value (NIS) X^- is calculated. For this purpose, for each $k_z, z = \overline{1, Z}$

$$a_z^- = \{\min a_{iz}^N, i = \overline{1, n}\} \tag{9}$$

is selected, and the following matrix is formed:

$$X^- = [a_z^-] = [(a_1^-, a_1^-, a_1^-, a_1^-), \ldots, (a_Z^-, a_Z^-, a_Z^-, a_Z^-)]. \tag{10}$$

Step 9. The distance of alternatives from PIS are calculated by formula (2) for the individual values of each partial criterion:

$$D_z^*(x_i, X^*) = \sqrt{\frac{1}{4}((a_{iz}^N - d_z^*)^2 + (b_{iz}^N - d_z^*)^2 + (c_{iz}^N - d_z^*)^2 + (d_{iz}^N - d_z^*)^2)} \tag{11}$$

Vector $[D^*] = [D_1^*, \ldots, D_Z^*]$ is formed, based on the obtained results.

Step 10. The distance of alternatives from NIS are calculated for the individual values of each partial criterion

$$D_z^-(x_i, X^-) = \sqrt{\frac{1}{4}((a_{iz}^N - a_z^-)^2 + (b_{iz}^N - a_z^-)^2 + (c_{iz}^N - a_z^-)^2 + (d_{iz}^N - a_z^-)^2)} \tag{12}$$

Vector $[D^-] = [D_1^-, \ldots, D_Z^-]$ is formed, based on the obtained results.

Step 11. The distance of each alternative from PIS is determined:

$$D^*(x_i) = \sqrt{\sum_{z=1}^{Z} (D_z^*(x_i, X^*))^2} \tag{13}$$

Step 12. The distance of each alternative from NIS is determined:

$$D^-(x_i) = \sqrt{\sum_{z=1}^{Z} (D_z^-(x_i, X^*))^2} \tag{14}$$

Step 13. The integral indicator (proximity coefficient) is calculated for each compared alternative, as the correlation between its calculated distance from the negative ideal solution, and the sum of distances between the best and the worst solutions:

$$D(x_i) = D^*(x_i) + D^-(x_i)$$
$$\varphi(x_i) = \frac{D^-(x_i)}{D(x_i)}. \tag{15}$$

The value of the proximity coefficient $\varphi(x_i)$ allows ranking alternatives. For example, the closer the value of the proximity coefficient $\varphi(x_i)$ to 1, the more preferable the compared alternative.

7 Application of the Suggested Method for Solving Personnel Selection and Recruitment Tasks

The suggested instrumental approach was tested during the solution of tasks of selecting and recruiting with a view to evaluating candidates. Experiments were conducted with a view to evaluating the applicants for the vacancy at the Human Resource Management Department. For this purpose, the following actions were taken:

1. A criteria system was formed with the participation of four experts, with a view to recruiting personnel to the HRM Department. The coefficients of criteria's relative importance were calculated, based on pairwise comparison [26]. The task of detecting contradictions in expert evaluations [26, 31] was also considered. The obtained results helped determine the coefficients of criteria's relative importance, and the weights of partial criteria, with which the latter will enter the calculation of the K integral criterion.
2. The obtained integral indicator (the proximity coefficient of compared alternatives) $\varphi(x_i)$ was expressed by a certain value (on the [0, 1] interval) of the probability of recruitment for each candidate x_i. The values of this variable allow making the final decision regarding each alternative candidate.
3. The coefficients of the competence of experts, who participate in the evaluation of candidates, were calculated by pairwise comparison [26, 31], based on the linguistic expression of "slight superiority of expert 1 and expert 4 over expert 2 and expert 3": $v_1 = 0.375$, $v_2 = 0.125$, $v_3 = 0.125$, $v_4 = 0.375$.
4. With the participation of four experts, based on seven-level linguistic variables, the authors evaluated the level of satisfaction (membership) of 12 partial criteria by three candidates for the job, who passed all necessary stages of selection.
5. A $3 \times 12 \times 4$ generalized matrix of fuzzy trapezoidal numbers was built, based on the evaluations of four experts on the basis of formulas (1) and (2).
6. Taking into account the competence of experts based on the formula (3), the matrix of trapezoidal fuzzy numbers was built and aggregated in accordance with formula (4) trapezoidal fuzzy numbers were defined.

Table 2 Distance of compared alternatives from PIS and NIS, the coefficient of their proximity to the ideal solution and respective ranks

Alternatives	X*	X⁻	X* + X⁻	$\varphi_K(x_i)$	Ranks	Solution
x_1	1.7208	1.6287	3.3495	0.486	3	Reception of the candidate is associated with a great risk
x_2	1.7344	1.6719	3.4063	0.491	2	Reception the candidate is associated with a great risk
x_3	1.6619	1.6651	3.3270	0.501	1	Reception of the candidate is associated with a small risk

7. The elements of the matrix of aggregated fuzzy trapezoidal numbers were multiplied by the weights of partial criteria according to formula (5), and the results were normalized.
8. The integral matrix of fuzzy positive (best) ideal solutions and fuzzy negative (worst) ideal solutions was formed in accordance with expressions (7)–(10).
9. The results of the calculations of distances between alternatives and PIS (X*) were calculated, based on formula (12) for the value of each partial criterion.
10. The results of the calculations of distances between alternatives and NIS (X⁻) were calculated, based on formula (11) for the value of each partial criterion.
11. The distances from each alternative to PIS and DIS were calculated in accordance with formulas (13) and (14), respectively. Formula (15) was used to calculate the values of the integral indicator, which expresses the proximity of each compared alternative to the ideal solution. The ranks of each alternative were determined in accordance with the results (Table 2).

8 Conclusion

The suggested methodological approach to the solution of HRM task with the use of TOPSIS-based multi-objective optimization allows improving the adequacy of made decisions by means of prioritization by the proximity to the ideal solution, ensures the objectiveness and transparency of managerial decisions, and provides opportunities for expanding the applicability of multi-objective optimization methods. The use of the described methodological approach as the mathematical basis for the computer system of decision-making support in HRM tasks can become an effective instrument for preparing and making efficient decisions in human resource management.

References

1. G.A. Cole, *Personnel and Human Resource Management* (Thomson Copyright, 2002), 511 pp.
2. M.L. Spencer, S.M. Spencer, Competence at Work: Models for Superior Performance (Business & Economics, 1993), 372 pp.
3. M. Armstrong, *Armstrong's Handbook of Strategic Human Resource Management* (Kogan Page Publishers, 2006), 216 pp.
4. S.V. Mikoni, Multicriteria selection on the final alternative set, in *Student Handbook* (Lan Publishing, 2009), 270 p.
5. M. Mammadova, *Decision-making, based on a knowledge database with a fuzzy relational structure* (Elm, Baku, 1997), p. 296
6. L.A. Zadeh, Fuzzy sets. Inf. Control **8**, 338–353 (1965)
7. E.A. Trachtengertz, Capabilities and realization of computer decision making support systems, in *Management theory and Systems*. vol. 3, (News of Academy of Sciences of Russia, 2001) pp. 86–113
8. A.R. Afshari, M. Nikolić, D. Ćoćkalo, Applications of fuzzy decision making for personnel selection problem—a review. J. Eng. Manag. Compet. **4**, 68–77 (2014)
9. C.F. Chien, L.F. Chen, Data mining to improve personnel selection and enhance human capital: a case study in high-technology industry. Expert Syst. Appl. **34**(2), 280–290 (2008)
10. Z. Gungor, G. Serhadlıoglu, S.E. Kesen, A fuzzy AHP approach to personnel selection problem. Appl. Soft Comput. **9**, 641–649 (2009)
11. P.C. Chen, A fuzzy multiple criteria decision making model in employee recruitment. IJCSNS Int. J. Comput. Sci. Netw. Secur. **9**(7), 113–117 (2009)
12. M. Varmazyar, B. Nouri, A fuzzy AHP approach for employee recruitment. Decis. Sci. Lett. **3**, 27–36 (2014)
13. C.L. Hwang, K. Yoon, *Multiple Attributes Decision Making: Methods And Applications* (Springer, Heidelberg, 1981)
14. C.T. Chen, Extensions of the TOPSIS for group decision-making under fuzzy environment. Fuzzy Sets Syst. **114**, 1–9 (2000)
15. M. Dursun, E. Karsak, A fuzzy MCDM approach for personnel selection. Expert Syst. Appl. **37**, 4324–4330 (2010)
16. S. Saghafian, S.R. Hejazi, Multi-criteria group decision making using a modified fuzzy TOPSIS procedure, in *International Conference on Computational Intelligence for Modelling, Control and Automation and International Conference on Intelligent Agents, Web Technologies and Internet Commerce (CIMCA-IAWTIC'06)* (IEEE, 2005), pp. 215–221
17. A. Kelemenis, D. Askounis, A new TOPSIS-based multi-criteria approach to personnel selections. Expert Syst. Appl. **37**, 4999–5008 (2010)
18. A. Kelemenis, K. Ergazakis, D. Askounis, Support managers' selection using an extension of fuzzy TOPSIS. Expert Syst. Appl. **38**, 2774–2782 (2011)
19. P.V. Polychroniou, I. Giannikos, A fuzzy multicriteria decision making methodology for selection of human resources in a Greek private bank. Career Dev. Int. **14**, 372–387 (2009)
20. M. Mammadova, Z. Jabrailova, S. Nobari, Application of TOPSIS Method in Decision-Making Support of Personnel Management Problems. *4th International Conference on Problems of Cybernetics and Informatics* (PCI), (2012)
21. M.H. Mammadova, Z.Q. Jabrayilova, F.R. Mammadzada, Fuzzy multi-scenario Approach to Decision-Making Support in Human Resource Management. Recent Developments and new direction in soft-computing foundations and applications. **342**, 19–36 (2016)
22. E. Akhlagh, A rough-set based approach to design an expert system for personnel selection. World Acad. Sci. Eng. Technol. **54**, 202–205 (2011)
23. M.H. Mamedova, Z.G. Djabrailova, Methods of family income estimation in the targeting social assistance system. Appl. Comput. Math. **1**, 80–87 (2007)

24. M.H. Mammadova, Z.Q. Jabrayilova, F.R. Mammadzada, Application of fuzzy situational analysis for IT-professionals labor market management. *2nd International conference on information science and control engineering ICISCE 2015*, 143–146 (2015)
25. M.H. Mammadova, Z.G. Jabrayilova, Managing the IT labor market in conditions of fuzzy information. Autom. Cont. Comp. Sci. **2**, 88–93 (2015)
26. T.L. Saaty, Y. Cho, The decision by the US Congress on China's trade status: a multicriteria analysis. Socioecon. Plann. Sci. **35**, 243–252 (2001)
27. V.P. Karelin, Models and methods of presenting knowledge and elaborating decisions in intelligent information systems with fuzzy logic. Taganrog Inst. Manag. Econ. Bull. **1**, 75–82 (2014)
28. L.A. Zadeh, Fuzzy logic = computing with words. IEEE Trans. Fuzzy Syst. **4**, 103–111
29. L.A. Zadeh, The concept of a linguistic variable and its application to approximate reasoning. Inf. Sci. **8**, 199–249 (1975)
30. H.M. Hsu, C.T. Chen. Fuzzy credibility relation method for multiple criteria decision-making problems. Inf. Sci. (Ny), **96**, 79–91 (1997)
31. Z.Q. Jabrayilova, S. Nobari, Processing methods of information about the importance of the criteria in the solution of personnel management problems and contradiction detection. Probl. Inf. Technol. **2**, 57–66 (2011)

Fuzzy Management of Imbalance Between Supply and Demand for IT Specialists

M. H. Mammadova, Z. G. Jabrayilova and F. R. Mammadzada

Abstract The levels of modeling the processes of interaction between supply and demand on the labor market of IT specialists identified. Different types of imbalance between supply and demand for IT specialists and identified, the main areas, models and methods of their coordination are defined. The management method by a supply and demand on IT specialists at the micro-level, based on fuzzy situation analysis and fuzzy pattern recognition is proposed. The method and algorithm of an estimation of the imbalance degree between supply and demand on IT-specialists labor market at the macro-level, based on fuzzy mismatch scale is offered.

1 Introduction

Everyone recognizes the special role of information and communication technologies (ICT) in the development of the world and national economies, the acceleration of the processes of transiting to a knowledge-based economy, increasing the efficiency, competitiveness and innovation potential of economic branches and enterprises. The use of ICT in various domains of human activity, the expansion of the population's access to various information resources and transformation of information into a global resource have caused a sharp increase in demand for IT specialists on the labor market, which far exceeds their supply in many countries. Thus, the deficit of IT specialists is recorded in practically every country of the European Union today [1, 2].

The rising demand for IT specialists and their insufficient supply is recorded in such developed countries as the USA and Canada [3, 4]. CIS member states, including Azerbaijan, which are actively integrating in the global information space, have also encountered the problem of the discrepancy of supply and demand of IT specialists [5–7].

M. H. Mammadova (✉) · Z. G. Jabrayilova · F. R. Mammadzada
Institute of Information Technology of National Academy Science of Azerbaijan,
Baku c., B. Vahabzadeh str., 9, AZ1141 Baku, Republic of Azerbaijan
e-mail: depart15@iit.ab.az

© Springer International Publishing AG, part of Springer Nature 2018
L. A. Zadeh et al. (eds.), *Recent Developments and the New Direction in Soft-Computing Foundations and Applications*, Studies in Fuzziness and Soft Computing 361, https://doi.org/10.1007/978-3-319-75408-6_18

Today the elaboration of adequate technologies and methods of support for managerial decisions on the coordination of supply and demand for IT specialists which takes into account the specifics of this sector and its main actors, i.e. IT specialists (supply) and employers (demand), remain open.

2 Literature Review

At present the labor market and the market of education have different mechanisms and time frames of functioning. The education system develops on the basis of long-term policies and strategies, whereas the IT industry must respond to fast changing market demands and technological innovation in order to remain competitive on the global, national and local markets [8, 9]. The political measures taken today to address the problem of imbalance between supply and demand (SD) for IT specialists (ITS), within the framework of which new technologies and instruments are implemented, have not produced the expected effect yet [4, 5, 10–14]. The discordance of SD on the market of ITS can be explained by: (1) quantitative imbalance as a result of lack or oversupply of ITS, including in terms of certain IT jobs and specializations; (2) qualitative imbalance which is accounted for by (a) structural misalignment of occupational training of ITS as a result of obsolescence of some IT skills and emergence of other, principally new IT jobs and specializations and the education system's delayed response to some of the labor market demands; (b) the discrepancy between the knowledge and skills as well as practical experience of IT specialists formed by educational institutions and those required by the labor market; (3) various combinations of the indicated kinds of imbalance [15–17]. The decisions taken to coordinate supply and demand for IT specialists will be different for each type of imbalance or a combination of them.

The approaches and models of assessing supply and demand on the labor market suggested in individual works [18–23] do not allow one to manage its separate segments efficiently, since the labor market is not an aggregate of the labor markets in specific branches of the economy. Specific features of every segment of the national economy, the conditions, factors and rates of development of the economic branches account for their various contributions to the general state of the labor market and necessitate the research of individual segments.

3 The Aim and Tasks of the Research

The aim of this study is the development of models and methods of assessment of the imbalance between supply and demand at the micro- and macro-levels on the basis of a conceptual approach to intelligent control of the IT specialists' labor market which is proposed by the authors. The following tasks are aimed at achieving this objective: to define the notion of intelligent control of supply and demand on the labor market of IT specialists; to identify the components of the intelligence system of management of the IT specialists' labor market; to elaborate a transaction scheme of the system of intelligent control of supply and demand on the IT specialists' labor market; to develop a method for assessing the qualitative imbalance of supply and demand; to develop a method for assessing the quantitative imbalance of supply and demand.

The study describes the authors' approach to fulfilling the above tasks.

4 Intelligent Control of Supply and Demand on the Labor Market of IT Specialists

At present demand plays the main role in the supply-and-demand interconnection in many economic sectors, including IT. It is quite robust and presents certain challenges to supply. This suggests the necessity of new strategic approaches to IT specialist training including closer contacts with employers who are now obliged to carry out the training, retraining and further training of IT experts on their own [13]. At the same time, employers need to participate more actively in the processes of developing the professional knowledge and skills, the advancement in training and further training of IT specialists [14].

We regard the elaboration of an intelligent system of managing the coordination of ITS as one of the ways of reducing the imbalance between them and quick adjustment of IT experts' supply to the fast-changing needs of the economy.

By intelligent control of SD of the market of ITS the authors mean managerial decision-making on the coordination (balancing) of the SD of ITS which comes down to: (1) generating alternative policy options (measures, strategies, tactics) in accordance with a specific task and the supply and demand at a given moment (period) of time; (2) making a managerial decision in keeping with the aim and terms of the task set as well as with the needs and preferences of the main actors of the labor market (employers and IT specialists) on the one hand and ensuring minimal difference between supply and demand on the other. The aim of intelligent control of SD of ITS consists in making well-grounded managerial decisions on staffing various branches of the national economy with qualified specialists. In order to achieve this aim, a system of intelligent control of the ITS labor market is designed, one of the basic components of which is a help module of the process of exercising control as regards the coordination of both quantitative and qualitative components of SD of the market of ITS.

5   Conceptual Approach to Investigating the Labor Market of IT Specialists

The shift of emphasis onto the human factor and its intellectual component determines the need to review the content of the main components of the labor market. According to the concept advanced by the authors [24], the labor market is seen as an intellectual space (environment or system) in which its subjects who represent demand (employers) and supply (IT experts) operate. The latter's competences which embody their personal intellectual potentials expressed by knowledge, skills and aptitudes as well as "soft" skills and idiosyncrasies are seen as the product of the IT segment of the labor market.

Interpreting the content of the main subjects of the labor market of ITS in keeping with the realities of knowledge-based economy allows one to identify the following basic structural components of intelligent control of the coordination of the labor market:

(1) demand for IT experts from the perspective of the needs of enterprises (employers) which is reflected in IT job requirements;
(2) supply, i.e. applicants for IT jobs who offer their intellectual potentials formed by means of continuous education services;
(3) mechanisms of identifying the correlation between supply and demand on the IT labor market which embody the degree of coordination (imbalance) of demand for IT specialists and their supply (models and methods of assessing the balance of supply and demand);
(4) mechanisms of managing the coordination of SD on the market of ITS (methods of support of managerial decision-making which come down to elaboration of policies in the sphere of labor and employment and adjustment of IT education and training to labor market needs) [25].

6   Modeling of the Processes of Interaction and Evaluation of the Imbalance of SD on the Market of ITS

In the examination of the processes of coordination of S and D it is also necessary to clearly identify the level at which the imbalance is assessed. Thus, the task of modeling the processes of interaction of S and D on the labor market of ITS and their management can be examined at the micro- and macro-levels. At the macro-level the task of identifying the states of S and D is examined from the viewpoints of individual subjects of the labor market (IT experts and employers) and their behavioral strategies. Enterprise is the smallest unit at this level. It is at the level of enterprise that the structure and volume of demand for ITS as well as expectations of their professional competencies and personal intelligences are specified. Given this, the establishment of the degree of coordination between S and

D at the micro-level boils down to elaboration of mechanisms of effective selection and recruitment of ITS. At the macro-level the modeling of the processes of interaction of S and D of ITS comes down to balancing the relevant supply and demand within the bounds of a territorial or geographical entity (at the levels of economic branches, regions, country as a whole etc.).

In formal terms, the problem of identification of supply and demand conditions can be defined by three components $D = \langle V, S, R \rangle$, where V is the set of vacancies;

S is the set of IT specialists; R is the set of rules defining the relationship between the elements of sets V and S, i.e., rules helping to compare the descriptions of actual conditions of IT specialists with all reference conditions of the demand side.

The recognition and evaluation of supply and demand take the form of the mapping $F: D \rightarrow Z$, where Z is the solution of the problem D set with the intelligent system as a particular target condition meeting the purpose of recognition and evaluation in a particular situation.

At the micro-level these are rules that allow to correlate the descriptions of the real states of IT specialists and all the reference states of demand (qualification criteria set by the employer for job applicants) and reveal the difference between them. At the macro-level these are rules that allow to establish the composite supply and demand of IT specialists in various IT jobs and specializations as well as the degree of their balance which reflects the conditions of the market and IT specialists.

A. *Managing the discrepancy between supply and demand of IT specialists at the micro-level*

Let $M_V = \{V, K, G, Q, U^p\}$ be a model of demand for IT professionals, defining requirements for the competence of the applicant to a particular workplace. The latter one is a system of employers' preference regarding specific job applicant, which are expressed by a set of desired competencies of the candidate, and it forms a search image of IT professional. Here V is a set of vacancies expressed with demands of employers for IT professionals, i.e. applicants for vacancies, $K = (L, C)$ is a set of core competencies characterizing IT professional, which is formed by L—a set of personal competencies needed to work in the IT sector, and by C—a set of professional competence, reflecting the necessary ability of functionality to be engaged in a particular position. G is a system of employer's preferences regarding owning individual indicators, $Q: V \cdot K \cdot U^p \rightarrow G$ is a decisive rule (assessment model) to display the system of employer's preferences to the set of competencies, U^p is a set of conditions, proposed to the candidates for IT profile position.

Proposal model $M_S = \{S, K, W, Q^*, U^s\}$ reflects the actual values of competencies and preferences of each individual IT professionals, identifying real search image (professional portrait) of IT professionals. Here S is a set of IT professionals seeking a job and applying for a particular job, $K = (L, C)$ is a set of personal characteristics and professional competencies of a specific IT specialist, i.e. a potential applicant for a specific job, W is a system of preferences of IT professionals, $Q^*: S \cdot K \cdot U^s \to W$ displays the system of preferences of IT specialist to a set of competencies, U^s is the requirements of IT professionals for IT-profile work place. During the interaction of a set of standard states of demand for IT professionals and the set of real states, that define their supply, many unique semi-structured (fuzzy) situations are formed.

The goal of the management problem of supply and demand in the labor market of IT professionals is the identification (recognition) of real search images of IT professionals and etalon query search images of the exact same pair, the conformity (proximity) degree of the elements of which has the greatest value from both the preference position (reference requirements) of the employer, and from the standpoint of the applicant claims.

B. *Solution of the management problem of supply and demand in the labor market of IT professionals*

Demand model $V = (L, C)$ can be described by three matrices $V_L = \left\| l_{ij} \right\|_{kn}$, $V_C = \left\| c_{ir} \right\|_{km}$, and $V_U = \left\| u_{iz} \right\|_{kp}$, where each row (V_i) describes a separate position in the IT labor market; the columns (l_n, c_m) display constantly expanding base of personality characteristics and competencies; the elements l_{kn}, c_{km} are the level of possessing separate indicators needed to fill a vacancy at time t; u_{kp} is the values of parameters characterizing the conditions offered for the applicant to fill a particular vacancy. Satisfaction rate of V_i vacancy for the indicators l_{ij} and c_{ir} is defined as fuzzy sets with membership functions $\mu_{l_{ij}}(V_i): V \times L \to [0, 1]$, $\mu_{c_{ir}}(V_i): V \times C \to [0, 1]$, which express the rate of owning individual competencies required by the given employers to fill a vacancy.

At the same time, the conditions offered to applicants are described by matrix $V_U = \left\| u_{iz} \right\|_{kp}$, where the membership functions $\mu_{u_{iz}}(V_i): V \times U \to [0, 1]$ are fuzzy measures of indicator intensity, characterizing the conditions of employment. $S = (l, c)$—the demand model is also described by three matrices $S_L = \left\| l_{ij} \right\|_{kn}$, $S_C = \left\| c_{ir} \right\|_{km}$ and $S_U = \left\| u_{iz} \right\|_{qp}$, where each row (S_q) describes a separate candidate for the job in the IT labor market; the columns (l_n, c_m) reflect the constantly expanding base of personality characteristics and competencies; the elements l_{qn}, c_{qm} are the rate of possessing different characteristics needed to fill a vacancy; u_{qp} is the indicator value, describing the requirements of IT—specialist to the vacancies. The rate of possession of specific IT professional competence S_l is determined by a separate membership function $\mu_{l_{ij}}(S_i): S \times L \to [0, 1]$, $\mu_{c_{ir}}(S_i): S \times C \to [0, 1]$.

Requirements of IT professionals for a vacancy are expressed by matrix $S_U = \left\| u_{iz} \right\|_{cp}$, and $\mu_{u_{cz}}(S_i): S \times U \to [0, 1]$ is a fuzzy measure of requirements for intensity of IT professionals. Actually, there are two sets of fuzzy conditions, describing the state of supply $\tilde{V}_k$ and demand $\tilde{S}_q$ in the labor market of IT professionals:

$$\tilde{S}_q = \left\{ \langle \mu_{l_{ij}}(S_q) \rangle, \langle \mu_{c_{ir}}(S_q) \rangle, \langle \mu_{u_{iz}}(S_q) \rangle \right\} = \left\{ \mu_{S_q}(y)/y \right\}$$

$$\tilde{V}_k = \left\{ \langle l_{ij}(V_k) \rangle, \langle \mu_{c_{ir}}(V_k) \rangle, \langle \mu_{u_{iz}}(V_k) \rangle \right\} = \left\{ \mu_{V_k}(y)/y \right\}$$

Here $\tilde{S}_q = \left\{ \mu_{S_q}(y)/y \right\}$ is a description of the set of fuzzy etalon situations, and $\tilde{V}_k = \left\{ \mu_{V_k}(y)/y \right\}$ is the set of description of fuzzy real situations [26, 27].

C. *Managing the imbalance between S and D on the market of labor of ITS at the macro-level*

The education services market is one of the main sources of workforce inflow onto the labor market and an infrastructure element of its regulation, including various occupational-skills groups [28]. The market of education services is the source of continuous inflow of specialists for specific jobs in the IT sector. This provides an opportunity to elaborate a model of interaction between the system of vacancies (demand), the IT workforce (supply), the institutions influencing the processes and mechanisms of managing the S and D of ITS and to establish the composite D (S) at the level of any territorial and geographical entity and country as a whole. We should also take into account the characteristic feature of the problems relevant to the assessment of the imbalance between S and D at the micro-level which consists in the fact that information for their resolution is derived from observing the states of S and D at various intervals and from different sources.

The task of managing the imbalance between S and D at the macro-level is the elaboration of approaches and methods: (1) calculation of the composite demand and composite supply for IT jobs and qualifications; (2) the evaluation of the degree of occupational-skill imbalance between S and D which reflects conditions of ITS labor market.

Let $[t_1, t_2]$ be a specified amount of time. The volume and structure of the demand for specialists in various IT jobs and specializations will be described by vector $V[t_1, t_2] = \{V_1[t_1, t_2], V_2[t_1, t_2], \ldots, V_N[t_1, t_2]\}$ which presents a set of IT jobs broken down by sectors of a national economy within time interval $[t_1, t_2]$, where N is the number of IT jobs and specializations on the labor market. The volume and the structure of supply of ITS in a specified time interval in terms of IT jobs and specializations can be described by the vector of supply $S[t_1, t_2] = \{S_1[t_1, t_2], S_2[t_1, t_2], \ldots, S_N[t_1, t_2]\}$.

The interaction of S and D on the labor market of ITS and the movement of resource flows occur predominantly on the basis of three sources. Thus, the overall number of ITS with a certain occupational-skill structure which seek employment independently (through friends, relatives etc.) can be described by vector $S^1[t_1, t_2] = \{S_1^1[t_1, t_2], S_2^1[t_1, t_2], \ldots, S_N^1[t_1, t_2]\}$. Let us designate the number of employed ITS from this category over time $[t_1, t_2]$ as $H^+[t_1, t_2]$, whereas the number of job seekers as $H^-[t_1, t_2]$. These vectors can be described also in terms of the number of the employed as $H^+[t_1, t_2] = \{H_1^+[t_1, t_2], H_2^+[t_1, t_2], \ldots, H_N^+[t_1, t_2]\}$ and those seeking employment, as $H^-[t_1, t_2] = \{H_1^-[t_1, t_2], H_2^-[t_1, t_2], \ldots, H_N^-[t_1, t_2]\}$ in every IT job and specialization. The overall number and make up of ITS who seek employment via the Internet and recruitment agencies will be designated as vector $S^2[t_1, t_2]$. Among them $W^+[t_1, t_2]$ is the number of the employed and $W^-[t_1, t_2]$ is the number of the unemployed. In terms of the occupational-skill structure the number of the employed and the unemployed IT experts is described by the following vectors: $W^+[t_1, t_2]$; $W^-[t_1, t_2]$. The general number and the makeup of ITS who seek employment through the market of education services will be described by vector $S^3[t_1, t_2]$. Among them $Q^+[t_1, t_2]$ is the number and the occupational-skill makeup of the employed who entered the labor market through forms of training and retraining, whereas $Q^-[t_1, t_2]$ is the number and the occupational-skill makeup of the unemployed from this category who apply for IT jobs. The constructed system of vectors allows one to describe the overall number and makeup of ITS on the labor market.

The unemployed part of ITS constitutes the vector of supply at a given moment t which can be described as follows: $S(t) = \{H^-[t_1, t_2] + W^-[t_1, t_2] + Q^-[t_1, t_2]\}$. The overall number of the satisfied cases of S and D of ITS over certain period of time $[t_1, t_2]$ according to all the sources of interaction of S and D can be presented as $VIS[t_1, t_2] = \{H^+[t_1, t_2] + W^+[t_1, t_2] + Q^+[t_1, t_2]\}$. Then the unsatisfied D at current moment in time t is calculated in the following way:

$$V(t) = V[t_1, t_2] - VIS[t_1, t_2], \tag{1}$$

and S can be calculated using the following equation:

$$S(t) = S[t_1, t_2] - VIS[t_1, t_2]. \tag{2}$$

D. *The conjuncture of the labor market of ITS*

The interaction between S and D of ITS forms the conditions of the respective segment of the labor market. Let us consider market conditions in the context of the IT segment. Thus, (1) if at moment in time t vector of demand $V(t)$ exceeds vector of supply $S(t)$, i.e. the composite demand for ITS in various IT professions and specializations is higher than the composite supply in terms of the specified jobs and specializations $V(t) > S(t)$, then there is a deficit of IT specialists on the market; (2) if at moment in time t vector of supply $S(t)$ exceeds vector of demand $V(t)$, i.e. the composite supply of specialists in various IT jobs and specializations is higher than the composite demand for ITS in terms of the IT jobs and specializations in question $V(t) < S(t)$, then the labor market has an excess of ITS; (3) if at moment in time t vectors of supply $S(t)$ and demand $V(t)$ coincide, i.e. the number and makeup of S of ITS are equal to the number of IT vacancies for specialists of appropriate qualifications and specializations are sought, then we can speak of an ideal situation of balance $V(t) = S(t)$ on the labor market. The description of the labor market's conjuncture by means of the three above mentioned states does not allow to reveal the entire range of interrelations between S and D of ITS. For instance, a gap between S and D can be quite substantial (critical) or, on the contrary, insignificant. It is also necessary to note the character of the tendencies of growth and diminishment of imbalance from the perspective of supply and demand, i.e. to uncover both the dynamics of S and D trends. The variety of possible states of S and D generates a respective set of conjuncture situations on the labor market of ITS.

7 Method of Assessing the Degree of Imbalance Between S and D Based on Fuzzy Mismach Scale

In order to obtain generalized characteristics of the correlation of S and D on the market of ITS within time interval $[t_1, t_2]$, let us introduce the notion of "indicator of quantitative imbalance", which represents the ratio of the num-ber of the unem-ployed and job-seeking ITS (supply) to the number of IT jobs: $\delta = S[t_1, t_2]/V[t_1, t_2]$, if $S[t_1, t_2] < V[t_1, t_2]$, and $\delta = V[t_1, t_2]/S[t_1, t_2]$ if $S[t_1, t_2] > V[t_1, t_2]$.

Let the indicators which describe the labor market of ITS be data-based and the above formulas (1) and (2) provide for calculating the values of S and D. As a rule, factual data for assessing supply and demand come from different information sources. These data are not ideal, far from comprehensive and not free from sub-jectivism and contradictions. Therefore it appears logical to examine the indicator of the "imbalance between supply and demand" as a linguistic variable [26, 29]. It is proposed to describe the range of variation of the imbalance of S and D of ITS as a scale of discordance consisting of two segments, one of which we will call a region of positive demand (in case of the latter's dominance over supply) and the other a region of positive supply in the contrary case. Figure 1 illustrates a scale of discordance which shows imbalance between S and D.

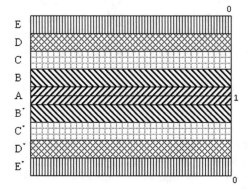

A – region with normative (optimal) value of the correlation between S and D;

B, B* – regions of minimal imbalance (deviation) between S and D;

C, C* – regions of acceptable imbalance between S and D;

D, D* – regions of critical imbalance between S and D;

E, E* – regions of complete imbalance between S and D.

Fig. 1 A graphic illustration of the degree of imbalance between S and D

As can be seen in the Figure, each of the specified regions of changes in the imbalance $[E, A]$ and $[E^*, A]$ is an interval [0, 1] divided in conformity with a possible degree of discordance between S and D into several segments, which are regions of deviation from the standard value of the imbalance. In order to formalize the parameters determining the degree of imbalance between S and D at a certain moment in time, let us use functions $s(S_j)$ described as fuzzy measures and being a real number which an expert assigns to every event S_j. These functions were introduced in the work [30] for assessing the measure of uncertainty. For $\forall i$, functions $s(S_j)$ characterize the degree of the expert's confidence that $s(S_j) \subset \Psi_S$. The scope of changes in the imbalance depending on the degree of its expressiveness can be divided, by expert assessment, into several fuzzy intervals representing the region of the change of functions of membership of fuzzy sets of verbal gradations of the linguistic variable "imbalance of S and D" δ_i, specified for the set of real numbers R_δ in the shape of $\mu_{\delta_i} : R_\delta \to [0, 1]$. Let us express the guideline value of imbalance (optimal correlation between supply and demand) at the moment in time t^m (or in certain time period) by $\delta^{t_m}_{norm} = S^{t_m}_{norm} / V^{t_m}_{norm}$ if $S^{t_m}_{norm} < V^{t_m}_{norm}$ and $\delta^{t_m}_{norm} = V^{t_m}_{norm} / S^{t_m}_{norm}$ if $S^{t_m}_{norm} > V^{t_m}_{norm}$. If running values of supply and demand are known (as is their correlation (the running imbalance)), i.e. $\delta^{t_m}_{cur} = S^{t_m}_{cur} / V^{t_m}_{cur}$, then the membership functions of the current state of imbalance can be established using the following:

$$\mu_l(x) = 1 - \left| \delta^{t_m}_{cur} - \delta^{t_m}_{norm} \right|. \tag{3}$$

As is seen in Fig. 1, the imbalance between S and D may vary over a wide range: from the standard value of the correlation of S and D to their complete imbalance.

The closer the running imbalance is to the standard one, the more favorable the region of changes is in which the values of membership functions of the current state occur.

The proposed approach to the assessment of the situation on the labor market allows one to make a fuzzy classification of its states according to the degree of imbalance between S and D. An algorithm for fuzzy classification of the states of the labor market imbalance in the degree of its intensity at a certain time is offered. The open knowledge base, in which generated by experts production rules correspond to control actions that should be adopted to eliminate the discrepancies between supply and demand in particular the current situation in the labor market.

8   Conclusion

At present the Institute of Information Technology of the National Academy of Science of Azerbaijan is engaged in designing an intelligent system of managing the labor market of ITS which is based on the methodological and conceptual approaches advanced in this paper. The system is meant to reduce the quantitative and qualitative imbalance between S and D on the ITS' labor market by means of providing information and consultation support for decision-makers at various levels of administration as well as for employers and ITS, in identifying the current needs of the labor market and the education services in the IT sector, the adoption of executive decisions adequate to the situation as regards balancing S and D, the management of human resources in the IT sector, continuous adjustment of decision-making to the changing business environment.

References

1. E-Skills for Jobs in Europe [Electronic resource], www.eskills2014.eu
2. European Vacancy Monitor (EVM) (2012), http://ec.europa.eu/social/main.jsp?catId=955&langId=en
3. D.M. Wennergren, Forecast of Future IT Labor Supply and Demand U.S. (2007), http://www.dodcio.defense.gov/Home/Initiatives/NetGeneration
4. D. Ticol, Labour supply/demand dynamics of Canada's information and communications technology (ICT) sector. Final Report, Nordicity (2012), p. 30
5. M.H. Mammadova, Z.G. Jabrayilova, M.I. Manafli, in *Monitoring of the Need for IT Specialists*, ed. by M.H. Mammadova (Baku:IT publishing, 2009), p. 199
6. IT Personnel 2010, The number of the employed in the Russian economy in 2009 and the forecast of the relevant need in 2010–2015. REAL analytical center, AE CITT (2010), http://www.apkit.ru/committees/education/projects/itcadry2010
7. Shortage of IT specialists will make 170,000 people in (2015), http://www.unian.net/society/871034-defitsit-it-spesialistov
8. W. Bartlett, Skills anticipation and matching systems in transition and developing countries: conditions and challenges, Working paper for ETF (2012), www.etf.europa.eu
9. Forecast and anticipation for skills supply and demand in ETF partner countries. Working paper for the ETF (2013), www.etf.europa.eu
10. M. Mammadova, F. Mammadzadeh, Formation of supply and demand for IT specialists on the base of competency model, in *Proceedings of IV International Conference "Problems of Cybernetics and Informatics"*, Baku, vol. IV (2012), pp. 199–201. https://doi.org/10.1109/icpci.2012.6486486
11. Research Study on High-Level Skill Needs in NI ICT Sector [Text]: Final Report. Oxford Economics (2009), 129 pp.
12. P. Doucek, L. Nedomova, M. Maryska, Differences between offer and demand on the ICT specialist's Czech Labor market. Organizacija **45**(6), 261–275 (2012), https://doi.org/10.2478/v10051-012-0026-0
13. European Commission. EU Skills Panorama Analytical Highlight 'ICT Professionals', http://euskillspanorama.ec.europa.eu
14. H. Salzman, D. Kuehn, L. Lowell, Guestworkers in the high-skill US labour market an analysis of supply, employment, and wage trends (2013), http://www.epi.org/publication/bp359-guestworkers
15. B.I.J.M. Van der Heijden, Employability management needs analysis for the ICT sector in Europe: the case of small and medium-sized enterprises. J. Cent. Cathedr. Bus. Econ. Res. J. **3**(2), 182–200 (2010)
16. L. Feiler, A. Fetsi, T. Kuusela, G. Platon, Anticipating and matching supply and demand of skills in ETF partner countries. Working paper for the European Training Foundation, European Training Foundation (2013), www.etf.europa.eu/webatt.nsf/0/FBEF620E5BFEB10

17. N. 'Vincent, N. Tremblay-Côté, Imbalances between Labour Supply and demand—2011–2020. Canadian Occupational Projection System, Employment and Social Deveplopment Canada (2011), http://www23.hrsdc.gc.ca/l.3bd.2t.1ilsht

18. M.H. Mammadova, Z.Q. Jabrayilova, F.R. Mammadzada, Fuzzy multi-scenario approach to decision-making support in human resource management, in *Recent Development and New Direction in Soft-computing Foundations and Applications*, vol. 342 (2016), pp. 19–36

19. T.V. Azarnova, T.V. Popova, A.N. Leontyev, Analysis algorithm of dynamics of change of quality of functioning of the labour market at realization of various strategy of quality management, in *Proceedings of Voronezh State University, Series: Systems Analysis and Information Technologies,* vol. 2 (2013), pp. 79–86

20. D.A. Gainanov, R.R. Galliamov, Model of minimization of structural imbalance of the labor market. Vestnik UGATU **8**(2), 89–92 (2006)

21. V.N. Vasilyev, *The labor market and the educational services market in the subjects of the Russian Federation* (Technosphere, Moscow, 2007), p. 680

22. S.V. Sigova, State regulation of the balance of the labor market: Synopsis of the thesis of the Doctor of Economics, Moscow (2011), www.dissers.ru/avtoreferati-dissertatsii-ekonomika/a176.php

23. K. Poulikas, A Balancing Act at Times of Austerity: Matching the Supply and Demand for Skills in the Greek Labour Market. Cedefop and IZA (2014), 43 pp., http://www.ftp.iza.org/dp7915.pdf

24. M.H. Mammadova, F.R. Mamedzade, Elaboration of the conceptual framework for intelligent control of supply and demand on the labor market of IT specialists. East.-Eur. J. Enterp. Technol. **4**(3)(76), 53–67 (2015)

25. M.H. Mammadova, Z.G. Jabrayilova, F.R. Mammadzada, Managing the IT labor market in conditions of fuzzy information. Autom. Control Comput. Sci. **2**(49), 88–93 (2015)

26. L.A. Zadeh, The concept of a linguistic variable and its application to approximate reasoning. I. J. Inf. Sci. **8**(3), 199–249 (1975)

27. A.N. Melikhov, L.S. Bernshtein, S.Y. Korovin, *Situatsionnye sovetuyushchie sistemy s nechetkoi logikoi (Situation Consulting Systems with Fuzzy Logic* (Nauka, Moscow, 1990)

28. E.M. Ilyin, M.A. Klupt, B.S. Lisovik et al., *Prediction of the Labour Market* (Levsha, St. Petersburg, 2001), p. 458

29. R.R. Yager, Fuzzy Set and Possibility Theory. Recent Developments, vol. XIII (Pergamon Press, New York, 1982), 633 pp.

30. M. Sugeno, Theory of fuzzy integral and its application. Ph.D. thesis, Tokyo Institute of Technology, Japan (1974), http://www.lamsade

Part VI
Experts, Games and Game-inspired Methods

Discrimination of Electroencephalograms on Recognizing and Recalling Playing Cards—A Magic Without Trick

T. Yamanoi, H. Toyoshima, H. Takayanagi, T. Yamazaki, S. Ohnishi and M. Sugeno

Abstract Authors measured electroencephalograms (EEGs) as participants recognized and recalled 13 playing card images (from ace to king of club) presented on a CRT monitor. During the experiment, electrodes were fixed on the scalps of the participants. Four EEG channels located over the right frontal and temporal cortices (Fp_2, F_4, C_4 and F_8 according to the international 10–20 system) were used in the discrimination. Sampling data were taken at latencies between 400 and 900 ms at 25 ms intervals for each trial. Thus, data were 84 dimensional vectors (21 time point X 4 channels). The number of objective variables was 13 (the number of different cards), and the number of explanatory variates was thus 84. Canonical discriminant analysis was applied to these single trial EEGs. Results of the

T. Yamanoi (✉)
Department of Life Science and Technology,
Hokkai-Gakuen University, Sapporo, Hokkaido 064-0926, Japan
e-mail: yamanoi@hgu.jp

H. Toyoshima
Japan Technical Software Co. Ltd, Sapporo, Hokkaido 001-0021, Japan
e-mail: toyoshima@jtsnet.co.jp

H. Takayanagi
Center for University-Society Relations and Collaboration,
Future University Hakodate, Hakodate, Hokkaido 041-8655, Japan
e-mail: taka@s2c.jp

T. Yamazaki
Faculty of Computer Science and System Technology,
Kyushu Institute of Technology, Iizuka, Fukuoka 820-8502, Japan
e-mail: t-ymz@bio.kyutech.ac.jp

S. Ohnishi
Department of Electronics and Information Engineering,
Hokkai-Gakuen University, Sapporo, Hokkaido 064-0926, Japan
e-mail: ohnishi@hgu.jp

M. Sugeno
Tokyo Institute of Technology, Yokohama, Kanagawa, Japan
e-mail: michio.sugenoi@gmail.com

© Springer International Publishing AG, part of Springer Nature 2018
L. A. Zadeh et al. (eds.), *Recent Developments and the New Direction in Soft-Computing Foundations and Applications*, Studies in Fuzziness and Soft Computing 361, https://doi.org/10.1007/978-3-319-75408-6_19

canonical discriminant analysis were obtained using the jack knife method and were 100% of nine participants. We could perform playing card estimation magic without a trick. This fact is sub production based on our series of precedent research.

1 Introduction

In the human brain, the primary processing of a visual stimulus occurs in areas V1 and V2 in the occipital lobe. Initially, a stimulus presented to the right visual field is processed in the left hemisphere and a stimulus presented to the left visual field is processed in the right hemisphere. Next, processing moves on to the temporal associative areas [1].

Higher order processing in the brain is associated with laterality. For example, language processing in Wernicke's area and Broca's area is located in the left hemisphere in 99% of right-handed people and 70% of left-handed people [2, 3]. Language is also processed in the angular gyrus (AnG), the fusiform gyrus (FuG), the inferior frontal gyrus (IFG), and the prefrontal cortex (PFC) [4].

Using equivalent current dipole localization techniques [5] applied to summed and averaged electroencephalograms (EEGs), we previously reported that for input stimuli comprised of the arrow symbol, equivalent current dipoles (ECDs) can be localized to the right middle temporal gyrus, and estimated in areas related to working memory for spatial perception, such as the right inferior or the right middle frontal gyrus. Further, using Chinese characters (Kanji) as stimuli, ECDs were also localized to the prefrontal cortex and the precentral gyrus [6, 7].

However, in the case of silent reading, spatiotemporal activities were observed in the same areas at around the same latencies regardless of the stimulus (Kanji or arrow). ECDs were localized to the Broca's area which is said to be the language area that control speech. And also after on the right frontal lobe, the spatiotemporal activities go to so-called working memory region. As in our previous studies, we found that latencies of main peak were almost the same, but that the polarities of potentials were opposite in the frontal lobe during higher order processing [6].

Research into executive function using the functional magnetic resonance imaging indicates that the middle frontal lobe is related to the central executive system, including working memory. Functions of the central executive system include to selecting information from the outer world, to holding it temporarily in memory, to ordering subsequent actions, to evaluating these orders and making decisions, and finally erasing temporarily stored information. Indeed, this art of the frontal cortex is the headquarters of higher order functions in the brain.

Previously, we compared signal latencies at each of three channels of EEG, and found that the channel 4 (F4), 6 (C4) and 12 (F8) were effective in discriminating EEGs during silent reading for four types of arrows and Kanji characters. Each discrimination ratio was more than 80% [6].

When the data were tested with the jack knife (cross validation) statistical technique, their discriminant ratios generally decreased. Thus, for the recent study,

Fig. 1 Electrodes placement over the right hemisphere

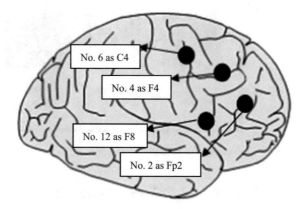

No. 6 as C4

No. 4 as F4

No. 12 as F8

No. 2 as Fp2

we have improved the technique by adding another EEG channel (channel 2: Fp$_2$) (Fig. 1). With these changes, discriminant analysis with the jack knife method resulted in means of discriminant ratios were greater than 95%.

2 Measurement of EEGs on Recognition and Recall

Participants were nine 22-year-old university students, who had normal visual acuity, and were right-handed. The participants were nine undergraduate students. We denote each experiment as YI, YT, YY, SH, RE, MK, CS, HM and TH. The participants wore an electrode cap with 19 active electrodes and attended visual stimuli that were presented on a 21-inch PC monitor placed 30 cm in front of them.

Participants kept their heads steady by placing their chins on a chin rest fixed to a table. Electrode positions were set according to the international 10–20 system and two other electrodes were fixed on the upper and lower eyelids for eye blink monitoring. Impedances were adjusted to less than 50 kΩ. Reference electrodes were put on both earlobes and the ground electrode was attached to the base of the nose.

EEGs were recorded on a multi-purpose portable bio-amplifier recording device (Polymate, TEAC). The frequency band was set between 1.0 and 2000 Hz. Output was transmitted to a recording computer. Analog outputs were sampled at a rate of 1 kHz and stored on computer hard disk.

During the experiment, participants were presented with 13 playing card images (the 13 Club). Each trial began with a 3000 ms fixation period, followed by the playing card image (encoding period) for 2000 ms, another fixation (delay) period for 3000 ms, and finally a 2000 ms recall period. During the recall period, participants imagined the playing card that had just been presented. Each playing card was presented randomly, and measurements were repeated several times for each playing card. Thus, the total number of experiment was about 100. We recorded EEGs the encoding and recall periods (Fig. 2) and these EEGs are used for inference of the playing card by use of the canonical discriminant analysis.

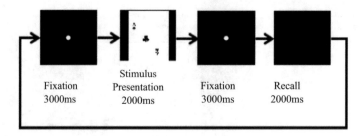

Fig. 2 Time chart of present experiment

3   Single Trial EEG and Sampling Period

Single-trial EEG data, recorded in the experiment with directional symbols were used in a type of multivariate analysis called canonical discriminant analysis. From the result of our past research [6], the silent reading pathway with directional symbols goes to the right frontal area at the latency after 400 ms. Therefore, in the current experiment, we sampled EEGs data from 400 to 900 ms. We also sampled data from 399 to 899 ms and from 398 to 898 ms. Each set of samples was collected 25 ms intervals, yielding 21data points from each channel for each sampling period.

For real-time applications, using a small number of EEG channels or sampling data is natural. Our previous work has investigated to what the minimal numbers of EEG channels and data samples are for the best results [6]. Especially, we wanted to determine the minimal sampling number necessary to obtain a perfect discriminant ratio (100%) at each channel for the same. In that set of experiments, we used the same time period, but the sampling interval was 50 ms. These results showed that by using EEGs, four types of order might be able to control. We must note that the discriminant analysis must be performed individually for each single trial of data. Thus, the discriminant coefficients are determined for each single data set. To improve the accuracy of single-trial discriminant ratios, we have adopted the jack knife (cross validation) method.

Of the 19 channels, we analyzed data from channels Fp_2 (No. 2), F_4 (No. 4), C_4 (No. 6), and F_8 (No. 12), according to the International 10–20 system (Fig. 1), because these points lie above the right frontal area. Although EEGs are time series data, we regarded them as vector values in a 84 dimensional space (4 channels X 21 time points).

4 Canonical Discriminant Analysis

Canonical discriminant analysis [8] is a dimension-reduction technique related to principal component analysis and canonical correlation. Given a classification objective variable and several explanatory variables, canonical discriminant analysis derives canonical variables (linear combinations of the explanatory variables) that summarize between-class variation in much the same way that principal components summarize total variation.

Given two or more groups of observations with measurements on several interval variables, canonical discriminant analysis derives a linear combination of the variables x_1, x_2, ..., and x_p that has the highest possible multiple correlation with the groups. This maximal multiple correlation is called the first canonical correlation. The coefficients of the linear combination are the canonical coefficients or canonical weights. The variable defined by the linear combination is the first canonical variable or canonical component. The second canonical correlation is obtained by finding the linear combination uncorrelated with the first canonical variable that has the highest possible multiple correlation with the groups. The process of extracting canonical variables can be repeated until the number of canonical variables equals the number of original variables or the number of classes minus one, whichever is smaller.

The expression of the fundamental canonical discriminant analysis is as follows:

$$y = a_1 x_1 + a_2 x_2 + \cdots + a_p x_p + b$$

where y is the canonical discriminant function, x_i is the discriminating variable, and a_i is the coefficient which produce the desired characteristics of the function, and b is the residual.

The first canonical correlation is at least as large as the multiple correlation between the groups and any of the original variables. If the original variables have high within-group correlations, the first canonical correlation can be large even if all the multiple correlations are small. In other words, the first canonical variable can show substantial differences among the classes, even if none of the original variables do.

For each canonical correlation, canonical discriminant analysis tests the hypothesis that it and all smaller canonical correlations are zero in the population. An F approximation is used that gives better small-sample results than the usual χ^2 approximation. The variables should have an approximate multivariate normal distribution within each class, with a common covariance matrix in order for the probability levels to be valid. The new variables with canonical variable scores in canonical discriminant analysis have either pooled within-class variances equal to one; standard pooled variance, or total-sample variances equal to one; standard total variance. By default, canonical variable scores have pooled within-class variances equal to one.

5   Results of Canonical Discrimination

We collected the EEGs data from each single trial and used them as learning data.
The data were resampled three times, in three types of sample timing; sampling data
1 are taken from latency of 400–900 ms at 25 ms interval (21 sampling points),
sampling data 2 are taken from latency of 399–899 ms at 25 ms interval and
sampling data 3 are taken from latency of 398–898 ms at 25 ms interval. Each
criterion variable has thirteen type indices, from Ace to King. We had tried so
called the jack knife statistical method, we took one sample to discriminate, and we
used the other samples left as learning data, and the method was repeated.

The participants were seven undergraduate students so the total number of
experiments was seven.

We tried to discriminate the thirteen types by EEG samples using the canonical
discriminant analysis. Each canonical discriminant coefficient was determined by
each participant and by each series about 100 experiments. As results, the discriminant ratios were 100%. Two results are shown in Tables 1 and 2.

Each result of discrimination for participants is presented on a computer screen
by software developed us. From this each result of discrimination of EEG is presented on the computer screen with the presented card and the corresponding
discriminated card (Fig. 3). And results are presented also on the computer screen
(Fig. 4).

Table 1 Example of result of canonical discriminant analysis for playing card recognition
(subject: YI, discriminant ratio: 100%)

Obs.	Pred.												
	A	2	3	4	5	6	7	8	9	10	J	Q	K
A	8	0	0	0	0	0	0	0	0	0	0	0	0
2	0	8	0	0	0	0	0	0	0	0	0	0	0
3	0	0	8	0	0	0	0	0	0	0	0	0	0
4	0	0	0	8	0	0	0	0	0	0	0	0	0
5	0	0	0	0	8	0	0	0	0	0	0	0	0
6	0	0	0	0	0	8	0	0	0	0	0	0	0
7	0	0	0	0	0	0	8	0	0	0	0	0	0
8	0	0	0	0	0	0	0	8	0	0	0	0	0
9	0	0	0	0	0	0	0	0	8	0	0	0	0
10	0	0	0	0	0	0	0	0	0	8	0	0	0
J	0	0	0	0	0	0	0	0	0	0	8	0	0
Q	0	0	0	0	0	0	0	0	0	0	0	8	0
K	0	0	0	0	0	0	0	0	0	0	0	0	8

Obs Observation, *Pred* Prediction

Table 2 Example of result of canonical discriminant analysis for playing card recognition (subject: YT, discriminant ratio: 100%)

Obs.	Pred.												
	A	2	3	4	5	6	7	8	9	10	J	Q	K
A	8	0	0	0	0	0	0	0	0	0	0	0	0
2	0	8	0	0	0	0	0	0	0	0	0	0	0
3	0	0	8	0	0	0	0	0	0	0	0	0	0
4	0	0	0	8	0	0	0	0	0	0	0	0	0
5	0	0	0	0	8	0	0	0	0	0	0	0	0
6	0	0	0	0	0	8	0	0	0	0	0	0	0
7	0	0	0	0	0	0	8	0	0	0	0	0	0
8	0	0	0	0	0	0	0	8	0	0	0	0	0
9	0	0	0	0	0	0	0	0	8	0	0	0	0
10	0	0	0	0	0	0	0	0	0	8	0	0	0
J	0	0	0	0	0	0	0	0	0	0	8	0	0
Q	0	0	0	0	0	0	0	0	0	0	0	8	0
K	0	0	0	0	0	0	0	0	0	0	0	0	8

Obs Observation, *Pred* Prediction

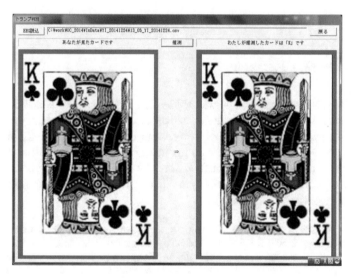

Fig. 3 Example of result on computer screen by canonical discriminant analysis for single-trial EEGs (right: presented card, left: discriminated result)

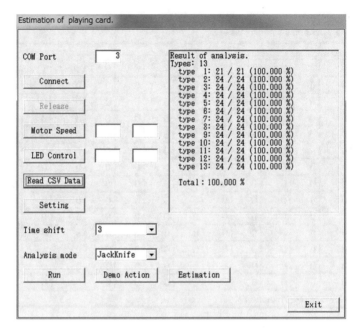

Fig. 4 Example of result on computer screen by canonical discriminant analysis for single-trial EEGs

6 Concluding Remarks

In this study, we recorded single-trial EEGs starting after 400 ms, and determined effective sampling latencies for the canonical discriminant analysis of 13 different images. We triple sampled EEG data from four channels (Fp_2, F_4, C_4, and F_8) at 25 ms intervals between 400 and 900 ms just after image presentation. Discriminant analysis using jack knife method for 13 objective variates yielded perfect (100%) discriminant rate for the nine participants. This attempt is original and none of simular research was found.

Acknowledgements This research was partially supported by a grant from the Ministry of Education, Culture, Sports, Science and Technology for the national project of the High-tech Research Center of Hokkai-Gakuen University in March 2013. This work was supported by JSPS KAKENHI Grant Number 16H02852. The experiment was approved by the ethical review board of Hokkaido University.

References

1. R.A. McCarthy, E.K. Warrington, *Cognitive Neuropsychology: A Clinical Introduction* (Academic Press, San Diego, 1990)
2. N. Geschwind, A.M. Galaburda, Cerebral lateralization, in *The Genetical Theory of Natural Selection* (Clarendon Press, Oxford, 1987)
3. K. Parmer, P.C. Hansen, M.L. Kringelbach, I. Holliday, G. Barnes, A. Hillebrand, K.H. Singh, P.L. Cornelissen, Visual word recognition: the first half second. NeuroImage **22**(4), pp. 1819–1825 (2004)
4. M. Iwata, M. Kawamura, M. Otsuki et al., Mechanisms of writing, Neurogrammatology (in Japanese), IGAKU-SHOIN Ltd, (2007), pp. 179–220
5. T. Yamazaki, K. Kamijo, T. Kiyuna, Y. Takaki, Y. Kuroiwa, A. Ochi, H. Otsubo, PC-based multiple equivalent current dipole source localization system and its applications. Res. Adv. Biomed. Eng. **2**, 97–109 (2001)
6. T. Yamanoi, T. Yamazaki, J.-L. Vercher, E. Sanchez, M. Sugeno, Dominance of recognition of words presented on right or left eye -Comparison of Kanji and Hiragana, in *Modern Information Processing From Theory to Applications*, ed. by B. Bouchon-Meunier, G. Coletti, R.R. Yager (Elsevier Science B.V., 2006) pp. 407–416
7. T. Yamanoi, H. Toyoshima, T. Yamazaki, S. Ohnishi, M. Sugeno, E. Sanchez, Micro robot control by use of electroencephalograms from right frontal area. J. Adv. Comput. Intell. Intell. Inf. **13**(2), 68–75 (2009)
8. J.F. Hair Jr, W.C. Black, B.J. Babin, R.E. Anderson, *Multivariate Data Analysis*, 7th edn. (Pearson India, 2009)

How to Describe Measurement Uncertainty and Uncertainty of Expert Estimates?

Nicolas Madrid, Irina Perfilieva and Vladik Kreinovich

Abstract Measurement and expert estimates are never absolutely accurate. Thus, when we know the result $M(u)$ of measurement or expert estimate, the actual value $A(u)$ of the corresponding quantity may be somewhat different from $M(u)$. In practical applications, it is desirable to know how different it can be, i.e., what are the bounds $f(M(u)) \leq A(u) \leq g(M(u))$. Ideally, we would like to know the tightest bounds, i.e., the largest possible values $f(x)$ and the smallest possible values $g(x)$. In this paper, we analyze for which (partially ordered) sets of values such tightest bounds always exist: it turns out that they always exist only for complete lattices.

1 Formulation of the Problem

How can we describe measurement uncertainty: formulation of the problem. We want to know the actual values of different quantities. To get these values, we perform measurements.

Measurements are never absolutely accurate, there is always measurement uncertainty, in the sense that the actual value $A(u)$ of the corresponding physical quantity is, in general, different from the measurement result $M(u)$; see, e.g., [1].

This uncertainty means that the actual value $A(u)$ can be somewhat different from the measurement result $M(u)$. It is therefore desirable to describe what are the possible values of $A(u)$. This will be a perfect way to describe uncertainty: for each measurement result $M(u)$, we describe the set of all possible values of $A(u)$.

How can we attain this description?

N. Madrid · I. Perfilieva
Institute for Research and Applications of Fuzzy Modeling, University of Ostrava,
Ostrava, Czech Republic
e-mail: nicolas.madrid@osu.cz

I. Perfilieva
e-mail: Irina.Perfilieva@osu.cz

V. Kreinovich (✉)
Department of Computer Science, University of Texas at El Paso, El Paso, TX 79968, USA
e-mail: vladik@utep.edu

© Springer International Publishing AG, part of Springer Nature 2018
L. A. Zadeh et al. (eds.), *Recent Developments and the New Direction
in Soft-Computing Foundations and Applications*, Studies in Fuzziness
and Soft Computing 361, https://doi.org/10.1007/978-3-319-75408-6_20

Important remark: in practice, we do not know the actual values. Ideally, for different situations u, we should compare the measurement result $M(u)$ with the actual value $A(u)$. The problem is that we do not know the actual value—if we knew the actual value, we would not need to perform any measurements. So how do practitioners actually gauge the accuracy of measuring instruments?

A usual approach (see, e.g., [1]) is to compare the measurement result $M(u)$ with the result $S(u)$ of measuring the same quantity by using a much more accurate ("standard") measuring instrument. If the standard measuring instrument is indeed much more accurate than the one whose accuracy we are gauging, then, for the purpose of this gauging, we can:

- assume that $S(u) = A(u)$, and
- compare the results $M(u)$ and $S(u)$ of measuring the same quantity by two different measuring instruments.

In general, all we have is measurement results, so all we can do to gauge accuracy is to compare two measurement results. So, from the practical viewpoint, the above problem can be reformulated as follows:

- we know the measurement result $M(u)$ corresponding to some situation u,
- we would like to describe the set of possible values $S(u)$ that we would have obtained if we apply a standard measuring instrument to these same situation u.

Let us first list typical situations. Before we consider the general case, let us first describe several typical situations.

Case of absolute measurement error. In some cases, we know the upper bound Δ on the absolute value of the measurement error $M(u) - A(u)$, i.e., we know that

$$|M(u) - A(u)| \leq \Delta.$$

In this case, once we know the measurement result $M(u)$, we can conclude that the (unknown) actual value $A(u)$ satisfies the inequality

$$M(u) - \Delta \leq A(u) \leq M(u) + \Delta.$$

In other words, we conclude that $A(u)$ belongs to the *interval* $[M(u) - \Delta, M(u) + \Delta]$; see, e.g., [2–4].

In more general terms, we can describe the corresponding bounds as

$$f(M(u)) \leq A(u) \leq g(M(u)),$$

where

$$f(x) \stackrel{\text{def}}{=} x - \Delta \text{ and } g(x) \stackrel{\text{def}}{=} x + \Delta.$$

Case of relative measurement error. In some other cases, we know the upper bound δ on the *relative* measurement error:

$$\frac{|M(u) - A(u)|}{|A(u)|} \leq \delta.$$

In this case, for positive values,

$$(1 - \delta) \cdot A(u) \leq M(u) \leq (1 + \delta) \cdot A(u).$$

Thus, once we know the measurement result $M(u)$, we can conclude that the actual (unknown) value $A(u)$ of the corresponding physical quantity satisfies the inequality

$$\frac{M(u)}{1 + \delta} \leq A(u) \leq \frac{M(u)}{1 - \delta}.$$

In other words, we have

$$f(M(u)) \leq A(u) \leq g(M(u))$$

for

$$f(x) \stackrel{\text{def}}{=} \frac{x}{1 + \delta} \text{ and } g(x) \stackrel{\text{def}}{=} \frac{x}{1 - \delta}.$$

In some cases, we have both types of measurement errors. In some cases, we have both:

- *additive* measurement errors, i.e., errors whose absolute value does not exceed Δ, and
- *multiplicative* measurement errors, i.e., errors whose relative value does not exceed δ and thus, whose absolute value does not exceed $\delta \cdot |A(u)|$.

In this case, for positive values, we get

$$A(u) - \Delta - \delta \cdot A(u) \leq M(u) \leq A(u) + \Delta + \delta \cdot A(u).$$

The left inequality can be reformulated as

$$A(u) \cdot (1 - \delta) - \Delta \leq M(u),$$

hence

$$A(u) \cdot (1 - \delta) \leq M(u) + \Delta$$

and thus,

$$A(u) \leq \frac{M(u) + \Delta}{1 - \delta}.$$

Similarly, the right inequality can be reformulated as

$$M(u) \leq A(u) \cdot (1 + \delta) + \Delta,$$

hence

$$A(u) \cdot (1 + \delta) \geq M(u) - \Delta$$

and thus,

$$A(u) \geq \frac{M(u) - \Delta}{1 + \delta}.$$

In this case, we have

$$f(M(u)) \leq A(u) \leq g(M(u)),$$

where

$$f(x) \overset{\text{def}}{=} \frac{x - \Delta}{1 + \delta} \text{ and } f(x) \overset{\text{def}}{=} \frac{x + \Delta}{1 - \delta}.$$

Towards a general case. The above formulas assume that parameters Δ and δ describing measurement accuracy are the same for the whole range. In reality, measuring instruments have different accuracies in different ranges. Hence, the resulting functions $f(x)$ and $g(x)$ are non-linear.

It should be mentioned that all the above functions $f(x)$ and $g(x)$ are monotonic, and this is usually true for all measuring instruments: when the measurement result is larger, this usually means that the bounds on possible values of the actual quantity also increase (or at least do not decrease).

To describe the accuracy of a general measuring instrument, it is therefore reasonable to use:

- the largest of the monotonic functions $f(x)$ for which $f(M(u)) \leq A(u)$ and
- the smallest of the monotonic functions $g(x)$ for which $A(u) \leq g(M(u))$.

Similarly, to describe the relative accuracy of a measuring instrument $M(u)$ in comparison to a standard measuring instrument $S(u)$, it is reasonable to use:

- the largest of the monotonic functions $f(x)$ for which $f(M(u)) \leq S(u)$ and
- the smallest of the monotonic functions $g(x)$ for which $S(u) \leq g(M(u))$.

From measurements to expert estimates. While measurement are very important, a large part of our knowledge comes from *expert estimates*. Expert estimates are extremely important in areas such a medicine.

In contrast to measurements that always result in numbers, expert estimates often can also result in "values" from a partially ordered set. For example, when a medical doctor is asked how probable is a certain diagnosis, the doctor may provide an approximate probability, or an interval of possible probabilities, or a natural-language term like "somewhat probable" or "very probable".

Such possibilities are described, e.g., in different generalizations and extensions of the traditional [0, 1]-based fuzzy logic; see, e.g., [5–7]; see also [8]. What is in common for all such extensions is that on the corresponding set of value L, there is always an *order* (sometimes partial), so that $\ell < \ell'$ means that ℓ' represents a stronger expert's degree of confidence.

Need to describe uncertainty of expert estimates. Some experts are very good, in the sense that based on their estimates, we make very effective decisions. These experts can be viewed as analogs of standard measuring instruments.

Other experts may be less accurate. It is therefore desirable to gauge the uncertainty of such experts in relation to the "standard" (very good) ones. If a regular expert provides an estimate $M(u)$ for a situation u, then, to make a good decision based on this estimate, we would like to know what would the perfect expert conclude in this case, i.e., what are the bounds on the perfect expert's estimates $S(u)$? In general, we may have several functions $f(x)$ and $g(x)$ for which

$$f(M(u)) \leq S(u) \leq g(M(u)).$$

It is desirable to find:

- the largest of the monotonic functions $f(x)$ for which $f(M(u)) \leq S(u)$ and
- the smallest of the monotonic functions $g(x)$ for which $S(u) \leq g(M(u))$.

What is known and what we do in this paper. For the case when the set L is an interval – e.g., the interval $[0, 1]$ – the existence of the largest $f(x)$ and smallest $g(x)$ was proven in [9] (see also [10]).

In this paper, we analyze for which partially ordered sets such largest $f(x)$ and smallest $g(x)$ exist. It turns out that they exist for complete lattices – and, in general, do not exist for more general partially ordered sets. To be more precise,

- the largest $f(x)$ exists for complete lower semi-lattices (precise definitions are given below), while
- the smallest $g(x)$ exists for complete upper semi-lattices.

2   Main Result: For Lattices, It Is Possible to Describe Uncertainty in Terms of The Bounding Functions $f(x)$ and $g(x)$

Definition 1   Let L be a (partially) ordered set, and let U be any set. We say that a function $F : U \to L$ is *smaller* that a function $G : U \to L$ is $F(u) \leq G(u)$ for all $u \in U$. We will denote this by $F \leq G$.

Definition 2   We say that a function $L \to L$ is *monotonic* if $x \leq y$ implies $f(x) \leq f(y)$.

Notation 1   For every ordered set L, by M_L, we denote the set of all monotonic functions $f : L \to L$.

Definition 3   Let $f \in M_L$. We say that a function $F : U \to L$ is f-*smaller* than a function $G : U \to L$ if $f(F(u)) \leq G(u)$ for all $u \in U$. We will denote this by $F \leq_f G$.

Notation 2 By $\mathscr{F}(F, G)$ we will denote the set of all functions $f \in M_L$ for which

$$F \leq_f G.$$

Definition 4 Let L be an ordered set, and let $S \subseteq L$ be its subset.

- We say that an element x is a *lower bound* for the set S if $x \leq s$ for all $s \in S$.
- An ordered set is called a *complete lower semi-lattice* if for every set S, among all its lower bounds, there exists the largest one. This largest lower bound is denoted by $\bigwedge S$.
- We say that an element x is an *upper bound* for the set S if $s \leq x$ for all $s \in S$.
- An ordered set is called a *complete upper semi-lattice* if for every set S, among all its upper bounds, there exists the smallest one. This smallest upper bound is denoted by $\bigvee S$.
- An ordered set L is called a *complete lattice* if it is both a complete lower semi-lattice and a complete upper semi-lattice.

Proposition 1 *If L is a complete lower semi-lattice, then for every two functions $F, G : U \to L$, the set $\mathscr{F}(F, G)$ has the largest element $f_{F,G}$ for which*

$$\mathscr{F}(F, G) = \{f \in M_L : f \leq f_{F,G}\}.$$

Proof We will prove that the function

$$f_{F,G}(x) \overset{\text{def}}{=} \bigwedge \{G(u) : x \leq F(u)\}$$

is the desired function. In other words, we will prove:

- that the function $f_{F,G}$ belongs to the class $\mathscr{F}(F, G)$ and
- that the function $f_{F,G}$ is the largest function in this class.

Let us first prove that $f_{F,G} \in \mathscr{F}(F, G)$, i.e., that for every u, we have

$$f_{F,G}(F(u)) \leq G(u).$$

Indeed, for $x = F(u)$, we have $x \leq F(u)$, and thus, the element $G(u)$ belongs to the set $\{G(u) : x \leq F(u)\}$. Thus, this element $G(u)$ is larger than or equal to the largest lower bound

$$f_{F,G}(x) = \bigwedge \{G(u) : x \leq F(u)\}$$

for this set, i.e., indeed

$$f_{F,G}(F(u)) = f_{F,G}(x) \leq G(u).$$

Let us now prove that the function $f_{F,G}$ is the largest function in the class $\mathscr{F}(F, G)$, i.e., that if $f \leq \mathscr{F}(F, G)$, then $f \leq f_{F,G}$. Indeed, let $f \in \mathscr{F}(F, G)$. By definition of this class, this means that f is monotonic and $f(F(u)) \leq G(u)$ for all u. Let us pick some $x \in L$ and show that $f(x) \leq f_{F,G}(x)$. Indeed, for every value $u \in U$ for which $x \leq F(u)$, we have, due to monotonicity, $f(x) \leq f(F(u))$. Since $f(F(u)) \leq G(u)$, we this conclude that $f(x) \leq G(u)$. So, the value $f(x)$ is smaller than or equal to all elements of the set $\{G(u) : x \leq F(u)\}$, i.e., $f(x)$ is a lower bound for this set. Every lower bound is smaller than or equal to the largest lower bound

$$f_{F,G}(x) = \bigwedge \{G(u) : x \leq F(u)\},$$

so indeed $f(x) \leq f_{F,G}(x)$.

Let us now prove that $\mathscr{F}(F, G) = \{f \in M_L : f \leq f_{F,G}\}$. We have shown that every function $f \in \mathscr{F}(F, G)$ is $\leq f_{F,G}$, i.e., that

$$\mathscr{F}(F, G) \subseteq \{f \in M_L : f \leq f_{F,G}\}.$$

Vice versa, if $f \leq f_{F,G}$, then for every u, from $f_{F,G}(F(u)) \leq G(u)$ and

$$f(F(u)) \leq f_{F,G}(F(u)),$$

we conclude that $f(F(u)) \leq G(u)$, i.e., that indeed $f \in \mathscr{F}(F, G)$.

The proposition is proven.

Discussion. A similar result can be obtained for the upper bounds.

Definition 5 Let $f \in M_L$. We say that a function $G : U \to L$ is *g-larger* than a function $F : U \to L$ if $F(u) \leq g(G(u))$ for all $u \in U$. We will denote this by $G \geq_g F$.

Notation 3 By $\mathscr{G}(F, G)$ we will denote the set of all functions $g \in M_L$ for which

$$G \geq_g F.$$

Proposition 2 *If L is a complete upper semi-lattice, then for every two functions $F, G : U \to L$, the set $\mathscr{G}(F, G)$ has the smallest element $g_{F,G}$ for which*

$$\mathscr{G}(F, G) = \{g \in M_L : g \geq g_{F,G}\}.$$

Proof Is similar to the proof of Proposition 1.

3 The Main Result Cannot Be Extended Beyond Complete Lower Semi-Lattices

Let us prove that this result cannot be extended beyond complete semi-lattices.

Proposition 3 *Let L be an ordered set for which, for every two functions $F, G : U \to L$, the set $\mathscr{F}(F, G)$ has the largest element. Then the set L is a complete lower semi-lattice.*

Proof Let us assume that the ordered set L has the above property. Let us prove that L is a complete lower semi-lattice. Indeed, let $S \subseteq L$ be any subset of L. Let us take $U = S$, and take $G(u) = u$ for all $u \in S$. Let us also pick any element $x_0 \in L$ and take $F(u) = x_0$ for all $u \in S$. Because of our assumption, the set $\mathscr{F}(F, G)$ of all the functions for which $f(F(u)) \leq G(u)$ for all u has the largest element $f_{F,G}$.

Because of our choice of the functions $F(u)$ and $G(u)$, the inequality

$$f(F(u)) \leq G(u)$$

simply means that $f(x_0) \leq u$ for all $u \in S$, i.e., that $f(x_0)$ is the lower bound for the set S. The fact that there is the largest of such functions $f \in \mathscr{F}(F, G)$ means that there is the largest of the lower bounds – which is exactly the definition of the complete lower semi-lattice. The proposition is proven.

Proposition 4 *Let L be an ordered set for which, for every two functions F, G : $U \to L$, the set $\mathscr{G}(F, G)$ has the smallest element. Then the set L is a complete upper semi-lattice.*

Proof Is similar to the proof of Proposition 3.

4   Auxiliary Results: What if There Is No Bias?

Comment about bias. In some practical situations, measuring instrument has a *bias* (shift): a clock can be regularly 2 min behind, a thermometer can regularly show temperatures which are 3 degrees higher, etc. Bias means that we get the measurement result $M(u)$, then this value *cannot* be equal to the actual value of the measured quantity: there is always a non-zero shift $A(u) - M(u)$.

Bias can easily be eliminated by re-calibrating the measuring instrument: for example, if I move to a different time zone, I can simply add or subtract the corresponding time difference and get the exact local time.

It is therefore reasonable to assume that the bias has already been eliminated, and that, thus, $A(u) = M(u)$ is one of the possible actual values. For this value $A(u) = M(u)$, our inequality

$$f(M(u)) \leq A(u) \leq g(M(u))$$

implies that

$$f(x) \leq x \leq g(x).$$

So, it makes sense to only consider functions $f(x)$ and $g(x)$ for which $f(x) \leq x$ and $x \leq g(x)$. It turns out that similar results hold when we thus restrict the functions $f(x)$ and $g(x)$.

Notation 4 For every ordered set L, by Ω_L, we denote the set of all monotonic functions $f : L \to L$ for which $f(x) \leq x$ for all $x \in L$.

Notation 5 By $\mathscr{F}_u(F, G)$ we will denote the set of all functions $f \in \Omega_L$ for which

$$F \leq_f G.$$

Proposition 5 *If L is a complete lower semi-lattice, then for every two functions $F, G : U \rightarrow L$, the set $\mathscr{F}_u(F, G)$ has the largest element $f_{F,G}$ for which*

$$\mathscr{F}_u(F, G) = \{f \in \Omega_L : f \leq f_{F,G}\}.$$

Proof We will prove that the function

$$f_{F,G}(x) \stackrel{\text{def}}{=} \bigwedge \{G(u) \wedge x : x \leq F(u)\}$$

is the desired function. In other words, we will prove:

- that the function $f_{F,G}$ belongs to the class $\mathscr{F}_u(F, G)$ and
- that the function $f_{F,G}$ is the largest function in this class.

Let us first prove that $f_{F,G} \in \mathscr{F}_u(F, G)$, i.e., that for every u, we have $f_{F,G}(F(u)) \leq G(u)$. Indeed, for $x = F(u)$, we have $x \leq F(u)$, and thus, the element $G(u)$ belongs to the set $\{G(u) : x \leq F(u)\}$. Thus, this element $G(u)$ is larger than or equal to the largest lower bound

$$f_{F,G}(x) = \bigwedge \{G(u) : x \leq F(u)\}$$

for this set, i.e., indeed

$$f_{F,G}(F(u)) = f_{F,G}(x) \leq G(u).$$

Since $G(u) \wedge x \leq x$, we conclude that $f_{F,G}(x) \leq x$. Thus, indeed, $f_{F,G} \in \Omega_L$.

Let us now prove that the function $f_{F,G}$ is the largest function in the class $\mathscr{F}_u(F, G)$, i.e., that if $f \leq \mathscr{F}_u(F, G)$, then $f \leq f_{F,G}$. Indeed, let $f \in \mathscr{F}_u(F, G)$. By definition of this class, this means that f is monotonic, $f(x) \leq x$ or all x, and $f(F(u)) \leq G(u)$ for all u. Let us pick some $x \in L$ and show that $f(x) \leq f_{F,G}(x)$. Indeed, for every value $u \in U$ for which $x \leq F(u)$, we have, due to monotonicity, $f(x) \leq f(F(u))$. Since $f(F(u)) \leq G(u)$, we this conclude that $f(x) \leq G(u)$. So, the value $f(x)$ is smaller than or equal to all elements of the set $\{G(u) : x \leq F(u)\}$, i.e., $f(x)$ is a lower bound for this set. Moreover, as $f(x) \leq x$, we have

$$f(x) \leq \bigwedge \{G(u) \wedge x : x \leq F(u)\} = f_{F,G}(x),$$

so indeed $f(x) \leq f_{F,G}(x)$.

Let us now prove that

$$\mathscr{F}_u(F, G) = \{f \in \Omega_L : f \leq f_{F,G}\}.$$

We have shown that every function $f \in \mathscr{F}_u(F, G)$ is $\leq f_{F,G}$, i.e., that

$$\mathscr{F}(F, G) \subseteq \{f \in \Omega_L : f \leq f_{F,G}\}.$$

Vice versa, if $f \leq f_{F,G}$, then for every u, from $f_{F,G}(F(u)) \leq G(u)$ and

$$f(F(u)) \leq f_{F,G}(F(u)),$$

we conclude that $f(F(u)) \leq G(u)$, i.e., that indeed $f \in \mathscr{F}_u(F, G)$.

The proposition is proven.

Notation 6 For every ordered set L, by Θ_L, we denote the set of all monotonic functions $g : L \rightarrow L$ for which $x \leq g(x)$ for all $x \in L$.

Notation 7 By $\mathscr{G}_u(F, G)$ we will denote the set of all functions $g \in \Theta_L$ for which

$$G \geq_g F.$$

Proposition 6 *If L is a complete upper semi-lattice, then for every two functions $F, G : U \rightarrow L$, the set $\mathscr{G}_u(F, G)$ has the smallest element $g_{F,G}$ for which*

$$\mathscr{G}_u(F, G) = \{g \in \Theta_L : g \geq g_{F,G}\}.$$

Proof Is similar to the proof of Proposition 5.

Acknowledgements This work was supported in part by the National Science Foundation grants HRD-0734825 and HRD-1242122 (Cyber-ShARE Center of Excellence) and DUE-0926721, and by an award "UTEP and Prudential Actuarial Science Academy and Pipeline Initiative" from Prudential Foundation. The authors are thankful to all the participants of IFSA'2015, especially to Enric Trillas and Francesc Esteva, and to the anonymous referees for valuable suggestions.

References

1. S.G. Rabinovich, Theory and practice, in *Measurement Errors and Uncertainty* (Springer, Berlin, 2005)
2. L. Jaulin, M. Kieffer, O. Didrit, E. Walter, *Applied Interval Analysis* (Springer, London, 2001)
3. V. Kreinovich, Interval computations and interval-related statistical techniques: tools for estimating uncertainty of the results of data processing and indirect measurements, in *Data Modeling for Metrology and Testing in Measurement Science*, ed. by F. Pavese, A.B. Forbes (Birkhauser-Springer, Boston, 2009), pp. 117–145
4. R.E. Moore, R.B. Kearfott, M.J. Cloud, *Introduction to Interval Analysis* (SIAM Press, Philadelphia, Pennsylvania, 2009)
5. G. Klir, B. Yuan, *Fuzzy Sets and Fuzzy Logic* (Prentice Hall, Upper Saddle River, New Jersey, 1995)
6. H.T. Nguyen, E.A. Walker, *A First Course in Fuzzy Logic* (Chapman and Hall/CRC, Boca Raton, Florida, 2006)
7. L.A. Zadeh, Fuzzy sets. Info. Control **8**, 338–353 (1965)
8. H. Bustince, E. Barrenechea, M. Pagola, J. Fernandez, Z. Xu, B. Bedregal, J. Montero, H. Hagras, F. Herrera, B. De Baets, A historical account of types of fuzzy sets and their relationships. IEEE Trans. Fuzzy Syst. **24**(1), 179–194 (2016)

9. N. Madrid, M. Ojeda-Aciego, I. Perfilieva, *f*-inclusion indexes between fuzzy sets, in *Proceedings of the 16th World Congress of the International Fuzzy Systems Association (IFSA) and the 9th Conference of the European Society for Fuzzy Logic and Technology (EUSFLAT)*, Gijon, Asturias, Spain, 30 June–3 July 2015, pp. 1528–1533

10. E.H. Ruspini, On the semantics of fuzzy logic. Int. J. Approx. Reason. **5**(1), 45–88 (1991)

Jagdambika Method for Solving Matrix Games with Fuzzy Payoffs

Tina Verma and Amit Kumar

Abstract Li (IEEE Trans Cybern 43:610-621, 2013) [1] recently proposed a method for solving matrix games with fuzzy payoffs and claimed that the obtained minimum expected gain of Player I and maximum expected loss of Player II, will be identical. Chandra and Aggarwal (Eur J Oper Res 2015. https://doi.org/10.1016/j.ejor.2015.05.011) [2], in their recent paper, pointed out the shortcomings of Li's approach and overcome the shortcomings of Li's approach. Chandra and Aggarwal, transformed the fuzzy mathematical programming problem into a multiobjective programming problem and obtained its result by using GAMS software. In this paper, it is pointed out that Chandra and Aggarwal have not considered some necessary constraints for the value of game to be a fuzzy number. Further, a new method (named as Jagdambika method) is proposed to overcome the limitations of existing method and to obtain the solution of matrix games with fuzzy payoffs. To illustrate the proposed Jagdambika method, an existing numerical problem of matrix games with fuzzy payoffs is solved by the proposed Jagdambika method.

1 Introduction

Game theory [3] is a mathematical tool to describe strategic interactions among multiple decision makers who behave rationale. It has many applications in broad areas such as strategic welfares, economic or social problems, animal behavior, political voting systems etc. Although, the concept of game theory was started with Von Neumanns study on zero-sum games [4], in which he proved the famous minimax

T. Verma (✉) · A. Kumar
Indian Institute of Technology Ropar, Rupnagar, Punjab 140001, India
e-mail: verma.tina21@gmail.com

A. Kumar
e-mail: amitkdma@gmail.com

© Springer International Publishing AG, part of Springer Nature 2018
L. A. Zadeh et al. (eds.), *Recent Developments and the New Direction in Soft-Computing Foundations and Applications*, Studies in Fuzziness and Soft Computing 361, https://doi.org/10.1007/978-3-319-75408-6_21

theorem for zero-sum games. It was also basis for the book Theory of Games and Economic Behavior by Von Neumann and Morgenstern [5]. But, Game theory was significantly advanced at Princeton University through the work of Nash [6].

Games are roughly be classified into two major categories: Cooperative games and Non-cooperative games [3]. Cooperation of players may be assumed in games, since it often exists in reality, though in many, if not most cases, the existence of non-cooperation is more attractive because it is often more realistic, especially in the presence of competition between players.

In non-cooperative games, an important, from a conceptual and applications point of view, class of games are matrix games.

In the classical (or crisp) matrix games, usually the payoffs of players are represented by crisp values i.e., real numbers. But in real life, there is need to represent the players' payoffs by their subjective judgments (or opinions) about competitive situations (or outcomes) instead of real numbers.

In the classical (or crisp) matrix games, usually the payoffs of players are represented by crisp values i.e., real numbers. But in real life, there is need to represent the players' payoffs by their subjective judgments (or opinions) about competitive situations (or outcomes) instead of real numbers. For example, the decision problem in which two companies try to improve some products sales in some target market may be regarded as a game problem. In this scenario, the payoffs of players (i.e., companies) are represented by the company managers subjective judgments (or opinions) of the product shares in the target market at various situations. Such subjective judgments may be expressed with terms of linguistic variables such as "very large", "larger", "medium" and "small" as well as "smaller". Obviously, these judgments usually involve some fuzziness or uncertainty due to the bounded rationality of players and behavior complexity. Usually, when some of the data are only approximately known, the most likely values or average is used to find a crisp solution of matrix games. Since, only one value is obtained so, some valuable information is lost sometimes.

One way to describe impreciseness in the payoffs is to represent the payoffs by fuzzy numbers [7].

In the literature, such matrix games in which payoffs are represented by fuzzy numbers are named as matrix games with fuzzy payoffs.

2 Preliminaries

In this section, some basic definitions, arithmetic operations of fuzzy numbers and method for comparing fuzzy numbers is presented [1].

2.1 Definitions

In this section, some basic definitions are reviewed

Definition 1 A fuzzy number $\tilde{A} = (a^L(0), a(1), a^R(0))$ is called a triangular fuzzy number if its membership function $\mu_{\tilde{A}}$ is given by

$$\mu_{\tilde{A}} = \begin{cases} \frac{x-a^L(0)}{a(1)-a^L(0)} & a^L(0) \leq x < a(1) \\ \frac{x-a^R(0)}{a(1)-a^R(0)} & a(1) \leq x < a^R(0) \\ 0 & x < a^L(0), x \geq a^R(0) \end{cases}$$

Further, the α−cut of the triangular fuzzy number $\tilde{A}$ is the closed interval $[a^L(0) + \alpha(a(1) - a^L(0)), a^R(0) + \alpha(a(1) - a^R(0))]$.

Definition 2 A fuzzy number $\tilde{A} = (a^L(0), a^L(1), a^R(1), a^R(0))$ is called a trapezoidal fuzzy number if its membership function $\mu_{\tilde{A}}$ is given by

$$\mu_{\tilde{A}} = \begin{cases} \frac{x-a^L(0)}{a^L(1)-a^L(0)} & a^L(0) \leq x < a^L(1) \\ 1 & a^L(1) \leq x \leq a^R(1) \\ \frac{x-a^R(0)}{a^R(1)-a^R(0)} & a^R(1) < x \leq a^R(0) \\ 0 & x < a^L(0), x > a^R(0) \end{cases}$$

Further, the α−cut of the trapezoidal fuzzy number $\tilde{A}$ is the closed interval $[a^L(0) + \alpha(a^L(1) - a^L(0)), a^R(0) + \alpha(a^R(1) - a^R(0))]$.

Definition 3 A fuzzy number $\tilde{A} = (a^L(0), a^L(1), a^R(1), a^R(0))$ is said to be non-negative if $a^L(0) \geq 0$ and atleast $a^R(0) > 0$.

2.2 Arithmetic Operations of Trapezoidal Fuzzy Numbers

In this section, some arithmetic operations of two trapezoidal fuzzy numbers, defined on universal set of real numbers $\mathfrak{R}$, are presented. Let $\tilde{A} = (a^L(0), a^L(1), a^R(1), a^R(0))$ and $\tilde{b} = (b^L(0), b^L(1), b^R(1), b^R(0))$ be two trapezoidal fuzzy numbers then

1. $\tilde{A} + \tilde{B} = (a^L(0) + b^L(0), a^L(1) + b^L(1), a^R(1) + b^R(1), a^R(0) + b^R(0))$.
2. $\tilde{A} - \tilde{B} = (a^L(0) - b^R(0), a^L(1) - b^R(1), a^R(1) - b^L(1), a^R(0) - b^L(0))$.
3. If λ is a real number then

$$\lambda \tilde{A} = \begin{cases} (\lambda a^L(0), \lambda a^L(1), \lambda a^R(1), \lambda a^R(0)) & \lambda \geq 0 \\ (\lambda a^R(0), \lambda a^R(1), \lambda a^L(1), \lambda a^L(0)) & \lambda < 0 \end{cases}$$

2.3 Comparison of Fuzzy Numbers

If a and b are two distinct real numbers then it can be easily verified that $a > b$ or $a < b$. However, if $\tilde{A}$ and tildeB are two fuzzy numbers then there is no unique way to verify $\tilde{A} > \tilde{B}$ or $\tilde{A} < \tilde{B}$. In this section, the method, used by the authors [1, 2] to compare two trapezoidal fuzzy numbers, is presented. Let $\tilde{A} = (a^L(0), a^L(1), a^R(1), a^R(0))$ and $\tilde{b} = (b^L(0), b^L(1), b^R(1), b^R(0))$ be two trapezoidal fuzzy numbers. Then,

1. $\tilde{A} \geq \tilde{B} \Leftrightarrow a^L(0) \geq b^L(0), a^L(1) \geq b^L(1), a^R(1) \geq b^R(1), a^R(0) \geq b^R(0)$.
2. $\tilde{A} \leq \tilde{B} \Leftrightarrow a^L(0) \leq b^L(0), a^L(1) \leq b^L(1), a^R(1) \leq b^R(1), a^R(0) \leq b^R(0)$.

3 Existing Mathematical Formulation of Matrix Games with Payoffs of Piecewise Linear Fuzzy Numbers

Let Player I and Player II be two players and $A = (a_{ij})$ be the payoff matrix for Player I. Let $S_1 = \{\delta_1, \delta_2, ..., \delta_m\}$ and $S_2 = \{\eta_1, \eta_2, ..., \eta_n\}$ be set of course of actions available to Player I and Player II respectively and $x_1, x_2, ..., x_m$ and $y_1, y_2, ..., y_n$ be the probabilities for selecting the course of action $\delta_1, \delta_2, ..., \delta_m$ and $\eta_1, \eta_2, ..., \eta_n$ respectively. Let $X = \left\{ x = (x_1, x_2, ..., x_m) | \sum_{i=1}^{m} x_i = 1, x_i \geq 0, i = 1, 2, ..., m \right\}$ and $Y = \left\{ y = (y_1, y_2, ..., y_n) | \sum_{j=1}^{n} y_j = 1, y_j \geq 0, j = 1, 2, ..., m \right\}$ be the set of strategies for Player I and Player II respectively. Then, the optimal value of Problem 3.1 and Problem 3.2 represents the minimum expected gain of Player I and maximum expected loss of Player II respectively [3]. Further, the optimal solution $\{x_1, x_2, ..., x_m\}$ and $\{y_1, y_2, ..., y_n\}$ of Problem 3.1 and Problem 3.2 represents the optimal strategies of Player I and Player II respectively.

Problem 3.1 Maximize$\{v\}$ Subject to
$$\sum_{i=1}^{m} a_{ij}x_i \geq v, j = 1, 2, ..., n;$$
$$\sum_{i=1}^{m} x_i = 1;$$
$$x_i \geq 0, i = 1, 2, ..., m.$$

Problem 3.2 Minimize$\{\omega\}$ Subject to
$$\sum_{j=1}^{n} a_{ij}y_j \leq \omega, i = 1, 2, ..., m;$$
$$\sum_{j=1}^{n} y_j = 1;$$
$$y_j \geq 0, j = 1, 2, ..., n.$$

On the same direction, Li [1] assumed that if payoffs are represented by trapezoidal fuzzy numbers instead of real numbers then to find the minimum expected gain of Player I and maximum expected loss of Player II, it is equivalent to find the optimal value of Problem 3.3 and Problem 3.4 respectively. Further, the optimal solution $\{y_1, y_2, ..., y_m\}$ and $\{z_1, z_2, ..., z_n\}$ of Problem 3.3 and Problem 3.4 represents the optimal strategies of Player I and Player II respectively.

Problem 3.3 Maximize$\{\tilde{v}\}$ Subject to

$$\sum_{i=1}^{m} \tilde{a}_{ij}x_i \geq \tilde{v}, j = 1, 2, ..., n;$$

$$\sum_{i=1}^{m} x_i = 1;$$

$$x_i \geq 0, i = 1, 2, ...m.$$

Problem 3.4 Minimize$\{\tilde{\omega}\}$ Subject to

$$\sum_{j=1}^{n} \tilde{a}_{ij}y_j \leq \tilde{\omega}, i = 1, 2, ..., m;$$

$$\sum_{j=1}^{n} y_j = 1; y_j \geq 0, j = 1, 2, ..., n.$$

4 Review of Existing Methods

Li [1] split the Problem 3.3 into four independent crisp problems and obtained the minimum expected gain of Player I with the help of these problems. Similarly, Problem 3.4 into four independent crisp problems and obtained the maximum expected loss of Player II. However, it is pointed out by Chandra and Aggarwal [2], that it is not correct way to transform the fuzzy programming problem i.e., Problem 3.3 into four problems as these are independent problems so different optimal solutions will be obtained. Therefore, it is not genuine to split it in four problems as Li [1] did. So, Problem 3.3 and Problem 3.4 can be transformed into multiobjective linear programming problem i.e., Problem 4.1 and Problem 4.2 respectively.

Problem 4.1 Maximize$\{(v^L(0), v^L(1), v^R(1), v^R(0))\}$
Subject to

$$\sum_{i=1}^{m} a_{ij}^L(0)x_i \geq v^L(0), j = 1, 2, ..., n;$$

$$\sum_{i=1}^{m} a_{ij}^L(1)x_i \geq v^L(1), j = 1, 2, ..., n;$$

$$\sum_{i=1}^{m} a_{ij}^R(1)x_i \geq v^R(1), j = 1, 2, ..., n;$$

$$\sum_{i=1}^{m} a_{ij}^{R}(0)x_i \geq v^{R}(0), j = 1, 2, \ldots, n;$$

$$\sum_{i=1}^{m} x_i = 1; x_i \geq 0, i = 1, 2, \ldots, m.$$

Problem 4.2 Minimize$\{(\omega^{L}(0), \omega^{L}(1), \omega^{R}(1), \omega^{R}(0))\}$
Subject to

$$\sum_{j=1}^{n} a_{ij}^{L}(0)y_j \leq \omega^{L}(0), i = 1, 2, \ldots, m;$$

$$\sum_{j=1}^{n} a_{ij}^{L}(1)y_j \leq \omega^{L}(1), i = 1, 2, \ldots, m;$$

$$\sum_{j=1}^{n} a_{ij}^{R}(1)y_j \leq \omega^{R}(1), i = 1, 2, \ldots, m;$$

$$\sum_{j=1}^{n} a_{ij}^{R}(0)y_j \leq \omega^{R}(0), i = 1, 2, \ldots, m;$$

$$\sum_{j=1}^{n} y_j = 1; y_j \geq 0, j = 1, 2, \ldots n.$$

5   Flaws of the Existing Methods

Chandra and Aggarwal [2] have used the relation $(a^{L}(0), a(1), a^{R}(0)) \geq (b^{L}(0), b(1), b^{R}(0)) \Rightarrow a^{L}(0) \geq b^{L}(0), a(1) \geq b(1), a^{R}(0) \geq b^{R}(0)$, for transforming the fuzzy constraints of Problem 3.3 into crisp constraints of Problem 4.1. As the constraints $v^{L}(0) \leq v(1) \leq v^{R}(0)$ are not considered in Problem 4.1 so for the optimal solution of Problem 4.1, the condition $v^{L}(0) \leq v(1) \leq v^{R}(0)$ may or may not be satisfied i.e., it is not always possible to obtain a triangular fuzzy number, representing the optimal value of Problem 4.1, by using the optimal solution of Problem 4.1. e.g., it can be verified that $v^{L}(0) = \frac{3725}{24}, v(1) = 0, v^{R}(0) = 0, x_1 = \frac{19}{24}, x_2 = \frac{5}{24}$ is an efficient solution of existing Problem 5.1 [1]. However, in the efficient solution the inequalities $v^{L}(0) \leq v(1)$ and $v(1) \leq v^{R}(0)$ are not satisfying i.e., $\tilde{v} = (v^{L}(0), v(1), v^{R}(0)) = (\frac{3725}{24}, 0, 0)$, the value of game is not a triangular fuzzy number.

Problem 5.1 Maximize$\{(v^{L}(0), v(1), v^{R}(0))\}$
Subject to
$175x_1 + 80x_2 \geq v^{L}(0); 150x_1 + 175x_2 \geq v^{L}(0);$
$190x_1 + 100x_2 \geq v^{R}(0); 158x_1 + 190x_2 \geq v^{R}(0);$
$180x_1 + 90x_2 \geq v(1); 156x_1 + 180x_2 \geq v(1);$
$x_1 + x_2 = 1, x_1 \geq 0, x_2 \geq 0.$

6 Proposed Jagdambika Method

In this section, a new method (named as Jagdambika method), is proposed to find minimum expected gain of Player I, maximum expected loss of Player II and the corresponding optimal strategies of Player II. The minimum expected gain of Player I, maximum expected loss of Player II and the corresponding optimal strategies can be obtained as follows:

Step 1: Transform the Problem 4.1 and Problem 4.2 into Problem 6.1 and Problem 6.2 respectively.

Problem 6.1 Maximize$\{v^L(0)\}$
Subject to
$$\sum_{i=1}^{m} a_{ij}^L(0)x_i \geq v^L(0), j = 1, 2, ..., n;$$
$$\sum_{i=1}^{m} a_{ij}^L(1)x_i \geq v^L(1), j = 1, 2, ..., n;$$
$$\sum_{i=1}^{m} a_{ij}^R(1)x_i \geq v^R(1), j = 1, 2, ..., n;$$
$$\sum_{i=1}^{m} a_{ij}^R(0)x_i \geq v^R(0), j = 1, 2, ..., n;$$
$$v^L(1) - v^L(0) \geq 0; v^R(1) - v^L(1) \geq 0; v^R(0) - v^R(1) \geq 0; \quad \sum_{i=1}^{m} x_i = 1; x_i \geq 0, i = 1,$$
$2, ...m.$

Problem 6.2 Maximize$\{\omega^L(0)\}$
Subject to
$$\sum_{j=1}^{n} a_{ij}^L(0)y_j \leq \omega^L(0), i = 1, 2, ..., m;$$
$$\sum_{j=1}^{n} a_{ij}^L(1)y_j \leq \omega^L(1), i = 1, 2, ..., m;$$
$$\sum_{j=1}^{n} a_{ij}^R(1)y_j \leq \omega^R(1), i = 1, 2, ..., m;$$
$$\sum_{j=1}^{n} a_{ij}^R(0)y_j \leq \omega^R(0), i = 1, 2, ..., m;$$
$$\omega^L(1) - \omega^L(0) \geq 0; \omega^R(1) - \omega^L(1) \geq 0; \omega^R(0) - \omega^R(1) \geq 0; \quad \sum_{j=1}^{n} y_j = 1; y_j \geq 0, j = 1,$$
$2, ...n.$

Step 2: Find the optimal solution $\{x_i, v^L(0), v^L(1), v^R(1), v^R(0), i = 1, 2, ..., m\}$ and $\{y_j, \omega^L(0), \omega^L(1), \omega^R(1), \omega^R(0), j = 1, 2, ..., n\}$ of Problem 6.1 and Problem 6.2 respectively.

Step 3: Using the optimal solution of Problem 6.1 and Problem 6.2, obtain the solution $\{x_i, v^L(0), v^L(1), v^R(0), i = 1, 2, ..., m\}$ and $\{y_j, \omega^L(0), \omega^L(1), \omega^R(0)j = 1, 2, ..., n\}$ of Problem 6.3 and Problem 6.4 respectively.

Problem 6.3 Maximize$\{v^L(1)\}$
Subject to
Constraints of Problem 6.1 with additional constraint $v^L(0) =$ optimal value of Problem 6.1.

Problem 6.4 Maximize$\{\omega^L(1)\}$
Subject to
Constraints of Problem 6.1 with additional constraint $\omega^L(0) =$ optimal value of Problem 6.1.

Step 4: Using the optimal solution of Problem 6.3 and Problem 6.4, obtain the solution $\{x_i, v^L(0), v^L(1), v^R(1), i = 1, 2, ..., m\}$ and $\{y_j, \omega^L(0), \omega^L(1), \omega^R(1), j = 1, 2, ..., n\}$ of Problem 6.5 and Problem 6.6 respectively.

Problem 6.5 Maximize$\{v^R(1)\}$
Subject to
Constraints of Problem 6.3 with additional constraint $v^L(1) =$ optimal value of Problem 6.3.

Problem 6.6 Maximize$\{\omega^R(1)\}$
Subject to
Constraints of Problem 6.4 with additional constraint $\omega^L(1) =$ optimal value of Problem 6.4.

Step 5: Using the optimal solution of Problem 6.5 and Problem 6.6, obtain the solution $\{x_i, v^L(0), v^L(1), v^R(1), v^R(0), i = 1, 2, ..., m\}$ and $\{y_j, \omega^L(0), \omega^L(1), \omega^R(1), \omega^R(0), j = 1, 2, ..., n\}$ of Problem 6.7 and Problem 6.8 respectively.

Problem 6.7 Maximize$\{v^R(0)\}$
Subject to
Constraints of Problem 6.5 with additional constraint $v^R(1) =$ optimal value of Problem 6.5.

Problem 6.8 Maximize$\{\omega^R(0)\}$
Subject to
Constraints of Problem 6.6 with additional constraint $\omega^R(1) =$ optimal value of Problem 6.6.

Step 6: Using the optimal solution $\{x_i, v^L(0), v^L(1), v^R(1),$
$v^R(0), i = 1, 2, ..., m\}$ and $\{y_j, \omega^L(0), \omega^L(1), \omega^R(1), \omega^R(0),$
$j = 1, 2, ..., n\}$ of Problem 6.7 and Problem 6.8, the minimum expected gain of
Player I and maximum expected loss of Player II is $(v^L(0), v^L(1), v^R(1), v^R(0))$ and
$(\omega^L(0), \omega^L(1), \omega^R(1), \omega^R(0))$ respectively. The corresponding optimal strategies for
Player I and Player II are $\{x_i^*, i = 1, 2, ..., m\}$ and $\{y_j^*, j = 1, 2, ..., n\}$ which are optimal solution of Problem 6.7 and Problem 6.8 respectively.
Step 7: Similarly, the Pareto optimal solutions can be obtained by changing the order
of the objectives.

7 Numerical Example

In this section, matrix games with fuzzy payoffs
$$\tilde{A} = \begin{bmatrix} (175, 180, 190) & (150, 156, 158) \\ (80, 90, 100) & (175, 180, 190) \end{bmatrix}, \text{ chosen by Li [1], is solved by the proposed}$$
Jagdambika method.
The mathematical programming problem for Player I and Player II will be Problem
7.1 and Problem 7.2 respectively.

Problem 7.1 Maximize$\{(v^L(0), v(1), v^R(0))\}$
Subject to
$(175, 180, 190)x_1 + (80, 90, 100)x_2 \geq (v^L(0), v(1), v^R(0))$;
$(150, 156, 158)x_1 + (175, 180, 190)x_2 \geq (v^L(0), v(1), v^R(0))$;
$x_1 + x_2 = 1; x_1 \geq 0, x_2 \geq 0.$

Problem 7.2 Maximize$\{(\omega^L(0), \omega(1), \omega^R(0))\}$
Subject to
$(175, 180, 190)y_1 + (150, 156, 158)y_2 \leq (\omega^L(0), \omega(1), \omega^R(0))$;
$(80, 90, 100)y_1 + (175, 180, 190)y_2 \leq (\omega^L(0), \omega(1), \omega^R(0))$;
$y_1 + y_2 = 1; y_1 \geq 0, y_2 \geq 0.$
Step 1: Transform the Problem 7.1 and Problem 7.2 into Problem 7.3 and Problem
7.4 respectively.

Problem 7.3 Maximize$\{v^L(0)\}$
Subject to
$175x_1 + 80x_2 \geq v^L(0); 180x_1 + 90x_2 \geq v(1)$;
$190x_1 + 100x_2 x_1 \geq v^R(0); 150x_1 + 175x_2 \geq v^L(0)$;
$156x_1 + 180x_2 \geq v(1); 158x_1 + 190x_2 \geq v^R(0))$;
$x_1 + x_2 = 1; x_1 \geq 0, x_2 \geq 0.$

Problem 7.4 Maximize$\{\omega^L(0)\}$

Subject to

$175y_1 + 150y_2 \leq \omega^L(0); 180y_1 + 156y_2 \leq \omega^{(}1);$
$190y_1 + 158y_2 \leq \omega^R(0); 80y_1 + 175y_2 \leq \omega^L(0);$
$90y_1 + 180y_2 \leq \omega(1); 100y_1 + 190y_2 \leq \omega^R(0);$
$y_1 + y_2 = 1; y_1 \geq 0, y_2 \geq 0.$

Step 2: The obtained solution of Problem 7.3 and Problem 7.4 is $\{x_1 = \frac{19}{24}, x_2 = \frac{5}{24}, v^L(0) = \frac{3725}{24}, v(1) = \frac{3725}{24}, v^R(0) = \frac{3725}{24}\}$ and $\{y_1 = \frac{5}{24}, y_2 = \frac{19}{24}, \omega^L(0) = \frac{3725}{24}, \omega(1) = \frac{645}{4}, \omega^R(0) = \frac{685}{4}\}$ respectively.

Step 3: Using the optimal solution of Problem 7.3 and Problem 7.4, the obtained solution of Problem 7.5 and Problem 7.6 is $\{x_1 = \frac{19}{24}, x_2 = \frac{5}{24}, v^L(0) = \frac{3725}{24}, v(1) = 161, v^R(0) = 161\}$ and $\{y_1 = \frac{5}{24}, y_2 = \frac{19}{24}, \omega^L(0) = \frac{3725}{24}, \omega(1) = \frac{645}{4}, \omega^R(0) = \frac{685}{4}\}$ respectively.

Problem 7.5 Maximize$\{v^L(1)\}$

Subject to

Constraints of Problem 7.3 with additional constraint $v^L(0) = \frac{3725}{24}$.

Problem 7.6 Maximize$\{\omega^L(1)\}$

Subject to

Constraints of Problem 7.4 with additional constraint $\omega^L(0) = \frac{3725}{24}$.

Step 4: Using the optimal solution of Problem 7.5 and Problem 7.6, the obtained solution of Problem 7.7 and Problem 7.8 is $\{x_1 = \frac{19}{24}, x_2 = \frac{5}{24}, v^L(0) = \frac{3725}{24}, v(1) = 161, v^R(0) = \frac{494}{3}\}$ and $\{y_1 = \frac{5}{24}, y_2 = \frac{19}{24}, \omega^L(0) = \frac{3725}{24}, \omega(1) = \frac{645}{4}, \omega^R(0) = \frac{685}{4}\}$ respectively.

Problem 7.7 Maximize$\{v^R(0)\}$

Subject to

Constraints of Problem 7.5 with additional constraint $v(1) = 161$.

Problem 7.8 Maximize$\{\omega^R(0)\}$

Subject to

Constraints of Problem 7.6 with additional constraint $\omega(1) = \frac{645}{4}$.

Step 6: Using the solution $\{x_1 = \frac{19}{24}, x_2 = \frac{5}{24}, v^L(0) = \frac{3725}{24}, v(1) = 161, v^R(0) = \frac{494}{3}\}$ and $\{y_1 = \frac{5}{24}, y_2 = \frac{19}{24}, \omega^L(0) = \frac{3725}{24}, \omega(1) = \frac{645}{4}, \omega^R(0) = \frac{685}{4}\}$ of Problem 7.7 and Problem 7.8, the minimum expected gain of Player I and maximum expected loss of Player II is $(v^L(0), v(1), v^R(0)) = (\frac{3725}{24}, 161, \frac{494}{3})$ and $(\omega^L(0), \omega(1), \omega^R(0)) = (\frac{3725}{24}, \frac{645}{4}, \frac{685}{4})$ respectively. The corresponding optimal strategies for Player I and Player II are $\{x_1 = \frac{19}{24}, x_2 = \frac{5}{24}\}$ and $\{y_1 = \frac{5}{24}, y_2 = \frac{19}{24}\}$ which are optimal solution of Problem 7.7 and Problem 7.8 respectively.

Step 7: The Pareto optimal solutions for minimum expected gain of Player I by changing the order of the objectives are $(v^L(0), v(1), v^R(0)) = (155, \frac{3060}{19}, \frac{3130}{19})$, $(v^L(0), v(1), v^R(0)) = (\frac{9155}{61}, \frac{9540}{61}, \frac{10150}{61})$ and the corresponding optimal strategies are $\{x_1 = \frac{15}{19}, x_2 = \frac{4}{19}\}$, $\{x_1 = \frac{45}{61}, x_2 = \frac{16}{61}\}$ respectively. Similarly, Pareto optimal solutions for maximum expected loss of Player II by changing the order of the objectives are $(\omega^L(0), \omega(1), \omega^R(0)) = (\frac{2950}{19}, \frac{3060}{19}, \frac{3250}{19})$, $(\omega^L(0), \omega(1), \omega^R(0)) = (\frac{9550}{61}, \frac{9900}{61}, \frac{10150}{61})$ and the corresponding optimal strategies are $\{y_1 = \frac{4}{19}, y_2 = \frac{15}{19}\}$, $\{y_1 = \frac{16}{61}, y_2 = \frac{45}{61}\}$ respectively.

8 Conclusion

On the basis of the present study, it can be concluded that some necessary constraints are missing in the existing methods [1, 2] so, the value of a game of matrix games in which payoffs are represented by fuzzy numbers, may or may not be fuzzy number. Also, it can be concluded that in the Jagdambika method, proposed in this paper, necessary constraints are considered and hence, Jagdambika method should be used for solving these type of problems.

Acknowledgements The authors would like to acknowledge the adolescent inner blessings of Mehar (lovely daughter of Dr. Amit Kumar's cousin). They believe that Mata Vaishno Devi has appeared on the earth in the form of Mehar and without her blessings it was not possible to think the ideas presented in this paper. The first author also acknowledge the financial support given to her by Department of Science and Technology under INSPIRE Programme for research students [IF130759] to complete Doctoral studies.

References

1. D.F. Li, An effective methodology for solving matrix games with fuzzy payoffs. IEEE Trans. Cybern. **43**, 610–621 (2013)
2. S. Chandra, A. Aggarwal, On solving matrix games with pay-offs of triangular fuzzy numbers: Certain observations and generalizations. Eur. J. Oper. Res. **246**, 575–581 (2015)
3. G. Owen, *Game Theory*, 2nd edn. (Academic Press, New York, 1982)
4. J. Von Neumann, Zur theorie der Gesellschaftssoiele. Math. Ann. **100**, 295–320 (1928)
5. J. Von Neumann, O. Morgenstern, *Theory of Games and Economic Behavior* (Princeton University Press, 1944)
6. J. Nash, Non-cooperative games. Ann. Math. **54**, 286–295 (1951)
7. L.A. Zadeh, Fuzzy sets. Inf. Control **8**, 338–353 (1965)

Part VII
Fuzzy Control

Structural and Parametric Optimization of Fuzzy Control and Decision Making Systems

Yuriy P. Kondratenko and Dan Simon

Abstract This paper analyzes various methods of structural and parametric optimization for fuzzy control and decision-making systems. Special attention is paid to hierarchical structure selection, rule base reduction, and reconfiguration in the presence of incomplete data sets. In addition fuzzy system parameter optimization based on gradient descent, Kalman filters, H-infinity filters, and maximization of envelope curve values, are considered for unconstrained and constrained cases. Simulation results show the validity of the proposed methods.

1 Introduction

Many systems require the automation of decision making and control processes for the efficient functioning of complex coupled objects in the presence of uncertainty. Such systems include mobile robots, marine objects, economic enterprises, and others. These systems include a sparsity of information, and a corresponding need for methods to decrease the degree and impact of uncertainty. Factors that impact uncertainty are: (a) non-stationary disturbances with characteristics that are not possible to measure in real time; (b) difficulties in creating accurate mathematical models; (c) human factors and the subjectivity of human evaluations and decisions.

One popular approach to the design of efficient control and decision-making systems in uncertain environments is fuzzy sets and fuzzy logic, first suggested by L. A. Zadeh [29] as a control method based on linguistic rules. Fuzzy logic allows the formation of linguistic models of processes and control methods based on

Y. P. Kondratenko (✉)
Department of Intelligent Information Systems, Petro Mohyla Black
Sea National University, Mykolaiv 54003, Ukraine
e-mail: yuriy.kondratenko@chmnu.edu.ua; y_kondrat2002@yahoo.com

Y. P. Kondratenko · D. Simon
Department of Electrical Engineering and Computer Science,
Cleveland State University, Cleveland, OH 44115, USA
e-mail: d.j.simon@csuohio.edu

© Springer International Publishing AG, part of Springer Nature 2018
L. A. Zadeh et al. (eds.), *Recent Developments and the New Direction in Soft-Computing Foundations and Applications*, Studies in Fuzziness and Soft Computing 361, https://doi.org/10.1007/978-3-319-75408-6_22

human experts. New applications of fuzzy sets and fuzzy logic often require new theoretical investigations, new approaches for the optimisation of fuzzy systems, and new design methods for the hardware realisation of fuzzy systems, all while taking into account requirements related to embedded systems and real-time decision making.

2 Related Works and Problem Statement

Many examples of fuzzy systems (FS) applications are given in the literature [2, 7, 15]: control of asynchronous, direct current, and thermoacoustic drives, vehicles, ships, ecopyrogenesis plants, intelligent robots; decision making in uncertainty, including route planning in transport logistics and intelligent robotics; and many others.

Approaches to the optimization of designing fuzzy controllers (FC) are considered in [2, 8, 14], especially for the optimization of membership function (MF) parameters of linguistic terms (LT) [8, 18, 25], weights of fuzzy rules [18], selection of defuzzification methods [17, 18], etc. Special attention is paid to fuzzy rule base reduction [5, 17] based on rule base interpolation [6], supervised fuzzy clustering [19], combining rule antecedents [5], linguistic 3-tuple representation [1], orthogonal transforms [20], multi-objective optimization [3], and evolutionary algorithms [16, 26]. Publications [4, 15, 27, 30] show that researchers continue to develop approaches for structural and parametric optimization (SPO). Recent research deals with applications of FS, design requirements, increasing levels of uncertainty (incomplete input data, random disturbances, and unknown parameters), etc. Choosing specific methods for FS SPO is in most cases based on the analysis of comparative modeling results and the designer's experience.

The aim of this paper is to provide an overview of some of the proposed methods of FS structural and parametric optimization and their abilities for increasing the efficiency of real-time control and decision making processes.

3 Optimization of Fuzzy Systems

3.1 Structural Optimization of Fuzzy Systems

Structural optimization should be applied at different stages of the design of a fuzzy logic system [8]:

- selection of the type of fuzzy inference engine (Mamdani, Takagi-Sugeno, modifications, etc.);

- selection of the most informative inputs; for example, the type of FC (PI, PD, PID, etc.) depends on the input signals: error, integral, derivative, specific measured disturbances, etc.;
- selection of MF the number K and H of LTs $L_{x_i} = \left\{ L_{x_i}^1 \ldots L_{x_i}^K \right\}$ for input signals $x_i (i = 1 \ldots m)$, and LTs $L_G = \left\{ L_G^1 \ldots L_G^H \right\}$ for output signals G, where K and H affect the flexibility and memory of the FS, as well as the computation time during fuzzy information processing;
- optimization and reduction of the rules number in rule base $\{ R_1(\nu_1) \ldots R_R(\nu_R) \}$ and the weights $\nu_j (\nu_j \in [0, 1])$; for example, if $m = 2$, then the j-th rule $R_j(\nu_j), (j \in [1, R])$ can be written as

$$\textbf{IF } (x_1 = L) \textbf{ AND } (x_2 = H) \textbf{ THEN } \left(G_j = LM, \right) \tag{1}$$

where L, H, LM are "Low", "High", and "Low-Medium";

- selection of fuzzy processing algorithms for aggregation, accumulation, and defuzzification; for example: (a) the AND operator can be realized by t-norm operators or by parameterized mean operators [17]; (b) the OR operator can be realized by the parameterized union mean operator, algebraic sum operator, arithmetic sum operator, or by s-norm operators [17]; (c) different methods [17, 18] can be used for defuzzification $G^* = Defuz \left(\mu_{res} \left(\underset{\sim}{G} \right) \right)$ of the fuzzy set $\mu_{res} \left(\underset{\sim}{G} \right) = \underset{G \in R^+}{\sup} \left\{ \mu_{L_G^1} \left(\underset{\sim}{G} \right) \ldots \mu_{L_G^H} \left(\underset{\sim}{G} \right) \right\}$.

All the above-mentioned structural types, algorithms, operators, and analytic models are candidates for structural optimization in fuzzy control and decision-making systems.

3.2 Parametric Optimization of Fuzzy Systems

Structural and parametric FS optimization can be done on the basis of minimization of the RMS criterion, for example,

$$J(\varepsilon, \bar{B}, \bar{\nu}) = \left(\sum_{i=0}^{N_{max} - 1} (\varepsilon[i])^2 \right) / T_{max}, \tag{2}$$

where $\varepsilon[i]$ is the error of the control system at time $t_i, t_i = i \cdot \Delta t$, $(i = 0 \ldots N_{max}); T_{max} = \Delta t \cdot N_{max}; \bar{B}$ is a matrix of LT parameters; and $\bar{\nu}$ is a vector of

Fig. 1 Fuzzy numbers: 1—
triangular
$C = (c - b^-, c, c + b^+)$; 2,3
—bell-shape from Eq. (4)
with various parameters of
MFs

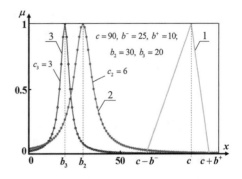

rule weights. Objective (2) can be used for evaluation of FS under the different
kinds of non-stationary disturbances and different system parameters to achieve
robustness. Some FS parameters that can be optimized task include the following.

(a) MF parameters of LTs $L_{x_i} = \left\{ L_{x_i}^1 \dots L_{x_i}^K \right\}$ for FS inputs [21, 23, 24] (triangular
   (3), trapezoidal, Gaussian, polynomial, or harmonic [17]). For example, three
   parameters (b^-, c, b^+) are optimized (Fig. 1) for triangular LTs
   $(c - b^-, c, c + b^+)$ [7, 17]:

$$\mu_{\underset{\sim}{A}}(x) = \begin{cases} 0, \forall (x \leq c - b^-) & \cup (x \geq c + b^+) \\ (x - c + b^-)/(b^-), & \forall (c - b^- < x \leq c) \\ (c + b^+ - x)/(b^+), & \forall (c < x < c + b^+) \end{cases} \tag{3}$$

Two parameters (b, c) are optimized for bell-shaped LTs [18]:

$$\mu_{\underset{\sim}{A}}(x) = 1 / \left(1 + ((x - b)/c)^2 \right) \tag{4}$$

(b) MF parameters of LTs $L_G = \left\{ L_G^1 \dots L_G^H \right\}$ for FS outputs G (for Mamdani-type
   FS) and parameters $\bar{K}^j = \left(k_P^j, k_I^j, k_D^j \right)$ of consequents $G_j \left(x_1, x_2, x_3, \bar{K}^j \right)$,
   $j \in [1 \dots R]$ (Takagi-Sugeno FS). For example, the contribution G_j of rule R_j to
   output signal G^* of a fuzzy PID controller [8] can be presented as:

$$G_j \left(x_1, x_2, x_3, \bar{K}^j \right) = k_P^j x_1 + k_I^j x_2 + k_D^j x_3 \tag{5}$$

(c) Weight coefficients ν_j of the fuzzy rules R_j [17, 18].

4 Hierarchical Systems and Incomplete Input Data

4.1 Rational Selection of the Hierarchical Structure

For multi-input FS it is appropriate to use a hierarchical structure, where the output of a subsystem is supplied to the input of another subsystem. Such a hierarchical approach allows reduction of the size of the FS rule base and an increase in the sensitivity to the input variables [9, 17]. The design process for a hierarchical FS depends on the selection of its structure (grouping, number of hierarchical levels, etc.).

Let us detail the main steps of the selection of the structure.

Step 1. Synthesis of the set $\bar{D}$ of alternative variants $D_i, (i = 1 \ldots n)$ for different groupings of input signals $x_j, (j = 1 \ldots m)$:

$$\bar{D} = \{D_1, D_2, \ldots, D_i, \ldots, D_n\} \tag{6}$$

For example, for a fuzzy system with 9 input signals (Fig. 2a) the set $\bar{D}$ of alternatives can be presented as $\bar{D} = \{D_1, D_2, D_3\}$, where

$$D_1 = (\{x_1, x_2, x_3\}, \{x_4, x_5, x_6\}, \{x_7, x_8, x_9\}) \tag{7}$$

$$D_2 = (\{x_1, x_2\}, \{x_3, x_4, x_5\}, \{x_6, x_7, x_8, x_9\}) \tag{8}$$

$$D_3 = (\{x_1, x_2, x_3, x_4\}, \{x_5, x_6, x_7, x_8\}, \{x_9\}) \tag{9}$$

Step 2. Synthesis of the set $\bar{St}$ of alternative structures $St_i(D_i), (i = 1 \ldots n)$ of the hierarchical FS based on (6):

$$\bar{St} = \{St_1(D_1), St_2(D_2), \ldots, St_i(D_i), \ldots, St_n(D_n)\} \tag{10}$$

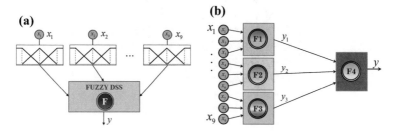

Fig. 2 FS structures with nine input signals. **a** Single-level FS; **b** hierarchical FS

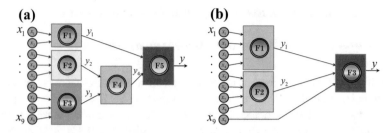

Fig. 3 Alternative structures of hierarchical fuzzy systems: **a** $St_2(D_2)$ based on (8); **b** $St_3(D_3)$ based on (9)

For example, three different structures of a hierarchical FS, based on (7), (8), and (9), are presented in Figs. 2b and 3.

Step 3. Evaluation of each alternative structure $St_i(D_i)$ using some suitable criterion $K(St_i), i = \{1, \ldots, n\}$, which quantifies the goodness of the FS output signals (accuracy, response time, etc.). For example, the structure of an FS for gait intent recognition during human ambulation can be evaluated by the number of correctly classified cases $K(St_i)$.

Step 4. The selection of the optimal configuration St_{opt} of a fuzzy system by solving the optimization problem

$$St_{opt} = Arg \underset{St_i}{Max} K(St_i), (i = 1 \ldots n). \tag{11}$$

Figure 4a presents the hierarchical structure of an FS for transport logistics [12] and Fig. 4b presents a FS for model-oriented support of university / IT-company cooperation [9], with 19 and 27 inputs, respectively.

The hierarchical FS of Fig. 4b with discrete outputs (seven alternative models) consists of 11 subsystems:

$$St_s = \begin{cases} y_1 = f_1(x_1, x_2, x_3), y_2 = f_2(x_4, x_5, x_6, x_7), \\ y_3 = f_3(x_8, \ldots, x_{13}), y_4 = f_4(x_{14}, \ldots, x_{17}), \\ y_5 = f_5(x_6, x_{18}, x_{19}), y_6 = f_6(x_{18}, \ldots, x_{23}), \\ y_7 = f_7(x_{24}, x_{25}, x_{26}, x_{27}), y_8 = f_8(y_1, y_2), \\ y_9 = f_9(y_3, y_4), y_{10} = f_{10}(y_5, y_6), \\ y = f_{11}(y_7, y_8, y_9, y_{10}) \end{cases}. \tag{12}$$

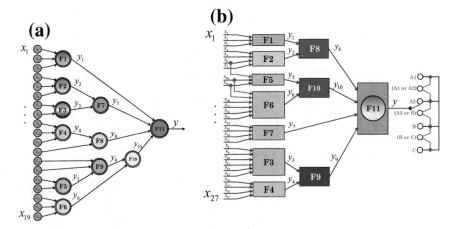

Fig. 4 Hierarchical structures of FSs for **a** transport logistics and **b** "university-IT-company" cooperation

4.2 Fuzzy System Reconfiguration with Incomplete Data

FS reconfiguration is required in the case of incomplete input data [11]. In some cases some input data may not be important according to human judgment. Suppose r is the set of informative inputs, and NI is the set of uninformative or uninteresting inputs. For example, if an FS operates with 16 inputs ($N = 16$), but we have data on only 11 input signals ($N_r = 11$), we exclude the other 5 inputs ($N_{NI} = 5$), so $N_r + N_{NI} = N$, and the dimension of the FS input vector decreases from 16 to 11. The FS rule base can be reconfigured [11] according to the following steps for a Mamdani-type FS.

Step 1. Automatic exclusion of all input variables $x_i \in NI$, $i \in \{1, 2, \ldots, N\}$ from the antecedents of all FS rules.

Step 2. Determination of the input dimension N_r for the newly-reconfigured antecedents: $N_r = N - N_{NI}$.

Step 3. Calculation of $Quant_j$ for each fuzzy rule:

$$Quant_j = \sum_{k=1}^{N_r} eval(ILT_{jk}), \tag{13}$$

where $eval(ILT_{jk})$ corresponds to the number of input linguistic terms ILT_{jk} for the k-th input signal x_k (Fig. 5) of the j-th fuzzy rule; $eval(ILT_{jk}) \in \{1, 2, \ldots, N_{LT_k}\}$. For example, if $ILT_{jk} \in \{L, M, H\}$, then $N_{LT_k} = 3$ and $eval(L) = 1$; $eval(M) = 2$; $eval(H) = 3$.

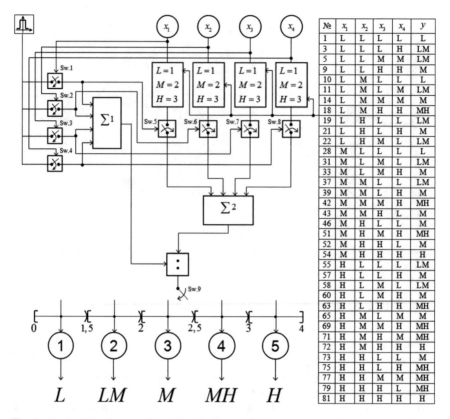

№	x_1	x_2	x_3	x_4	y
1	L	L	L	L	L
3	L	L	L	H	LM
5	L	L	M	M	LM
9	L	L	H	H	M
10	L	M	L	L	L
11	L	M	L	M	LM
14	L	M	M	M	M
18	L	M	H	H	MH
19	L	H	L	L	LM
21	L	H	L	H	M
22	L	H	M	L	LM
28	M	L	L	L	L
31	M	L	M	L	LM
33	M	L	M	H	M
37	M	M	L	L	LM
39	M	M	L	H	M
42	M	M	M	H	MH
43	M	M	H	L	M
46	M	H	L	L	M
51	M	H	M	H	MH
52	M	H	H	L	M
54	M	H	H	H	H
55	H	L	L	L	LM
57	H	L	L	H	M
58	H	L	M	L	LM
60	H	L	M	H	M
63	H	L	H	H	MH
65	H	M	L	M	M
69	H	M	M	H	MH
71	H	M	H	M	MH
72	H	M	H	H	H
73	H	H	L	L	M
75	H	H	L	H	MH
77	H	H	M	M	MH
79	H	H	H	L	MH
81	H	H	H	H	H

Fig. 5 An algorithm for reconfiguration of a fuzzy rule base

Step 4. Calculation of *Div* (Fig. 5):

$$Div = Quant_j/N_r = U_{out}(\textstyle\sum 2)/U_{out}(\textstyle\sum 1) \qquad (14)$$

which determines the new output linguistic term OLT_j in the consequent of the *j*-th rule. For example, if (as in Fig. 5) $OLT_j \in \{L, LM, M, MH, H\}$ and $Div \subset \{Int_1, Int_2, \ldots, Int_5\}$, then $OLT_j = L$, if $Div \subset Int_1$; $\ldots;OLT_j = H$, if $Div \subset Int_5$, where $Int_1 = [0, 1.5)$; $Int_2 = [1.5, 2.0)$; $Int_3 = [2.0, 2.5)$; $Int_4 = [2.5, 3.0)$; $Int_5 = [3.0, 4.0]$.

The application of the proposed method to the reconfiguration of an FS for transport logistics with various numbers of input signals confirms its effectiveness [11, 12].

5 Fuzzy Rule Base Reduction

5.1 Rule Base Reduction via Singular Value Decomposition

The reduction of the size of a fuzzy rule base via singular value decomposition (SVD) [28] is an efficient method of FS optimization. One of its successful implementations is rule base reduction of a fuzzy filter for the estimation of motor currents [23]. For a FS with a single output r and two inputs (a, b) with a corresponding number of LTs n_a, n_b, the fuzzy rule base can be described using the following rules:

$$IF f_{i1}(a) \; AND f_{j2}(b), \; THEN \; r_{ij}, \; (i = 1 \ldots n_a; j = 1 \ldots n_b) \tag{15}$$

which can be represented with the $(n_a \times n_b)$ matrix

$$R = \begin{bmatrix} r_{11} & r_{12} & \cdots & r_{1n_b} \\ r_{21} & r_{22} & \cdots & r_{2n_b} \\ . & . & \cdots & . \\ r_{n_a1} & r_{n_a2} & \cdots & r_{n_an_b} \end{bmatrix}. \tag{16}$$

The SVD of R can be represented as

$$R = U \Sigma V^T, \tag{17}$$

where U is $n_a \times n_a$, V is $n_b \times n_b$ and the singular values in Σ indicate the relative importance of the corresponding columns in U and V in the decomposition of R [22]. Initially seven LTs are used for two inputs $(f_{i1}(a), f_{j2}(b))$ and one output signal, so $n_a = n_b = 7$ in the fuzzy filter [23]. The initial rule base with 49 fuzzy rules is reduced to 9 fuzzy rules after the implementation of SVD. The reduced MFs for the first input a, the second input b, and the output are presented in Fig. 6.

5.2 Rule Base Reduction Based on the Evaluation of Each Rule Contribution

Reduction based on the evaluation of each rule contribution to the FS output signal is proposed [10] for the optimization of the rule base of a Sugeno-type PID FC with 27 initial fuzzy rules (Fig. 7). The minimal number of rules can be determined to maintain control quality within acceptable limits.

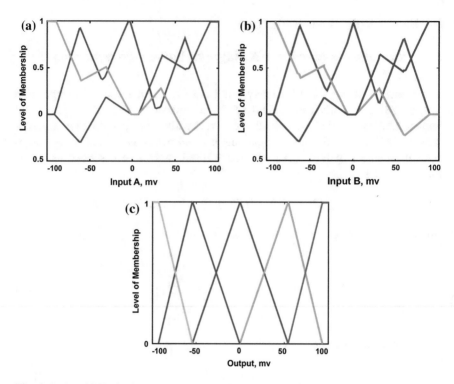

Fig. 6 Reduced MFs for input *a* (**a**), input *b* (**b**), and output (**c**)

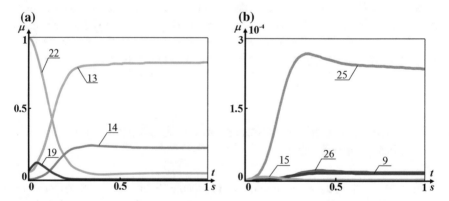

Fig. 7 Evaluation of fuzzy rule contributions $\mu(t)$: **a** rules 13, 14, 19, 22—$\mu(t) \in [0.1, 1]$ (large contribution); **b** rules 9, 15, 25, 26—$\mu(t) \in [0, 0.00027]$ (small contribution)

6 Parametric Optimization of Linguistic and Analytical Components in Fuzzy Rule Base Design

6.1 Gradient Descent Based on Sum Normal Constraints

One of the most popular methods for FS parameter optimization is gradient descent (GD) [18, 24]. The following iterative procedure can be used for optimization of the three parameters as c, b^- and b^+ of a triangular MF:

$$c(k+1) = c(k) - \eta \frac{\partial E}{\partial c}\bigg|_{c(k)}, \tag{18}$$

$$b^-(k+1) = b^-(k) - \eta \frac{\partial E}{\partial b^-}\bigg|_{b^-(k)} \tag{19}$$

$$b^+(k+1) = b^+(k) - \eta \frac{\partial E}{\partial b^+}\bigg|_{b^+(k)}, \tag{20}$$

where k is the iteration number, E is the FS output error function, and η is a gradient descent step size. E can be formed as

$$E = \left(\sum_{n=1}^{N} g_n (y_n - \hat{y}_n)^2 \right)/2N, \tag{21}$$

where $y_n, \hat{y}_n$ are the target and actual outputs of the FS; g_n is a time-dependent weighting function depending on user preference; and N is the number of training samples. Various methods can be used with gradient descent for avoiding local minima. One problem that arises with this method is that the family of optimized MFs are not sum normal; that is, the resulting MF values do not add up to 1 at each point in the FS domain. In [24] the author proposes to use gradient descent with additional constraints to enforce sum normality:

$$c_1 + b_1^+ = c_1 + b_2^- = c_2; \; c_2 + b_2^+ = c_2 + b_3^- = c_3; \; \ldots; \tag{22}$$

$$c_{v-1} + b_{v-1}^+ = c_{v-1} + b_v^- = c_v \tag{23}$$

where v is the number of LTs. This approach provides sum normal parametric optimization of triangular MFs for corresponding input and output LTs:

$$\left\{ (c_1 - b_1^-, c_1, c + b_1^+), \ldots, (c_i - b_i^-, c_i, c_i + b_i^+), \ldots \right\} \tag{24}$$

6.2   Kalman Filtering for Parametric MF Optimization

The parameter optimization problem can be formulated as a nonlinear filtering problem which can be solved using a Kalman filter (KF) or an H_∞ filter (HiF). The nonlinear system model to which the filter can be applied is:

$$\mathbf{x}_{n+1} = \mathbf{f}(\mathbf{x}_n) + \mathbf{w}_n$$
$$\mathbf{d}_n = \mathbf{h}(\mathbf{x}_n) + \mathbf{v}_n, \tag{25}$$

where $\mathbf{d}_n$ is the observation vector; $\mathbf{f}(\mathbf{x}_n), \mathbf{h}_n(\mathbf{x}_n)$ are non-linear vector functions of the state $\mathbf{x}_n$ at time step n; $\mathbf{w}_n, \mathbf{v}_n$ are artificially added noise processes. The use of a state estimator for optimization of triangular MFs requires the formation of the state vector $\mathbf{x}$, which consists of all MF parameters arranged in a column vector:

$$\mathbf{x} = \begin{bmatrix} b_{11}^- \ b_{11}^+ \ c_{11} \ b_{21}^- \ b_{21}^+ \ c_{21} & \dots & b_{v_11}^- \ b_{v_11}^+ \ c_{v_11} \ \dots \\ b_{1r}^- \ b_{1r}^+ \ c_{1r} \ b_{2r}^- \ b_{2r}^+ \ c_{2r} & \dots & b_{v_r r}^- \ b_{v_r r}^+ \ c_{v_r r} \\ b_{10}^- \ b_{10}^+ \ c_{10} \ b_{20}^- \ b_{20}^+ \ c_{20} & \dots & b_{v_00}^- \ b_{v_00}^+ \ c_{v_00} \end{bmatrix}^T, \tag{26}$$

where b_{ij}^-, b_{ij}^+, c_{ij} are the parameters of the i-th triangular LT for the j-th input, $(i = 1 \dots v_j; j = 1 \dots r)$;   and $b_{i0}^-, b_{i0}^+, c_{i0}$ are   the   output LT parameters $(i = 1 \dots v_0; j = 0)$. The estimate $\hat{\mathbf{x}}_n$ can be obtained using the extended Kalman filter [21, 22]:

$$F_n = \frac{\partial \mathbf{f}(x)}{\partial \mathbf{x}}\bigg|_{\mathbf{x}=\mathbf{x}_n}, \; H_n = \frac{\partial \mathbf{h}(x)}{\partial \mathbf{x}}\bigg|_{\mathbf{x}=\mathbf{x}_n},$$
$$K_n = P_n H_n^T \left(R + H_n P_n H_n^T\right)^{-1},$$

$$\hat{\mathbf{x}}_n = \mathbf{f}(\hat{\mathbf{x}}_{n-1}) + K_n[\mathbf{d}_{n-1} - \mathbf{h}(\hat{\mathbf{x}}_{n-1})],$$
$$P_{n+1} = F_n(P_n - K_n H_n P_n)F_n^T + Q, \tag{27}$$

where F_n is the identity matrix; H_n is the partial derivative of the fuzzy output with respect to the MF parameters; K_n is the Kalman gain; and Q, R are the covariance matrices of $\{\mathbf{w}_n\}$ and $\{\mathbf{v}_n\}$, respectively. Comparative results of the parametric optimization for a fuzzy automotive cruise control [21, 24] are presented in Fig. 8 for unconstrained and sum normal (constrained) cases.

6.3   Parametric Optimization Based on H_∞ estimation

If we consider a two-input, one-output fuzzy system, then the nonlinear model to which the H_∞ filter can be applied is

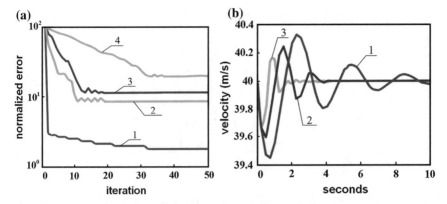

Fig. 8 MF optimization: **a** gradient descent and Kalman filter optimization (1—unconstrained KF, 2—constrained KF, 3—unconstrained GD, 4—constrained GD); **b** FC transients for the target speed 40 m/s (1—default FC, 2—GD, 3—KF)

$$\mathbf{x}_{n+1} = \mathbf{x}_n + B\mathbf{w}_n + \delta_n$$
$$\mathbf{d}_n = \mathbf{h}(\mathbf{x}_n) + \mathbf{v}_n, \tag{28}$$

where B is a tuning parameter which is proportional to the magnitude of the artificial noise process; and δ_n is an arbitrary noise sequence. The augmented noise vector $\mathbf{e}_n$ and the estimation error $\tilde{\mathbf{x}}_n$ can be defined [25] as

$$\mathbf{e}_n = \begin{bmatrix} \mathbf{w}_n^{\mathrm{T}} & \mathbf{v}_n^{\mathrm{T}} \end{bmatrix}^{\mathrm{T}}, \; \tilde{\mathbf{x}}_n = \mathbf{x}_n - \hat{\mathbf{x}}_n. \tag{29}$$

An H_∞ filter for optimization of MF parameters (26) $\hat{\mathbf{x}}_n$ satisfies the condition

$$\|G_{\tilde{\mathbf{x}}\mathbf{e}}\|_\infty < \gamma, \tag{30}$$

which is the infinity norm of the transfer function from the augmented noise vector $\mathbf{e}$ to the estimation error $\tilde{\mathbf{x}}$, which is bounded by a user-specified value γ. The desired estimate $\hat{\mathbf{x}}_n$ can be obtained [25] with an H_∞ estimator:

$$F_n = \left.\frac{\partial \mathbf{f}(x)}{\partial \mathbf{x}}\right|_{\mathbf{x}=\mathbf{x}_n}, \; H_n = \left.\frac{\partial \mathbf{h}(x)}{\partial \mathbf{x}}\right|_{\mathbf{x}=\mathbf{x}_n}, \; Q_0 = E\left(\mathbf{x}_0\mathbf{x}_0^{\mathrm{T}}\right),$$

$$Q_n\left(I - H^{\mathrm{T}}HP_n\right) = \left(I - Q_n/\gamma^2\right)P_n,$$
$$Q_{n+1} = FP_nF^{\mathrm{T}} + BB^{\mathrm{T}}, K_n = FP_nH^{\mathrm{T}},$$

$$\hat{\mathbf{x}}_{n+1} = F\hat{\mathbf{x}}_n + K_n[\mathbf{d}_{n-1} - H(\hat{\mathbf{x}}_{n-1})], \tag{31}$$

with the assumption that $\{Q_n\}$ and $\{P_n\}$ are nonsingular matrices. Comparative results for the optimization of MF parameters of a fuzzy automotive cruise control

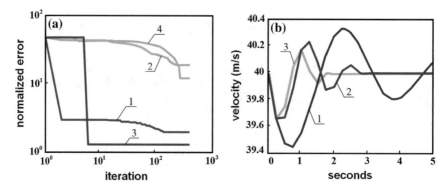

Fig. 9 MF optimization: **a** state estimation processes (1—unconstrained KF, 2—constrained KF, 3—unconstrained HiF, 4—constrained HiF) and **b** control transients (1—default FC, 2—KF, 3—HiF)

system [21, 25], using H_∞ and Kalman filters are presented in Fig. 9 for both unconstrained and constrained (sum normal) cases.

6.4   The Envelope Curve of Instantaneous MF Values

Analyzing (for desired transients of a control system) the instantaneous values of MF grades $\mu_i(\mathbf{X}(t)), (i = 1 \ldots R)$ at the consequents of all FC rules makes it is possible to build a corresponding envelope curve (Fig. 10a). If the vector of instantaneous inputs is $\mathbf{X}(t) = \{x_1^*(t), x_2^*(t), \ldots, x_n^*(t)\}$, then the envelope curve characterizes the instantaneous maximal value of the MF grade $\mu_{max}(\mathbf{X}(t))$ at time t. Figure 10a shows that in some cases this maximal value $\mu_{max}(\mathbf{X}(t)) < 1$, but in some cases, for example, for time-intervals $t \in [t_1, t_2]$ and $t \in [t_3, t_4]$, this value does

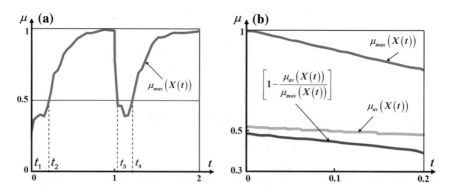

Fig. 10   a The envelope curve; and **b** $C(\mathbf{K})$ from Eq. (32)

not exceed 0.5. We introduce the criterion $C(\mathbf{K})$ for solving the MF parameter optimization problem [13] (Fig. 10b):

$$C(\mathbf{K}) = \frac{1}{T} \int_0^T \left[1 - \frac{\mu_{av}(\mathbf{X}(t))}{\mu_{max}(\mathbf{X}(t))} \right] dt, \tag{32}$$

where $\mu_{max}(\mathbf{X}(t)) = \overset{m}{\underset{i=1}{\sup}}[\mu_i(\mathbf{X}(t))]$ is the maximal value of the MF grade for all fuzzy rules; $\mu_{av}(\mathbf{X}(t)) = \left(\sum_{i=1}^m [\mu_i(\mathbf{X}(t))] \right)/m$ is the average value of $\mu_i(\mathbf{X}(t))$; and $\mathbf{K}$ is a vector of optimized MF parameters. If we maximize the criterion $\mathrm{MAX}C(\mathbf{K})$ we can achieve $\mu_{\lim} < \mu_{max}(\mathbf{X}(t)) \le 1$ and $\mu_{max}(\mathbf{X}(t)) \gg \mu_{av}(\mathbf{X}(t))$, where $\mu_{\lim}$ is the minimum acceptable value.

7 Conclusion

Various solutions to decrease memory and computational time requirements in FS require new optimisation methods at all stages of the design processes. The SPO methods discussed in this paper can be recommended for the design of fuzzy control and decision-making systems, in particular for ship docking, mobile robotics, transport logistics, etc. The authors' experience in fuzzy logic as well as numerous simulation and experimental results confirm the effectiveness of the proposed methods, algorithms, and approaches. In the future we plan to apply FS multi-objective optimization based on the combination of these methods and evolutionary algorithms.

Acknowledgements This work is partially supported by NSF Grant # 1344954 and Fulbright Program (USA).

References

1. R. Alcalá, J. Alcalá-Fdez, M.J. Gacto, F. Herrera, Rule base reduction and genetic tuning of fuzzy systems based on the linguistic 3-tuples representation. Soft. Comput. **11**(5), 401–419 (2007)
2. D. Driankov, H. Hellendoorn, M. Reinfrank, *An introduction to fuzzy control* (Springer Science & Business Media, 2013)
3. H. Ishibuchi, T. Yamamoto, Fuzzy rule selection by multi-objective genetic local search algorithms and rule evaluation measures in data mining. Fuzzy Sets Syst. **141**(1), 59–88 (2004)
4. M. Jamshidi, V. Kreinovich, J. Kacprzyk (eds.), *Advance Trends in Soft Computing* (Springer, Cham, 2013)
5. B. Jayaram, Rule reduction for efficient inferencing in similarity based reasoning. Int. J. Approximate Reasoning **48**(1), 156–173 (2008)

6. L.T. Koczy, K. Hirota, Size reduction by interpolation in fuzzy rule bases. IEEE Trans. Syst. Man Cybern. B Cybern. **27**(1), 14–25 (1997)

7. G.V. Kondratenko, Y.P. Kondratenko, D.O. Romanov, Fuzzy models for capacitive vehicle routing problem in uncertainty, in *Proceedings 17th International DAAAM Symposium on "Intelligent Manufacturing and Automation: Focus on Mechatronics & Robotics"*, 2006, pp. 205–206

8. Y.P. Kondratenko, E.Y.M. Al Zubi, The optimization approach for increasing efficiency of digital fuzzy controllers, in *Annals of DAAAM for 2009 & Proceeding of the 20th Inernational DAAAM Symposium on Intelligent Manufacturing and Automation*, 2009, pp. 1589–1591

9. Y.P. Kondratenko, G.V. Kondratenko, Ie.V. Sidenko, V.S. Kharchenko, Cooperation models of universities and IT-companies: decision-making systems based on fuzzy logic, in *Kharkiv: NASU "KhAI"*, ed. by Y.P. Kondratenko (2015) (in Ukrainian)

10. Y.P. Kondratenko, L.P. Klymenko, E.Y.M. Al Zu'bi, Structural optimization of fuzzy systems' rules base and aggregation models. Kybernetes **42**(5), 831–843 (2013)

11. Y.P. Kondratenko, Ie.V. Sidenko, Method of actual correction of the knowledge database of fuzzy decision support system with flexible hierarchical structure, in *Computational Techniques in Modeling and Simulation*, ed. by V. Krasnoproshin, A.M. Gil Lafuente, C. Zopounidis (Nova Science Publishers, New York, 2013), pp. 55–74

12. Y.P. Kondratenko, S.B. Encheva, E.V. Sidenko, Synthesis of inelligent decision support systems for transport logistic, in *Proceedings of 6th IEEE International Conference on Intelligent Data Acquisition and Advanced Computing Systems: Technology and Applications*, vol. 2 (2011), pp. 642–646

13. Y.P. Kondratenko, S.A. Sydorenko, Multi-objective optimization of embedded computer components of fuzzy control systems, *Technical News*, no. 1(29), 2(30), 2009, pp. 98–101. (in Ukrainian)

14. W.A. Lodwick, J. Kacprzhyk (eds.), *Fuzzy Optimization*, STUDFUZ 254 (Springer, Berlin, Heidelberg, 2010)

15. J.M. Merigo, A.M. Gil-Lafuente, R.R. Yager, An overview of fuzzy research with bibliometric indicators. Appl. Soft Comput. **27**, 420–433 (2015)

16. W. Pedrycz, K. Li, M. Reformat, *Evolutionary reduction of fuzzy rule-based models, in Fifty Years of Fuzzy Logic and its Applications, STUDFUZ 326* (Springer, Cham, 2015), pp. 459–481

17. A. Piegat, *Fuzzy Modeling and Control*, vol. 69 (Physica, 2013)

18. A.P. Rotshtein, H.B. Rakytyanska, *Fuzzy Evidence in Identification, Forecasting and Diagnosis*, vol. 275 (Springer, Heidelberg, 2012)

19. M. Setnes, Supervised fuzzy clustering for rule extraction. IEEE Trans. Fuzzy Syst. **8**(4), 416–424 (2000)

20. M. Setnes, R. Babuška, Rule base reduction: some comments on the use of orthogonal transforms. IEEE Trans. Syst. Man Cybern. Part C Appl. Rev. **31**(2), 199–206 (2001)

21. D. Simon, Training fuzzy systems with the extended Kalman filter. Fuzzy Sets Syst. **132**, 189–199 (2002)

22. D. Simon, *Optimal State Estimation: Kalman, H-infinity, and Nonlinear Approaches* (Wiley, 2006)

23. D. Simon, Design and rule base reduction of a fuzzy filter for the estimation of motor currents. Int. J. Approx. Reason. **25**, 145–167 (2000)

24. D. Simon, Sum normal optimization of fuzzy membership functions. Intern. J. Uncertain. Fuzziness Knowl. Based Syst. **10**, 363–384 (2002)

25. D. Simon, H∞ estimation for fuzzy membership function optimization. Int. J. Approx. Reason. **40**, 224–242 (2005)

26. D. Simon, *Evolutionary Optimization Algorithms: Biologically Inspired and Population-Based Approaches to Computer Intelligence* (Wiley, 2013)

27. D.E. Tamir, N.D. Rishe, A. Kandel (eds.), *Fifty Years of Fuzzy Logic and its Applications*. STUDFUZ 326 (Springer, Cham, 2015)

28. Y. Yam, P. Baranyi, C.-T. Yang, Reduction of fuzzy rule base via singular value decomposition. IEEE Trans. Fuzzy Syst. **7**(2), 120–132 (1999)
29. L.A. Zadeh, Fuzzy Sets. Inf. Control **8**, 338–353 (1965)
30. L.A. Zadeh, A.M. Abbasov, R.R. Yager, S.N. Shahbazova, M.Z. Reformat (eds.), *Recent Developments and New Directions in Soft Computing, STUDFUZ 317* (Springer, Cham, 2014)

Statistical Comparison of the Bee Colony Optimization and Fuzzy BCO Algorithms for Fuzzy Controller Design Using Trapezoidals MFs

Leticia Amador-Angulo and Oscar Castillo

Abstract This paper focuses on a statistical comparison with a proposed Fuzzy BCO based on an Interval Type-2 Fuzzy System and the Original BCO algorithm using Trapezoidal Membership Functions. The Fuzzy Bee Colony Optimization method applied for tuning the parameters of the Fuzzy Logic Controller is presented. The objective of the work is based on the main reasons for the statistical comparison of BCO and Fuzzy BCO algorithm that is to find the optimal design in the fuzzy logic controller for two problems in fuzzy control. We added perturbations in the model with band-limited noise so that the Interval Type-2 Fuzzy Logic System is better analyzed under uncertainty and to verify that the Fuzzy BCO shows better results than the Original BCO.

Keywords Interval type-2 fuzzy logic system · Fuzzy controller
Fuzzy bee colony optimization · Perturbation · Uncertainty

1 Introduction

The nonlinear characteristics of ill-defined and complex modern plants make classical controllers inadequate in whose cases for such systems. However, using fuzzy sets and fuzzy logic principles [1] has enabled researchers to understand better and hence control, complex systems that are difficult to model mathematically. These newly developed fuzzy logic controllers [2–4] have given control systems a certain degree of intelligence.

Interval Type-2 fuzzy models have emerged as an interesting generalization of Type-1 fuzzy models based upon fuzzy sets. In fact, these models are also called Interval Type-2 fuzzy models. There have been a number of claims put forward as

L. Amador-Angulo (✉) · O. Castillo
Division of Graduate Studies, Tijuana Institute of Technology, Tijuana, Mexico
e-mail: leticia.amadorangulo@yahoo.com.mx

O. Castillo
e-mail: ocastillo@tectijuana.mx

© Springer International Publishing AG, part of Springer Nature 2018
L. A. Zadeh et al. (eds.), *Recent Developments and the New Direction in Soft-Computing Foundations and Applications*, Studies in Fuzziness and Soft Computing 361, https://doi.org/10.1007/978-3-319-75408-6_23

to the relevance of Type-2 fuzzy sets being regarded as generic building constructs of fuzzy models. Fuzzy Controllers have the advantage that they can be adaptive when disturbances in the plant are present. In an Interval Type-2 fuzzy system, the membership functions can now return a range of values for the membership degrees, which vary depending on the uncertainty involved in not only the inputs, but also in the same membership functions [5, 6]. When we consider noise presence for the IT2FLC, the results show that the Interval Type-2 Fuzzy Logic System has better stability characteristics. Interval Type-2 Fuzzy Sets have shown to handle uncertainty in the field of Fuzzy Control [6].

The Bee Colony Optimization (BCO) metaheuristic has been successfully applied to various engineering and management problems by Teodorović et al. [7–9]. This study focuses on using a new Optimization technique called the BCO to find the optimal Fuzzy membership functions of controlling the trajectory in an autonomous mobile robot and the filling in the water tank. The term intelligence is defined as the ability to make the right decision [10] of human or bio-systems. By the definition, the BCO mimics the honey foraging behavior of bees [11] and uses mechanisms, such as, the waggle move to optimally locate honey sources and to find new source could be a choice of intelligence control system.

This work focuses on a statistical comparison of Fuzzy BCO and BCO algorithm on tuning fuzzy controllers for two problems. Section 2 describes a briefly describes of Interval Type-2 Fuzzy Logic System. Section 3 describes the studied cases. Section 4 describes the BCO and the Fuzzy BCO algorithm. Section 5 describes the comparative results with the two techniques. Section 6 describes the statistical comparison, Finally, Sect. 7 offers some conclusions of this work.

2 Interval Type-2 Fuzzy Logic System

Based on Zadeh's ideas [12, 13], in 1972 Mizumoto et al. [14] presented the mathematical definition of a type-2 fuzzy set. Since then, several authors have studied these sets, for example in [15, 16] Mendel, John and Mouzouris defined these sets as follows:

An Interval Type-2 Fuzzy Set $\tilde{A}$, denoted by $\underline{\mu}_{\tilde{A}}(x)$ and $\bar{\mu}_{\tilde{A}}(x)$ is represented by the lower and upper membership functions of $\mu_{\tilde{A}}(x)$. Where $x \in X$. In this case, Eq. (1) shows a sample IT2FS [16–18].

$$\tilde{A} = \{((x, u), 1) | \forall x \in X, \forall u \in J_x \subseteq [0, 1]\} \tag{1}$$

where X is the primary domain, J_x is the secondary domain. All secondary degrees $(\mu_{\tilde{A}}(x, u))$ are equal to 1. Figure 1 shows the architecture of an IT2FLS.

The output processor includes a type-reducer and defuzzifier; it generates a Type-1 fuzzy set output (from the type-reducer) or a crisp number (from the defuzzifier) [15–18].

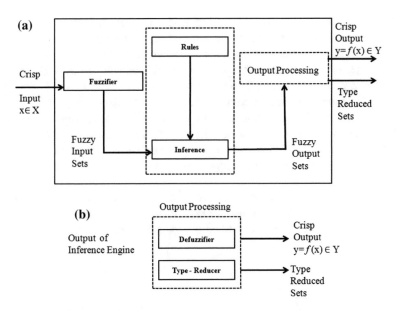

Fig. 1 Architecture of an interval type-2 fuzzy logic system

In this paper the statistical comparison with BCO and Fuzzy BCO in the optimization of the parameter values of the membership functions applied to design fuzzy controllers is presented. The two control problems and the fuzzy BCO algorithm are detailed in the following sections.

3 Studied Cases

3.1 Water Tank Controller

The first problem to be considered is known as the water tank controller, which aims at controlling the water level in a tank, therefore, based on the actual water level in the tank the controller has to be able to provide the proper activation of the valve. To evaluate the valve opening in a precise way we rely on fuzzy logic, which is implemented as a fuzzy controller that performs the control on the valve that determines how fast the water enters the tank to maintain the level of water in a better way [19]. The process of filling the water tank is presented as a differential equation for the height of water in the tank, H, and is given by Eq. (2).

$$\frac{d}{dt} Vol = A \frac{dH}{dt} = bV - a\sqrt{H} \tag{2}$$

Where **Vol** is the volume of water in the tank, **A** is the cross-sectional area of the tank, **b** is a constant related to the flow rate into the tank, and **a** is a constant related to the flow rate out of the tank. The equation describes the height of water, H, as a function of time, due to the difference between flow rates into and out of the tank. The Membership functions are for the two inputs to the fuzzy system: the first is called **level,** which has three Trapezoidal membership functions with linguistic values of *low, okay* and *high*. The second input variable is called **rate** with three membership functions corresponding to the linguistic values of *negative, none* and *positive*. The fuzzy logic controller has an output called **valve,** which is composed of five Trapezoidal membership functions with the following linguistic values: *closefast, closeslow, nochange, openslow* and *openfast*, and we show in Fig. 2 the representations of the input and output variables. Figure 3 represents the Fuzzy Controller in the model.

The rules are based on the behavior of the water tank to be filled. All the combinations of rules were taken from experimental knowledge according to how the process is performed in a tank filled with water, which are detailed in Table 1.

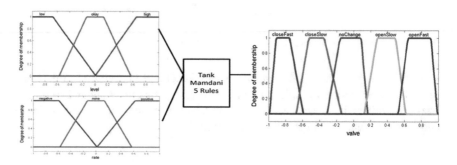

Fig. 2 Characteristics of the type-1 FLC

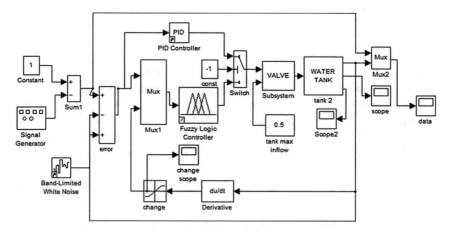

Fig. 3 Block diagram for the simulation of the FLC

Table 1 Rules for the first studied case

# Rules	Input 1 level	Input 2 rate	Output valve
1	okay	—	nochange
2	low	—	openfast
3	high	—	closefast
4	okay	positive	closeslow
5	okay	negative	openslow

3.2 Mobile Robot Controller

The second studied case is for controlling the trajectory of the unicycle mobile robot [20]. We added levels of perturbation (noise) in the model with the goal of analyzing the behavior under uncertainty. The model consisting of two driving wheels located on the same axis and a front freewheel, and Fig. 4 shows a graphical description of the robot model.

The robot model assumes that the motion of the free wheel can be ignored in its dynamics, as shown by Eqs. (3)–(4).

$$M(q)\dot{v} + C(q, \dot{q})v + Dv = \tau + P(t) \tag{3}$$

where,

$q = (x, y, \theta)^T$ is the vector of the configuration coordinates,

$v = (v, w)^T$ is the vector of velocities,

$\tau = (\tau_1, \tau_2)$ is the vector of the torques applied to the wheels of the robot where τ_1 and τ_2 denote the torques of the right and left wheel,

$P \in R^2$ is the uniformly bounded disturbance vector,

$M(q) \in R^{2 \times 2}$ is the positive-definite inertia matrix,

$C(q, \dot{q})\vartheta$ is the vector of centripetal and Coriolis forces, and

$D \in R^{2 \times 2}$ is a diagonal positive-definite damping matrix.

The kinematic system is represented by Eq. (4).

Fig. 4 Mobile robot model

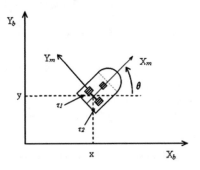

$$\dot{q} = \underbrace{\begin{bmatrix} \cos\theta & 0 \\ \sin\theta & 0 \\ 0 & 1 \end{bmatrix}}_{J(q)} \underbrace{\begin{bmatrix} v \\ w \end{bmatrix}}_{\upsilon} \qquad (4)$$

where,

(x, y) is the position in the X – Y (world) reference frame,
θ is the angle between the heading direction and the x-axis, v and w are the linear and angular velocities.

Furthermore, Eq. (5) shows the non-holonomic constraint which this system has, which corresponds to a no-slip wheel condition preventing the robot from moving sideways.

$$\dot{y}\cos\theta - \dot{x}\sin\theta = 0 \qquad (5)$$

The Membership functions are for the two inputs to the fuzzy system: the first is called **ev (angular velocity)**, which has three membership functions with linguistic values of N, Z and P. The second input variable is called **ew (linear velocity)** with three membership functions with the same linguistic values. The type-1 fuzzy logic controller has two outputs called **T1 (Torque 1)**, and **T2 (Torque 2)**, which are composed of three trapezoidal membership functions with the following linguistic values, respectively: N, Z, P, and in Fig. 5 we show the representations of the input and output variables.

The combination of the rules is shown in Table 2 and the model of the Fuzzy Logic Controller can be found in Fig. 6.

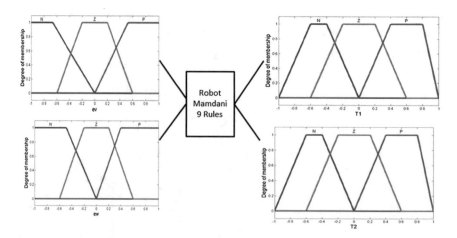

Fig. 5 Inputs and outputs variables of the type-1 FLC of the mobile robot controller

Table 2 Fuzzy rules used by the fuzzy controller of mobile robot

# Rules	Input 1 ev	Input 2 ew	Output T1	Output T2
1	N	N	N	N
2	N	Z	N	Z
3	N	P	N	P
4	Z	N	Z	N
5	Z	Z	Z	Z
6	Z	P	Z	P
7	P	N	P	N
8	P	Z	P	Z
9	P	P	P	P

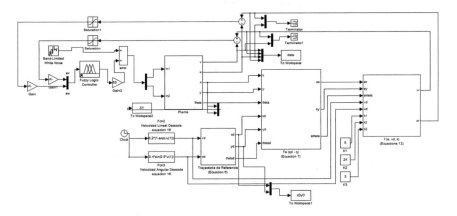

Fig. 6 Fuzzy controller of the mobile robot

The test criteria are a series of Performance Indices; where the Integral Square Error (ISE), Integral Absolute Error (IAE) and Root Mean Square Error (RMSE) are used, respectively shown in Eqs. (6)–(8).

$$ISE = \int_0^\infty e^2(t)dt \tag{6}$$

$$IAE = \int_0^t (|\Delta P(t)| + |\Delta Q(t)|)dt \tag{7}$$

$$\varepsilon = \sqrt{\frac{1}{N}\sum_{t=1}^N (X_t - \hat{X}_t)^2} \tag{8}$$

4 Bee Colony Optimization Algorithm

4.1 Original BCO

The communication between individual insects in a colony of social insects has been well known. The BCO is inspired by the bees' behavior in the nature. The basic idea behind the BCO is to create the multi agent system (colony of artificial bees) capable to successfully solve difficult combinatorial optimization problems. The artificial bee colony behaves partially alike, and partially differently from bee colonies in nature [7, 19, 21]. The basic step of the BCO algorithm is shown in Table 3. The BCO algorithm is based on Eqs. (9)–(12):

$$P_{ij,n} = \frac{[\rho_{ij,n}]^\alpha \cdot [\frac{1}{d_{ij}}]^\beta}{\sum\limits_{j \in Ai,n} [\rho_{ij,n}]^\alpha \cdot [\frac{1}{d_{ij}}]^\beta} \tag{9}$$

$$D_i = K \cdot \frac{Pf_i}{Pf_{colony}} \tag{10}$$

$$Pf_i = \frac{1}{L_I}, L_i = \text{Tour Length} \tag{11}$$

$$Pf_{colony} = \frac{1}{N_{Bee}} \sum_{i=1}^{N_{Bee}} Pf_i \tag{12}$$

Table 3 Basic steps of the BCO algorithm

Pseudocode of BCO
1. Initialization: an empty solution is assigned to every bee;
2. For every bee: //the forward pass
a) Set k = 1; //counter for constructive moves in the forward pass;
b) Evaluate all possible constructive moves;
c) According to evaluation, choose on move using the roulette wheel;
d) k = k + 1; if k ≤ NC goto step b.
3. All bees are back to the hive; //backward pass stars.
4. Evaluate (partial) objective function value for each bee;
5. Every bee decide randomly whether to continue its own exploration and become a recruiter, or to become a follower;
6. For every follower, choose a new solution from recruiters by the roulette wheel;
7. If solutions are not completed goto step 2;
8. Evaluate all solutions and find the best one;
9. If stopping condition is not met goto step 2;
10. Output the best solution found.
B: Represents the number of bees in the hive.
NC: Represents the number of constructive moves during one forward pass.

Equation (9) represents the probability of a bee k located on a node i selects the next node denoted by j, where, N_i^k is the set of feasible nodes (in a neighborhood) connected to node i with respect to bee k, ρ_{ij} is the probability to visit the following node. Note that the β is inversely proportional to the city distance; d_{ij} represents the distance of node i until node j, for this algorithm indicate the total the dance that a bee have in this moment. Finally, α is a binary variable that is used to find better solutions in the algorithm. Equation (11) represents that a waggle dance will last for certain duration, determined by a linear function, where K denotes the waggle dance scaling factor, Pf_i denotes the profitability scores of bee i as defined in Eq. (12) and Pf_{colony} denotes the bee colony's average profitability as in Eq. (13) and is updated after each bee completes its tour. For this research the waggle dance is represented by the mean square error (MSE) that the model to find once that is done the simulation in the iteration of the algorithm. In the BCO algorithm, a bee represents the values of the distribution of the membership functions. The design of the T1FLS for the mobile robot controller has Trapezoidal membership functions in the inputs and outputs (see Fig. 5), obtaining a total of 48 values. The variable Pos represents the size for each bee that for this studied case is of 48 values, and for the second studied case is of 44 values.

4.2 Fuzzy BCO

In the BCO algorithm the waggle dance represents the intensity with which a bee finds a possible good solution. If the intensity of the waggle dance is large this means that the solution found by the bee is the best of all the population. For this work the waggle dance is represented by the mean square error that all models find once the simulation in the iteration of the algorithm is done [19–21]. For measuring the iterations of the algorithm, it was decided to use the percentage of iterations as a variable, i.e. when starting the algorithm the iterations will be considered "low", and when the iterations are completed it will be considered "high" or close to 100%. We represent this idea using Eq. (13) [19–21]:

$$\text{Iteration} = \frac{\text{Current Iteration}}{\text{Maximum of Iterations}} \tag{13}$$

The diversity measure is defined by Eq. (14), which measures the degree of dispersion of the bee, i.e. when the bees are closer together; there is less diversity as well as when bees are separated then the diversity is higher. As the reader will realize the equation of diversity can be considered as the average of the Euclidean distances between each bee and the best bee [19, 20].

$$Diversity(S(t)) = \frac{1}{n_s} \sum_{i=1}^{n_x} \sqrt{X_{ij}(t) - \bar{X}_j(t))^2} \tag{14}$$

The fitness function in the BCO algorithm is represented with the Mean Square Error shown in Eq. (15). For each Follower Bee for N Cycles, the Type-1 FLS design for the BCO algorithm is evaluated and the objective is to minimize the error.

$$MSE = \frac{1}{n} \sum_{i=1}^{n} (\bar{Y}_i - Y_i)^2 \qquad (15)$$

The distribution of the membership functions in the inputs and outputs is realized in a symmetrical way. The design of the input and output variables can be appreciated in Fig. 7 and Fig. 8 for the Type-1 FLS and Interval Type-2 FLS, respectively. The rules are shows in Table 4.

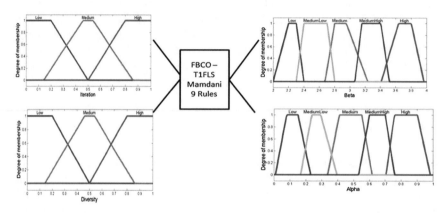

Fig. 7 Fuzzy BCO T1FLS

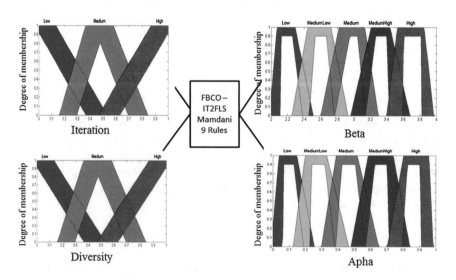

Fig. 8 Fuzzy BCO IT2FLS

Table 4 Rules for the FBCO with dynamic beta and alpha

# Rules	Input 1 iteration	Input 2 diversity	Output beta	Output alpha
1	Low	Low	High	Low
2	Low	Medium	MediumHigh	Medium
3	Low	High	MediumHigh	MediumLow
4	Medium	Low	MediumHigh	MediumLow
5	Medium	Medium	Medium	Medium
6	Medium	High	MediumLow	MediumHigh
7	High	Low	Medium	High
8	High	Medium	MediumLow	MediumHigh
9	High	High	Low	High

5 Comparative Results

Experimentation was performed with perturbations. We used the specific noise generators with Band-Limited noise with a value of 0.5 and delay of 1000. The configuration of the each experiment was with the following parameters; *Population* of 50, *Follower Bees* of 30, *MaxCycle* of 20, *Alpha* and *Beta* of 0.5 and 2.5 for the Original BCO and Dynamic for the Fuzzy BCO with T1FLS and IT2FLS. Simulation results in Table 5 shows the best experiment without perturbation in the fuzzy logic controller for each bio-inspired algorithm.

Table 5 shows that when the perturbation is not applied in the model the Fuzzy BCO with Interval Type-2 FLS is better with a simulation error of **0.00002.**

Table 6 shows in the studied case of the mobile robot controller that when we used the Interval Type-2 FLS in the model the simulation error is better. Figure 9 shows a comparison with the results with the three methods considered in the work for the first and Fig. 10 for the second studied cases.

The behavior of the Fuzzy BCO with Interval Type-2 FLS, Type-1 FLS and the Original BCO is presented in Figs. 11, 12 and 13 for the mobile robot controller.

In Fig. 14, a comparison of the proposed method applied a bee colony optimization with parameter dynamics against other nature inspired techniques that

Table 5 Average or 30 experiments for the water tank controller without perturbation

Index	Methods		
	Original BCO	FBCO with T1FLS	FBCO with IT2FLS
ISE	1.6909	1.7545	1.7769
IAE	4.0853	4.0933	4.3245
MSE	0.0520	0.0006	0.0003
RMSE	0.0122	0.0086	0.0531
SD	0.0116	0.0008	0.0005
Best	0.0076	0.0001	**0.00002**
Worst	0.0541	0.0047	0.0025

Table 6 Average or 30 experiments for the mobile robot controller without perturbation

Index	Methods		
	Original BCO	FBCO with T1FLS	FBCO with IT2FLS
ISE	7.8503	7.5676	7.9283
IAE	19.1775	18.5116	19.3201
MSE	7.4609	11.2145	7.9848
RMSE	13.3005	18.2449	13.4967
SD	10.2270	13.9492	10.0397
Best	0.0086	0.0606	**0.0059**
Worst	46.9678	41.8009	44.9853

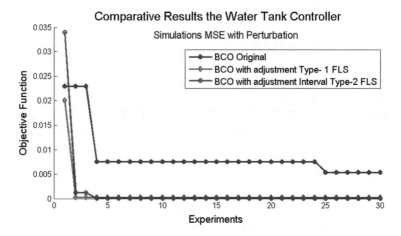

Fig. 9 Comparative results of water tank controller with perturbation

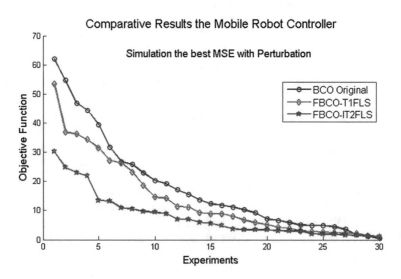

Fig. 10 Comparative results mobile robot controller

Fig. 11 Behavior of the original BCO

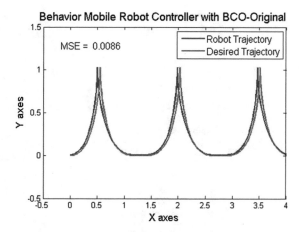

Fig. 12 Behavior of the fuzzy BCO-T1FLS

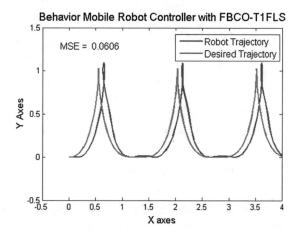

Fig. 13 Behavior of the fuzzy BCO-IT2FLS

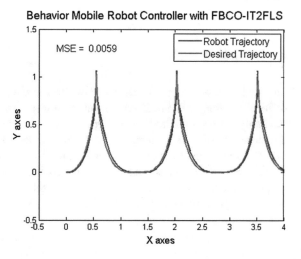

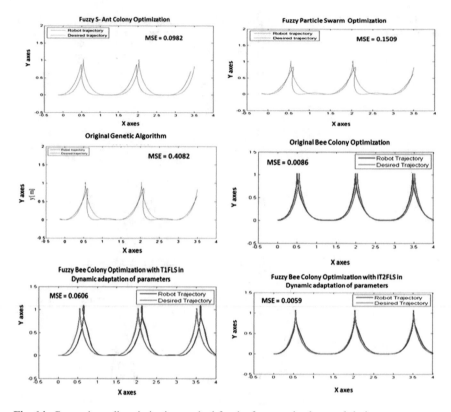

Fig. 14 Comparison all optimization method for the fuzzy optimal control designs

have been applied to solve the same problem, of mobile robot controller is shown [22–24]. The tracking error of the proposed method clearly outperforms the other techniques.

6 Statistical Test

The statistical test used for result comparison is the z-test, whose parameters are defined in Table 7. We realized the statistical test with a sample of 30 experiments randomly for each method, obtaining the results contained in Table 8. In applying the z-test statistic, with a significance level of 0.05, and the alternative hypothesis says that the results found of the proposed method (IT2FBCO) is lower than the of the Original BCO, and of course the null hypothesis tells us that the results of the proposed method are greater than or equal to the results of Original BCO, with a rejection region for all values that fall below −1.645. Tables 8 shows the statistical test results.

Table 7 Parameters for the statistical z-test

Parameter	Value
Level of significance	95%
Alpha	0.05%
Ha	$\mu_1 < \mu_2$
Ho	$\mu_1 \geq \mu_2$
Critical value	-1.645

Table 8 Results of applying the statistical z-test

Studied Case	Noise	μ_1	μ_2	Z value	Evidence
Tank controller	Not	FBCO-IT2FLS	Original BCO	-5.7539	Significant
	Not	FBCO-T1FLS	Original BCO	-5.7171	Significant
	Not	FBCO-IT2FLS	FBCO-T1FLS	-3.7836	Significant
	Yes	FBCO-IT2FLS	Original BCO	-8.6446	Significant
	Yes	FBCO-T1FLS	Original BCO	-3.7984	Significant
	Yes	FBCO-IT2FLS	FBCO-T1FLS	-21.850	Significant
Mobile robot controller	Not	FBCO-IT2FLS	Original BCO	-1.2546	NotSignificant
	Not	FBCO-T1FLS	Original BCO	1.1886	NotSignificant
	Not	FBCO-IT2FLS	FBCO-T1FLS	-1.0293	NotSignificant
	Yes	FBCO-IT2FLS	Original BCO	-2.9857	Significant
	Yes	FBCO-T1FLS	Original BCO	-5.912	Significant
	Yes	FBCO-IT2FLS	FBCO-T1FLS	-2.0732	Significant

7 Conclusions

This paper presents a statistical comparison of the Original BCO and Fuzzy BCO algorithm to tune membership functions in fuzzy controller design. With the intention of demonstrating, statistically, that the robustness of the Fuzzy BCO with IT2FLS is better than the Original BCO for the stabilization of the problems in Fuzzy Control. Results obtained shown that finding the optimal *alpha* and *beta* values through fuzzy sets significantly improve the performance of the BCO algorithm therefore, it is shown that these parameters; *alpha* representing the exploitation and *beta* representing exploration, they are a good technique in optimizing parameters of fuzzy logic controller.

The future work includes the realization of the optimization with FBCO of the Interval Type-2 FLS in both Fuzzy Controllers.

References

1. D.T. Pham, A. Haj Darwish, E.E. Eldukhiri, S. Otri, Using the bees algorithm to tune a fuzzy logic controller for a robot gymnast, in *Intelligent Production Machines and Systems* (2007)

2. K.M. Passino, S. Yurkovich, Fuzzy Control (Addison-Wesley Longman, Menlo Park, CA, 1998), p. 475
3. E.H. Mamdani, S. Assilian, An experiment in linguistic synthesis with a fuzzy logic controller. Int. J. Man-Mach. Stud. 1–13 (1975)
4. M. Sugeno, Industrial Applications of Fuzzy Control (Elsevier Science, 1985), pp. 269
5. J.M. Mendel, *Uncertain Rule-Based Fuzzy Logic System: Introduction and New Directions* (Practice Hall, New Jersey, 2001)
6. J.M. Mendel, G.C. Mouzouris, Type-2 fuzzy logic system. IEEE Trans. Fuzzy Syst. **7**(6), 642–658 (1999)
7. P. Lučić, D. Teodorović, Vehicle routing problem with uncertain demand at nodes: the bee system and fuzzy logic approach, in *Fuzzy Sets in Optimization* ed. by J.L. Verdegay (Springer, Heidelberg, Berlin, 2003), pp. 67–82
8. D. Teodorović, "Transport modeling by multi-agent systems": a swarm intelligence approach. Transp. Plan. Technol. **26**(4), 289–312 (2003)
9. D. Teodorović, Swarm intelligence systems for transportation engineering: principles and applications. Transp. Res. Part. C Emerg. Technol. **16**(6), 651–782 (2008)
10. M.T. Jones, Artificial Intelligence: A Systems Approach, 1st edn. (Jones & Bartlett Learning, 2007), p. 500
11. D.T. Pham, A. Ghanbarzadeh, E. Koc, S. Otri, S. Rahim et al, The Bees Algorithm—A Novel Tool for Complex Optimisation Problems (Cardiff University, 2006)
12. L. A. Zadeh, The concept of a lingüistic variable and its application to approximate reasoning, part II. Inf. Sci. **8**, 301–357 (1975)
13. L. A. Zadeh, Fuzzy Sets, Inf. Control, **8**, 338–353 (1965)
14. M. Mizumoto, J. Toyoda, K. Tanaka, General formulation on formal grammars. Inf. Sci. **4**, 87–100 (1972)
15. J.M. Mendel, Computing derivatives in interval type-2 fuzzy logic systems. IEEE Trans. Fuzzy Syst. **12**(1), 84–98 (2004)
16. J.M. Mendel, R.I.B. John, Type-2 fuzzy sets made simple. IEEE Trans. Fuzzy Syst. **10**(2), 117–127 (2002)
17. J.M. Mendel, A quantitative comparison of interval type-2 and type1fuzzy logic systems: first results, in *Proceedings of the IEEE International Conference on Fuzzy Systems (FUZZ)* (2010), pp. 1–8
18. J.M. Mendel, On KM algorithms for solving type-2 fuzzy set problems. IEEE Trans. Fuzzy Syst. **21**(3), 426–446 (2013)
19. L. Amador-Angulo, O. Castillo, A new algorithm based in the smart behavior of the bees for the design of mamdani-style fuzzy controllers using complex non-linear plants, in *Design of Intelligent Systems based on Fuzzy Logic, Neural Network and Nature-Inspired Optimization* (2015), pp. 617–637
20. L. Amador-Angulo, O. Castillo, Statistical analysis of type-1 and interval type-2 fuzzy Logic in dynamic parameter adaptation of the BCO, in *IFSA-EUSFLAT* (2015), pp. 776–783
21. P. Melin, F. Olivas, O. Castillo, F. Valdez, J. Soria, M. Valdez, Optimal design of fuzzy classification systems using PSO with dynamic parameter adaptation through fuzzy logic. Expert Syst. Appl. **40**(8), 1–12 (2013)
22. O. Castillo, R. Martinez-Marroquin, P. Melin, F. Valdez, J. Soria, Comparative study of bio-inspired algorithms applied to the optimization of type-1 and type-2 fuzzy Controllers for an autonomous mobile robot. Inf. Sci. **192**, 19–38 (2012)
23. O. Castillo, R. Martinez, P. Melin, Bio-inspired optimization of fuzzy logic controllers for robotic autonomous system with PSO and ACO. Fuzzy Inf. Eng. 119–143 (2010)
24. R. Martinez, O. Castillo, L. Aguilar, Optimization of type-2 fuzzy logic controllers for a perturbed autonomous wheeled mobile robot using genetic algorithms. Inf. Sci. **179**(13), 2158–2174 (2009)

Load Frequency Control of Hydro-Hydro System with Fuzzy Logic Controller Considering Non-linearity

K. Jagatheesan, B. Anand, Nilanjan Dey, Amira S. Ashour and Valentina E. Balas

Abstract The current work handles Automatic Generation Control (AGC) of an interconnected two area hydro-hydro system. The proposed system is integrated with conventional Proportional Integral (PI) as well as Fuzzy Logic Controller (FLC). Since, the conventional PI controller does not offer sufficient control performance. Thus, non-linearities such as the Generation Rate Constraint (GRC) and Governor Dead Band (GDB) are included in the system in order to overcome this drawback with employing Fuzzy Logic Controller (FLC) in the system. The results reported the time domain simulation that used to study the performance, when 1% step load disturbance is given in either area of the system. Furthermore, the conventional PI controller simulation results are compared to fuzzy logic controller. The simulation results depicted that the FLC achieved superior control performance.

K. Jagatheesan (✉)
Department of EEE, Mahendra Institute of Engineering & Technology, Namakkal, Tamilnadu, India
e-mail: jaga.ksr@gmail.com

B. Anand
Department of EEE, Hindusthan College of Engineering & Technology, Coimbatore, Tamilnadu, India
e-mail: b_anand_eee@yahoo.co.in

N. Dey
Department of Information Technology, Techno India College of Technology, Kolkata, India
e-mail: neelanjandey@gmail.com

A. S. Ashour
Department of Electronics and Electrical Communications Engineering, Tanta University, Tanta, Egypt
e-mail: amirasashour@yahoo.com

V. E. Balas
Aurel Vlaicu University of Arad, Arad, Romania
e-mail: balas@drbalas.ro

© Springer International Publishing AG, part of Springer Nature 2018
L. A. Zadeh et al. (eds.), *Recent Developments and the New Direction in Soft-Computing Foundations and Applications*, Studies in Fuzziness and Soft Computing 361, https://doi.org/10.1007/978-3-319-75408-6_24

Keywords Load frequency control · Generation rate constraint
Automatic generation control · Hydro-Hydro power system · Fuzzy logic
controller

Nomenclature

Δ	Deviation
i	Subscript referred to area (1, 2)
f	Nominal system frequency
Kpi	Gain constant of generator
Tpi	Time constant of a generator
Pri	Rated area power
T1, T3	Time constants of hydro governor
T2	Mechanical governor reset time constant
Tw	Water starting time
Kdc	Gain associated with dc link
Tdc	Time constant of dc link
T12	Synchronizing coefficient
Ptie	Tie line power
Pdi	Load disturbance
Ri	Governor speed regulation parameter
Bi	Frequency bias constant
KP	Proportional controller gain
Ki	Integral controller gain
a12	Pr1/Pr2
ACE	Area Control Error
LFC	Load Frequency Control
J	Cost index
T	Sampling time period

1   Introduction

For successful interconnected electric power system operation, matching of the total
generation with total load demand, in addition to the system losses reduction is
required. Since, in the electric power system, the operating point may change with
respect to time. This leads to undesirable effects [1], which can be overcome by
controllers or automatic generation controllers. To perform this task, the area
control error (ACE) is properly adjusted to comprise the system frequency and tie
line power exchanges, which referred to as tie line bias control. Typically, the AGC
studies are carried out using simulation models. Numerous research works introduced different models for the interconnected thermal and hydro-thermal system

[2–13]. The system frequency deviation and tie line power exchange are controlled through ACE in conventional control process as follows:

$$ACE = Df \cdot B + \Delta Ptie \tag{1}$$

Here, Δf is frequency deviation, B is frequency bias constant, and ΔP tie is the change in tie line power. Consequently, the present study introduced the performance of two area interconnected hydro-hydro system with conventional PI and fuzzy logic controller. Since, the conventional PI control approach does not provide satisfactory control performance, when using 1% step load disturbance in either area of the system. Thus, an optimal fuzzy logic controller is proposed in this paper. The complexity to obtain the optimum settling time of the controller is mitigated by using FLC.

The structure of the remaining sections is as follows. Section 2 included the system under investigation. The fuzzy logic controller is explained in Sect. 3. Section 4 illustrated the results along with the discussion. Finally, the conclusion is presented in Sect. 5.

2 System Studied

The block diagram of two area interconnected hydro-hydro power system employed in the proposed study is illustrated in Fig. 1. In the conventional procedure, each area within the system, which operates at a different frequency and tie line power exchanges between the areas is computed as the product of the tie line constant and the angular difference between the areas [14]. Traditional turning methods are considered unsuccessful for hydro-hydro interconnections. In the proposed system, the frequency of both the areas is proposed to be same difference between the area frequencies which can be neglected. Also tie line computation block is absent. The system dynamics and system performance is improved significantly with the inclusion of incremental DC link power flow. The frequency control of interconnected power systems through the DC link is concerned. In order to obtain tie line power, the following power balance equation is used [15].

$$Ptie + Pgi - Pdi = Hi[d(f)/dt] \tag{2}$$

where, *Ptie* is the tie line power flow, *Pgi* is the area generation of *i*th area, *Pdi* is the *i*th area load disturbance, H_i is the inertia constant of *i*th area and *f* is the system frequency. This model approach is not unjustified. Divya and Rao [15] stated that "in AGC studies, the momentary difference between the frequencies of different areas can be ignored". For all AGC purpose, the frequency used for one area to compute ACE should be the same as used in the other areas so long as they remain interconnected.

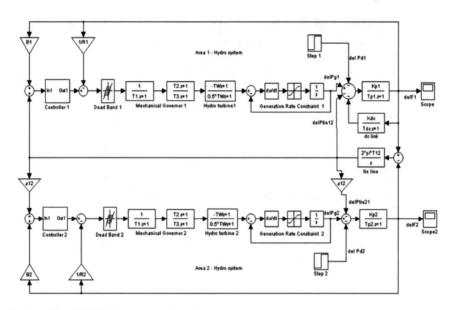

Fig. 1 Block diagram of two area hydro-hydro power system

2.1   Conventional PI Controller

The PI controller generates an output signal which consists of two terms, namely the proportional to error signal and the Integral of error signal [2], given be:

$$u(t) = \mu[e(t) + \int e(t)\, dt]$$

(3)

$$u(t) = K_p\, e(t) + K_i \int e(t)\, dt$$

(4)

Here, K_p is the proportional controller gain and K_i is the integral controller gain. By applying the Laplace transform, the following expression is obtained.

$$u(s) = K_p E(s) + K_i(s)/E(s)$$

(5)

$$U(S)/E(S) = K_p + K_i/s$$

(6)

The integral square error (ISE) technique is adopted in both the areas to find the optimum PI value. Figure 2 demonstrated the performance index curve. For ISE technique, the used objective function is given by:

Fig. 2 Performance index curve

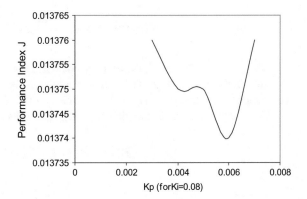

$$J = \int \left(\Delta f_1^2 + \Delta f_2^2 + \Delta Ptie^2 \right) \tag{7}$$

where, ΔT is the small time interval during sample, Δf is the changing frequency and $\Delta Ptie$ is the change in tie line power.

2.2 Governor Dead Band

The Governor dead band is known as the total magnitude of a sustained speed change, where no resulting change in the valve position is produced. A continuous sinusoidal oscillation which has natural period of about 2 s is produced using the Backlash non-linearity [16, 17]. The speed governor dead band has major influence on the dynamic performance of load frequency control system. The hysteresis of the non-linearities is given by $y = f(x, \dot{x})$ rather than $y = F(x)$.

Assume that the variable x is sufficiently close to a sinusoidal in order to solve the non-linear problem, thus:

$$x \approx A \, Sin \, \omega_0 t \tag{8}$$

where, A and ω_0 are the oscillation amplitude and frequency; respectively, which expressed by:

$$\omega_0 = 2\pi f_0 = \pi \tag{9}$$

The variable function is periodic, complex and can be developed as a Fourier series [16], which expressed by:

$$F(x, \dot{x}) = F^0 + N_1 x + \frac{N_2}{\omega_0} \dot{x} + \ldots \tag{10}$$

Since, the backlash nonlinearity is symmetrical about the origin, thus F^0 is zero. For simplicity, higher order terms can be neglected in Eq. (10), where the Fourier co-efficient are derived as $N_1 = 0.8$ and $N_2 = -0.2$. Substituting these values in Eq. (10), thus the transfer function for GDB is expressed as follows:

$$F(x, \dot{x}) = 0.8x - \frac{0.2}{\pi} \dot{x} \tag{11}$$

2.3 Generation Rate Constraint

Due to thermodynamic and mechanical constraint in the practical steam turbine system, there is a boundary to the rate at which its output power (dp1/dt) can be changed. This limit is known as the generation rate constraint. The rate limits are impressed to avoid a broad variation in the process variables, such as the temperature and pressure for the equipment safety. Generation rate constraint of 30% p.u. MW/min are usually applied to non-reheat turbines [18]. Typically, it was established from the literatures that fuzzy logic controller can be applied in several applications including AGC [19–30].

To prevent the extreme moment, a limiter is also added to the rate limit on valve position. These constraints yield to nonlinear system. The frequency control provides larger peak overshoots and larger settling time in the dynamic responses with the presence of GRC than those without considering GRC.

3 Fuzzy Logic Controller

Fuzzy logic control is based on a logical system, which inspirited by the human thinking and natural language system. Recently, fuzzy logic is employed in almost all sectors of industry and science. Since, the main goal of load frequency control in interconnected power system is to preserve the balance between production and consumption. Moreover, the conventional control techniques may not provide satisfactory solutions due to the complexity and multi variable conditions of the power system, [28–31]. Conversely, their robustness and reliability make fuzzy controllers useful in solving an extensive range of control problems. Therefore, the artificial intelligence based gain scheduling is an alternative technique generally used in designing controllers for non-linear systems. The fuzzy logic controller is comprised of four main components; namely the fuzzification, inference engine, rule base and defuzzification. The fuzzification transforms the numeric/crisp value into fuzzy sets [32]. The *Fuzzification* process is used to convert a numerical variable (real number or crisp variables) into a linguistic variable (fuzzy number).

To design the fuzzy controller, it is required to decide which state variables represent the system dynamic performance must be taken as the input signal to the controller. System variables, which are used as the fuzzy controller inputs comprises states error, state error derivative or state error integral. In the power system, Area Control Error (ACE) and its derivative (d(ACE)/dt) are chosen to be the input signals of fuzzy AGC.

The membership function (MF) is a graphical illustration of the magnitude of participation of each input. There are different MFs associated with each input and output response. In this study, the triangular MF for input and output variables is used. The number of MFs determines the quality of control which can be achieved using fuzzy controller, where as the number of MF increases, the control quality improves. As the number of linguistic variables increases, the computational time and required memory increases. Consequently, a compromise between the control quality and computational time is needed to select the number of linguistic variables.

Recently, fuzzy set theory based approach has emerged as a complement tool to mathematical techniques for solving power system problems. Fuzzy set theory and fuzzy logic establish the rules of a non-linear mapping, which obtained based on experiments of the process step response, error signal and its time derivative. Figure 3 demonstrated the fuzzy logic controller, which has two input signals, namely ACE and ACE. The output signal (y) is used for controlling the load frequency control in the interconnected power system.

The foremost component of the FLC is the inference engine, which performs all logic manipulations. The rule base consists of MFs and control rules [33]. In the proposed study, mamdani fuzzy inference engine is selected and the centroid method is used in defuzzification process. Seven triangular MFs are taken, thus 49 control rules are used in the proposed study. Ranges of the MFs are chosen from simulation results. The control rules are built from the If-then statement.

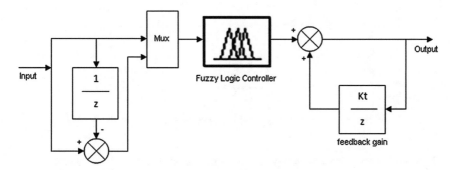

Fig. 3 Block diagram of fuzzy logic controller

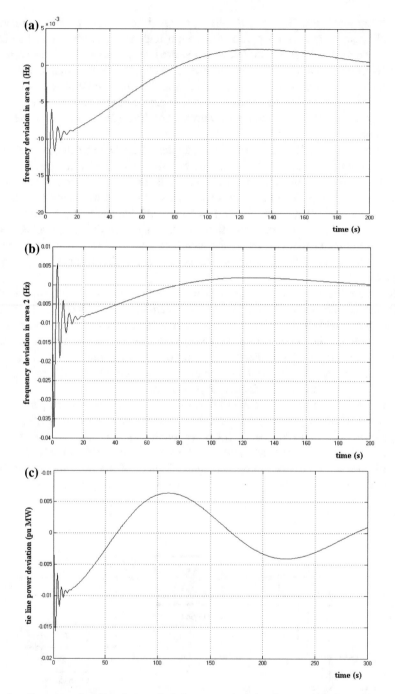

Fig. 4 **a** Frequency deviation in area 1, **b** frequency deviation in area 2, **c** change in tie line power, **d** frequency deviation in area 1 with PI and FLC, **e** frequency deviation in area 2 with PI and FLC, **f** change in tie line power with PI and FLC

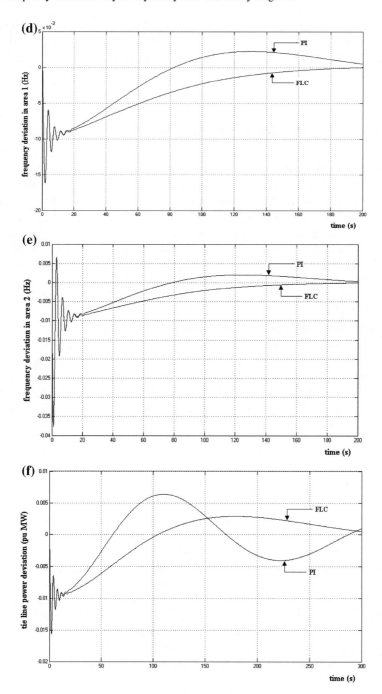

Fig. 4 (continued)

In a fuzzy scale, each MF of the seven linguistic states of triangular type are mapped into the values of Negative Large (NL), Negative Small (NS), Negative Medium (NM), Zero Error (ZE), Positive Small (PS), Positive Medium (PS) and Positive Large (PL).

4 Simulation Results and Discussion

The conventional PI and fuzzy logic controller are applied to two area interconnected hydro-hydro power system. Performance criteria, such as the settling time and overshoots are considered in the simulation for the system dynamic parameter, frequency deviation in both the area and the tie line power deviations. A system with load disturbance of 1% step is simulated in either area of the system. Frequency deviations in area 1 and 2 with conventional PI controller are illustrated in the Fig. 4a and b; respectively. In addition, the tie line power deviation is demonstrated in the Fig. 4c. Comparative results of PI and FLC for frequency deviation in area 1 and 2 is depicted in Fig. 4d, e. The tie line power deviation is shown in Fig. 4f.

Figure 4 established that the conventional PI controller does not provide good control performance. Furthermore, it takes more settling time to settle down the steady state error due to the fixed value of PI gains irrespective of the changing errors. However, the fuzzy logic controller provides satisfactory control performance compared to the conventional PI controller. The tie line waveform reported that the FLC yields very good performance, while the PI controller has high oscillations nearer to set point.

5 Conclusion

In this study, the fuzzy logic controller is employed for load frequency control. The proposed controller can handle two area hydro power systems, which achieved adequate control performance. The proposed controller effectiveness in increasing the damping of load and inter area modes of oscillations is demonstrated in the two area interconnected power system. The simulation results are compared to a conventional PI controller, which reported that the proposed intelligent controller improved the dynamic response. The presence of FLC in both areas and small step perturbation in either area provided better dynamic response than with conventional PI controller. Thus, the simulation results confirm that the fuzzy logic controller greatly reduces the overshoots as well as the settling time.

References

1. O.I. Elgerd, *Electric Energy Systems Theory: An Introduction* (Mc-Graw Hill, New York, 1983)
2. J. Nanda, A. Mangla, S. Suri, Some new findings on automatic generation control of an interconnected hydrothermal system with conventional controllers. IEEE Trans. Energy Convers. **21**(1), 187–193 (2006)
3. J. Nanda, A. Mangla, Automatic generation control of an interconnected hydro-thermal system using conventional Integral and fuzzy logic controller, in *2004 IEEE International Conference on Electric Utility Deregulation, Restructuring and Power Technologies* (Hongkong, Apr 2004), pp. 372–377
4. S.C. Tripathy, R. Balasubramanian, P.S.C. Nair, Effect of superconducting magnetic energy storage on automatic generation control considering governor deadband and boiler dynamics. IEEE Trans. Power Syst. **7**(3), 1266–1272 (1992)
5. IEEE Committee Report, Dynamic models for steam and hydro turbines in power system studies. IEEE Trans. Power Apparat. Syst. **92**(4), 1904–1911 (1973)
6. K. Jagatheesan, B. Anand, N. Dey, Automatic generation control of thermal-thermal-hydro power systems with PID controller using ant colony optimization. Int. J. Serv. Sci. Manage. Eng. Technol. **6**(2), 18–34 (2015)
7. K. Jagatheesan, B. Anand, N. Dey, A.S. Ashour, Ant colony optimization algorithm based PID controller for LFC of single area power system with non-linearity and boiler dynamics. World J. Model. Simul. **12**(1), 3–14 (2016)
8. K. Jagatheesan, B. Anand, B. Sourav Samanta, N. Dey, V. Santhi, A.S. Ashiur, V.E. Balas, Application of flower pollination algorithm in load frequency control of multi-area interconnected power system with non-linearity. Neural Comput. Appl. 1–14 (2016)
9. K. Jagatheesan, B. Anand, S. Samanta, N. Dey, A.S. Ashour, V.E. Balas, H. He, Design of proportional-integral-derivative controller for AGC of multi-area power thermal systems using firefly algorithm. IEEE/CAA J. Autom. Sinica (2016) (Accepted for Publication)
10. K. Jagatheesan, B. Anand, AGC of multi-area hydro-thermal power systems with GRC non-linearity and classical controller. J. Global Inf. Manage. (2016) (Accepted for Publication)
11. K. Jagatheesan, B. Anand, S. Samanta, N. Dey, A.S. Ashour, V.E. Balas, Particle swarm optimization based parameters optimization of PID controller for load frequency control of multi-area reheat thermal power systems. Int. J. Artif. Paradigm (2016) (Accepted for Publication)
12. K. Jagatheesan, B. Anand, N. Dey, A.S. Ashour, V.E. Balas, Conventional controller based AGC of multi-area Hydro-Thermal power systems, in *Proceedings of 2016 International Conference on Computer Communication and Informatics (ICCCI-2016)* (Coimbatore, India, 7–9 Jan 2016), pp. 560–565
13. K. Jagatheesan, B. Anand, V. Santhi, N. Dey, A.S. Ashour, V.E. Balas, Dynamic performance analysis of AGC of multi-area power system considering proportional-integral-derivative controller with different cost functions, in *Proceeding of IEEE International Conference on Electrical, Electronics, and Optimization Techniques (ICEEOT)-2016* (DMI College of Engineering, Chennai, Tamilnadu, 3–5 Mar 2016)
14. N. Cohn, Some aspects of tie-line bias control on interconnected power systems. AIEE Trans. **75**, 1415–1436 (1957)
15. K.C. Divya, P.S.N. Rao, A simulation model for AGC studies of hydro-hydro systems. Electr. Power Energy Syst. **27**, 335–342 (2005)
16. S.C. Tripathy, T.S. Bhatti, C.S. Jha, O.P. Malik, G.S. Hope, Sampled data automatic generation control analysis with reheat steam turbines and governor dead-band effects. IEEE Trans. Power Apparat. Syst. **PAS-103**(5) May 1984

17. M.F. Hossain, T. Takahashi, M.G. Rabbani, M.R.I. Sheikh, M.S. Anower, Fuzzy-proportional integral controller for an AGC in a single area power system, in *Fourth International Conference on Electrical and Computer Engineering ICECE 2006* (Dhaka, Bangladesh, 19–21 Dec 2006)

18. D. Das, J. Nanda, M.L. Kothari, D.P. Kothari, Automatic generation control of a hydrothermal system with new area control error considering generation rate constraint. Electr. Mach. Power Syst. **18**, 461 (1990)

19. M.E. Baydokhty, A. Zare, S. baochian, Performance of optimal hierarchical type 2 fuzzy controller for load-frequency system with production rate limitation and governor dead band. Alexndraia Eng. J. (2016) (Article in Press)

20. R.E. Precup, S. Preitl, M. Bălaş, V. Bălaş, Fuzzy controllers for tire slip control in anti-lock braking systems, in *2004 IEEE Proceedings, International Conference on Fuzzy Systems*, vol. 3 (2004), pp. 1317–1322

21. I. Filip, O. Prostean, I. Szeidert, V. Balas, G. Prostean, Adaptive fuzzy controller and adaptive self-tuning controller: comparative analysis for the excitation control of a synchronous generator, in *7th WSEAS International Conference on Automation and Information (ICAI'06) (Cavtat, Croatia, 2006)*, pp. 89–94

22. M. Madic, D. Markovic, M. Radovanovic, Performance comparison of meta-heuristic algorithms for training artificial neural networks in modeling laser cutting. Int. J. Adv. Intell. Paradigms **2**(4), 316–335 (2012)

23. E. Yesil, Interval type-2 fuzzy PID load frequency controller using big bang-big crunch optimization. Appl. Soft Comput. **15**, 100–112 (2014)

24. H. Bevrani, P.R. Daneshmand, Fuzzy logic-based load-frequency control concerning high penetration of wind turbines. IEEE Syst. J. **6**(1), 173–180 (2012)

25. S. Tyagi, K.l.K. Bharadwaj, A particle swarm optimization approach to fuzzy case-based reasoning in the framework of collaborative filtering. Int. J. Rough Sets Data Anal. (IJRSDA) **1**(1), 48–64 (2014)

26. H.A. Yousef, K.A.L. Kharusi, M.H. Albadi, N. Hosseinzadeh, Load frequency control of a multi-area power system: an adaptive fuzzy logic approach. IEEE Trans. Power Syst. **29**(4), 1822–1830 (2014)

27. N.A. Setiawan, Fuzzy decision support system for coronary artery disease diagnosis based on rough set theory. Int. J. Rough Sets Data Anal. (IJRSDA) **1**(1), 65–80 (2014)

28. N.A. Mhetre, A.V. Deshpande, P.N. Mahalle, Trust management model based on fuzzy approach for ubiquitous computing. Int. J. Ambient Comput. Intell. (IJACI) **7**(2), 33–46 (2016)

29. V. Bureš, P. Tučník, P. Mikulecký, K. Mls, P. Blecha, Application of ambient intelligence in educational institutions: visions and architecture. Int. J. Ambient Comput. Intell. (IJACI) **7**(1), 94–120 (2016)

30. S. Bersch, D. Azzi, R. Khusainov, I.E. Achumba, Artificial immune systems for anomaly detection in ambient assisted living applications. Int. J. Ambient Comput. Intell. (IJACI) **5**(3), 1–15 (2013)

31. T.J. Ross, *Fuzzy Logic with Engineering Applications* (Wiley, England, 2004)

32. L.A. Zadeh, Fuzzy sets. Inf. Control **8**, 338–353 (1965)

33. G.A. Chown, R.C. Hartman, Design and experience with a fuzzy logic controller for automatic generation control. IEEE Trans. Power Syst. **13**(3), 965–970 (1998)

Evolutionary Algorithm Tuned Fuzzy PI Controller for a Networked HVAC System

Narendra Kumar Dhar, Nishchal K. Verma and Laxmidhar Behera

Abstract Heating, ventilation and air-conditioning (HVAC) system is an important component of Smart Home. The HVAC system is connected to network for the transfer of measurement data and control action packets from sensors to controller and controller to actuators respectively. The HVAC system can therefore be categorized as a Cyber-Physical system (CPS). Such systems are prone to communication uncertainties like packet losses and delays. Such systems require integrated architecture of communication and control. An evolutionary algorithm tuned fuzzy PI controller design coupled in a communication framework is presented in this paper for performance improvements of HVAC system. The entire architecture considers relevant system objectives based on system states and actuator actions. The formulated problem has been solved through real time optimization approach using the designed controller following the communication protocol. The developed algorithm helps in obtaining optimal actuator actions and shows a fast convergence to the different desired temperature sets. The results also show that the system can recover from sudden burst packet losses.

Keywords Cyber-physical system (CPS) · Heating ventilation and air conditioning (HVAC) · Fuzzy proportional-integral (PI) · Evolutionary algorithm
Wireless networked control

1 Introduction

Smart Homes have been an active technical research area in decades. They are founded on a symbiosis of applications that is geared for central control. The smart home concept on a large scale consists of lot of physical systems which are to be monitored and controlled using communication networks and means [1, 2]. The whole system can be viewed as Cyber-Physical system (CPS). A proper control and

N. K. Dhar (✉) · N. K. Verma · L. Behera
Department of Electrical Engineering, IIT Kanpur, Kanpur, India
e-mail: dharnarendra@gmail.com

© Springer International Publishing AG, part of Springer Nature 2018
L. A. Zadeh et al. (eds.), *Recent Developments and the New Direction
in Soft-Computing Foundations and Applications*, Studies in Fuzziness
and Soft Computing 361, https://doi.org/10.1007/978-3-319-75408-6_25

coordination between the systems can provide improved convenience, comfort, energy efficiency, robustness, reliability and has promise of providing increased quality of life. The heating, ventilation and air-conditioning (HVAC) system stands out as the essential component of Smart Home.

CPS is the intersection between the computational platform, networks and physical processes. Hence effective system integration design has emerged as biggest challenge [3, 4]. The difficulty emerges due to the heterogeneity of components and interactions across different platforms during the design. The interactions between the two platforms affect each other. The compositionality across multiple domains is a big challenge because of the lack of orthogonality among the design principles. CPS therefore requires the understanding of the joint dynamics of physical, computational, and communication disciplines.

The HVAC system comprise of a number of mechanical and electrical components [5]. These include the cooling, and heating units (chillers, dehumidifier and boilers), the ventilation units (air handling unit and variable air volume unit), and different zones handled by air regulator units. The hardware components include sensors (e.g., humidity, temperature and flow rate), mechanical and electrical actuators (e.g., valves, coils and dampers) and controllers form the parts of each subsystem. All these components need to be coordinated properly using communication protocol and control algorithm.

Communication architecture forms an integral part of CPS. The sensor-actuator network and various nodes deployed in between are responsible for reliable data transmission during real time control [6]. The realization of such proper communication architecture is challenged by heterogeneous reliability requirements, delay bounds, node mobility and route failures, channel errors and the energy efficiency of nodes [7].

From the application point of view there is dearth of approaches for the framework which involves the aspects of control, computation and communication simultaneously. Various works have been done on HVAC system. Improvement in performance for HVAC system has been investigated by the implementation of multiple-input-multiple-output (MIMO) robust controllers in [8]. Internet-based monitoring and controls for HVAC system has been done in [9] but lacks the inclusion and effects of communication constraints on the performance of system. The work in [10] proposes a distributed architecture for HVAC sensor fault detection and isolation. The work in [11] considers networked HVAC system as case study and proposes an online optimization approach for both control and communication parameters. This paper attempts to provide a combined framework for HVAC system. The approach is summarized as follows.

(1) A joint framework is presented which integrates both control and communication. An overall optimization objective is considered which accounts for various system constraints such as dropped packets, time delay etc.
(2) A communication protocol for online control of CPS is considered. This protocol coordinates with the control of physical processes and helps in effective transmission of measurement and control input packets.

(3) The comfort level of different zones in terms of different temperature requirement in a limited number of sampling periods is being achieved through the given approach. The fuzzy PI controller auto tuned by evolutionary algorithm gives a good and fast convergence.

The rest of the paper is organized as follows. In Sect. 2, the paper presents the modeling of HVAC system, the mathematical relation of thermal level attained in the respective zones and the required preliminaries. Since the HVAC model is a networked system, Sect. 3 provides the communication protocol connecting the physical devices, sensors, actuators etc. Section 4 contains the fuzzy PI controller design. The fuzzy control parameters are tuned using evolutionary algorithm. The algorithm is given in Sect. 5. The case study results are shown in Sect. 6. Section 7 summarizes the main conclusion of the work presented in the paper.

2 HVAC System Modeling

The HVAC system consists of number of subsystems. The optimization of individual subsystem leads to suboptimal solutions. Therefore a joint framework approach is required to obtain the optimal solution for the whole system.

The thermal system [11] consisting of four zones as shown in Fig. 1 is considered for our analysis. The HVAC inlet lets the fresh air enter inside at temperature $T_0(t)$ and flow rate $f_0(t)$. This is mixed with a part of recirculated air. The mixed air passes through the heat exchanger coil where an amount of heat (positive for heating and negative for cooling) is exchanged with the mixed air. The resulting air temperature in heat exchanger is T_s. It is assumed that perfect mixing in heat exchanger takes place.

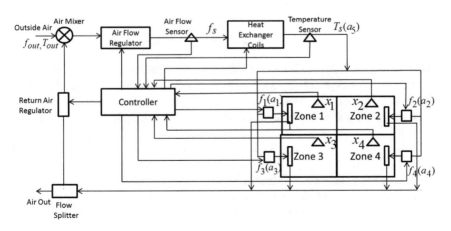

Fig. 1 HVAC system model

The conditioned air from the heat exchanger is passed into the thermal zones. Each zone has an actuator for controlling the air flow rate. The individual actuator action a can vary depending on the temperature requirement of each zone. It is assumed that the temperature throughout the thermal space is same as that of air entering and exiting out of it. The air returning from each zone is let out and the remaining air is allowed to mix with air from outside.

The controllers work in tandem to regulate the temperature of air in the heat exchanger coils as well as actuator actions to maintain the flow rate to attain desired temperature in each zone. It is assumed that there are no infiltration and exfiltration effects and thus the net flow rate is the sum of individual flow rates. The air humidity and transient effects in different parts of the system are neglected.

There are n_s wireless sensors deployed to measure the physical process of interest. The respective process state variables associated are $\{x_j | 1 \leq j \leq n_s\}$ in numbers. Similarly there are n_a actuators and the actions are defined as $a_i\{1 \leq i \leq n_a\}$. The sampling period length is fixed and is given as h.

The dynamics of the system states for HVAC are considered same as [11] and is given as:

$$\dot{x}(t) = f(a, t)x(t) + g(a, t) \tag{1}$$

where $f(a, t) \in \mathbb{R}^{n_a \times n}$ and $g(a, t) \in \mathbb{R}^{n \times n_a}$ are functions of $a(t)$. The control input is generated after each sampling period; hence the discrete system state is

$$x[k] = F(a[k - 1])x[k - 1] + G(a[k - 1]) \tag{2}$$

where, $F(a[k]) = \exp(\int_{h_k}^{h_{k+1}} f(a, t)dt)$, and $G(a[k]) = \exp(\int_{h_k}^{h_{k+1}} g(a, t) \exp(\int_{h}^{t} f(a, \tau)) d\tau)$.

The system state $X = \{x, a\}$ is defined for all the variables considered for system objective function. The objective function is given as

$$J(X) = J\{x, a\}$$
$$J(x, a) = \lambda_1 J_1 + \lambda_2 J_2 + \cdots + \lambda_n J_n \tag{3}$$

where, $\lambda_i, J_i\{i = 1, 2, \ldots, n\}$ are weighting factors and individual cost functions respectively. The objective function is constrained by limited abilities of actuator devices $a \in [\underline{a}, \overline{a}]$, dropped packets and time delay during communication.

The system equation is based on energy conservation principle and is given by

$$\rho_a c_p V \dot{x} = f \rho_a c_p (T_s - x) + Q \tag{4}$$

where, x is the temperature of thermal zone, ρ_a is air density, c_p is specific heat capacity, V is volume of the thermal zone, f is the air flow rate, T_s is the temperature of heat exchanger coil, and Q is the thermal load.

3 Communication Protocol

The network transmission of data packets are prone to time delays and packet losses [12–14]. The Fig. 2 shows the schematic of closed network between sensors, controller and actuators using wireless communication. The concept of communication model as in [11] has been employed in our work. The estimation and prediction [15] of system states and control decisions form an integral part of this.

The actuators are time-driven. The sensors and actuators are synchronized with the same sampling period h. The control packet with the latest timestamp is used for controlling the plant. The packets arriving at the actuators in incorrect sequence are considered as dropped packets. The packets which are corrupt are also considered as dropped packets. The communication model considers a number of control intervals θ_a. Each control interval θ_a indexed as k_a and its successive values consists of number of sampling periods k_s. The length of each sampling period is fixed but the control intervals have variable lengths. The algorithm obtains the optimal length of each control interval. Larger the length of each control interval better the decision for actuator actions. The actuator actions are decided based on the comparative value of objective function. The algorithm obtains the objective function value for the forthcoming control interval $[k_a + 2]$ at different sampling instants k_s. Since the sensors, controller and actuators are well synchronized each of them perform their tasks accordingly during the respective sampling periods. Basically the tasks are scheduled for each of the devices in the network.

The sensors report the measurement packets as well as other required packets to the controller in the initial phase of each sampling period. After the packets are transmitted the sensors go on to sleep mode for the rest of the sampling period. The controller receives the packets from the sensor as the first task in every sampling period. It then estimates and predicts the system states [16, 17]. The objective function value is determined based on all the estimated and predicted states and

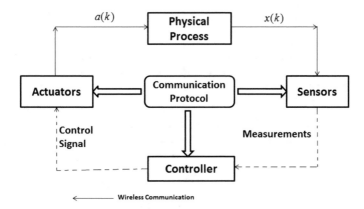

Fig. 2 Schematic of cyber physical system

control actions. It then waits for the next sampling period. The actuators continue to act as per the control commands received at the beginning of each control interval. The actuator has a task of receiving the command packets in the last sampling period. After the required packets are received successfully the actuator sends back the acknowledgments to the controller. If the packets are corrupt or get dropped due to inconsistency in the communication the actuators continue to perform as per the previous command packets.

The controller decides the required control actions in the last sampling period. It sends them to the actuators and waits for the acknowledgment. The control actions are sent in the form of a command packet $U[k_a + 1]$ having index as $k_a + 1$.

$$U[k_a + 1] = [u[k_a], u_{eq}[ka + 2]] \tag{5}$$

$u[k_a + 1]$ is control action for the next control interval. The actuator on receiving the command packet updates its control action as $u[k_a + 1]$. The action $u_{eq}[k_a + 2]$ stabilizes the system during $[k_a + 2]$th step.

The controller does the task of estimation and prediction based on the received packets. Similarly, the actuators also update their actions on successful reception of packets. Hence, the communication protocol takes care of the packet losses with the help of two parameters $v_i[k_s]$ and $w_j[k_s]$. v_i is designed for sensor to controller packet losses whereas $w_j[k_s]$ is for controller to actuator packet losses. On reception of packets by the controller from the ith sensor the v_ith parameter is set as 1 otherwise it is 0. The w_jth parameter is 1 when the jth actuator receives command packet from controller successfully otherwise $w_j = 0$ (Fig. 3).

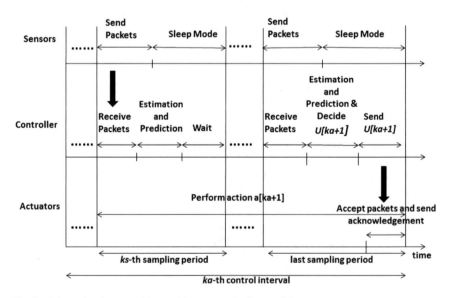

Fig. 3 Schematic of a control interval in communication model

4 Fuzzy PI Controller

Controllers consisting of proportional-integral (PI) units are generally used in commercial HVAC systems. According to Haines, PI controllers are the usually used method to control HVAC systems in order to improve accuracy and consumption of energy as compared to that of proportional control [18].

Automation devices typically implement the velocity or incremental algorithm. The incremental algorithm is based on the calculation of the change in control value. The conventional systems can be designed using fuzzy theory [19]. The PI controller designed for the HVAC system is basically a fuzzy PI controller [20]. The incremental fuzzy PI controller with r rules is as follows:

Rule i : IF $e(k)$ is F_i, THEN

$$\Delta u^i(k) = K_p(k)(e(k) - e(k-1)) + K_i(k)e(k) \tag{6}$$

where $e(k) = x_d - x(k)$, x_d is the desired temperature. $F_i(i = 1, \ldots, 2n-1)$ are fuzzy sets, $n > 1$. The number of fuzzy rules is $r = 2n - 1$, and K_p, K_i are PI parameters. The fuzzy sets F_j are represented by membership functions μ_j. Assuming the fuzzy set F_n accounts for $e(k) = 0$, F_j considers more negative $e(k)$ than F_{j+1} accounts for $1 \leq j \leq n - 1$. F_{j+1} considers more positive $e(k)$ than F_j accounts for $n \leq j \leq 2n - 2$. Figure 4 characterizes the membership functions μ_i with $n = 3$. $ZO(F_3)$ denotes zero, $NB(F_1)$ represents negative big, $NM(F_2)$ is negative medium, $PM(F_4)$ is positive medium, and $PB(F_5)$ is positive big. To improve the transient performances the gains are set as

$$K_p^1 \geq K_p^2 \geq \cdots \geq K_p^{n-1} \geq K_p^n \geq 0$$
$$K_p^{2n-1} \geq K_p^{2n-2} \geq \cdots \geq K_p^{n+1} \geq K_p^n \geq 0$$
$$K_i^1 \geq K_i^2 \geq \cdots \geq K_i^{n-1} \geq K_i^n \geq 0$$
$$K_i^{2n-1} \geq K_i^{2n-2} \geq \cdots \geq K_i^{n+1} \geq K_i^n \geq 0 \tag{7}$$

The final PI control input can be obtained using standard fuzzy inference method,

Fig. 4 Membership functions for fuzzy PI controller

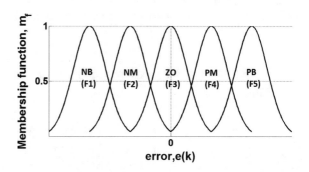

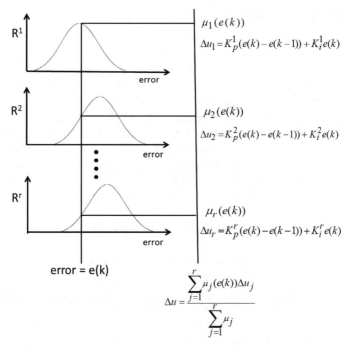

Fig. 5 Fuzzy inference for final PI control input

$$\Delta u(k) = - \sum_{i}^{r} h_i(e(k))[K_p^i(k)(e(k) - e(k-1)) + K_i^i(k)e(k)] \qquad (8)$$

where $R_1, R_2, \ldots, R_r$ are r rules and each IF-THEN rule has normalized weight $h_i = \mu_i / \sum_{j=1}^{r} \mu_j$, $h_i \geq 0$, $\sum_{i}^{r} h_i = 1$. The final PI control input is obtained after fuzzy inference as shown in Fig. 5. The input variable is $e(k)$. The number of the fuzzy rules is $r = 2n - 1$, and $\Delta u(k)$ is obtained as output variable.

The PI controller parameters generating control signal for actuators of HVAC system are not fixed. They are adjusted using evolutionary algorithm. The algorithm considers a population of individuals representing parameters of controller. The parameters of best individual are considered for the control input generation.

5 Evolutionary Algorithm for Tuning Fuzzy PI Controller Parameters

The fuzzy PI control law depends on the shape of membership function and K_p^j and K_i^j values. The fuzzy parameters need to be fine tuned. For this purpose evolutionary algorithm (EA) [20] has been used. EA has been widely used for optimization problems. EAs search for the parameters globally. The algorithm passes through three basic processes: selection, crossover and mutation. The solution is characterized by a chromosome containing several genes. These genes are the required variables. A chromosome containing N_v genes, is represented as an N_v element vector. The N_p chromosomes with N_v genes can be represented through $N_p \times N_v$ matrix. Every chromosome is ranked on the basis of fitness function value. The selection process considers highly ranked chromosomes for the mating pool. The mating takes place with the given crossover rate X_r. There are exchange of genes between the parent chromosomes from mating pool during crossover operation. The mutation process then changes gene values of chromosomes randomly. The required parameters of EA are G_m (generation count), N_P (number of individuals in population), X_r (rate of crossover), and X_m (rate of mutation). The basic steps followed in EA are.

5.1 Search Variables and Their Relation

The chromosomes in the population represent fuzzy PI parameters.

$$K_p^1 = K_p^{2n-1} = K_p^{max}, K_p^2 = K_p^{2n-2} = \alpha_1 K_p^{max}, \ldots, K_p^n = \prod_{j=1}^{n-1} \alpha_j K_p^{max}$$

$$K_i^1 = K_i^{2n-1} = K_i^{max}, k_i^2 = k_i^{2n-2} = \beta_1 K_i^{max}, \ldots, K_i^n = \prod_{j=1}^{n-1} \beta_j K_i^{max} \tag{9}$$

where K_p^{max} and K_i^{max} are maximum PI gains and $\alpha_j \in [0, 1], \beta_j \in [0, 1], (j = 1, \ldots, n)$. The Gaussian membership functions are chosen for fuzzy sets.

$$\mu_j(e(k)) = e^{-\sigma_j(e(k)-W_j)^2} \tag{10}$$

where σ_j and W_j are parameters shaping the membership functions.

$$W_{2n-1} = -W_1 = \delta_1 W_{max}; W_{2n-2} = -W_2 = \delta_1 \delta_2 W_{max}$$

$$W_{n+1} = -W_{n-1} = \prod_{j=1}^{n-1} \delta_j W_{max}; W_n = 0$$

$$\sigma_1 = \epsilon_1 \sigma_{max} > 0; \sigma_{n-1} = \epsilon_{n-1} \sigma_{max} > 0$$

$$\sigma_{2n-1} = \sigma_1; \sigma_{n+1} = \sigma_{n-1}; \sigma_n = \epsilon_n \sigma_{max} > 0 \tag{11}$$

where, $\sigma_j > 0, W_j > 0, \delta_j \in [0, 1]$, and $\epsilon_j \in [0, 1]$. There are $r = 2n - 1$ fuzzy rules and $4n - 3$ fuzzy parameters. A chromosome c_j therefore consists of $4n - 3$ genes, i.e. $N_v = 4n - 3$. For $n = 3$, a chromosome c_j has $N_v = 9$ genes.

$$c_j = [\alpha_j^1, \alpha_j^2, \beta_j^1, \beta_j^2, \delta_j^1, \delta_j^2, \epsilon_j^1, \epsilon_j^2, \epsilon_j^3]$$

The fuzzy membership function values for individual chromosome c_j are then realized as:

$$\mu_1(e(k)) = e^{-\epsilon_j^1 \sigma_{max}(e(k)+\delta_j^1 W_{max})^2}$$

$$\mu_2(e(k)) = e^{-\epsilon_j^2 \sigma_{max}(e(k)+\delta_j^1 \delta_j^2 W_{max})^2}$$

$$\mu_3(e(k)) = e^{-\epsilon_j^3 \sigma_{max} e(k)^2}$$

$$\mu_4(e(k)) = e^{-\epsilon_j^2 \sigma_{max}(e(k)-\delta_j^1 \delta_j^2 W_{max})^2}$$

$$\mu_5(e(k)) = e^{-\epsilon_j^1 \sigma_{max}(e(k)-\delta_j^1 W_{max})^2} \qquad (12)$$

5.2 Population Generation and Initialization

The population is generated with N_p individuals. The crossover and mutation operators X_r and X_m respectively are considered. The fuzzy parameters are tuned for generation or iteration size of G_m. The initial population individuals are randomly initialized between 0 and 1.

5.3 Fitness Evaluation and Ranking Individuals

The chromosomes or individual members of the population are ranked by evaluating the fitness function. The function for fitness evaluation considered is root mean square error (RMSE) between the desired temperature of the zones and the actual temperature. The best individuals are then considered for forming the mating pool. The crossover rate X_r helps in obtaining $X_r N_p$ parents.

5.4 Crossover Operation

The crossover operation takes two parents randomly from the mating pool to obtain an offspring. Let two parents be c_a and c_b and the offspring obtained be c_{off}.

$$c_{off} = c_a - \eta(c_a - c_b)$$

This process is repeated until $(1 - X_r)N_p$ offspring is obtained. The value of η is randomly considered between 0 and 1.

5.5 Mutation Operation

Some of the genes from offspring are randomly considered with a mutation rate X_m. The selected genes amount to $X_m(1 - X_r)N_pN_v$. These genes are randomly assigned values between 0 and 1. A new population is formed from the $(1 - X_r)N_p$ offspring and the X_rN_p individuals of the mating pool after mutation.

5.6 Final Parameters

The fuzzy PI parameters are tuned over iterations to obtain the desired state values. The final parameters are obtained in either of the two cases. One being the convergence of the fitness function value while the other is reaching the maximum generation number G_m.

6 Case Study

The proposed method is applied to evaluate the performance of HVAC system. The HVAC system has to regulate the air flow to four thermal zones. The supply air is a mixture of outdoor and return air. The air blown into each of the four zones is heated or cooled to temperature T_s. There are four flow rate regulators in each of the zones.

$$f_s = \sum_{i=1}^{4} f_i \tag{13}$$

f_s is the net flow rate and f_i is the flow rate in each zone. The objective function for the HVAC system is given as:

$$J(X) = \lambda_1 J_1 + \lambda_2 J_2 + \lambda_3 J_3$$

$$J_1 = \sqrt{\sum_{i=1}^{4} PMV_i^2/4}; J_2 \propto f_s^3$$

$$J_3 \propto f_s \rho_a c_p |T_s - T_o ut|; J_4 \propto 1/\theta_a \tag{14}$$

J_1 is based on the thermal comfort level in each zone. Predicted Mean Vote (PMV) is adopted as the thermal comfort level index [21]. J_2 and J_3 represent the energy consumption by flow rate regulators and the thermal exchanger coils. The sub-objective function J_4 depends on the control interval value. Larger the control step value better it is for determining appropriate control actions. The temperatures of each zones are defined as $x_i\{i = 1, \ldots, 4\}$. Each of the four zones has wireless temperature sensor which sends the measured temperature data to the controller. The parameter values given in [11] are considered, $\rho_a = 1.19\,\text{kg/m}^3$, $V_i = \{36, 40, 30, 35\}\,\text{m}^3$, $Q_i = 500\,\text{W}$, $f_s \in [0.04, 2]\,\text{m/s}$, $T_s \in [20, 36]\,°\text{C}$, $\theta_a \in [20, 180]$ ms, weights $\lambda_i i = 1, 2, 3 = [0.8, 0.1, 0.1]$, length of each sampling period $h = 20\,\text{ms}$. The measurement noise in sensor data has a covariance of 0.1. The initial temperatures for the zones are $\{20, 20, 20, 20\}\,°\text{C}$. The fuzzy PI controller has five gaussian membership functions {NB, NM, ZO, PM, PB}. The PI parameters K_p and K_i both have maximum values set as 0.5 as obtained from the basic tuning methods such as Ziegler-Nichols. The evolutionary algorithm considers a population size of 20 individuals. Each member is assigned random value between 0 and 1 initially. The maximum generation count is 500. The designed controller is implemented on HVAC system for $\{27, 24, 26, 25\}\,°\text{C}$. Figures 6 and 7 show the results for consistent and inconsistent (data losses nearly 25%) communication respectively. Various parameter and objective function values for both the cases are presented in Table 1. Table 2 shows the convergence and effective packet values for different data cases. It takes around 200 sampling instants for fulfilling the different temperature requirement of all the zones. The actuator actions are also optimized as can be seen from Table 1. The control step value tends to its extremum. It thereby incorporates a number of sampling periods for required task execution and also saves communication energy in frequent command packet transmission. The objective function values for both the cases are approximately same. It is little bit larger for inconsistent communication. Figure 8a shows that the system can recover from sudden burst packet losses. Figure 8b, c show the convergence for two communication scenarios when thermal zones are initially at different temperatures. The convergence of system is

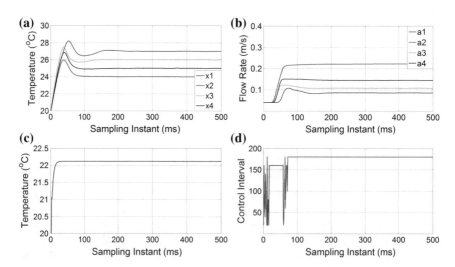

Fig. 6 Four zone temperatures $\{27, 24, 26, 25\}\,°\text{C}$ **a** real temperature **b** flow rate **c** coil temperature **d** control interval

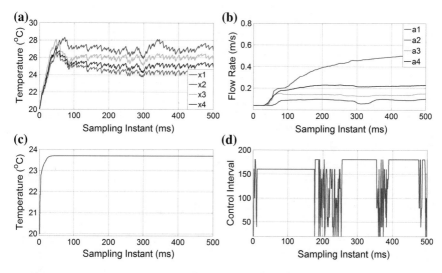

Fig. 7 Four zone temperatures {27, 24, 26, 25} °C **a** real temperature **b** flow rate **c** coil temperature **d** control interval

Table 1 Achieved state and other parameters at various iterations

{27, 24, 26, 25} °C	Iteration	Temperature (°C)				Flow rate regulators (m/s)				Coil temp.	Obj. fn.
		Zone-1	Zone-2	Zone-3	Zone-4	a_1	a_2	a_3	a_4		
No packet loss	50	28.2011	25.9360	27.4473	26.8338	0.0658	0.1160	0.0889	0.0955	22.1143	5.7335
	100	26.4713	24.0541	25.7428	24.9064	0.0954	0.2178	0.1134	0.1502	22.1147	5.5154
	200	27.0467	24.0043	26.0090	24.9966	0.0854	0.2209	0.1075	0.1453	22.1147	5.5531
Packet losses (25%)	50	27.3589	26.3859	27.8425	26.6327	0.0401	0.0889	0.0801	0.0959	23.6920	5.7450
	100	27.0496	25.0377	25.9504	25.2917	0.0878	0.2174	0.1408	0.1828	23.6931	5.5886
	200	26.7091	24.4014	26.0097	25.0780	0.0973	0.3948	0.1423	0.2292	23.6931	5.5830

Table 2 Effective packets and convergence of fuzzy PI controller for {27, 24, 26, 25} °C

Data packets (%)	Convergence (sampling instant)	Effective packets
100	200 (smooth)	500
73.6	350 (not so smooth)	368
52.2	435 (not smooth)	261

rather quick when desired temperature requirements are same and initially at different temperatures as shown in Fig. 8d.

The Simulated Annealing based optimization algorithm (SABA) developed in [11] applied on HVAC system looks for optimal state values. The objective function

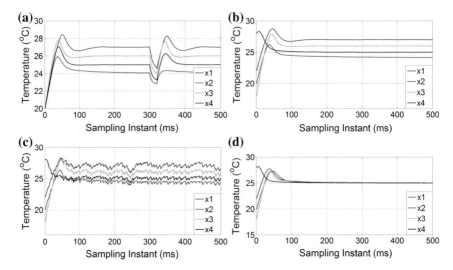

Fig. 8 Parameters for four zone temperatures $\{27, 24, 26, 25\}$ °C **a** burst packet losses **b** no packet loss **c** continuous packet losses **d** same desired temperature

Fig. 9 Optimal temperatures in thermal zones using SABA

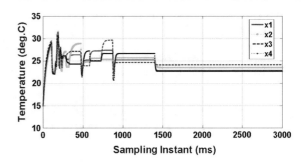

is similar. Minimization of the objective function on implementation of the algorithm given in the paper provided us a convergence around a common temperature for all the four thermal zones. The Fig. 9 shows convergence occurs around 2000 sampling instant whereas in this work the convergence is fast and it is around 200 instants for different required temperatures and around 120 instants for common temperatures. The evolutionary fuzzy PI controller obtains the desired temperatures and optimal control actions very quickly.

7 Conclusion

The HVAC system is a networked one and therefore requires an efficient communication protocol for control action generation. The paper considers a joint optimization framework consisting of communication protocol and online intelligent control. The control actions for the actuators are generated based on the principle of variable control steps and task scheduling of the devices in the network. The algorithm takes care of uncertainties as well as the physically constrained actuator actions. The fuzzy PI controller gives fast convergence. The controller parameters are not fixed but auto tuned using evolutionary algorithm. The whole methodology shows its efficacy in providing fast, optimized results for the networked HVAC system. The entire architecture and algorithm can be implemented on other networked systems also.

References

1. L. Mingfu, L.H. Ju, Design and implementation of smart home control system based on wireless sensor networks and power line communications. IEEE Trans. Ind. Electron. **62**(7), 4430–4442 (2015)
2. A. Zanella, N. Bui, A. Castellani, L. Vangelista, M. Zorzi, Internet of things for smart cities. IEEE Internet Things J. **1**(1), 22–32 (2014)
3. J. Sztipanovits, G. Karsai, P. Antsaklis, J. Baras, Toward a science of cyber-physical system integration. Proc. IEEE **100**(1), 29–44 (2012)
4. W. Zhang, M.S. Branicky, S.M. Pjillips, Stability of networked control systems. IEEE Control Syst. Mag. **1**, 84–99 (2001)
5. B. Sun, P.B. Luh, Q.S. Jia, Z. O'Neill, F. Song, Building energy doctors: an SPC and Kalman filter-based method for system-level fault detection in HVAC systems. IEEE Trans. Auto. Sci. and Eng. **11**(1), 215–229 (2014)
6. M.B.G. Cloosterman, N.D. Wouw, W.P.M.H. Heemels, H. Nijmeijer, Stability of networked control systems with uncertain time-varying delays. IEEE Trans. Auto. Control **54**(7), 1575–1580 (2009)
7. M. Nourian, A.S. Leong, S. Dey, D.E. Quevedo, An optimal transmission strategy for Kalman filtering over packet dropping links with imperfect acknowledgements. IEEE Trans. Control Netw. Syst. **1**(3), 259–271 (2014)
8. M. Anderson, M. Buehner, P. Young, D. Hittle, C. Anderson, J. Tu, D. Hodgson, MIMO robust control for HVAC systems. IEEE Trans. Control Syst. Tech. **16**(3), 475–483 (2008)
9. P.I.-H. Lin, H.L. Broberg, Internet-based monitoring and controls for HVAC applications. IEEE Ind. Appl. Mag. **8**(1), 49–54 (2002)
10. V. Reppa, P. Papadopouos, M.M. Polycarpou, C.G. Panayiotou, A distributed architecture for HVAC sensor fault detection and isolation. IEEE Trans. Control Syst. Tech. **23**(4), 1323–1337 (2015)
11. X. Cao, P. Cheng, J. Chen, Y. Sun, An online optimization approach for control and communication codesign in networked cyber-physical systems. IEEE Trans. Ind. Inf. **9**(1), 439–450 (2013)
12. A. Chiuso, N. Laurenti, L. Schenato, A. Zanella, LQG cheap control over SNR-limited lossy channels with delay, in *IEEE 52nd Annual Conference on Decision and Control (CDC)* (2013), pp. 3988–3993
13. S. Dey, A. Chiuso, L. Schenato, Remote estimation with noisy measurments subject to packet loss and quantization noise. IEEE Trans. Control Netw. Syst. **1**(3), 204–217 (2014)
14. H. Li, M.Y. Chow, Z. Sun, Optimal stabilizing gain selection for networked control systems with time delays and packet losses. IEEE Trans. Control Syst. Tech. **17**(5), 1154–1162 (2009)
15. Z.H. Pang, G.P. Liu, D. Zhou, M. Chen, Output tracking control for networked systems: a model-based prediction approach. IEEE Trans. Ind. Electron. **61**(9), 4867–4877 (2014)
16. L. Schenato, B. Sinopoli, M. Franceschetti, K. Poola, S.S. Sastry, Foundations of control and estimation over lossy networks. Proc. IEEE **95**(1), 1693–1698 (2007)
17. X. Cao, J. Chen, C. Gao, Y. Sun, An optimal control method over wireless sensor/actuator networks. Comput. Electr. Eng. **35**(5), 748–756 (2009)

18. J. Bai, X. Zhang, A new adaptive PI controller and its application in HVAC systems. Energy Convers. Manage. (ScienceDirect) **49**, 1043–1054 (2007)
19. N.K. Verma, M. Hanmandlu, From a Gaussian mixture model to nonadditive fuzzy systems. IEEE Trans. Fuzzy Syst. **15**(5), 809–826 (2007)
20. H.H. Choi, H.M. Yun, Y. Kim, Implementation of evolutionary fuzzy PID speed controller for PM synchronous motor. IEEE Trans. Ind. Inf. **11**(2), 540–547 (2015)
21. G. Ye, C. Yang, Y. Chen, Y. Li, A new approach for measuring predicted mean vote (PMV) and standaard effective temperature (SET). Build. Environ. **38**(1), 33–44 (2003)

Part VIII
Clustering, Classification, and Hierarchical Modeling

Study of Soft Computing Methods for Large-Scale Multinomial Malware Types and Families Detection

Lars Strande Grini, Andrii Shalaginov and Katrin Franke

Abstract There exist different methods of malware identification, while the most common is signature-based used by anti-virus vendors that includes one-way cryptographic hash sums to characterize each particular malware sample. In most cases such detection results in a simple classification into malware and goodware. In a modern Information Security society it is not enough to separate only between goodware and malware. The reason for this is increasingly complex functionality used by various malware families, in which there has been several thousand of new ones created during the last decade. In addition to this, a number of new malware types have emerged. We believe that Soft Computing (SC) may help to understand such complicated multinomial problems better. To study this we ensambled a novel large-scale dataset based on 400 k malware samples. Furthermore, we investigated the limitation of community-accepted Soft Computing methods and can clearly observe that the optimization is required for such non-trivial task. The contribution of this paper is a thorough investigation of large-scale multinomial malware classification by Soft Computing using static characteristics.

1 Introduction

Malicious software or malware are software that perform unwanted actions in targeted system. McAfee malware zoo [1] now includes 440,000,000 samples by Q2 2015. Malware pose a significant threat to every device connected to Internet in terms of both privacy and economical loses. The majority target MS Windows NT

L. S. Grini (✉) · A. Shalaginov · K. Franke
Norwegian Information Security Laboratory, Center for Cyber - and Information Security,
Norwegian University of Science and Technology, Trondheim, Norway
e-mail: lars.grini@gmail.com

A. Shalaginov
e-mail: andrii.shalaginov@ccis.no

K. Franke
e-mail: katrin.franke@ccis.no

© Springer International Publishing AG, part of Springer Nature 2018
L. A. Zadeh et al. (eds.), *Recent Developments and the New Direction in Soft-Computing Foundations and Applications*, Studies in Fuzziness and Soft Computing 361, https://doi.org/10.1007/978-3-319-75408-6_26

Operation Systems (OS) family that has been in use since the end of 1990th. There are two main approaches to analyze malware samples: static and dynamic analysis. *Static analysis* includes scanning files to collect relevant raw characteristics from the file, while *dynamic analysis* reveals behavior characteristics by executing them in an isolated environment as studied by Ravi et al. [2] for multiple malware families. As a result of obfuscation, different methods utilized by malware to avoid detection, some dynamic analysis methods require user interaction to trigger the malicious behaviour. Despite the fact that dynamic analysis could be comprehensive it requires much more time and resources than static analysis methods. From this we consider static analysis as more appropriate for our project taking in mind the large number of samples and the time constraint. The main goal of this paper is to study how static analysis of PE32 and classification by SC method can facilitate large-scale detection of malware families and malware types.

Static analysis is a term that covers a range of techniques to dissect malware samples and to gather as much information as possible without executing the file. One of the commonly used tools is the *VirusTotal* [3] online scanner described by Seltzer [4], which provides output from different scanning tools, and also checks the submitted sample against 65 anti-virus databases and then gives the detection ratio as well as the names that each vendor has labelled the sample with. Another is the *PEframe* [5] that is able to detect packers, anti-debug and anti-VM techniques as well as URLs and filenames used by the sample. *PEframe* and *strings* are somewhat redundant, the different tools can yield different information as, i.e. *PEframe* gives better structured output. Reverse engineering is a growing discipline to perform this dissection thoroughly in terms of using software to generate the assembly instructions and then to be able to determine the actions that sample performs on a system as given in the Malware Cookbook by Ligh et al. [6]. We will however not use this approach in our work. Due to the increasingly used and more complex obfuscation techniques, static malware analysis are becoming increasingly difficult to perform as studied by Gavrilut et al. [7] and Idika et al. [8]. The biggest advantage of static analysis methods however, are that they considerably quicker, making it scalable and environment-independent.

Most antivirus scanners use signature-based and heuristic-based detection methods, where they search for known patterns in executable code as well as hash sum of the file against known malicious files in a database. A limitation of signature-based methods are that the malware must be obtained and analyzed before the antivirus vendors can update their databases as described by Ye et al. [9] and Kolter et al. [10]. Far more flexibility provides ML-based methods such that nature-inspired SC methods that allows to build inexact models from incomplete and complex data. Das et al. [11] gave an overview of how SC methods can be used in different areas, including Information Security.

Our main motivation was to perform study of SC method on a class of malware detection problem, both types and families that are designed for MS Windows as Portable Executables (PE32) files. If we look on previous studies that involve PE32 file formats for MS Windows, we can see that majority only target differentiation between "*malicious*" and "*benign*" samples. However, our idea is to study how static analysis of PE32 files with help of SC can facilitate large-scale malware detection into families and types. We also done malware samples acquisition that results in more than 400k samples that were analysed afterwards. Further, we selected 328k for multinomial malware classification. The paper is organized as following. The Sect. 2 gives an overview of the different malware classification approaches, naming schemes used by anti-virus vendors, etc. Later, Sect. 3 provides an overview of the novel dataset was created, extracted features and feature selection process. Then, Sect. 4 gives a comprehensive overview of the accuracy comparison between SC and Hard Computing (HC) on such class of problems. Finally, Sect. 5 concludes the contribution of this study.

2 Related Works

Today there exists, to the authors' knowledge, no research which focuses on large-scale multinomial classification of malware into types and families based on features derived from static analysis using SC methods.

2.1 Malware Classification

Previous works mostly focus on differentiation of a file in benign or malicious. This is a *binary classification*, where a heap of malware samples is classified against a collection of goodware. Cohen [12] suggested in 1987 that there are no algorithm that will be able to confidently detect all computer viruses. This assertion also was strengthened by Chess et al. [13]. As result, we can assume that no methods can achieve 100% classification accuracy on large-scale sets. Bragen [14] applied Machine Learning (ML) on opcode sequences and achieved 95% accuracy with RandomForest method. Kolter et al. [10] used 1,971 malicious files and 1,651 benign, while Bragen used only 992 malicious and 771 benign. Markel et al. [15] used PE32 header data in malware and benign files detection on Decision Tree, Naive Bayes and Logistic Regression. Authors achieved 0.97 F-score on binary classification. Further, Shankarapani et al. [16] applied PE32 file parser to extract static features for similarity analysis. In overall 1,593 samples were acquired for binary classification. The limitations of mentioned researches are low number of files.

In contrary to binary, *multinomial classification* can be described as a detection of whether a malware belongs to a particular family or type. Rieck et al. [17] studied 14 different malware families extracted from 10,072 unique binaries. Authors achieved on average 88% accuracy of family detection using individual SVM for each one. Further, Zhang et al. [18] explored binary classification using binary sub-sets of 450 viruses and goodware based on the 2-gram analysis. So, not many works target the problem explored in this paper.

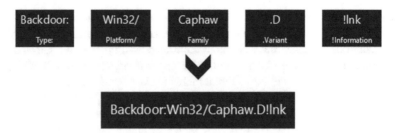

Fig. 1 CARO malware naming scheme [20]

2.2 Malware Naming by Anti-virus Vendors

We can state that there exist a number of malware types (like *trojan, backdoor*, etc.) and families (like *Poison, Ramdo*, etc.), which are commonly defined by Information Security community. In 1991 the Computer Antivirus Research Organization (CARO) proposed a standardized naming scheme for malware [19]. Although CARO states that this naming scheme is "widely accepted", we found that from all the vendors on *VirusTotal*, apparently Microsoft is the only one that complies with this. So, it is challenging to establish common pattern in scanners results across anti-virus databases. Example of CARO naming is given in the Fig. 1.

2.3 Soft Computing in Multinomial Classification

For the classification of viruses, static features can be automatically extracted from PE32 headers and the results used to build the classification model(s). To the authors knowledge, the application of SC for multinomial malware detection has not been studied. With respect to classification problems, SC encompasses a set of powerful methodologies that can produce generalized models, although with inexact solutions. Some of the existing methods, such as MLP and SVM based classifiers, were originally designed for binary problems. Other methods, such as Naive Bayes and Bayesian Networks, were designed to handle multinomial tasks. In a prominent study by Ou et al. [21], different models of multilayer ANN were studied with respect to multi-class problems. In a work by Shalaginov et al. [22] some thoughts regarding application of Neuro-Fuzzy for multi-class problems were presented with a number of improvements. However, used data sets were small.

3 Methodology

Below we will give an overview of how the malware collection was performed, what are the features extracted and which SC methods were selection for our study.

3.1 Malware Acquisition

There was a malware collection initiative that took place within the Testimon Research group[1] at NTNU Gjøvik during Summer 2015 using following sources: *maltrieve* [23] that extracts recent modern malware samples, 10 first archives available at *VirusShare* [24] starting from *VirusShare_00000.zip*, collection from *VxHeaven* and files that students share within the group. After thorough analysis and filtering, we derived PE32 files and removed other types of files. We ended up with 407,741 malware samples.

Since we targeted a static analysis it was decided to extract as much raw characteristic as possible to facilitate large-scale malware analysis. Two main sources that we used were *PEframe* [5] and *VirusTotal* [3]. *PEframe* presents comprehensive set of attributes that can be found in the PE32 header. There were some works before showing that it can be possible to identify malware using such headers information. *VirusTotal* presents scan results from over 65 anti-virus databases, information about possible packers and compressors in addition to basis PE32 headers data. All data from the *VirusTotal* were gathered using Private API. Moreover, we used standard Linux tools to retrieve various file characteristics, e.g. size of different sections, strings and entropy. Finally, we created a MySQL database with raw characteristics of all PE32 executables filtered out from the heap of gathered earlier samples.

3.2 Feature Extraction

Before meaningful automated analysis can be executed, we must extract numerical features by performing a preliminary manual processing of raw static characteristics. On this step we (i) transform gathered raw characteristics into numerical features. To accomplish this task, we processed *VirusTotal* and *PEframe* reports. The content from both sources was in a format of serialized JSON object. Therefore, we focused on extracting as many relevant numerical features as possible from this structure. Then, (ii) we extracted corresponding malware types and families names from *VirusTotal* output. Further, it was made feature selection to determine which features that contribute most to classification. Finally, a Python script was written to first extract the data, and then put it into a new database table. Extracted features are given in the Table 1.

3.3 Feature Selection Methods

Proper feature selection will contribute to a higher classification accuracy and reduction of the computational complexity, which in turn will increase the overall classi-

---

[1]https://testimon.ccis.no.

Table 1 Description of all 37 numerical features that were extracted from raw malware characteristics

From PEframe output	
pe_api	The number of suspicious API class
pe_debug	The number of opcodes recognized as common anti debug techniques
pe_packer	The number of packers discovered by PEframe
pe_library	The number of DLL calls in the file. Retrieved from PEframe
pe_autogen	The number of autogens discovered by PEframe
pe_object	The number of object calls discovered by PEframe
pe_executable	The number of .exe calls within the file
pe_text	The length of the text-field
pe_binary	The number of binary(.bin) files called by the file
pe_temporary	The number of .tmp files accessed by the file
pe_database	The number of .db files accessed by the file
pe_log	The number of log entries accessed by the file
pe_webpage	The number of web pages access by the file
pe_backup	The number of backups the file performs or accesses
pe_cabinet	The number of references to .cab files
pe_data	The number of .dat files accessed by file
pe_registry	The number of registry keys accessed or modified by the file
pe_directories	The number of directories accessed by the file
pe_dll	The number of DLL's accessed by the file
pe_detected	The number of suspicious sections in the file
From VirusTotal report	
vt_codesize	The size of the code in the file, retrieved from virustotal
vt_res_langs	The number of resource languages detected in the file
vt_res_types	The number of PE32 resource types detected in the file
vt_sections	The number of PE32 sections in the file
vt_entry_point	Decimal value of entry point, i.e. the location in code where control is transferred from the host OS to the file
vt_initDataSize	The size of the initialized data
vt_productName	The length of the field 'ProductName'. Can e.g. be "Microsoft(R) Windows(R) Operating System"
vt_originalFileName	The length of the original file name
vt_unitializedDataSize	The size of the part of the file that contain all global, uninitialized variables or variables initialized to zero
vt_legalCopyright	The length of the field LegalCopyRight. E.g. "(C) Microsoft Corporation. All rights reserved"

(continued)

Table 1 (continued)

From other Linux command line tools	
size_TEXT	The first output from the 'file' command. This is the size of the instructions
size_DATA	The second output with size of all declared/initialized variables
size_OBJ	The third output that contains the size of all unitialized data, i.e. the BSS-field
size_TOT	The fourth output that indicates the sum of the text, data and bss fields of the file
filesize	The total size of the file, including metadata

fication performance as suggested by Kononenko et al. [25]. After examination of current state of the art in feature selection methods we decided to use following methods:

- *Cfs* selects features with high correlation to classes and disregards features with a low correlation. Experiments by Hall [26] show that it can easily eliminate irrelevant, redundant, and noisy features.
- *Information Gain* ranks features by entropy with respect to each class as presented by Roobaert et al. [27].
- *ReliefF* ranks features from how well they help to separate classes of data samples that are close to each other. This is an extension from the two-class Relief algorithm. The algorithm was designed to perform on data with missing values as given by Kononenko [25].

Our main motivation also is to see how the feature's merit influences classification and whether we can judge about utility of the feature in malware analysis as compromise indicator.

3.4 Choice of Soft Computing Methods

As mentioned earlier SC is a set of nature-inspired methods that are designed to handle complex data and produce inexact solutions that can be also interpretable. Singh et al. [28] sketched how the variety of SC methods can be applied for malware detection. In this work we concentrate on the following SC methods in our experiments.

- *Naive Bayes* is a simple classifier build on an asssumpotion that features are independent from classes as described by Rish [29]. The method performs well on nominal features compared to other classifiers.
- *Bayesian Network* is based on conditional probabilities of features in a directed acyclic graph as presented by Friedman et al. [30]. Then, probabilities of the observable variables are computed.

- *MultiLayer Perceptron* is a type of Artificial Neural Network that simulates how the human brain neurons learn from observations according to Kononenko et al. [25]. This is achieved by using multiple hidden layers for better representation of non-linear data.
- *Support Vector Machine* learns from data by determining the optimal hyperplane to best separate the data. The better accuracy is achieved by using kernels in SVM that can project the data to a higher dimension feature space to create such a hyperplane according to Hearst et al. [31].

4 Experiments and Results

During this research we performed more than a hundred of different experiments. Since this is a novel work, we figured out that many of the methods could not be successfully performed due to the large model size in computer memory. Also we did not consider results of the experiments that took more than a week to execute. Below achieved accuracy is given for SC and HC methods on different sets of malware and features.

4.1 Experimental Setup

All the experiments were performed on a Virtual Server (Ubuntu 14.04) available at the Testimon Research group, NTNU Gjøvik in additional to other three laptops, including MacBookPro. The specifications of the server are as following. CPU is Intel(R) Core(TM) i7-3820 CPU @ 3.60 GHz with 2 cores (4 threads). Disk space was allocated on the SSD RAID storage with 8 GB of installed RAM Kingston PC-1600 memory. Files pre-processing were performed using *bash* scripts due to native support in Linux OS. To store extracted features MySQL 5.5.46 database engine was used. Characteristics harvesting and feature extraction was implemented in PHP 5.5.9 and run for many days without time limitation. Further, pre-processing, feature selection and classification was performed using opensource *Weka* v 3.7.13 package [32] that contains community-accepted ML methods.

4.2 Experimental Design

As mentioned in the Methodology Sect. 3, it was gathered a novel large-scale malware dataset. Initially (i) we reduced a number of files by eliminating duplicates and non-PE32 files. After thorough analysis and filtering, we derived all possible types of PE32 files designed for MS Windows and removed other types of files. In overall we ended up with 407,741 unique malware samples that are PE32 files.

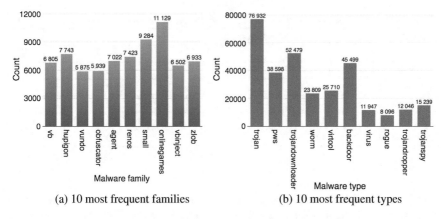

(a) 10 most frequent families   (b) 10 most frequent types

Fig. 2 Distribution in malware families and types datasets

The total size of all the executables is 136 GB despite the fact that all archives from *VirusShare* occupied 378 GB. Further, (ii) we performed raw characteristic acquisition from *PEframe* and *VirusTotal* and obtained a MySQL table of 19.5 GB, that later on resulted in 98.7 MB of numerical features of total 328,337 malware samples. The number is lower since we used only samples that contained CARO-formatted report from Microsoft anti-virus. So, after filtering and pre-processing we ended up with 328 k samples that consisted of 10,362 malware families and 35 types. Our preliminary study showed that such high number of families makes application of most ML methods infeasible. Therefore, it was decided to create following subsets limiting the most frequent types and families to decrease the number of highly-imbalanced classes.

(1) *10 families* containing 74,655 samples, 11.8 MB.
(2) *100 families* containing 227,191 samples, 36.3 MB.
(3) *500 families* containing 292,596 samples, 46.7 MB.
(4) *10 types* containing 310,355 samples, 50.5 MB.
(5) *35 types (full set)* containing 328,337 samples, 53.5 MB.

The Fig. 2 presents distribution of the smallest sets.

4.3 Analysis

All experiments on feature selection and classification were performed using cross-validation to see how well the methods can handle unlabelled data samples.

Feature selection resulted in a set of various features. The results from *Cfs* contain only the selection features and not the features merits. This is because the *Cfs* utilizes another search algorithm than *Information Gain* and *ReliefF*. Also we can say that *ReliefF* method shows considerably smaller average merit for each feature

Table 2 Number of selected features out of 27 initial features for each of the method

Dataset	Feature Selection methods		
	Cfs	InfoGain	ReliefF
10 families	15	15	25
100 families	16	17	27
500 families	16	13	27
10 types	12	11	23
35 types (full)	14	15	25

that it was achieved by *Information Gain*. It means that we have to use more features with much smaller threshold to be able to achieve good results. The numbers of extracted features per method are shown in the Table 2. For *ReliefF* we used only features with merit $\geq 10^{-3}$. For *Information Gain* we used features with merit ≥ 0.1 for types and ≥ 0.5 for families respectively. *Cfs* is not ranking-based methods and only generated final subsets of features.

Further, we can state that *ReliefF* in general gives a bigger number of features with not that significant range of feature's merits. For example, on a set of 500 most frequent families the merit range for *ReliefF* is 0.0–0.063 against 0.0–2.188 in *Information Gain*. The Table 3 shows the features selected by both *Cfs* and *Information Gain* on each data subset. The features in bold were selected by both methods for three subsets of families and two subsets of types respectively. An interesting observation considering the features selected for family classification is that the features selected converge with the increase in number of classes.

Two important features that we can highlight are "*pe_api*" and "*vt_sections*". Both features was selected by all three methods for feature selection. What we can derive from this is that the different malware families differs in the numbers of API-calls made in the file and the number of PE32-sections in the files. We have previously discussed the structure of a PE32 file. There are a number of sections mandatory for a PE32 file to run, while the total possible number of files is much bigger (maximum $2^{16} - 1 = 65,535$) as studied by Kath [33]. The number of API calls within the different families has a certain variety as well, from which we can tell that similar malware families will have a similar number of API calls. Since the different families have different capabilities and functionality, our assumption that API calls will be of interest in labelling malware is strengthened after this observation.

Classification performance was estimated using 5-fold cross validation. Also experiments with the resampling filtering method showed an improvement in accuracy, yet we did not include it in the final results. Along with SC method we decided to compare accuracy of a set of HC methods such that symbolic reasoning as was also studied by Abraham [34]. The comparison of achieved accuracy are shown in the Table 4. Highest accuracy for each subset is highlighted with respect to feature selection methods. To give more wide coverage, we only considered *Accuracy* as performance metric for all methods. *Bayes Network* shows considerably good perfor-

Table 3 Commonly selected features for malware families and types datasets using different feature selection methods

Malware families			Malware types	
10 families	1000 families	500 families	10 types	35 types
vt_res_langs	vt_codesize	vt_codesize	**vt_codesize**	**vt_codesize**
vt_res_types	**vt_res_langs**	**vt_res_langs**	**vt_entry_point**	**vt_entry_point**
vt_sections	**vt_sections**	**vt_sections**	**vt_initDataSize**	**vt_initDataSize**
vt_entry_point	**vt_entry_point**	**vt_entry_point**	**vt_productName**	**vt_productName**
vt_initDataSize	**vt_initDataSize**	**vt_initDataSize**	**vt_unitializedDataSize**	**vt_unitializedDataSize**
vt_productName	**vt_productName**	**vt_productName**	**vt_legalCopyright**	**vt_legalCopyright**
vt_originalFileName	**pe_api**	**pe_api**	pe_api	pe_api
vt_unitializedDataSize	pe_packer	pe_packer	**pe_dll**	**pe_dll**
pe_api	pe_library	pe_library	**size_TEXT**	**size_TEXT**
pe_debug	pe_dll	pe_executable	**size_DATA**	size_DATA
pe_dll	size_TEXT	**pe_dll**	size_OBJ	**size_OBJ**
size_DATA	**size_DATA**	**size_DATA**	**filesize**	size_TOT
filesize	**filesize**	**filesize**		**filesize**
	entropy	entropy		

Table 4 Accuracy of Soft Computing and selected Hard Computing methods, in %

Dataset	Feature selection	Soft computing				Hard computing	
		Naive bayes	Bayes network	MLP	SVM	C4.5	Random forest
10 families	Full set	23.9408	**72.6462**	42.3133	32.9824	83.2871	88.7965
	Cfs	20.8760	70.8285	40.7206	32.7962	83.9180	88.1990
	InfoGain	23.9354	**72.6462**	**51.5947**	38.0952	**84.2100**	**88.9572**
	ReliefF	**31.7835**	65.9996	41.0703	**41.2283**	82.7821	87.2279
100 families	Full set	20.9062	61.6056	11.8116	–	76.6250	81.3078
	Cfs	21.0008	**61.6851**	**18.1451**	–	77.8019	**81.9535**
	InfoGain	20.1905	60.9452	15.9034	–	**77.8204**	81.8329
	ReliefF	**23.9406**	57.3495	17.4743	–	75.3771	79.4565
500 families	Full set	16.1718	**57.2137**	9.1806	–	72.4210	–
	Cfs	**16.7610**	56.8853	**11.5360**	–	**73.2847**	–
	InfoGain	13.7654	54.1846	6.8251	–	72.4487	–
	ReliefF	2.8466	52.2410	11.3481	–	70.7730	–
10 types	Full set	**10.4316**	**51.3618**	28.4829	–	73.6789	**78.6000**
	Cfs	7.3857	51.1862	25.9661	–	**73.8200**	77.1639
	InfoGain	4.9063	49.9312	26.9623	–	73.6099	77.5177
	ReliefF	9.9650	47.2420	27.5317	–	72.4954	77.0888
35 types (full)	Full set	2.3576	48.4487	27.2826	–	72.6272	77.8797
	Cfs	1.9894	**48.6290**	25.1683	–	73.7480	77.3324
	InfoGain	1.9785	48.4140	25.7842	–	**73.8254**	**77.9687**
	ReliefF	**3.3517**	44.9438	**28.0675**	–	71.7842	76.4943

mance among SC methods. *MLP* performs much better on smaller sets, which might indicate a need to use a higher non-linearity, for example Deep Neural Network. Further, *Naive Bayes* gives a huge error rate on all sets, meaning that the method only works well on nominal-valued features. On the other hand, we can see that in general HC methods such that tree-based *C4.5* and *Random Forest* tend to have a better performance on defined problems of malware classification. However, for 500 families dataset *C4.5* can result in 32,904 leaves with a total size of tree equal to 65,807. For 35 malware types this method produces 33,952 leaves with a total size of the tree equal to 67,903. Yet, in real life this model is not practical.

Note that we were able to train *SVM* only on the smallest dataset (10 families). Also *Random Forest* exhausted available memory on the largest dataset (500 families).

5 Discussions and Conclusions

In this paper we investigated a large-scale malware detection using Soft Computing methods based on the static features extracted from PE32. It is important not only to distinguish between benign and malware files, yet also to understand what kind of malicious file it is, meaning malware families and types. To explore this problem we created a novel datasets of malware features with respect to families and types using publicly available sources and tools. At this point we can see that among SC methods Bayes Network performs much better than other, while Naive Bayes is the worse one. HC can produce reasonable results and better accuracy, yet resulting model is incredibly large for C4.5 and Random Forest. The last also failed to train on the largest dataset. Feature selection mostly cannot provide a consistent improvement. Moreover, SVM is not scalable at all. This extensive study contributes to the area of Soft Computing and Information Security also giving an insight for malware analyst on how to detect malware types and families fasted from static analysis.

Acknowledgements The authors would like to acknowledge help and support by Karl Hiramoto from VirusTotal. Special thanks to Carl Leichter for valuable and critical comments. Also we are grateful for sponsorship and support from COINS Research School of Computer and Information Security.

References

1. McAfee. Part of Intel Security., "Threats report," McAfee., Technical Report, Aug 2015. Accessed 19 Sep 2015
2. S. Ravi, N. Balakrishnan, B. Venkatesh, Behavior-based malware analysis using profile hidden markov models, in *2013 International Conference on Security and Cryptography (SECRYPT)*, July 2013, pp. 1–12
3. "Virustotal," https://www.virustotal.com/. Accessed 10 Aug 2015
4. L. Seltzer, *Tools for Analyzing Static Properties of Suspicious Files on Windows*, 2014
5. G. Amato, "Peframe," https://github.com/guelfoweb/peframe, Nov 2015. Accessed 17 June 2015
6. M. Ligh, S. Adair, B. Hartstein, M. Richard, *Malware Analyst's Cookbook and DVD: Tools and Techniques for Fighting Malicious Code* (Wiley Publishing, 2010)
7. D. Gavrilut, M. Cimpoesu, D. Anton, L. Ciortuz, Malware detection using machine learning, in *International Multiconference on Computer Science and Information Technology, 2009. IMCSIT '09*, Oct 2009, pp. 735–741
8. N. Idika, A.P. Mathur, *A Survey of Malware Detection Techniques*, vol. 48 (Purdue University, 2007)
9. Y. Ye, D. Wang, T. Li, D. Ye, Imds: Intelligent malware detection system, in *Proceedings of the 13th ACM SIGKDD International Conference on Knowledge Discovery and Data Mining*, ser. KDD '07. New York, NY, USA (ACM, 2007), pp. 1043–1047
10. J.Z. Kolter, M.A. Maloof, Learning to detect and classify malicious executables in the wild. J. Mach. Learn. Res. **7**, 2721–2744 (2006)
11. S.K. Das, A. Kumar, B. Das, A. Burnwal, On soft computing techniques in various areas. Comput. Sci. Inf. Technol. 59 (2013)
12. F. Cohen, Computer viruses: theory and experiments. Comput. Secur. **6**(1), 22–35 (1987)

13. D.M. Chess, S.R. White, An undetectable computer virus, in *Proceedings of Virus Bulletin Conference*, vol. 5, 2000
14. S. R. Bragen, Malware detection through opcode sequence analysis using machine learning, *Gjvik University College*, 2015
15. Z. Markel, M. Bilzor, Building a machine learning classifier for malware detection, in *Second Workshop on Anti-malware Testing Research (WATeR)*, vol. 2014 (IEEE, 2014), pp. 1–4
16. M. Shankarapani, K. Kancherla, S. Ramammoorthy, R. Movva, S. Mukkamala, Kernel machines for malware classification and similarity analysis, in *The 2010 International Joint Conference on Neural Networks (IJCNN)* (IEEE, 2010), pp. 1–6
17. K. Rieck, T. Holz, C. Willems, P. Düssel, P. Laskov, Learning and classification of malware behavior, in *Proceedings of the 5th International Conference on Detection of Intrusions and Malware, and Vulnerability Assessment*, ser. DIMVA '08 (Springer, Berlin, Heidelberg, 2008), pp. 108–125
18. B. Zhang, J. Yin, J. Hao, D. Zhang, S. Wang, Malicious codes detection based on ensemble learning, in *Proceedings of the 4th International Conference on Autonomic and Trusted Computing*, ser. ATC'07 (Springer, Berlin, Heidelberg, 2007), pp. 468–477
19. Naming scheme-caro-computer antivirus research organization (2015), www.caro.org/naming/scheme.html. Accessed 20 Aug 2015
20. Microsoft Malware Protection Center, "Naming malware"
21. G. Ou, Y.L. Murphey, Multi-class pattern classification using neural networks. Pattern Recogn. **40**(1), 4–18 (2007)
22. A. Shalaginov, K. Franke, Towards improvement of multinomial classification accuracy of neuro-fuzzy for digital forensics applications, in *15th International Conference on Hybrid Intelligent Systems (HIS 2015)*, vol. 420, no. 1 (Springer Publishing, 2015), p. 1
23. K. Maxwell, "maltrieve," May 2015. Accessed 10 June 2015
24. "Virusshare," https://virusshare.com/. Accessed 08 May 2015
25. I. Kononenko, M. Kukar, *Machine Learning and Data Mining: Introduction to Principles and Algorithms.* (Horwood Publishing Limited, 2007)
26. M.A. Hall, Correlation-based feature selection for machine learning, Ph.D. dissertation, The University of Waikato, 1999
27. D. Roobaert, G. Karakoulas, N.V. Chawla, Information gain, correlation and support vector machines, in *Feature Extraction* (Springer, 2006), pp. 463–470
28. R. Singh, H. Kumar, R. Singla, Review of soft computing in malware detection. Spec. Issues IP Multimed. Commun. **1**, 55–60 (2011)
29. I. Rish, An empirical study of the naive bayes classifier, in *IJCAI, Workshop on Empirical Methods in Artificial Intelligence*, vol. 3, no. 22 (IBM, New York, 2001), pp. 41–46
30. N. Friedman, D. Geiger, M. Goldszmidt, Bayesian network classifiers. Mach. Learn. **29**(2–3), 131–163 (1997)
31. M.A. Hearst, S.T. Dumais, E. Osman, J. Platt, B. Scholkopf, Support vector machines. Intell. Syst. Appl. IEEE **13**(4), 18–28 (1998)
32. Weka 3: Data mining software in java. Accessed: 15 Dec 2015
33. R. Kath, *The portable executable file format from top to bottom* (Microsoft Corporation, MSDN Library, 1993)
34. A. Abraham, Hybrid soft and hard computing based forex monitoring systems, in *Fuzzy Systems Engineering* (Springer, 2005), pp. 113–129

Automatic Image Classification for Web Content Filtering: New Dataset Evaluation

V. P. Fralenko, R. E. Suvorov and I. A. Tikhomirov

Abstract The paper presents experimental evaluation of image classification in the field of web content filtering using bag of visual features and convolutional neural networks approach. A more difficult data set than traditionally used ones was built from very similar types of images in order to make conditions closer to real world practice. F1-measure of classifiers that are based on bags of visual features was significantly lower than that reported in previously published papers. Convolutional neural networks performed much better. Also, we measured and compared training and prediction time of various algorithms.

Keywords Web content filtering · Automatic image classification
Scalability · Convolutional neural networks · Bag of visual features

1 Introduction

The problem of access restriction for certain groups of users in the modern Web is being widely and actively discussed. Companies are interested in deploying a content filtering solution to force their employees to spend more time on their primary work instead of surfing social networks and entertainment websites. Individuals are interested in preventing their children from watching pornography and learning about suicide and drugs. Nowadays, it is no doubt that in order to

V. P. Fralenko (✉)
Aylamazyan Program Systems Institute of the Russian Academy of Sciences,
Moscow, Russia
e-mail: alarmod85@hotmail.com

R. E. Suvorov · I. A. Tikhomirov
Federal Research Center "Computer Science and Control",
Russian Academy of Sciences, Moscow, Russia
e-mail: rsuvorov@isa.ru

I. A. Tikhomirov
e-mail: tih@isa.ru

© Springer International Publishing AG, part of Springer Nature 2018
L. A. Zadeh et al. (eds.), *Recent Developments and the New Direction
in Soft-Computing Foundations and Applications*, Studies in Fuzziness
and Soft Computing 361, https://doi.org/10.1007/978-3-319-75408-6_27

robustly control the Internet usage, it is necessary to analyze multiple types of content simultaneously and on-the-fly. Text-based content filtering is well studied [1, 2]. Utilizing interlinked nature of web documents is a less investigated and a more challenging problem, but there are some approaches to it [2]. Numerous research has been conducted in the field of automatic image classification. However, this research area is very dynamic and thus it makes sense to revise applicability of the most recent methods to the web content filtering problem.

The core idea of the presented research is to evaluate image classifiers in complicated real world conditions by constructing training and testing data sets from hardly separable images. We evaluate two methods: a promising approach using convolutional neural network (CNN) and a traditional well-known approach using bags-of-visual-features (BOVF). We consider BOVF-classifier as a baseline. By making the evaluation dataset harder, we make the difference between "good" and "very good" results more significant. In other words, it is impossible to objectively compare two or more methods in simple conditions (in this case, even the worst method would give F1 about 0.99 and there would not be any difference in quality metrics). To our knowledge, there are no other papers presenting such experiments.

During the experiments we solved a traditional binary classification problem, which consists in labeling "appropriate" and "inappropriate" images in test set. Usually, pictures of cars, animals, nature etc. are considered to be "appropriate" and pornography pictures are "inappropriate". In this research we built the "inappropriate" set of explicitly pornographic pictures from image hostings and photoblogs such as Tumblr. We decided to use such Web sites as a data source, because their pages are almost text-free (thus they are a hard nut to crack for text-based content filters). The "appropriate" collection was built of erotic pictures with many skin-colored pixels but without explicit sexual exposure. This makes the image classification task much harder.

The rest of the paper is organized as follows: in Sect. 2 "Related work" we review the state of the art in the field of image-based web content filtering, in Sect. 3 "Methods" we briefly describe the methods, in Sect. 4 "Experimental evaluation" we explain how experimental dataset was built and how the quality and speed was measured. Finally, in Sect. 5 "Conclusion" we summarize the work done and discuss the future work.

2 Related Work

This paper presents results mainly in the fields of automatic image classification and web content filtering (or adult content detection). In this section, we review the closest papers and compare the presented results.

The majority of papers [3–5] focus on various skin detection techniques, including non-typical multi-agent learning [6]. The decision is based on how much of image area is occupied by skin-colored pixels. Some researches [5, 7] follow information retrieval approach: images are mapped to a relatively low-dimensional space and the most frequent category from the found images is assigned to the query image.

There are a number of researches [8, 9] based on bag-of-visual-features approach with features extracted using SIFT [10], SURF [11], FREAK [12] or custom discrete cosine transformation-based technique [9]. The most interesting approaches [13–15] rely not only on skin detection, but also on shape and texture.

Shape is described via Fourier transform [14], statistics on various parts of image (e.g. skin-color area percent or intensity gradient direction in each block of 3×3 grid) [7] or approximation using ellipses [16].

Multistep approach with fast preliminary filtering and slower but more accurate final categorization is very popular [5, 17]. Researchers usually employ such decision functions as C4.5, SVM with RBF kernel, neural networks or k-nearest neighbors etc.

There are also papers that analyze motion in a video stream to detect obscene content [18].

Despite convolutional neural networks are a very promising framework, there are quite a few works that apply CNN to web content filtering. One [19] is quite close to this paper: it presents an evaluation of ConvNet-based classifiers on a dataset that contains images that are thematically similar and harder to distinguish.

Skin detection-based approach is somewhat restricted, because there are many resources in the Web that contain adult images with changed colors (e.g. with shifted hue or in gray scale) or even painted pictures. Most of the reviewed papers use manually crafted rules or special algorithms to extract features or make decisions. Most of these rules suit well the problem of adult content detection, but are not applicable to recognition of other content types (of other topics).

The quality measures were cross-validated or held-out precision, recall and F1, which highly depend on the used dataset. Some papers present evaluation on rather small datasets or do not present it at all. Usually, datasets are built of images on very different topics (e.g. pornography and nature), which are easy to distinguish, except [19].

3 Methods

3.1 Bag of Visual Features

Bag of Visual Features is one of the most frequently used methods for image-to-vector mapping, classification [8] and search [20]. Its core idea is to treat each image as a set of descriptors (like words in texts), which represent some interesting image fragments (such as eyes or distinctive shapes of contrast and color). They are extracted using some sub-algorithm (like SIFT [10] or SURF [11]). In this work, we decided to represent descriptors as points in 128-dimensional space. After all descriptors in the dataset are extracted, they are clustered in order to get unique and significant features (i.e. dimensionality reduction). Thus, each descriptor is identified by its cluster and each image is represented using a vector of size equal to the clusters number. In this vector, i-th element equals to a number of descriptors that belong to i-th cluster and are found in the image. These vectors are normalized and gathered into a matrix, representing a dataset. After that a classification method such as Naive Bayes or SVM is applied to determine categories of images. The classifier implementation is based on OpenCV framework [21].

3.2 Convolutional Neural Network

Multi-layer artificial neural networks (MLANN) are widely adopted in the fields of computer vision, object recognition and image classification. Nowadays, convolutional neural networks, a special type of MLANN, are becoming more popular. CNN aims on addressing major drawbacks of MLANN, including inability to deal with noisy, skewed, stretched, rotated images. This goal is achieved by a special CNN architecture that tries to model structure of visual cortex. Roughly speaking, the core idea is that different neurons are sensitive to different shapes and patterns (e.g. some are more sensitive to vertical lines, others—to horizontal ones) [22, 23].

Before being submitted to CNN input, images were preprocessed as follows. Image was extended with blue pixels so that it is of square shape. Blue color was used as it is considered to be the least visible color. Then, each image was scaled to the size of 68 × 68 pixels. We also tried scaling to larger sizes with no significant effect on classification quality. Input layer of CNN has 68 × 68 pixels. Then, a number of convolutional and subscaling layers follow one after another. In these layers, each neuron is connected only with a part of neurons of the previous layer, which represent some local area. Convolutional neurons aim on recognizing specific shapes (according to a given set of convolution kernels). Each convolution kernel represents input weights. Subscaling layers reduce size of image by mapping 4 adjacent pixels into one. At the end of CNN, there may be a number of fully-connected layers, trained using traditional backpropagation method. The implementation is based on nnForge framework [24].

The exact functions of the used CNN layers are as follows:

(1) Local contrast enhancement using 9 × 9 mask.
(2) Convolution layer with 120 5 × 5 masks and *abs(tanh())* or *MaxOut* function.
(3) Subsampling layer that maps 2 × 2 area to one pixel using *avg* or *max* function.
(4) Convolution layer with 38 5 × 5 masks and *abs(tanh())* or *MaxOut* function.
(5) Subsampling layer that maps 2 × 2 area to one pixel using *avg* or *max* function.
(6) Convolution layer with 5 × 5 masks and *abs(tanh())* (23 masks) or *MaxOut* function (22 masks).

(7) Subsampling layer that maps 2 × 2 area to one pixel using *avg* or *max* function.
(8) Convolution layer with 5 × 5 masks and *abs(tanh())* (19 masks) or *MaxOut* function (18 masks).
(9) Subsampling layer that maps 2 × 2 area to one pixel using *avg* or *max* function.
(10) Fully-connected layer with 2 neurons and *tanh* activation function.

4 Experimental Evaluation

4.1 Classification Quality

The core idea of the conducted experiments is to evaluate and compare quality using various classification methods and their combinations in conditions that are close to real conditions of web content filtering software.

To build the evaluation data set, we collected 1000 web pages in semi-automatic manner. Then, we took all the pictures linked to these pages and selected only those of them which were wider and higher than 41 pixel. Animated pictures and videos (GIF, MP4) were converted to a set of randomly chosen frames. All the pictures were converted to PNG. Thus, we collected about 30000 pictures. Then, we randomly subsampled them to create a set of datasets of various sizes. The goal of this is to estimate, how much the classification quality depends on the training dataset size. The final datasets contained from 200 to 8000 images (from 100 to 4000 images in each category).

We calculated accuracy on training data to estimate discriminative power of classifiers and cross-validated F1 to measure the classification quality. The metrics were evaluated over 5 cross-validated runs and then were macro-averaged. We used the same splits to evaluate each classifier. We decided to feed the classifiers with full images without interesting area detection and cropping (e.g. using skin detection), because we found that it is very popular to change color scale or slightly perturb colors to trick content filters.

We conducted a series of experiments with various combinations of the described methods:

- BOVF + NormalBayesClassifier. Input data is treated as a Gaussian mixture.
- BOVF + CvANN_MLP. 4-layer feed-forward neural network trained using Rprop algorithm [25]. Sizes of internal layers were 200, 100, 50, 2. We also tried reducing number of layers, leading to much worse results.
- BOVF + RBF-C_SVC,   BOVF + Poly-C_SVC,   BOVF + RBF-NuSVC, BOVF + Poly-NuSVC. Classical SVM [26, 27] and Nu-SVM [28] with Radial Basis Function or Polynomial kernel with hyper parameters tuned using 5-cross validated grid search.

Table 1 Experimental results

Method	Accuracy on training data	Cross-validated F1
BOVF + NormalBayesClassifier	0.61	0.47
BOVF + CvANN MLP	0.99	0.64
BOVF + RBF-C SVC	0.76	0.72
BOVF + Poly-C SVC 0.72 0.71	0.72	0.71
BOVF + RBF-NuSVC	0.99	**0.78**
BOVF + Poly-NuSVC	0.99	0.75
CNN-Avg	0.99	0.86
CNN-Avg-2	0.99	0.86
CNN-Max	0.99	0.86
CNN-Max-2	0.99	**0.87**

- CNN-Avg. Eight-layer CNN (3 iterations of convolution-subscaling, 1 convolution and fully-connected layer). Fully-connected layer was trained using stochastic diagonal Levenberg-Marquardt algorithm [29]. We set 120, 38, 23, 19 convolution kernels for convolutional layers. To improve classification quality, we preprocessed the training dataset by randomized transformations. From each source image we generated 5 images with uniformly-sampled transformations, including rotation (max $\pm 15°$), scale (max $\pm 10\%$), shift (max $\pm 2px$), contrast and brightness adjustment (max $\pm 50\%$).
- CNN-Avg-2. The same as CNN-Avg, but with 240, 76, 46 and 39 convolution kernels.
- CNN-Max and CNN-Max-2. The same as CNN-Avg and CNN-Avg-2 but with max-pooling layers [30] instead of subscaling.

Joint results are present in Table 1 "Experimental results". Additionally, we present a chart (Fig. 1 "Dependency of classification quality on size of training data

Fig. 1 Dependency of classification quality on size of training data set

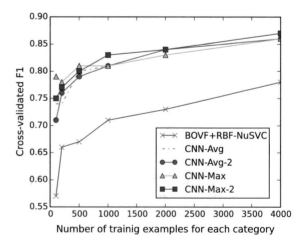

Fig. 2 Training time of various algorithms (log10 scale)

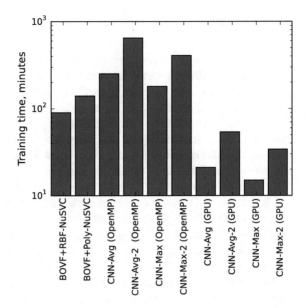

set") that shows how classification quality depends on size of the training data set. In average, the best results with BOVF were achieved when descriptors were clustered into 120 or 350 groups. We also tried increasing the number of clusters up to 1024 with no significant effect on classification quality.

4.2 Processing Time

Also, we measured amount of time that it takes to train a model and to predict a label for an image. Benchmarking was done on a computer with two Intel Xeon E5410 CPU (4 cores, 2.33 GHz) and two GPU: Nvidia Tesla K20c (Kepler architecture) and Nvidia 980 GTX (Maxwell architecture). Average training and prediction time is present on Fig. 2 "Training time of various algorithms (log10 scale)" and in Table 2 "Average prediction time for a single image". Training and prediction time of BOVF-* classifiers was measured only on CPU with dictionary of 350 features. For CNN-* classifiers, time was almost the same using both GPU.

Table 2 Average prediction time for a single image

	BOVF + RBF-NuSVC	CNN-Max (1 CPU)	CNN-Max (GPU)
Time, seconds	0.017	0.060	0.003

5   Conclusion

In this work, we conducted a series of experiments with two families of classification methods (bag of visual features and convolutional neural networks). The main peculiarity of the evaluation is "hardened" data set, built from hard-to-distinguish images (erotic and pornography pictures), often distorted or with altered colors. These conditions made such common techniques like skin detection unusable. This is the first time such evaluation was done. **However, making this dataset public needs some legal investigation. Thus, the dataset currently is not publicly available, but we are working on it. If you are interested in conducting new experiments using it, please contact us.**

The best F1 (0.78) using the bag of visual features approach were achieved using NuSVC (also tried C_SVC, multilayer perceptron and Gaussian mixture-based Naive Bayes). This F1 is much lower than that reported in previous researches of other authors. This is due to a more difficult dataset. Convolutional neural network performed much better (F1 = 0.87), which confirms CNN strength.

From Fig. 1 one can conclude that difference in classification quality is most significant when the model is trained on small data sets. Obviously, classifiers perform better on larger data. CNN-Max-* classifiers performed slightly better than CNN-Avg-*, especially on small training datasets. CNN-*-2 are also more stable than CNN-*, yet slower. If the training set is large enough, a relatively small number of convolution kernels is sufficient. Further increase of convolution kernels number did not gain more classification accuracy. Training data set size is more important than number of convolution kernels.

Main direction of the future work is linking all the work done [1, 2, 31] into a comprehensive software system suitable for real-time content classification that inspects text, images, links between documents and user's behavior. Furthermore, output produced by CNN can be converted to normalized likelihood and thus it is possible to build a comprehensive decision function based on methods from the fields of fuzzy logic and soft computing. It is also worth comparing with other methods and systems (including [19]) on The Pornography database [32].

Acknowledgements   The research is supported by Russian Foundation for Basic Research (Grant 15-37-20360).

References

1. R. Suvorov, I. Sochenkov, I. Tikhomirov, Method for pornography filtering in the web based on automatic classification and natural language processing, in *Speech and Computer* (Springer, 2013), pp. 233–240
2. R. Suvorov, I. Sochenkov, I. Tikhomirov, Training datasets collection and evaluation of feature selection methods for web content filtering, in *Artificial Intelligence: Methodology, Systems, and Applications* (Springer, 2014), pp. 129–138

3. A.F. Drimbarean, P.M. Corcoran, M. Cuic, V. Buzuloiu, Image processing techniques to detect and filter objectionable images based on skin tone and shape recognition, in *ICCE. International Conference on Consumer Electronics, 2001* (IEEE, 2001), pp. 278–279

4. M. Hammami, Y. Chahir, L. Chen, Webguard: A web filtering engine combining textual, structural, and visual content-based analysis. IEEE Trans. Knowl. Data Eng. **18**(2), 272–284 (2006)

5. B.-B. Liu, J.-Y. Su, Z.-M. Lu, Z. Li, Pornographic images detection based on cbir and skin analysis, in *Fourth International Conference on Semantics, Knowledge and Grid* (IEEE, 2008), pp. 487–488

6. A. Zaidan, H.A. Karim, N. Ahmad, B. Zaidan, M.M. Kiah, Robust pornography classification solving the image size variation problem based on multi-agent learning. J. Circuits Syst. Comput. **24**(02), 1550023 (2015)

7. J.-L. Shih, C.-H. Lee, C.-S. Yang, An adult image identification system employing image retrieval technique. Pattern Recogn. Lett. **28**(16), 2367–2374 (2007)

8. T. Deselaers, L. Pimenidis, H. Ney, Bag-of-visual-words models for adult image classification and filtering, in *19th International Conference on Pattern Recognition, 2008. ICPR 2008* (IEEE, 2008), pp. 1–4

9. A. Ulges, A. Stahl, Automatic detection of child pornography using color visual words, in *2011 IEEE International Conference on Multimedia and Expo (ICME)* (IEEE, 2011), pp. 1–6

10. D.G. Lowe, Object recognition from local scale-invariant features, in *The proceedings of the seventh IEEE international conference on Computer vision, 1999*, vol. 2 (IEEE, 1999), pp. 1150–1157

11. H. Bay, T. Tuytelaars, L. Van Gool, Surf: speeded up robust features, in *Computer Vision–ECCV 2006* (Springer, 2006), pp. 404–417

12. S.H. Yaghoubyan, M.A. Maarof, A. Zainal, M.F. Rohani, M.M. Oghaz, Fast and effective bag-of-visual-word model to pornographic images recognition using the freak descriptor. J. Soft Comput. Decis. Support Syst. **2**(6), 27–33 (2015)

13. Y. Fu, W. Wang, Fast and effectively identify pornographic images, in *2011 Seventh International Conference on Computational Intelligence and Security (CIS)* (IEEE, 2011), pp. 1122–1126

14. S.M. Kia, H. Rahmani, R. Mortezaei, M.E. Moghaddam, A. Namazi, *A Novel Scheme for Intelligent Recognition of Pornographic Images*, 2014. arXiv:1402.5792

15. J. Polpinij, C. Sibunruang, S. Paungpronpitag, R. Chamchong, A. Chotthanom, A web pornography patrol system by content-based analysis: in particular text and image, in *IEEE International Conference on Systems, Man and Cybernetics, 2008. SMC 2008* (IEEE, 2008), pp. 500–505

16. U. Sayed, S. Sadek, B. Michaelis, Two phases neural network based system for pornographic image classification in *Proceedings of the IEEE International Conference on Sciences of Electronic, Technologies of Information and Telecommunications*, 2009, pp. 1–6

17. J.A.M. Basilio, G.A. Torres, G. S´anchez, E. Hernadez, Explicit content image detection. Signal Image Process. Int. J. (SIPIJ) **1**(2), 47–58 (2010)

18. R. Mustafa, D. Zhu, A novel method for sensing obscene videos using scene change detection. TELKOMNIKA Indonesian J. Electr. Eng. **13**(2), 300–304 (2015)

19. M. Moustafa, in *Applying Deep Learning to Classify Prnographic Images and Videos*, 2015. arXiv:1511.08899

20. H. J´egou, M. Douze, C. Schmid, Improving bag-of-features for large scale image search. Int. J. Comput. Vis. **87**(3), 316–336 (2010)

21. G. Bradski, A. Kaehler, *Learning OpenCV: Computer Vision with the OpenCV Library* (O'Reilly Media, Inc., 2008)

22. Y. LeCun, Y. Bengio, Convolutional networks for images, speech, and time series, in *The Handbook of Brain Theory and Neural Networks*, vol. 3361, no. 10, 1995

23. D.H. Hubel, T.N. Wiesel, Receptive fields, binocular interaction and functional architecture in the cat's visual cortex. J. Physiol. **160**(1), 106 (1962)

24. M. Milakov, Convolutional and fully-connected neural networks C++ framework. http://milakov.github.io/nnForge/
25. M. Riedmiller, H. Braun, A direct adaptive method for faster backpropagation learning: the Rprop algorithm, in *IEEE International Conference on Neural Networks, 1993*, (IEEE, 1993), pp. 586–591
26. B.E. Boser, I.M. Guyon, V.N. Vapnik, A training algorithm for optimal margin classifiers, in *Proceedings of the Fifth Annual Workshop on Computational Learning Theory* (ACM, 1992), pp. 144–152
27. C. Cortes, V. Vapnik, Support-vector networks. Mach. Learn. **20**(3), 273–297 (1995)
28. P.-H. Chen, C.-J. Lin, B. Schölkopf, A tutorial on ν-support vector machines. Appl. Stoch. Models Bus. Ind. **21**(2), 111–136 (2005)
29. D.C. Ciresan, U. Meier, J. Masci, L. Maria Gambardella, J. Schmidhuber, Flexible, high performance convolutional neural networks for image classification, in *IJCAI Proceedings-International Joint Conference on Artificial Intelligence*, vol. 22, no. 1, 2011, p. 1237
30. L.-H. Lee, Y.-C. Juan, W.-L. Tseng, H.-H. Chen, Y.-H. Tseng, Mining browsing behaviors for objectionable content filtering. J. Assoc. Inf. Sci. Technol. **66**(5), 930–942 (2015)
31. S. Avila, N. Thome, M. Cord, E. Valle, A.D.A. Araujo, Pooling in image representation: the visual codeword point of view. Comput. Vis. Image Underst. **117**(5), 453–465 (2013)
32. S. Becker, Y. Le Cun, Improving the convergence of backpropagation learning with second order methods, in *Proceedings of the 1988 Connectionist Models Summer School* (Morgan Kaufmann, San Matteo, CA, 1988), pp. 29–37

V. P. Fralenko, Ph.D. He is a senior researcher at Aylamazyan Program Systems Institute of Russian Academy of Sciences Pereslavl, Russia. Graduated from the Program Systems Institute—Pereslavl University in 2007. He is an author of 56 papers. His research interests include artificial neural networks, machine learning, distributed computing, network security and software development.

R. E. Suvorov. He is a Ph.D. student and research engineer at Federal Research Center "Computer Science and Control" of the Russian Academy of Sciences, Moscow, Russia. Graduated from the Rybinsk State Aviation Technology University in 2012. He is an author of 17 papers. His research interests include data mining, machine learning, information extraction, Web content processing, graph mining, distributed and high-load systems.

I. A. Tikhomirov, Ph.D. He is the head of laboratory "Intelligent Technologies and Systems" at Federal Research Center "Computer Science and Control" of the Russian Academy of Sciences, Moscow, Russia. Graduated from the Rybinsk State Aviation Technology University in 2002. He is an author of more than 50 papers. His research interests include natural language processing, search engines, artificial intelligence, web systems.

Differentiations in Hierarchical Fuzzy Systems

Begum Mutlu and Ebru A. Sezer

Abstract Hierarchical fuzzy systems are one of the most popular solutions for the curse of dimensionality problem occurred in complex fuzzy rule based systems with a large number of input parameters. Nevertheless these systems have a hidden inaccuracy and instability problem. In detail, the outputs of hierarchical systems, based on Mamdani style inference, differ from the outputs of equivalent single system. Moreover they are not stable in any variation of system modeling even if the rules and membership functions do not expose any differentiation. This paper revisits inaccuracy and instability problems of hierarchical fuzzy inference systems. It investigates the differentiation in systems' behaviors against the variations in system modeling, and provides a pattern to identify the magnitude of this differentiation.

1 Introduction

Besides their high approximation capability, fuzzy systems provide transparency and plausibility to the inference models since they directly use the expert knowledge (instead of historical data) for their reasoning process. Therefore they have been successfully implemented in lots of applications from various research areas [1]. However a fuzzy inference system suffers from *curse of dimensionality* [2] when it has to deal with complex rule based problems containing a lot of inputs and/or fuzzy sets, and this is an important bottleneck for the implementation of fuzzy inference systems. Wang et al. [1] clearly defined this problem as *rule*, *parameter* and *data dimensionality*. The *rule dimensionality* is about exponential increase in the number of fuzzy rules. *Parameter dimensionality* addresses the increase in complexity of mathematical expressions of fuzzy systems. And *data dimensionality* emphasizes

B. Mutlu (✉)
Department of Computer Engineering, Gazi University, Ankara, Turkey
e-mail: begummutlu@gazi.edu.tr

E. A. Sezer
Department of Computer Engineering, Hacettepe University, Ankara, Turkey
e-mail: ebruakcapinarsezer@gmail.com

© Springer International Publishing AG, part of Springer Nature 2018
L. A. Zadeh et al. (eds.), *Recent Developments and the New Direction in Soft-Computing Foundations and Applications*, Studies in Fuzziness and Soft Computing 361, https://doi.org/10.1007/978-3-319-75408-6_28

that the expert needs much more information to identify a fuzzy system with a lot of input parameters. As a consequence, the system becomes less interpretable and more complex, and occupies too much memory and time for computation. These drawbacks make the implementation of such systems infeasible and inapplicable.

Hierarchical Fuzzy Systems (HFSs) are popular and reasonable solutions to the curse of dimensionality problem. Instead of a high dimensional single fuzzy inference system, lower dimensional sub-fuzzy systems (SFSs) are generated and linked hierarchically, and each SFS collaborates with its successor to contribute the final decision. How this collaboration is done determines the hierarchic structure.

Hierarchic structures have been classified as incremental, aggregated and cascaded in [3]. In incremental structure, a tentative decision is made in the lowest layer SFS by using some of the original inputs. Then some of the other SFSs contribute to this tentative decision in the subsequent SFS until all of the inputs are linked to the SFSs in different layers. In aggregated structure, the large scale problem is separated into lower dimensional ones whose decision is made in different SFSs in the lowest layer. Then outputs provided from lowest layer SFSs are merged in subsequent layers to make the final decision. In cascaded structure, on the other hand, the decision is made in the lowest layer SFS with all of the original inputs, and it is revised in the subsequent SFSs. More recently Benitez and Casillas [4] have made a similar classification of hierarchic structures as serial (as incremental), parallel (as aggregated) and hybrid (a customized structure contains both incremental and aggregated decision making strategies).

Main expectation from HFSs is that they have to keep their approximation ability. In other words, if any system is modeled in both single and complex fuzzy system and hierarchical fuzzy system; output of these two models should be same. Therefore, an HFS should provide same behaviors with the single system which is actually the solution before input-space separation [5]. Additionally, output of HFS should be stable against any differentiation in HFS modeling parameters (such as the hierarchic structure, the order of inputs, links between the inputs and SFSs and/or SFSs' each other).

Regarding hierarchical fuzzy inference based on Takagi-Sugeno-Kang (TSK) [6, 7] fuzzy systems, it has been studied that these systems are universal approximators, in other words, it is possible to build an HFS that approximates a desired function [8–11], even if this function is based on single system solution [5]. On the other hand, the approximation capability of Mamdani style HFSs is still a disputable topic. Moreover preliminary researches showed that, this type of HFSs' behaviors are highly likely to differ from corresponding single system's behaviors, and even a tiny variation in the hierarchical system modeling causes significant differentiation in the HFS behaviors [12]. The reason behind this outcome is clearly explained in [13] as follows: In conventional Mamdani style hierarchical inference, each SFS finalizes the whole reasoning process (from fuzzification to defuzzification), and transfers the provided crisp value to the subsequent layer. Here each defuzzification step causes a degeneration in the propagated fuzzy information [1, 14], and it provokes uncertainty in the final decision. In detail, once a defuzzification is applied on the output of aggregation step (which actually refers a fuzzy set), and this output fuzzy set is transformed into a crisp value by any defuzzifier, reforming the same fuzzy set in the subsequent SFS is not possible for most of the cases. At this point, aforementioned degeneration in the fuzziness occurs; spreads towards the whole hierarchy, and finally causes differentiation from single system's behaviors. It also reasons *instability* in differently structured HFS outputs. Consequently, *accuracy*, and *stability* properties and directly approximation capability of this kind of HFSs become a questionable topic. Surprisingly this issue has been gathered little attention, and a lot of applications have been still performed without considering its inaccuracy [15–18].

In literature there are three prominent studies whose most important motivation is to provide a solution to conventional HFSs inaccuracy. First Rattasiri and Halgamuge proposed Hierarchical Classifying-Type Fuzzy System: HCTFS [19] relies on Alternative Model of Hierarchical Fuzzy System: AHFS [20] and Classifying-Type Fuzzy System. Although this approach criticizes the redundantly repeated defuzzification steps' in the inner layers, it does not aim to provide single system outputs, but it mostly investigates the favorable outcome of proposed inference method on cost of computation against the single system. In addition, it insists the use of AHFS. Even though AHFS minimizes the total number of rules, it may be inapplicable for some problems because of the difficulties in rule generation. Especially in the inner layer rules, it may be really challenging to plausibly merge the resulting outputs of low layer SFSs. After a long while Joo and Sudkamp proposed a method to convert a single system into a two-layer HFS in [5]. The most important point achieved here is the proposed HFS's ability in generating same outputs with the corresponding single system. In the proposed hierarchy, some of the original inputs are linked to the first layer and the others cooperate with the outputs of this layer in the subsequent layer. First the rules of both layers are generated from the high dimensional rule matrix of single system, and second rule reduction is applied to reduce the number of rules and simplify each one. However proposed two layer structure is a customized version of HFSs, and the inference process strictly relies on it. Briefly both of these proposed methods depend on specific hierarchic structures. Additionally their reasoning is related with the TSK type fuzzy inference. However the inaccuracy problem of HFSs is caused by the degeneration of fuzziness in the inner layers, and for Mamdani style HFSs, this problem has precedence over TSK based HFSs.

Addressing the inaccuracy in Mamdani style HFSs, Mutlu et al. proposed Defuzzification-Free Hierarchical Fuzzy Systems: DF-HFS which performs the single system behaviors without restricting the use of any customized hierarchic structure [13]. Shortly, it prevents the degeneration in fuzziness and propagates the fuzzy information from the lowest layer to the top-most one. However DF-HFS is evaluated only on one conceptual case and a real world problem with maximum 9 inputs in [13]. In fact, it is necessary to evaluate DF-HFS behaviors and its boundaries for larger scale problems. Additionally, although there is not a restriction about the hierarchic structure for DF-HFS, AHFS is applied in [13] in order to make a fair comparison between DF-HFS and HCTFS accuracy. Moreover the effect of the variations in hierarchic structure is not investigated sufficiently by neither empirical nor mathematical proofs.

Two motivations encourage this study: (i) Examining the magnitude of propagated degeneration in fuzziness during Mamdani style HFS inference process, and determining a relation between conventional HFSs' accuracy and the chosen hierarchic structure regarding the number of SFSs. (ii) Testing the accuracy and stability of DF-HFS in large scale experiments, and verify its flexibility on both chosen inference methods and hierarchic structures. Addition operation on fuzzy numbers is chosen as the case study of this paper. There are several reasons for this problem selection: (a) Implementation is easy. (b) It is possible to employ very large scale experiments with huge number of inputs. (c) It is reasonable and fair to compare the behavioral differences if and only if the systems rules are modeled consistently. In fuzzy addition problem, there is no restriction to generate the rules base of HFSs which is equivalent with single system's, as addition is a decomposable function with commutativity and associativity properties. In line with the aforementioned objectives, various experiments are performed on fuzzy addition operation with changing number of inputs on different hierarchic structures which are the variations of incremental and aggregated HFSs. The intervals of number of inputs and number of applied SFSs are [4,512] and [2,255] respectively. 100 randomly generated input tuples are evaluated in the experiments. Since defuzzification is the most vital point for a fuzzy systems outputs, experiments are repeated two times by using different defuzzification strategies: Center of Area (COA) and Weighted Average. Provided test results are evaluated with Root Mean Square Error (RMSE) by basing the differentiation of corresponding HFS behaviors with the single systems'. R-Square (R^2) is also calculated to observe the distribution of systems' behaviors. Results of experiments have been proved the accuracy and stability of HF-HFS. Therefore, in such cases where the number of inputs are so high to be modeled by single system, DF-HFS outputs are based to complete the tests. Moreover, obtained results have been showed that a pattern can be extracted between the magnitude of reflection of degeneration in fuzziness to HFS outputs and total number of SFSs. Finally, the magnitude of this degeneration for unseen experiments can be approximately predicted by using this pattern.

2   Data and Method

2.1   Data: Fuzzy Numbers

In this study, addition operation on fuzzy numbers is employed as an example experimental case study.

In [21], Chen described a generalized trapezoidal fuzzy number $\tilde{A}$ as $\tilde{A} = (a, b, c, d; w)$ where a, b, c, d and w are real numbers, and $0 < w \leq 1$. A fuzzy set $\tilde{A}$ of the real line R with membership function $\mu_{\tilde{A}} : R \to [0, 1]$ is called a generalized fuzzy number if the following conditions are satisfied:

- $\mu_{\tilde{A}}(x) = 0$ where $-\infty < x \leq a$

- $\mu_{\tilde{A}}(x)$ is increasing on $[a, b]$
- $\mu_{\tilde{A}}(x) = w$ where $b \le x \le c$
- $\mu_{\tilde{A}}(x)$ is decreasing on $[c, d]$
- $\mu_{\tilde{A}}(x) = 0$ where $d \le x < \infty$

Note that $\tilde{A}$ is a triangular fuzzy number where $b = c$, and it is identified as $\tilde{A} = (a, b, d; w)$.

Let $\tilde{A}_1$ and $\tilde{A}_2$ are two trapezoidal fuzzy numbers. The addition of these numbers is calculated by (1).

$$\begin{aligned} \tilde{A}_1 \oplus \tilde{A}_2 &= (a_1, b_1, c_1, d_1; w_1) \oplus (a_2, b_2, c_2, d_2; w_2) \\ &= (a_1 + a_2, b_1 + b_2, c_1 + c_2, d_1 + d_2; min(w_1, w_2)) \end{aligned} \tag{1}$$

Basic properties of fuzzy addition operation are detailed in [22] and some of them are as follows:

- Commutativity: $\tilde{A}_1 \oplus \tilde{A}_2 = \tilde{A}_2 \oplus \tilde{A}_1$
- Associativity: $(\tilde{A}_1 \oplus \tilde{A}_2) \oplus \tilde{A}_3 = \tilde{A}_1 \oplus (\tilde{A}_2 \oplus \tilde{A}_3)$

It is known that some functions which are non-decomposable should not built by an HFS [9]. Surely in such cases where the desired function of single system is non-decomposable, the resulting HFS cannot provide the same behaviors with single system. Since associativity property ensures decomposability to the fuzzy addition operation, it ensures the construction of HFSs without any limitation on hierarchic structure or the fuzzy rules. Here, this flexibility is the most significant point for the reliability and correctness of implementations. Commutativity, on the other hand, lets variations in the order of input parameters, and provide elasticity to the HFS modeling. Briefly, these two properties make differently structured HFSs be equivalent in terms of fuzzy rules, and annihilate any reason which causes behavioral differentiation of HFSs for both each other and single system. And if this behavior differs somehow, these properties ensures that this outcome does not raise from the modeling issues, but the inference process inside.

2.2 Hierarchical Fuzzy Inference

In Mamdani style fuzzy inference, there are four basic steps: fuzzification, rule fitting, aggregation and defuzzification. Even though there are several different methods for defuzzification, all of these methods aim to generalize the aggregated fuzzy output to single crisp value. It is highly challenging and even mathematically impossible to extract that crisp value which determines the whole fuzzy data for all kinds of membership functions. Therefore this step affects fuzziness permanently.

In conventional implementation of HFSs, each SFS performs all of the Mamdani inference steps one by one, and provide a crisp output to its successor. This process is plausible and easy to implement. However the defuzzification steps in the inner layer SFSs cause a degeneration in the corresponding system's output [1].

In DF-HFS the redundantly repeated defuzzification steps are eliminated from the hierarchical inference process, and the fuzzy data provided from aggregation step in a SFS is directly transferred to the subsequent layer as its input. Since this input is already a fuzzy variable, the fuzzification step is also eliminated. Consequently fuzziness in the lowest layer is propagated without any deformation until the top-most SFS which procures the final decision of the corresponding HFS.

Since DF-HFS aims to preserve the fuzziness from the lowest layer to the top-most one, it pinpoints the inaccuracy problem in HFSs. In other words, DF-HFS aims to provide the same outputs with the corresponding single system if HFS rules are able to be generated equally with the single system's. If this objective can be satisfied, it means that DF-HFS is stable and robust against the variations on structural differences of HFS.

3 Experiments and Results

Experiments are performed on fuzzy addition operation with changing number of inputs on .NET platform by using FuzzyNet library [23]. These experiments are repeated for 100 randomly created input tuples. Each crisp input is determined by two triangular fuzzy numbers. One of them is the greatest integer less than this input and the other one is the smallest integer greater than it. These integers specifies the cores x of corresponding fuzzy numbers $\tilde{X}$, and the intervals are specified by $[x - 1, x + 1]$. For example if the input is 3.14, fuzzy numbers $\tilde{3} = (2, 3, 4; 1)$ and $\tilde{4} = (4, 5, 6; 1)$ are specified and processed during the fuzzy inference.

The utilized hierarchic structures are aggregated and incremental. Surely it is challenging to construct all of the variations for each hierarchic structure. Therefore a small but sufficient subset of differently constructed HFSs are implemented here. Each row in Table 1 and Table 4 corresponds the implemented cases one by one for aggregated and incremental structure respectively.

The HFSs based on aggregated structure are constructed by an up-down strategy. Suppose that n is the number of inputs, and s is the lowest divisor of n. The input space is separated into s sub-spaces which are then determined in different SFSs containing n/s inputs. Input space separation proceeds $m - 1$ times for the construction of m-layer aggregated structure. Regarding the incremental HFSs, n inputs are distributed into l layers as homogeneous as possible. Since each layer has only one SFS, totally l SFSs are generated in the hierarchy.

Table 1 and Table 4 show the experimental results containing the behavioral differences of HFSs for aggregated and incremental structure respectively. Here first four columns determine the identity number of related experimental case, the number of inputs, layers and SFSs. From the 5th column to the 8th, RMSE and R^2 values of conventional HFS and DF-HFS are presented for the use of COA defuzzifier, and the others corresponds RMSE and R^2 values for weighted average defuzzifier. Since

Table 1 Results of experiments on aggregated structure

Case ID	# Inputs	# Layers	# SFSs	COA Defuzzifier						Weighted Average Defuzzifier					
				Conventional HFS		DF-HFS		Conventional HFS		DF-HFS					
				RMSE	R^2	RMSE	R^2	RMSE	R^2	RMSE	R^2				
1	4	2	3	0.036787	0.994257	0	1	0.068476	0.984937	0	1
2	8	2	3	0.038951	0.997463	0	1	0.070294	0.991370	0	1
3	8	3	7	0.059919	0.995449	0	1	0.143252	0.980482	0	1
4	12	2	3	0.047189	0.997732	0	1	0.080159	0.993262	0	1
5	12	3	7	0.073952	0.994444	0	1	0.162874	0.977873	0	1
6	16	2	3	0.050347	0.997812	0	1	0.076693	0.993981	0	1
7	16	3	7	0.071838	0.995940	0	1	0.129873	0.985303	0	1
8	16	4	15	0.094425	0.993898	0	1	0.247285	0.962990	0	1
9	18	2	3	0.058454	0.997004	0	1	0.085736	0.990885	0	1
10	18	3	9	0.109431	0.989399	0	1	0.234238	0.945546	0	1
11	18	4	27	0.128135	0.989314	0	1	0.244058	0.943045	0	1

(continued)

Table 1 (continued)

Case ID	# Inputs	# Layers	# SFSs	COA Defuzzifier				Weighted Average Defuzzifier			
				Conventional HFS		DF-HFS		Conventional HFS		DF-HFS	
				RMSE	R^2	RMSE	R^2	RMSE	R^2	RMSE	R^2
12	32	3	7	0.089669	0.996898	0	1	0.131928	0.990476	0	1
13	32	4	15	0.116394	0.994977	0	1	0.228046	0.977126	0	1
14	32	5	31	0.146781	0.993268	0	1	0.395523	0.950357	0	1
15	64	4	15	0.132230	0.995973	–	–	0.183991	0.990289	–	–
16	64	5	31	0.176528	0.993350	–	–	0.325536	0.974514	–	–
17	64	6	63	0.218939	0.991166	–	–	0.553455	0.945563	–	–
18	128	5	31	0.197000	0.996090	–	–	0.274842	0.990780	–	–
19	128	6	63	0.291209	0.992445	–	–	0.459772	0.980108	–	–
20	128	7	127	0.318850	0.992704	–	–	0.797419	0.954666	–	–
21	256	6	63	0.290863	0.995620	–	–	0.424994	0.987770	–	–
22	256	7	127	0.360638	0.993861	–	–	0.671068	0.980171	–	–
23	256	8	255	0.472737	0.992252	–	–	1,155739	0.958046	–	–
24	512	7	127	0.345949	0.996169	–	–	0.476808	0.990993	–	–
25	512	8	255	0.520015	0.992462	–	–	0.840230	0.976258	–	–
26	512	9	511	0.651571	0.989821	–	–	1,523419	0.947324	–	–

the most important motivation for HFS performance is providing the single system behaviors besides handling the curse of dimensionality, RMSE and R^2 values are calculated by basing the single system outputs.

It is clearly seen from Tables 1 and 4 that conventional process of HFSs cannot provide the same inputs with corresponding single system. On the other hand, DF-HFS satisfies this objective for all of the cases without being effected from the chosen hierarchic structure. This observation is evaluated as an accuracy and stability test of DF-HFS. In the cases with more than 32 input parameters (*ID* > 14 in Table 1 and *ID* > 37 in Table 4), single system solution becomes infeasible due to the curse of dimensionality bottleneck. Therefore the outputs provided from DF-HFS is used as a base in order to see the behaviors of conventional process for larger scale experiments.

In Tables 1 and 4, it can be seen that, for a constant number of inputs, the difference between single system's and HFS's output is increased when the number of layers are increased. In fact this issue is related with the number of SFSs which deter-

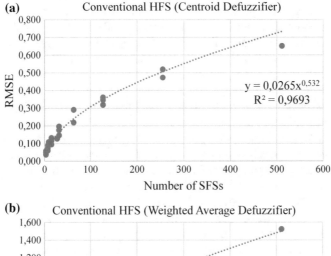

Fig. 1 Effect of number of SFSs on HFS outputs for aggregated HFS. **a** Experiments on COA, and **b** weighted average defuzzifier

mines the number of defuzzification steps performed in the inner layers. In a single system solution, only one defuzzification is required, and it is plausible since a crisp output should be provided to finalize the reasoning process. However for an HFS, it is repeated in every SFSs, and each one causes an amount of deformation until the root SFS which is in fact the only SFS providing a crisp value. Consequently, when the number of inputs keeps constant, the use of more SFSs increases the degeneration in transferred fuzzy data, and the HFS behaviors staidly diverge from the behavior provided by single system.

To make the experimental results in Table 1 more comprehensible, these results are projected into two charts in Fig. 1 which illustrate the differentiation between the single system's and the HFS's outputs for aggregated structure. Figure 1a, b shows the differentiation of RMSE values by the increase in number of SFSs for COA and weighted average defuzzifier. It can be easily concluded that when the number of SFSs (indirectly the number of performed defuzzification steps) increases, RMSE value exposes an exponential increase.

Summarizing Table 1 and Table 4, RMSE values are post-processed for aggregated (see Table 2 and Table 3) and incremental (see Table 5 and Table 6) structures respectively. Herein it is possible to see the minimum, maximum and average RMSE values with standard deviation for both the overall HFS's and individual SFS's perspective. The effect of individual SFS on these RMSE values are provided from dividing the corresponding RMSE value into the number of SFSs.

From all these findings of experiments, a new discussion can be coined here: *Is this possible to predict the magnitude of degeneration on fuzziness before the implementation of single system?* In detail, now it is known that the conventional process of hierarchical inference fails because of its inaccuracy arisen from the degeneration of fuzziness. In this study it is easy to measure this degeneration. Because it is possible to construct the single system for most of the cases. However it is clear that, the construction of single system for more complex problems may not be applicable. Therefore the behavior of single system remains unknown. Unless the system designer uses DF-HFS for hierarchical reasoning, (s)he should at least approximately anticipate the magnitude of possible differentiation in the HFS's outputs from the single system. Here it can be derived from the graphics in Fig. 1 that, the differentiation in the system's output can be stated by a specific function with very high precision. By using it, this differentiation can be predicted for an unknown case. For example if an aggregated HFS with 3 SFSs are required, that HFS's RMSE value can be calculated from Fig. 1a as $0.0265 \times 3^{0.532} = 0.047541664$. Resulting value can be verified by checking the minimum and maximum intervals in Table 2. If it is inside the related interval, it is verified. All of these cases can be verified by using the same perspective. It is concluded from this calculation that the degeneration in system's fuzziness is not random or occasional, and has a consistent pattern which can be extracted by some empirical studies.

Table 2 The boundaries of RMSE values on aggregated structure for COA defuzzifier

# Layers	# SFSs	Conventional HFS				Individual SFS Effect of Conventional HFS			
		Minimum	Maximum	Average	Std. Dev.	Minimum	Maximum	Average	Std. Dev.
2	3	0.036787	0.058454	0.046346	0.007866	0.012262323	0.019484780	0.015448546	0.002622098
3	7	0.059919	0.089669	0.073844	0.010587	0.008559853	0.012809873	0.010549207	0.001512495
3	9	0.109431	0.109431	0.109431	0	0.012159047	0.012159047	0.012159047	0
4	15	0.094425	0.132230	0.114350	0.015501	0.006295029	0.008815351	0.007623322	0.001033419
4	27	0.128135	0.128135	0.128135	0	0.004745726	0.004745726	0.004745726	0
5	31	0.146781	0.197000	0.173436	0.020618	0.004734856	0.006354834	0.005594712	0.000665103
6	63	0.218939	0.291209	0.267004	0.033987	0.003475229	0.004622369	0.004238157	0.000539476
7	127	0.318850	0.360638	0.341812	0.016661	0.002510633	0.002839667	0.002691435	0.000136288
8	255	0.472737	0.520015	0.496376	0.023639	0.00185387	0.002039274	0.001946572	9,27021E-05

Table 3 The boundaries of RMSE values on aggregated structure for weighted average defuzzifier

# Layers	# SFSs	Conventional HFS				Individual SFS Effect of Conventional HFS			
		Minimum	Maximum	Average	Std. Dev.	Minimum	Maximum	Average	Std. Dev.
2	3	0.068476	0.085736	0.076272	0.006346	0.022825360	0.028578727	0.025423919	0.002115406
3	7	0.129873	0.162874	0.141982	0.013094	0.018553237	0.023267735	0.020283104	0.001870576
3	9	0.234238	0.234238	0.234238	0	0.026026488	0.026026488	0.026026488	0
4	15	0.183991	0.247285	0.219774	0.026493	0.012266034	0.016485634	0.01465158	0.001766232
4	27	0.244058	0.244058	0.244058	0	0.009039195	0.009039195	0.009039195	0
5	31	0.274842	0.395523	0.331967	0.049477	0.008865867	0.012758821	0.010708612	0.001596048
6	63	0.424994	0.553455	0.479407	0.054251	0.006745931	0.008785001	0.007609636	0.000861124
7	127	0.476808	0.990993	0.975276	0.146258	0.003754396	0.00627889	0.005105761	0.001038297
8	255	0.840230	1,155739	0.997984	0.157754	0.003295021	0.004532309	0.003913665	0.000618644

Table 4 Results of experiments on incremental structure

Case ID	# Inputs	# Layers	# SFSs	COA Defuzzifier				Weighted Average Defuzzifier			
				Conventional HFS		DF-HFS		Conventional HFS		DF-HFS	
				RMSE	R^2	RMSE	R^2	RMSE	R^2	RMSE	R^2
1	4	2	2	0.028717	0.996469	0	1	0.064010	0.984323	0	1
2	8	2	2	0.038764	0.997529	0	1	0.081687	0.987678	0	1
3	8	3	3	0.046637	0.996387	0	1	0.096921	0.984345	0	1
4	8	4	4	0.052070	0.995835	0	1	0.119565	0.982121	0	1
5	12	2	2	0.045762	0.997813	0	1	0.104096	0.985586	0	1
6	12	3	3	0.064994	0.995658	0	1	0.134886	0.979973	0	1
7	12	4	4	0.070285	0.994889	0	1	0.155077	0.972539	0	1
8	12	5	5	0.078362	0.994111	0	1	0.166661	0.974581	0	1
9	12	6	6	0.081887	0.993509	0	1	0.194280	0.967745	0	1
10	16	2	2	0.049038	0.997922	0	1	0.109626	0.987121	0	1
11	16	3	3	0.060394	0.996935	0	1	0.136976	0.980525	0	1
12	16	4	4	0.075907	0.995260	0	1	0.151014	0.979941	0	1
13	16	5	5	0.089621	0.993170	0	1	0.181772	0.969104	0	1
14	16	6	6	0.080637	0.995290	0	1	0.175889	0.973936	0	1
15	16	7	7	0.086780	0.994631	0	1	0.191379	0.971666	0	1
16	16	8	8	0.093175	0.994084	0	1	0.221633	0.965792	0	1

(continued)

Table 4 (continued)

Case ID	# Inputs	# Layers	# SFSs	COA Defuzzifier				Weighted Average Defuzzifier			
				Conventional HFS		DF-HFS		Conventional HFS		DF-HFS	
				RMSE	R^2	RMSE	R^2	RMSE	R^2	RMSE	R^2
				RMSE	R^2	RMSE	R^2	RMSE	R^2	RMSE	R^2
17	18	2	2	0.058549	0.996946	0	1	0.117211	0.982213	0	1
18	18	3	3	0.076667	0.994623	0	1	0.154707	0.971134	0	1
19	18	4	4	0.082160	0.994814	0	1	0.150565	0.972878	0	1
20	18	5	5	0.086594	0.994007	0	1	0.192762	0.962982	0	1
21	18	6	6	0.094939	0.992444	0	1	0.213972	0.957809	0	1
22	18	7	7	0.093017	0.992175	0	1	0.189334	0.960788	0	1
23	18	8	8	0.103867	0.990621	0	1	0.218357	0.953183	0	1
24	18	9	9	0.107708	0.990400	0	1	0.238723	0.947106	0	1
25	32	4	4	0.087859	0.996508	0	1	0.175507	0.981711	0	1
26	32	5	5	0.093706	0.996162	0	1	0.195659	0.976106	0	1
27	32	6	6	0.123807	0.993728	0	1	0.238183	0.970520	0	1
28	32	7	7	0.117788	0.993506	0	1	0.229857	0.969735	0	1
29	32	8	8	0.129103	0.992199	0	1	0.251574	0.967509	0	1
30	32	9	9	0.128926	0.993495	0	1	0.268276	0.962607	0	1
31	32	10	10	0.138429	0.992605	0	1	0.285447	0.961568	0	1
32	32	11	11	0.122082	0.993289	0	1	0.250077	0.968002	0	1
33	32	12	12	0.129864	0.992523	0	1	0.266844	0.962627	0	1
34	32	13	13	0.127179	0.992439	0	1	0.264066	0.965374	0	1
35	32	14	14	0.135145	0.991475	0	1	0.268882	0.964416	0	1
36	32	15	15	0.144803	0.990200	0	1	0.288336	0.960780	0	1
37	32	16	16	0.147986	0.989813	0	1	0.306402	0.956212	0	1

(continued)

Table 4 (continued)

Case ID	# Inputs	# Layers	# SFSs	COA Defuzzifier				Weighted Average Defuzzifier			
				Conventional HFS		DF-HFS		Conventional HFS		DF-HFS	
				RMSE	R^2	RMSE	R^2	RMSE	R^2	RMSE	R^2
38	64	8	8	0.127988	0.995863	–	–	0.265505	0.978517	–	–
39	64	16	16	0.189721	0.991278	–	–	0.328125	0.968386	–	–
40	64	32	32	0.221121	0.988808	–	–	0.420184	0.955378	–	–
41	128	16	16	0.219842	0.995153	–	–	0.389128	0.980231	–	–
42	128	32	32	0.258150	0.994399	–	–	0.530891	0.971298	–	–
43	128	64	64	0.303855	0.990921	–	–	0.641261	0.957631	–	–
44	256	32	32	0.290031	0.995330	–	–	0.576159	0.977882	–	–
45	256	64	64	0.376943	0.992999	–	–	0.706476	0.972023	–	–
46	256	128	128	0.419523	0.991299	–	–	0.894822	0.958731	–	–
47	512	64	64	0.420490	0.994022	–	–	0.824880	0.969824	–	–
48	512	128	128	0.478786	0.992494	–	–	1,014838	0.961012	–	–
49	512	256	256	0.750131	0.984825	–	–	1,190337	0.952961	–	–

Table 5 The boundaries of RMSE values on incremental structure for COA defuzzifier

# Layers	# SFSs	Conventional HFS				Individual SFS Effect of Conventional HFS			
		Minimum	Maximum	Average	Std. Dev.	Minimum	Maximum	Average	Std. Dev.
2	2	0.028717	0.058549	0.044166	0.010010	0.014358256	0.029274303	0.022083025	0.005005056
3	3	0.046637	0.076667	0.062173	0.010754	0.015545779	0.025555789	0.020724394	0.003584566
4	4	0.052070	0.087859	0.073656	0.012300	0.013017487	0.021964674	0.018414004	0.00307494
5	5	0.078362	0.093706	0.087071	0.005626	0.01567238	0.018741156	0.017414172	0.001125166
6	6	0.080637	0.123807	0.095318	0.017376	0.013439479	0.020634512	0.015886253	0.002895993
7	7	0.086780	0.117788	0.099195	0.013391	0.012397088	0.016826828	0.014170706	0.001913071
8	8	0.093175	0.129103	0.113533	0.015486	0.011646885	0.016137868	0.014191666	0.00193575
9	9	0.107708	0.128926	0.118317	0.010609	0.011967571	0.014325121	0.013146346	0.001178775
10–11–12	10–11–12	0.122082	0.138429	0.130125	0.006676	0.010822023	0.013842943	0.01192112	0.00136361
13–14–15	13–14–15	0.127179	0.144803	0.135709	0.007206	0.009653212	0.009783019	0.009696598	6,11092E-05
16	16	0.147986	0.219842	0.185850	0.029463	0.009249101	0.013740147	0.011615608	0.001841428
32	32	0.221121	0.290031	0.256434	0.028159	0.006910031	0.009063477	0.008013569	0.000879958
64	64	0.303855	0.420490	0.367096	0.048122	0.004747739	0.006570159	0.005735878	0.000751912
128	128	0.419523	0.478786	0.449154	0.029632	0.00327752	0.003740514	0.003509017	0.000231497

Table 6 The boundaries of RMSE values on incremental structure for weighted average defuzzifier

# Layers	# SFSs	Conventional HFS				Individual SFS Effect of Conventional HFS			
		Minimum	Maximum	Average	Std. Dev.	Minimum	Maximum	Average	Std. Dev.
2	2	0.064010	0.117211	0.095326	0.019632	0.032005247	0.058605464	0.047663	0.009815897
3	3	0.096921	0.154707	0.130872	0.021061	0.032306842	0.051569105	0.043624118	0.007020191
4	4	0.119565	0.175507	0.150346	0.017908	0.029891371	0.043876743	0.037586471	0.00447688
5	5	0.166661	0.195659	0.184213	0.011381	0.033332269	0.039131751	0.036842699	0.002276201
6	6	0.175889	0.238183	0.205581	0.023144	0.029314852	0.039697231	0.034263528	0.003857389
7	7	0.189334	0.229857	0.203523	0.018639	0.027047676	0.032836666	0.029074709	0.002662776
8	8	0.218357	0.265505	0.239267	0.019925	0.027294583	0.033188101	0.029908391	0.002490653
9	9	0.238723	0.268276	0.253500	0.014776	0.026524825	0.029808490	0.028166658	0.001641833
10–11–12	10–11–12	0.250076	0.285446	0.267455	0.014446	0.022236974	0.028544683	0.024505304	0.002863478
13–14–15	13–14–15	0.264066	0.288336	0.273761	0.010492	0.019205863	0.020312766	0.019580344	0.000517945
16	16	0.306402	0.389128	0.341218	0.035019	0.019150117	0.024320489	0.021326137	0.002188673
32	32	0.420184	0.576159	0.509078	0.065518	0.013130758	0.018004963	0.015908684	0.00204743
64	64	0.641261	0.824880	0.724206	0.093085	0.010019699	0.012888752	0.011315712	0.001187554
128	128	0.894822	1,014838	0.954830	0.060008	0.006990795	0.007928423	0.007459609	0.000468814

4 Conclusion

Single system solutions are inapplicable for large scale problems with a lot of input parameters because of the problem called *curse of dimensionality*. Hierarchical fuzzy systems are proposed to handle this problem by separating the large input space into lower dimensional sub-spaces. However the conventional process of HFSs has a hidden problem: inaccuracy. It cannot provide the single system outputs, and the provided outputs differentiate in any tiny variation in the hierarchic structure. This issue reduces the stability of HFSs against these variations. The inaccuracy is a bigger problem for Mamdani style fuzzy inference, since it relies on linguistic fuzzy reasoning for both antecedent and consequent parts of fuzzy rules.

In this study, the inaccuracy and instability of Mamdani style hierarchical fuzzy inference process are investigated by an experimental manner on fuzzy addition problem. The experiments are performed on various hierarchic structures which are the variations of aggregated and incremental HFSs. Two area based defuzzifiers are implemented: center of area and weighted average. Basing single system outputs, RMSE and R^2 values are calculated to evaluate the accuracy of HFSs. For large scale experiments whose solution by single system is inapplicable because of the curse of dimensionality, Defuzzification-Free Hierarchical Fuzzy System: DF-HFS outputs are used as a base for the measurements, since it ensures to provide the same outputs with the single system.

The experiments showed that the conventional process of HFSs cannot provide the same outputs with single system, and any variation in hierarchic structure causes differentiation in HFS behaviors. However this differentiation is not occasional or random. On the contrary, there is a consistent pattern which can be expressed by a mathematical function. Therefore in such cases where the system designer should use this kind of HFSs, (s)he can predict the potential magnitude of this differentiation and take it into account as an error range. DF-HFS, on the other hand, always provides the same outputs with the single system and it is robust against the variations in the hierarchic structure.

Although there is no need to derive rules for addition operation, real world problems strictly require fuzzy rules for a complete fuzzy system implementation. Therefore subsequent studies will be focusing on automatic generation of HFS by partitioning the rules of single systems.

References

1. D. Wang, X. Zeng, J. Keane, A survey of hierarchical fuzzy systems. International Journal of Computational Conginiton **4**(1), 18–29 (2006)
2. G. Raju, J. Zhou, R. Kisner, Hierarchical fuzzy control. Int. J. Control **54**(5), 1201–1216 (1991). https://doi.org/10.1080/00207179108934205
3. F. Chung, J. Duan, On multistage fuzzy neural network modeling. IEEE Trans. Fuzzy Syst. **8**(2), 125–142 (2000)

4. A.D. Benitez, J. Casillas, Multi-objective genetic learning of serial hierarchical fuzzy systems for large-scale problems. Soft Comput. **17**(1), 165–194 (2013)
5. M.G. Joo, T. Sudkamp, A method of converting a fuzzy system to a two-layered hierarchical fuzzy system and its run-time efficiency. IEEE Trans. Fuzzy Syst. **17**(1), 93–103 (2009)
6. T. Takagi, M. Sugeno, Fuzzy identification of systems and its applications to modeling and control. IEEE Trans. Syst. Man Cybern. Syst. **15**(1), 116–132 (1985)
7. M. Sugeno, G. Kang, Structure identification of fuzzy model. Fuzzy Sets Syst. **28**(1), 15–33 (1988)
8. L.-X. Wang, Universal approximation by hierarchical fuzzy systems. Fuzzy Sets Syst. **93**, 223–230 (1998)
9. V. Torra, A review of the construction of hierarchical fuzzy systems. Int. J. Intel. Syst. **17** (2002)
10. M.G. Joo, J.S. Lee, Universal approximation by hierarchical fuzzy system with constraints on the fuzzy rule. Fuzzy Sets Syst. **130**, 175–188 (2002)
11. L. Cai, Z. Cui, H. Liu, Hierarchical fuzzy systems as universal approximators, in *IEEE International Conference on Information Theory and Information Security (ICITIS)* (2010)
12. B. Mutlu, E.A. Sezer, H.A. Nefeslioglu, Consequences of structural differences between hierarchical systems while fuzzy inference, in *Lecture Notes in Computer Science Advances, Computational Intelligence.* (2015) pp. 549–560
13. B. Mutlu, E.A. Sezer, H.A. Nefeslioglu, A defuzzification-free hierarchical fuzzy system (DF-HFS): Rock mass rating prediction. Fuzzy Sets Syst. (2016)
14. H. Maeda, A study on the spread of fuzziness in multi-fold multi-stage fuzzy reasoningeffect of two-dimensional multi-fold fuzzy membership function forms. Fuzzy Sets Syst. **80**, 133–148 (1996)
15. A. Daftaribesheli, M. Ataei, F. Sereshki, Assessment of rock slope stability using the Fuzzy Slope Mass Rating (FSMR) system. Appl. Soft Comput. **11**, 4465–4473 (2011)
16. S. Khanmohammadi, C. Dagli, F. Esfahlani, A fuzzy inference model for predicting irregular human behaviour during stressful missions. Procedia Comput. Sci. **12**, 265–270 (2012)
17. C. Qu R. Buyya, A cloud trust evaluation system using hierarchical fuzzy inference system for service selection, in *IEEE 28th International Conference on Advanced Information Networking and Applications,* (2014)
18. X. Zhang, E. Onieva, A. Perallos, E. Osaba, V.C. Lee, Hierarchical fuzzy rule-based system optimized with genetic algorithms for short term traffic congestion prediction. Transp. Res. Part C: Emerg. Technol. **43**(2014), 127–142 (2014)
19. W. Rattasiri, S. Halgamuge, Computationally advantageous and stable hierarchical fuzzy systems for active suspension. IEEE Trans. Ind. Electron. **50**(1), 48–61 (2003)
20. W. Rattasiri S. Halgamuge, Computational complexity of hierarchical fuzzy systems, in *Fuzzy Information Processing Society, NAFIPS. 19th International Conference of the North American,* vol. 2 (Atlanta, 2000), pp. 383–387
21. S.-J. Chen, S.-M. Chen, Fuzzy risk analysis based on similarity measures of generalized fuzzy numbers. IEEE Trans. Fuzzy Syst. **11**(1), 45–56 (2003)
22. M. Mzumoto, K. Tanaka, Some properties of fuzzy numbers. Ad. Fuzzy Set Theory Appl. **1**(2), 153–164 (1979)
23. D. Kaluzhny, *Fuzzy Net: Fuzzy Logic Library for Microsoft.NET* (2009)

Part IX
Image Analysis

A Fuzzy Shape Extraction Method

A. R. Várkonyi-Kóczy, B. Tusor and J. T. Tóth

Abstract This chapter presents an easily implementable method of fuzzy shape extraction for shape recognition. The method uses Fuzzy Hypermatrix-based classifiers in order to find the potential location of the target objects based on their colors, then determines the areas where the most densely occurring positive findings in order to restrict the area of operation thus speeding the process up. In these areas the edges are detected, the edges are mapped to tree structures, which are trimmed down to simple outline sequences using heuristics from the Fuzzy Hypermatrix. Finally, fuzzy information is extracted from the outlines that can be used to classify the shape with a fuzzy inference machine.

1 Introduction

Shape recognition is an important field in computer vision. Its numerous fields of application involve industrial manufacturing, traffic control, human hand posture recognition, and many other applications that involve the recognition or objects in the data of various (typically visual) sensors.

In order to be able to classify the shape, various descriptors have to be found in the images.

A. R. Várkonyi-Kóczy (✉)
Integrated Intelligent Systems Japanese-Hungarian Laboratory, Department of Mathematics and Informatics, J. Selye University, Komarno, Slovakia
e-mail: varkonyi-koczy@uni-obuda.hu

B. Tusor
Integrated Intelligent Systems Japanese-Hungarian Laboratory, Doctoral School of Applied Informatics and Applied Mathematics, Óbuda University, Budapest, Hungary
e-mail: balazs.tusor@gmail.com

B. Tusor · J. T. Tóth
Department of Mathematics and Informatics, J Selye University, Komarno, Slovakia
e-mail: tothj@selyeuni.sk

© Springer International Publishing AG, part of Springer Nature 2018
L. A. Zadeh et al. (eds.), *Recent Developments and the New Direction in Soft-Computing Foundations and Applications*, Studies in Fuzziness and Soft Computing 361, https://doi.org/10.1007/978-3-319-75408-6_29

There are many systems designed for shape detection and recognition. Panwar [1] presents a hand gesture recognition system that removes the background noise in pre-processing steps and applies K-means clustering for segmenting the hand from the rest of the background, so that only segmented significant cluster or hand object is to be processed in order to calculate shape based features like orientation, center of mass, status of fingers, etc. It does not consider skin color, thus avoiding problems caused by different lighting conditions.

Wang et al. [2] introduces robust feature descriptors for shape feature detection from intensity order information, considering the intensity relationships among all the neighboring sample points around a pixel, and exploiting the coarsely quantized overall intensity order of these sampling points.

Inoges et al. [3] proposes a shape detector based on inner-angles and inner-distances for hand detection on images of Near Infrared cameras.

Chen et al. [4] uses a fusion of calibrated observation data from two RGB-D sensors installed on the head and the hand of a humanoid tomato-harvesting robot. It applies a pointcloud model segmentation to obtain the primitive shape model of each fruit, and a probabilistic model to determine the picking order of the tomatoes within the branch.

Li et al.[5] proposes two kinds of neural networks for automatic lane boundary detection in traffic scenes: a multitask deep convolutional network for detecting the presence of the geometric attributes (location and orientation) of the target object with respect to the region of interest; and a recurrent neuron layer for structured visual detection.

In general, the first step in such systems is finding the appropriate descriptors that describe the shapes in the image efficiently and then comparing the shapes (based on these descriptors) to that of the known objects. For example, this step can be composed of segmenting the whole image with a suitable edge detection algorithm, then analyzing the resulting contours. However, this can be a lengthy process depending on the size of the image and its complexity (the amount of objects in it). If heuristics are available about the sought objects (e.g. color tones associated with them), the extraction process can be shortened by restricting the area of operation to certain parts of the image.

The system proposed in this paper is using this principle. It consists of the following steps: (a) detecting the areas associated with the sought objects based on color tones using a fast classifier, (b) constructing the contour of given objects in the resulting areas, and (c) extracting robust fuzzy descriptors that can be analyzed in order to identify the shapes.

In the first step, a 3D Fuzzy Hypermatrices-based (FHM) classifier [6] (developed by the authors in an earlier stage of this research) is used to detect the pixels of the image that have a similar color tone to the object that has to be recognized (e.g. human skin color areas). The resulting image is filtered with a density-based filter in order to get rid of the scarcely occurring false-positive findings (which, in this case is basically just noise), then using dilation (a simple operation of mathematical morphology, for more details see [7]). The results are the areas where the sought objects are.

On the resulting areas (so-called blobs) the system carries out edge detection (e.g. the Canny [8] or Russo's fuzzy edge detection [9] method), then it builds tree structures from the detected edges. The trees are trimmed (using heuristics) until they are reduced to simple lines. The resulting neighboring lines are combined and extended so they create a loop (in each blob).

After that, the analysis of the looping pixel-sequences can be started. This is done by finding points of interests (POIs) in the sequence. Finally, fuzzy features are extracted from the features of the POIs and from their patterns the shape of the detected object is determined.

Although the system can be used for any object, in this paper authors focus on human hand shapes, so the examples shown are from a human hand posture recognition problem. The implementation of the system has been tested on an average PC (HP ProBook 4540 s (Intel® Core™ i5-2450 M CPU, @2.50 GHz, 4 GB RAM)).

The rest of the chapter is, as follows. Section 2 describes the system through 7 subsections, shows the testing results, and provides comparison to other descriptors. Finally, Sect. 3 concludes the paper and presents future work.

2 Shape Detection

2.1 Determining the Area of Operation and Edge Detection

The first step is to determine the area of operation in order to restrict further processing, thus enhancing the speed of the system. For this, the system has to find all pixels that are similar to the known color tones of the object that is needed to be recognized. Fuzzy Hypermatrices are used, because they are among the quickest classifiers due to their minimalistic runtime operation. The principle behind them is the pre-calculation and storing of the problem space (in this case, the HSV color space) in multi-dimensional arrays (where each dimension corresponds to an input attribute), so the classification results are immediately available during the evaluation phase. Figure 1 shows the image where the hand shape is needed to be recognized (left) and the output of the FHM classifier (right). The image with resolution 640×480 pixels requires about 0.13 s to process.

The output of the classifier is filtered by a density based filtering (DBF) algorithm. The idea behind the method is dividing the image to $N \times M$ subimages and calculating the number of positively marked pixels in each subimage. A subimage is copied to the output only if (a) it has a sufficient number (ϑ_1) of marked pixels, or (b) it has at least ϑ_2 marked pixels and has a neighboring subimage with at least ϑ_1 marked pixels ($0 < \vartheta_2 < \vartheta_1$). Figure 2 (left) shows the output of the DBF algorithm ($M = 24$ and $N = 32$), thus dividing the image into subimages with size 20×20; using appropriately chosen thresholds $\vartheta_1 = 25$ and $\vartheta_2 = 5$. The filtering takes about 0.05 s.

Fig. 1 An input image (left) and the filtered human skin color areas using FHMs (right)

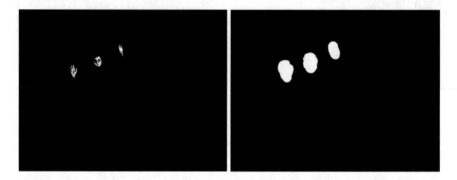

Fig. 2 The output of the density-based filter (left) and the dilated blobs (right)

After that, the resulting image is processed using dilation. Figure 2. shows the dilated image (right), using an ellipse shaped kernel with size 10. It takes 0.24 s to process the whole image. The results are the areas (so-called *blobs*) in the image where the rest of the analysis is carried out, making the process much faster. In order to find the frame of the subimage that contains the blob, a simple algorithm is used that operates on the same idea as the flood fill method [10]:

(1) While inspecting the image pixel by pixel, if the examined pixel is marked then check its neighbors, recursively mapping the connected area and noting the location and properties of the frame they fit in.
(2) After the whole connected area is mapped, then continue processing the image considering pixels that have not been examined yet.

The algorithm finds 3 blobs in the input image which takes about 0.034 s per blob. Similarly to the case of the density based filter, blobs that do not exceed an arbitrary size can be discarded.

Fig. 3 The results of the edge detection (using Canny method) for each blob

While the processing of the whole image (Fig. 1 left) with the Canny method in order to find the edges takes approximately 0.016 s, processing the blobs instead takes only 0.003 s. The results can be seen in Fig. 3.

2.2 The Mapping of Outline Trees

The edges detected in the previous step are stored in a directed tree structure, where each node corresponds to a pixel. Each neighboring pixels are directly connected in the tree; thus each node has one parental (except the root node) and 7 child nodes (because each pixel is surrounded by 8 other pixels).

The tree-building algorithm recursively maps the blob. If it finds a marked pixel, then appoints it as a root node, and sets all neighboring unprocessed pixels as its children, then continues the process through them until it runs out of such pixels. The tree building algorithm produces the directed trees for each blob in 0.0123 s on average.

The next step is the pruning of the tree down to a simple node sequence (a tree where all nodes have less than 2 children). Let us consider the following definitions:

- *branching nodes* are nodes with at least 2 children,
- *branches* are directed sequences of nodes from a branching node till a leaf that do not contain branching nodes (thus the parental branching node is not a part of the branches).

The pruning of tree is done by removing branches of the tree that are not likely to be part of the outline. A fitness function is used to determine which branches should be removed, thus a simple greedy algorithm (in this case, always choosing the option with the lowest fitness value) can be used.

One good way to determine the fitness value is calculating the number of marked pixels in the given branches, but it alone is not sufficient. Depending on the edge detection algorithm, not all edges that belong to the shape of the sought object are

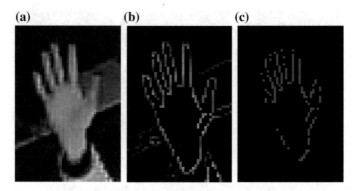

Fig. 4 **a** The original blob image, **b** the results of the edge detection and **c** the pixels of the results of the edge detection that have the sought color tone

actually part of the object on the image, as it can be seen in Fig. 4. As it can be seen, (a) shows the original subimage, (b) the edges found by the Canny method and (c) the edges that actually have the color tones of the human hand. This issue can be fixed by considering the colors of not only the pixels in the given branches, but the pixels *around* the branches too.

In this research, three qualifier variables are used to determine the fitness of a given branch:

- L: The number of pixels covered by the branch,
- Q: The number of *marked* pixels covered by the branch,
- Q_n: The number of marked pixels *around* the pixels covered by the branch.
  Using the combination of these qualifiers, rules can be set for the pruning. Such rules are e.g.: delete branches
- where $L = 1$ (thus removing all 1-pixel "dead ends").
- where L is much larger than $Q + Q_n$ (thus removing every branch with length much larger than all the marked pixels in and around the branch).
- where Q_n is much larger, than L (thus removing branches that are surrounded by too many marked pixels). This is the case of edges that are inside the sought shape, thus are not part of the outline.

The removal of branches that satisfy the removal rules is done until there is: (a) only one simple node sequence from node to leaf, or (b) only one branching node, which is the root node with two branches. In the first case, the algorithm is finished. In the second case, one of the branches is needed to be reversed in order to get a simple node sequence.

The change in the trees can be observed in Fig. 5 after one cycle of branch removal. As it can be seen, many trees have already been reduced to either to a single sequence or to their root node. The pixels with lighter color are the root nodes of the corresponding tree. The whole pruning phase results in only 3 sequences in the example that can be seen in Fig. 5b. The required time is 0.0356 on average per blob.

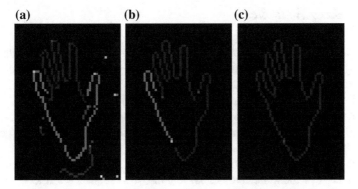

Fig. 5 The trees **a** after one pruning cycle, **b** after all the whole pruning and improvement phases; and **c** the united sequence

2.3 Node Sequence Improvement

The resulting node sequences should be improved by fixing outlying pixels (i.e. pixels where the direction of the sequence takes a sudden turn, only to turn back to the original direction by the next pixel, see Fig. 6a) that would otherwise make the further analysis steps more complex. Such outliers can be fixed by using a 3 pixels wide window on the sequence, changing the position of the middle pixel considering the position of the other two. This change can be summarized in two rules, (where x_i and y_i are the coordinates of pixel i):

- If $(x_{i-1} = x_{i+1})$ and $(y_{i-1} = y_i - 1)$ and $(y_{i-1} = y_i + 1)$ then $x_i = x_{i-1}$
- If $(y_{i-1} = y_{i+1})$ and $(x_{i-1} = x_i - 1)$ and $(x_{i-1} = x_i + 1)$ then $y_i = y_{i-1}$

Further improvements are made as well in order to remove the sharp (90°) turns in the sequence (see Fig. 6b). Similarly to the previous case, this also involves using a 3-pixel wide window. Whenever the algorithm encounters 3 pixels with a horizontal and vertical transitional direction, the middle node is removed from the sequence. The results of the sequence improvement phase can be seen in Fig. 5b, the processing time of each blob is ∼0.001 s.

Fig. 6 Sequence improvements: **a** fixing outliers and **b** removing sharp turns

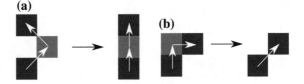

2.4 Uniting Sequences

Before the analyzation the outline of the shape, the sequences are needed to be joined by their ends. Firstly, the positions of the individual sequences are compared relative to each other. The sequences that have ends sufficiently close to each other (their distances are smaller than an arbitrary value) then they are joined, by adding pixels to bridge the distance between them. This can be done by simply making a straight line or a curve considering the average direction of the two sequences around their examined ends.

If one of the ends is the root and the other is a leaf, then the joining process is trivial. If both of them are roots or leaves, then one sequence has to be reversed. Finally, if the resulting sequence is not a closed loop, then it is discarded.

The results of the unification phase can be seen in Fig. 5c, the processing time of each blob is ~0.0036 s.

2.5 Sequence Analysis

In order to extract useful features of the shape, the node sequence has to be divided into *segments*, bounded by so-called *points of interests* (*POIs*).

Let us define a POI as a node, where a significant change occurs in the general direction of the sequence. Therefore, in order to appoint them in the sequence, the general direction is needed to be examined. In this implementation, two different direction measures are used: absolute and relative directions. Former marks the direction of the sequence considering the whole image on a $[0 \ldots 7]$ scale, as it can be seen in Fig. 7a. The latter marks the average direction of the segment since the last POI on a scale of $[-3 \ldots 3]$ (Fig. 7b), where 0 marks the straight direction, $[-3 \ldots 0)$ the left hand-side and $(0 \ldots 3]$ the right hand side deviation from the straight direction.

Latter is also used to detect if moving onto the next examined pixel in the sequence causes a larger change in the general direction. If this change is larger than an arbitrary threshold (δ), then the pixel is a POI. For this threshold either a static value or a dynamic value can be used. Latter values are calculated by a function that provides a high output in the beginning, then lowering considering the length of the segment.

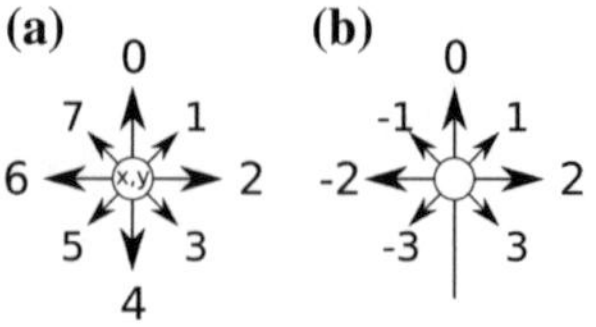

Fig. 7 a Absolute and **b** relative directions

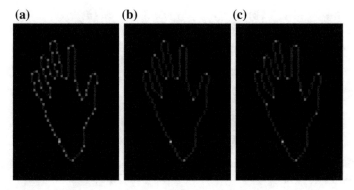

Fig. 8 Resulting points of interests using different threshold values: **a** δ = 1, **b** δ = 1.5 and **c** a dynamic thresholding function

Figure 8 shows the POIs considering different threshold values: (a) considering δ = 1 allowed relative deviation, (b) δ = 1.5 and (c) using a dynamic value ($T_1 = 0.5$, $T_2 = 0.5$, $q_1 = 10$, $q_2 = 20$). As it can be seen, the first threshold results in too many POIs, while the other two provides a much better segmentation, from which valuable information can be extracted considering the features of the hand. (In the figure, the light green pixel shows the starting point (the root node) of the sequence, which is always considered as a POI in the current implementation.) Figure 9 shows the results for the other two blobs, using dynamic thresholding. As it can be seen, although the hand posture is the same as in Fig. 4a, due to its angle and blurred image it is less informative of its true shape. Nevertheless, its shape can still be analyzed. The circle of the face can be clearly seen as well. The average processing time equals about 0.001 s for each blob.

2.6 POI Analysis: Fuzzy Shape Extraction

By analyzing the segments (the POIs and the paths (node sequences) between them) we can get information about the shape. Each POI can be described by the following:

- *Type: peak or valley*; this can be determined by noting the direction of the sequence with respect to the image. If the direction is clockwise, then every POI where the change in the relative direction is negative (thus taking a left-hand side turn) is a valley, while every positive change in the direction marks a peak. In case of counter-clockwise propagation the sides are swapped. For example, the POIs of fingertips of the hand are peaks and the POIs of the webbing between fingers are valleys.
- *The length of the path* from the POI to the next one.

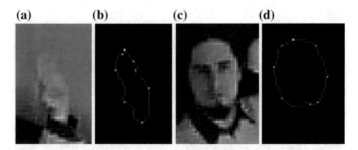

Fig. 9 The results of the shape extraction for the other two blobs: **a** and **c** are the original subimages, while **b** and **d** are the extracted shapes, respectively

- *The relative direction* of said path compared to the direction of the path leading to the POI.
- *The ratio between* the lengths of said *two paths* (length of the next one divided by the length of the previous one).

These descriptors can be used to describe a shape. Considering Fig. 8c, numerous patterns are noticeable of that particular hand shape. For example, the lengths of the paths leading to and from a peak of fingers are roughly the same, their directions are roughly the opposite of each other, and so on.

Remark: it is easy to determine if the sequence propagates in a clockwise or counter-clockwise manner. Take a pixel that is next to the root (but not part of the sequence), recursively examine its neighbors, then the neighbors of its neighbor, etc. until either there are no other non-sequence pixels to consider, or the margin of the image is reached. In the former case the initial pixel is an inner pixel (part of the object that the shape belongs to), while in the latter case it is not. If it is an inner pixel and its position is to the right of the direction from the root and the next pixel in the sequence, then the sequence is propagating in a clockwise manner.

The values of the descriptors can be represented with fuzzy linguistic variables as well. The length of each path can be compared to the diameter of the shape and described by fuzzy values such as VERY SHORT, SHORT, MEDIUM, LONG, VERY LONG. The relative change in direction can be described by FAR LEFT, LEFT, NEAR LEFT, STRAIGHT, NEAR RIGHT, RIGHT, FAR RIGHT, and so on.

Thus fuzzy rules can be formulated which can be evaluated by fuzzy inference machines, classifying as one of the known shapes that is the most similar to the detected one. One such system has been proposed by the authors in [11].

Table 1 summarizes the time required for the feature extraction. The whole process takes 0.6861 s, of which the feature extraction itself takes 0.2661 s, or 0.0887 s per blob. This shows how fast the system is, and pinpoints the areas that are needed to be improved in order to further enhance the operational speed.

Table 1 Time required for the feature extraction

Operation	For one blob on average	For all blobs
Pixel color filtering (FHM) (s)	0.13	
Density based filtering (s)	0.05	
Dilation (s)	0.24	
Blob mapping (s)	0.034	0.103
Edge detection (Canny) (s)	0.001	0.003
Tree building (s)	0.012	0.037
Tree pruning (s)	0.036	0.107
Sequence improvement (s)	0.001	0.003
Sequence unification (s)	0.0037	0.011
POI extraction (s)	0.001	0.003
Time required for blob operations (s)	0.0887	0.2661
All time required (s)	0.5087	0.6861

2.7 Comparison and Overall Analysis

The feature descriptor (proposed in subsection E) belongs to the group of so-called *spatial relationship features* [12]. These descriptors define shapes or contours using the geometric features of their pixels or curves (e.g. length, curvature, orientation, etc.). Here a comparison is made between the proposed descriptor and some of the existing ones.

In *shape context* analysis [13], the whole shape or contour is mapped from the viewpoint of each point it contains, thus gaining local information that can be used to match feature points. It is inherently translation invariant, and with some additional steps it can be made rotation and scale invariant as well. The disadvantage is the considerable computational and storage cost per feature point.

Shock graphs [14] create an axial stick-like skeleton for each shape they represent. They are rotation, scale, and translation invariant and impervious to noise on the shape boundary, but their computational complexity is very high.

Chain codes [15] describe the movement along the sequence of pixels by noting the direction for each pixels. The chain code descriptors in the literature are generally invariant to translation and have a low computational complexity, but they are not rotation and scale invariant, have high dimensions and weak against noise.

The proposed fuzzy descriptor is similar to chain codes, by encoding the general direction of the pixel sequences. However, by using relative directions the proposed descriptor is gains *rotational invariance*. Furthermore, by appointing feature points only at certain points (local extrema), the amount of feature points can be kept relatively low. With the usage of the dynamic thresholding function, it is possible to obtain more or less the same feature points regardless of the size of the shape.

Because of the noise-filtering effect of the Canny edge detector (due to its Gaussian blur step) and an additional sequence improvement step (Sect. 2.3) that removes outliers, the descriptor is also resilient against noise, while retaining a low computational complexity.

The downside of the proposed descriptor is the varying (although generally low) amount of feature points, which restricts the range of the applicable classifiers as most of them expect a constant number of inputs. Another disadvantage is that the starting point of the sequence is varying as well. Both problems can be solved by using fuzzy inference machines: for the former problem the fuzzy rules can be examined to any number of features, while for the latter the feature point sequence can be rotated along the loop until a fitting model is found.

3 Conclusions

In this chapter, a shape detection system is presented for shape feature extraction in camera images. It uses heuristics (the known color tones of the object) to find the areas in the image where the object is and applies the feature extraction only on them, thus making the process faster and more reliable.

The system uses a Fuzzy Hypermatrix-based classifier to find the areas of operation, combined with a density-based filter and morphological dilation. On the resulting areas (blobs) it carries out edge detection, maps the edges into directed tree structures, reduces the trees into simple sequences using heuristics, unites and improves the sequences to get the outline of the shape. Finally, points of interests are extracted from the outline, analyzed and converted into fuzzy values that can be classified by fuzzy inference machines.

The operation of the proposed system is illustrated on a hand recognition process.

The proposed system can work well on images with heterogeneous backgrounds and multiple skin regions, with the only restriction that the areas cannot be over-lapping. Furthermore, with the proposed descriptor virtually any 2D shape can be described, it is translation, rotation, and scale invariant and have a low computa-tional cost due to the reduced area of operation.

The accuracy of the system depends on multiple factors. One is the general accuracy of the hypermatrix based classifier, which can be easily improved by further training it with more color tones associated with the target object. The other one is the performance of the applied edge detector, since the contour construction is based on its output. In the current implementation, the Canny edge detector has problems with detecting edges of an object if the background has a similar color tone. There are further adjustable parameters as well that can influence the accuracy of the system (e.g. dynamic thresholding, the range in which contour lines can be united, etc.), which will be thoroughly investigated in future work.

In order to achieve an even faster operation, certain parts of the system will be improved. Dilation is one such part, which takes about 1/3 of the whole processing time, making it the longest process in the procedure. In future work, a better solution will be developed for it, as well as an improvement for a more reliable edge detection.

Acknowledgements This work has partially been sponsored by the Hungarian National Scientific Fund under contract OTKA 105846 and the Research & Development Operational Program for the project "Modernization and Improvement of Technical Infrastructure for Research and Development of J. Selye University in the Fields of Nanotechnology and Intelligent Space", ITMS 26210120042, co-funded by the European Regional Development Fund.

References

1. M. Panwar, Hand gesture recognition based on shape parameters, in *2012 International Conference on Computing, Communication and Applications (ICCCA)* (Dindigul, Tamilnadu, 2012), pp. 1–6
2. Z. Wang, B. Fan, G. Wang, F.C. Wu, Exploring local and overall ordinal information for robust feature description. IEEE Trans. Pattern Anal. Mach. Intell. **PP**(99), 1–1
3. A. Inoges, G. Lopez-Nicolas, C. Sagues, S. Llorente, Elastic hand contour matching in NIR images with a novel shape descriptor parametrization, in *EUROCON 2015—International Conference on Computer as a Tool (EUROCON)* (IEEE, Salamanca, 2015), pp. 1–6
4. X. Chen et al., Reasoning-based vision recognition for agricultural humanoid robot toward tomato harvesting, in *2015 IEEE/RSJ International Conference on Intelligent Robots and Systems (IROS)* (Hamburg, 2015), pp. 6487–6494
5. J. Li, X. Mei, D. Prokhorov, Deep neural network for structural prediction and lane detection in traffic scene. IEEE Trans. Neural Netw. Learn. Syst. **PP**(99), 1–14
6. A.R. Várkonyi-Kóczy, B. Tusor, J.T. Tóth, A fuzzy hypermatrix-based skin color filtering method, in *Proceedings of the 19th IEEE Inernational Conference on Intelligent Engineering Systems, INES2015* (Bratislava, Slovakia, 3–5 Sept 2015), pp. 173–178
7. J. Serra, *Image Analysis and Mathematical Morphology: Theoretical Advances* (Academic Press, USA, 1982), p. 610
8. J. Canny, A computational approach to edge detection. IEEE Trans. Pattern Anal. Mach. Intell. **8**(6), 679–698 (1986)
9. F. Russo, Edge detection in noisy images using fuzzy reasoning. IEEE Trans. Instrum. Meas. **47**(5), 1102–1105 (1998)
10. G. Law, Quantitative comparison of flood fill and modified flood fill algorithms. Int. J. Comput. Theory Eng. **5**(3), June 2013
11. A.R. Várkonyi-Kóczy, B. Tusor, Human-computer interaction for smart environment applications using fuzzy hand posture and gesture models. IEEE Trans. Instrum. Measure. **60**(5), 1505–1514 (2011)
12. M. Yang, K. Kpalma, J. Ronsin, A survey of shape feature extraction techniques, in *Pattern Recognition*, ed. by P.-Y. Yin (IN-TECH, 2008), pp. 43–90
13. S. Belongie, J. Malik, J. Puzicha, Shape matching and object recognition using shape contexts. IEEE Trans. Pattern Anal. Mach. Intell. **24**(24), 509–521 (2002)
14. K. Siddiqi, B.B. Kimia, A shock grammar for recognition, in *IEEE Computer Society Conference on Computer Vision and Pattern Recognition*, CVPR'96 (San Francisco, CA, June 1996), pp. 507–513
15. H. Freeman, L.S. Davis, A corner finding algorithm for chain coded curves. IEEE Trans. Comput. **C-26**(3), 297–303 March 1977

Pipelined Hardware Architecture for a Fuzzy Logic Based Edge Detection System

Aous H. Kurdi and Janos L. Grantner

Abstract Edge detection is a fundamental task for any image processing system. It serves as an entry point for a lot of major algorithms such as image identification, segmentation, and feature extraction. Consequently, a lot of different techniques have evolved to accomplish this task. The most commonly known methods include Sobel, Laplacian, Prewitt, and fuzzy logic based methodology. In this paper, a novel, pipelined architecture for a type-1 fuzzy edge detector system is discussed. It has been implemented on different Xilinx devices. The fuzzy system consists of modules as follows: preprocessing, fuzzification which creates four fuzzy input variables, inference and defuzzification resulting in a single crisp output. The hardware accelerator utilizes a pipeline of seven stages. Each stage requires just one clock cycle. The system can operate at a frequency range of 83–100 MHz depending on the speed grade of the FPGA device it is compiled to. Using the 1080P HD standard, the proposed architecture is capable of processing up to 45 fps which makes it feasible for real time applications. The system was developed using Xilinx Vivado and 7000-series FPGA devices. Simulations were carried out using ModelSim by Mentor Graphics.

1 Introduction

Knowledge representation, in the context of conventional approaches, is based upon bivalent logic. The major drawback is the limited capability of dealing with the problems of uncertainty and inaccuracy [1]. Consequently, conventional approaches do not provide suitable models to represent human knowledge and the uncertain reasoning cognitive function. Fuzzy logic, on the other hand, offers a

A. H. Kurdi (✉) · J. L. Grantner
Department of Electrical and Computer Engineering,
Western Michigan University, Kalamazoo, MI 49008, USA
e-mail: aoushammad.kurdi@wmich.edu

J. L. Grantner
e-mail: janos.grantner@wmich.edu

© Springer International Publishing AG, part of Springer Nature 2018
L. A. Zadeh et al. (eds.), *Recent Developments and the New Direction in Soft-Computing Foundations and Applications*, Studies in Fuzziness and Soft Computing 361, https://doi.org/10.1007/978-3-319-75408-6_30

mathematical framework to deal with the degree of truth rather being limited to the classic true, or false logic.

Fuzzy logic has evolved to be an essential tool for a wide spectrum of applications including systems control, intelligent systems, and image processing [2]. Edge detection is in the core of most image processing algorithms. It aims to distinguish a set of points in a digital image in which the intensity level changes sharply with respect to the surrounding neighbor points [3]. These points are grouped into curved line shapes called edges. Different methods have been developed to extract the edges in an image such as the Sobel operator, Laplacian, and Prewitt. These methods depend on specific parameters, such as a threshold, to realize the edge detection process [4]. A fuzzy logic based edge detection approach has the advantage of transferring human knowledge into a model that can adapt to change in the environment such as the presence of noise in the input feed rather than depending on a static threshold value. Fuzzy logic based systems work with a linguistic representation of knowledge in a way that describes uncertainty in a form of IF THEN rules. Complex systems can be modeled using those rules that are instinctively understandable to human beings [5]. To unleash the powerful capability of fuzzy logic based systems in real life applications, practical platforms with low energy consumption and high computing power are crucially needed to implement them.

Field Programmable Gate Arrays (FPGAs) [6] are one of the most eligible candidates for implementing fuzzy logic based systems. In the Xilinx 7 Series devices, the programmable elements organized in blocks called Configurable Logic Block (CLB). Each CLB consists of two slices and each slice equipped with a 6-input 1-output look-up table (LUT), distributed memory, shift register, high speed logic for arithmetic functionality, a wide multiplexer, and a switching matrix to facilitate the access to routing elements on the chip [7]. The synthesizer tool assigns the chip's resources, mainly the CLBs, depending on the designer input to implement sequential or combinational logic circuits.

This paper is organized as follows: Sect. 2 gives references to a couple edge detection algorithms using fuzzy logic. Section 3 presents the proposed fuzzy logic system. Section 4 describes the design of the hardware accelerator. Section 5 summarizes the experimental results and provides comparisons with other methods. Conclusions are given in Sect. 6.

2 Background

Fuzzy logic based edge detection has emerged as one of the hot areas for research in the field of image processing. In [8], an improved edge detection algorithm using fuzzy logic was proposed. In that study, the authors applied fuzzy technique on a 3×3 pixels mask. This mask is utilized in the process of examining each pixel's relation with its neighbors. Each pixel is considered as a fuzzy input resulting in a multi-input-single-output (MISO) fuzzy system. Another approach uses the pixels' gradient and standard deviation values as inputs to the fuzzy system [9]. In the system proposed in this paper the fuzzy system makes a decision on which pixel is considered an edge, or not, by carrying out inference calculations based upon a set of fuzzy IF THEN rules.

In the field of applying FPGAs to implement fuzzy logic, as in [5], the authors used a Hardware Description Language (HDL) to design and implement a Mamdani-type fuzzy system.

3 The Proposed Fuzzy Logic System

The proposed edge detector is a Mamdani-type [10] fuzzy system with four fuzzy inputs, one output and seven IF THEN rules. The first two inputs are the gradients with respect to the x-axis and the y-axis out of a kernel of 3 × 3 pixels. The corresponding fuzzy sets are denoted as GX, and GY, respectively. The third input is the output of a low-pass filter and the associated fuzzy set is named LF. The fourth one is the output of a high-pass filter and the fuzzy set representing this input is HF.

3.1 Preprocessing

In the preprocessing stage, a kernel of 3 × 3 pixels coming from the input image are processed to calculate the gradient in x-direction, gradient in y-direction, low pass filter, and high pass filter using Eqs. (1), (2), (3), and (4), respectively. I(x, y) stands for the kernel with the targeted pixel at the center.

$$GX = \begin{bmatrix} -1 & 0 & 1 \\ -2 & 0 & 2 \\ -1 & 0 & 1 \end{bmatrix} \cdot I(x, y) \tag{1}$$

$$GY = \begin{bmatrix} -1 & -2 & -1 \\ 0 & 0 & 0 \\ 1 & 2 & 1 \end{bmatrix} \cdot I(x, y) \tag{2}$$

$$LF = \begin{bmatrix} 1 & 1 & 1 \\ 1 & 1 & 1 \\ 1 & 1 & 1 \end{bmatrix} \cdot \frac{1}{9} \cdot I(x, y) \tag{3}$$

$$HF = \begin{bmatrix} -1 & -1 & -1 \\ -1 & 8 & -1 \\ -1 & -1 & -1 \end{bmatrix} \cdot \frac{1}{9} \cdot I(x, y) \tag{4}$$

3.2 Fuzzification

In fuzzification, crisp inputs converted into fuzzy quantities. Linguistic variables were used to represent the fuzzy qualities spanning over a practical range on crisp values. For each input, three fuzzy sets were defined over the universes of discourses. These fuzzy sets are LOW, MED, and HIGH.

3.3 The Inference System

For each input variable, three membership functions (MF) was defined, LOW, MED, and HIGH. LOW and HIGH are trapezoids MFs and MED is a triangle MF. The membership functions are distributed over the universe of discourse values from 0 to 255 (the intensity range in a grayscale image) as illustrated in Fig. 1.

Fuzzy union implemented as MAX function was adopted to pick the linguistic variable with the most membership function strength for each individual input.

The proposed system utilizes a Mamdani inference system. The knowledge base is made up of seven fuzzy IF THEN rules as shown in Table 1. All the rules have the weight of 1. In the implication process the MIN operator is used for fuzzy intersection and the MAX operator is used for fuzzy union, respectively.

Fig. 1 The membership functions of input GX

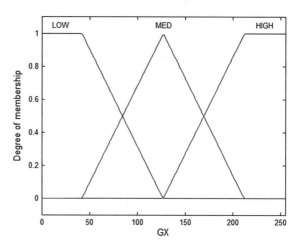

Table 1 Pseudo-code of the fuzzy rules

If GX = LI and GY = LI then E = LE
If GX = MI and GY = MI then E = ME
If GX = MI and GY = HI then E = HE
If GX = LI and HF = LI then E = ME
If GY = LI and HF = LI then E = ME
If GX = LI and LF = LI then E = LE
If GY = LI and LF = LI then E = LE

Fig. 2 Membership
functions for output E

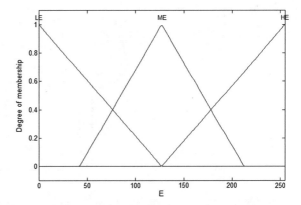

3.4 Defuzzification

For the single output variable E (Edge), three membership functions are defined,
LE, ME, and HE. All three membership functions are of triangular shape. They are
distributed over the universe of discourse values from 0 to 255 as shown in Fig. 2.
For defuzzification, the Mean of the Maxima (MOM) method was used.

4 Design of the Pipelined Hardware Accelerator

A pipelined hardware accelerator was designed and implemented on a Xilinx
7000-series device using the Xilinx Vivado Design Suite. The system consists of
four main blocks: Preprocessor, Fuzzifier, Inference System, and Defuzzifier
extending over seven pipeline stages. Each stage requires just one clock cycle of
execution time. The Preprocessor uses 3 stages, the inference system utilizes 2
stages and the other two blocks use one stage each.

4.1 Preprocessing

The Preprocessor Unit's functional block diagram is shown in Fig. 3. The hardware
architecture of the Preprocessor includes a Block RAM (BRAM) Module that is
configured as a Dual-Port RAM with asynchronous Read and Write cycles. The
memory organization is set up with parallel data width of 72 bits by 512 locations.
Each location contains the representation of the intensity of 9 pixels forming a
3 × 3 kernel window.

The first pipeline stage reads one location from memory. For calculating the
gradient in x-direction, the gradient in y-direction, the low pass filter and the high
pass filter, a dedicated block of hardware was implemented for each one of them.

Fig. 3 Preprocessor block
diagram

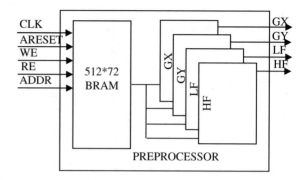

Each block utilizes two stages out of the system's seven pipeline stages and they work in parallel.

For GX and GY, two specialized sub-blocks were designed to execute the process of calculating the positive and negative parts of the gradient masks in Eqs. (1) and (2). These sub-blocks work in parallel and they form the second stage of the pipeline. The third stage performs the tasks as follows: the addition of the outputs of the previous sub-blocks, finding the magnitude value of the sum and scaling the value down to the established maximum value (255 in this case) if the results exceed this maximum.

For LF, three sub-blocks were designed to find the sum of each row in Eq. (3). These sub-blocks work in parallel and represent the second stage of the pipeline in the LF block. The third stage calculates the sum of the results of the previous stage and divides the result by 9 to find the average.

The last block is the HF. This block is also divided into two stages. The first stage consists of three sub-blocks. The first sub-block preforms the process of multiplying the first four elements of the HF mask in Eq. (4) with their corresponding input pixels and finds the sum. The second sub-block does the same as the first sub-block but for the last four elements of the HF mask. The third sub-block carries out the multiplication of the center element by 8. The second stage calculates the sum of the previous stage and then divides the sum over 9.

4.2   Fuzzifier

The conversion of the crisp inputs into fuzzy variables takes place in the fourth stage of the system pipeline in the fuzzification block. The Fuzzifier Unit's functional block diagram is depicted in Fig. 4. The Fuzzifier block consists of four identical sub-blocks working in parallel. These sub-blocks map the crisp inputs to linguistic labels in the corresponding fuzzy universes of discourses along with the degrees of consistency. The inputs of the Fuzzifier block are GX, GY, LF, and HF which are the outputs of the Preprocessor block, as well as CLK, and ARESET. The outputs of the Fuzzifier block are as follows: fuzzy variable 1(F-Var1), degree of consistency of

Fig. 4 Fuzzifier block diagram

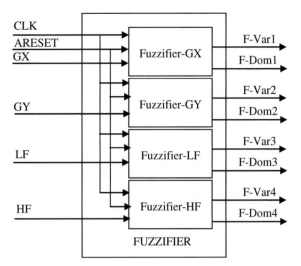

Fig. 5 Inference system block diagram

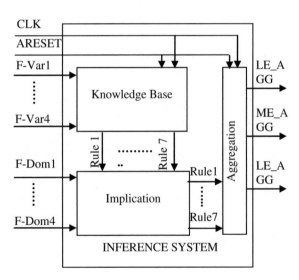

variable 1(F-Dom1), fuzzy variable 2(F-Var2), degree of consistency of variable 2 (F-Dom2), fuzzy variable 3(F-Var-3), degree of consistency of variable 3(F-Dom3), fuzzy variable 4(F-Var-4), and degree of consistency of variable 4(F-Dom4).

4.3   Inference System

The hardware design of the inference system uses two stages out of the system's seven pipeline stages. The first stage implements the knowledge base, which

consists of seven IF THEN rules, and calculates the implication for each rule. The
second stage preforms the aggregation the outcomes of the rules into three fuzzy
variables LE, ME, and HE, respectively. The Inference System block diagram is
given in Fig. 5.

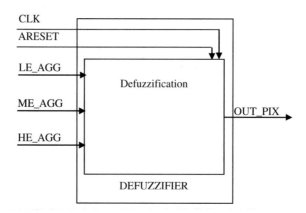

Fig. 6 Defuzzifier block diagram

Table 2 SNR using Lena's image

Noise type	SNR (dB)			
Salt and pepper	The proposed system	Sobel	Roberts	Marr-Hildreth
3%	11.3229	2.1172	0.3227	5.7949
2%	12.9713	2.838	0.5179	6.8407
1%	15.6542	4.4507	1.2663	8.8938
0.5%	19.0904	7.1571	2.6619	11.2930
0.01%	36.4372	22.3805	16.4128	28.4842
Poisson	5.770	4.8552	2.3705	8.3296
White Gaussian	4.9275	2.0647	0.4376	4.8246

Table 3 SNR using cameraman image

Noise type	SNR (dB)			
Salt and pepper	The proposed system	Sobel	Roberts	Marr-Hildreth
3%	11.7017	4.0597	0.435	7.1072
2%	13.3604	5.0934	0.8328	8.1922
1%	15.9836	7.6217	2.2562	9.7482
0.5%	20.2011	10.2151	4.4911	13.2041
0.01%	36.2842	28.811	19.7059	29.0869
Poisson	6.5649	5.7147	3.8914	9.7991
White Gaussian	5.3962	3.1791	0.8788	5.3655

4.4 Defuzzifier

In this stage, the aggregated fuzzy variables LE, ME, and HE are defuzzified using Mean of Maxima defuzzification method. The output of the Defuzzifier Unit is an 8 bit representation of a pixel intensity value in the output image. The Defuzzifier Unit's block diagram is shown in Fig. 6.

5 Results

In this research, a hardware accelerator for fuzzy logic based edge detector was designed, implemented, and tested. The system's performance has been compared, in the presence of noise, with two traditional edge detection techniques. The pipelined hardware accelerator was developed using Xilinx Vivado and simulated

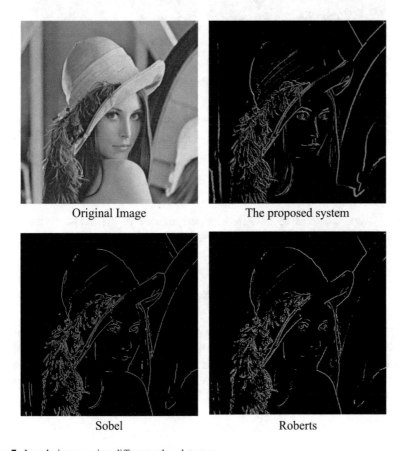

| Original Image | The proposed system |
| Sobel | Roberts |

Fig. 7 Lena's image using different edge detectors

Original Image   The proposed system

Sobel   Roberts

Fig. 8 Cameraman image using different edge detectors

using Mentor Graphics ModelSim. The performance of the hardware accelerator
was investigated using different Xilinx Artix7 devices.

5.1   The Performance of the Proposed Fuzzy System

The proposed system was tested using two benchmark grayscale images, Lena and
the Cameraman. We also compared the results with other edge detection techniques
such as Sobel's, Roberts's, and Marr-Hildreth edge detection methods using Signal
to Noise Ratio (SNR) as a quantitive measure. SNR is the physical indicator of the
imaging system sensitivity to noise [11]. It is calculated using Eq. (5).

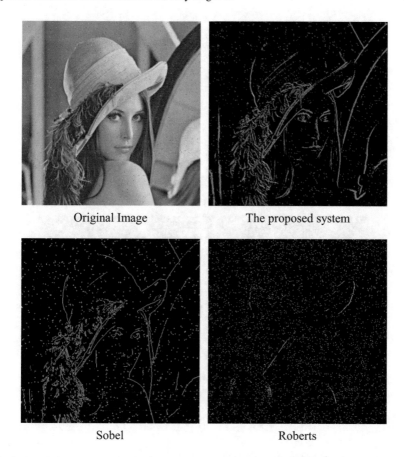

Original Image	The proposed system

| Sobel | Roberts |

Fig. 9 Lena's image with 0.05% salt and pepper noise using different edge detectors

$$SNR = 10 \log_{10} \frac{\sum_0^{n_x - 1} \sum_0^{n_y - 1} [r(x, y)]^2}{\sum_0^{n_x - 1} \sum_0^{n_y - 1} [r(x, y) - t(x, y)]^2} \tag{5}$$

where r is the reference image, t is the image to be tested, n_x is the width of the image in pixels, and n_y is the height of the image in pixels. The system showed better immunity to noise compare to the other methods in most of the cases. The noise test was carried out using artificial noise added to the original image as shown in Tables 2 and 3 and Figs. 7, 8, 9 and 10.

| Original Image | The proposed system |
| Sobel | Roberts |

Fig. 10 Cameraman image with 0.1% salt and pepper noise using different edge detectors

5.2 The Hardware Accelerator

The hardware accelerator design based upon a pipeline architecture of seven stages. Each stage requires just one clock cycle of execution time. The system was tested using three Xilinx Artix7 devices with different speed grades as indicated in Table 4. The system can work with a clock frequency rate of up to 100 MHz producing an output pixel in each 10 ns.

Simulation results using a 11 ns clock cycle have shown that the system needs 77 ns (7 cycles) to fill the pipeline and after that 176 ns (16 cycles) to produce 16 outputs as illustrated in Fig. 11. To compare the performance of the hardware

Table 4 Device utilization and maximum operating speed

Device name	FF	LUT	BRAM	Speed (MHz)
xc7a200tfbg676-1	269	860	1	83.3
xc7a200tfbg676-2	269	860	1	90.9
xc7a200tfbg676-3	269	860	1	100

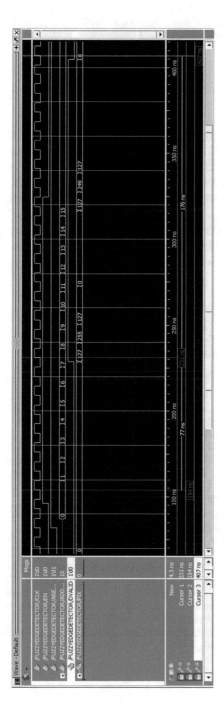

Fig. 11 System simulation using a 11 ns clock cycle

accelerator with its software counterpart, MATLAB was used to implement the system on a PC with Intel Core i7 processor and 8 GB of memory. The tests yield execution time of 1.3178 ms per pixel.

6 Conclusions

A novel seven-stage pipeline architecture for fuzzy edge detector was proposed and implemented using Xilinx Vivado and VHDL. The system performance was evaluated using three Xilinx Artix7 devices. The system's noise immunity was compared with other traditional methods using SNR as a quantitive measure. The system delivers better results in all of the comparisons. Working with a system clock of a 100 MHz, the system can process a frame with 1080P HD resolution within 0.020736 s or a 0.62208 s for a stream of 30 fps. It is equivalent to about 45 fps. By attaching suitable input and output peripheral devices to the proposed hardware accelerator it will make a good tool for real time applications.

References

1. L.A. Zedeh, Knowledge representation in fuzzy logic. IEEE Trans. Knowl. Data Eng. **1**(1), 89–100 (1989)
2. K. Tanaka, *An Introduction to Fuzzy Logic for Practical Applications* (Springer, 1997)
3. Scott E. Umbaugh, *Digital Image Processing and Analysis: Human and Computer Vision Applications with CVIP Tools*, 2nd edn. (CRC Press, Boca Raton, FL, 2010)
4. Y. Becerikli, T.M. Karan, A new fuzzy approach for edge detection, in *Computational Intelligence and Bioinspired Systems* (Springer, Berlin, Heidelberg, 2005), pp. 943–951
5. D.N. Oliveira, A.P. de Souza Braga, O. da Mota Almeida, Fuzzy logic controller implementation on a FPGA using VHDL, in *Fuzzy Information Processing Society (NAFIPS), 2010 Annual Meeting of the North American* (12–14 July 2010), pp. 1–6
6. C. Dick, F. Harris, FPGA signal processing using sigma-delta modulation. Sig. Process. Mag. IEEE **17**(1), 20–35 (2000)
7. Xilinx, 7 Series FPGAs Configurable Logic Block: User Guide (2014), http://www.xilinx.com/support/documentation/user_guides/ug474_7Series_CLB.pdf
8. A.B. Borker, M. Atulkar, Detection of edges using fuzzy inference system. Int. J. Innov. Res. Comput. Commun. Eng. (2013)
9. W. Barkhoda, F.A. Tab, O.-K. Shahryari, Fuzzy edge detection based on pixel's gradient and standard deviation values, in *International Multiconference on Computer Science and Information Technology, 2009. IMCSI'09* (12–14 Oct 2009), pp. 7–10
10. E.H. Mamdani, S. Assilian, An experiment in linguistic synthesis with a fuzzy logic controller. Int. J. Man Mach. Stud. **7**(1), 1–13 (1975)
11. R.C. Gonzalez, R.E. Woods, *Digital Image Processing*, 3rd edn. (Prentice Hall, 2008)

A Quantitative Assessment of Edge Preserving Smoothing Filters for Edge Detection

Huseyin Gunduz, Cihan Topal and Cuneyt Akinlar

Abstract Edge detection algorithms have traditionally utilized the Gaussian Linear Filter (GLF) for image smoothing. Although GLF has very good properties in removing noise and unwanted artifacts from an image, it is also known to remove many valid edges. To cope with this problem, edge preserving smoothing filters have been proposed and they have recently attracted increased attention. In this paper, we quantitatively compare three prominent edge preserving smoothing filters; namely, Bilateral Filter (BLF), Anisotropic Diffusion (AD) and Weighted Least Squares (WLS) with each other and with GLF in terms of their effects on the final detected edges using the precision/recall framework of the famous Berkeley Segmentation Dataset (BSDS 300). We conclude that edge preserving smoothing filters indeed improve the performance of the edge detectors, and of the filters compared, WLS yields the best performance with AD also outperforming the GLF.

Keywords Edge detection · Canny · Edge drawing (ED) · Edge preserving smoothing · Bilateral filter · Anisotropic diffusion · Weighted least squares (WLS)

1 Introduction

Edge detection is one of the fundamental tools of image processing and computer vision. It is usually performed as the first step of a processing pipeline and is especially used in feature detection and extraction. The importance of this problem

H. Gunduz (✉) · C. Topal · C. Akinlar
Department of Computer Engineering, Anadolu University, Eskisehir, Turkey
e-mail: huseyingunduz@anadolu.edu.tr

C. Topal
e-mail: cihant@anadolu.edu.tr

C. Akinlar
e-mail: cakinlar@anadolu.edu.tr

411

have led the researchers to develop many edge detection algorithms in the literature [1–4].

The first step of any edge detection algorithm is to remove noise and reduce the amount of detail and unwanted artifacts in the image. The easiest and the most widely used filter for this purpose is the Gaussian Linear Filter (GLF) [1]. This filter smooths out every pixel of the input image using the same Gaussian function and thus removes potential edges along with the unwanted image artifacts, which inversely affects the edge detection performance.

To overcome this problem, edge preserving smoothing filters have been proposed in the literature. The main goal of these filters is to remove unwanted artifacts from the image as in GLF while preserving valid edge crossings. These filters can be analyzed in two different categories: (1) Those that work locally and compute the output value for each pixel as some sort of an average of the local neighborhood, (2) Those that formulate a global optimization problem using the image pixel intensity values and smoothing coefficients, and solve this problem to obtain the final filter output.

The most important method in the first category is the Anisotropic Diffusion (AD) proposed by Perona and Malik [5]. In this method, the filter coefficients are variable as opposed to Gaussian Linear Filer (GLF) and changes depending on the structure of the input image. Shortly stated, using an anisotropic diffusion based equation, the smoothing rate is adjusted based on the gradient value at each pixel. While smoothing rate is reduced at the edge crossings, it is increased over the continuous, smooth regions of the image. Although this method works well in practice, it does not perform well especially on noisy images. Konishi proposes making use of different statistical methods in the determination of edge crossings instead of using a gradient operator [6]. In this method, the image is modelled as a random field and the relationships between the pixels are analyzed. In addition to the original AD proposed by Perona and Malik, many different AD variants have been proposed in the literature [7–11].

Another important and popular edge preserving smoothing filter from the first category is the Bileteral Filter (BLF) proposed by Tomasi and Manduchi [12]. This filter makes use the spatial and range differences around the center pixel to produce a weighted average around the neighborhood. The fact that the filter kernel size is not linear creates computational problems. For this reason, several methods have been proposed to speed up the filter and make it available for applications that require fast computation [13–15].

An important method from the second category is the Weighted Least Squares (WLS) [16, 17]. The idea with this method is to formulate image smoothing as a global optimization problem and solve a system of linear equations to obtain the output image.

2 Image Smoothing Filters

In this study, four popular image smoothing filters have quantitatively been compared. Here is a brief description of the filters to be compared.

2.1 Gaussian Linear Filter (GLF)

Given an input image f, and a Gaussian smoothing filter function g, the smoothed output image h is calculated at each pixel (x, y) by the following convolution equation:

$$h(x, y) = f(x, y)g(x, y),$$ (1)

Since this filter is based on the Gaussian function, the weights of the pixels closer to the center pixel would be higher than the weights of the pixels far from the center as follows:

$$G(x, y) = \frac{1}{2\pi\sigma^2} \exp\left(-\frac{x^2 + y^2}{2\sigma^2}\right).$$ (2)

2.2 Bilateral Filter (BLF)

Contrary to GLF, where the weights of the filter coefficients are the same for each pixel of the input image, the bilateral filter adjusts the weights of the filter coefficients based on the intensity values of the current pixel region, thus trying to preserve boundary crossings. If the difference between the intensity of the center pixel and the intensity of the pixels in the neighborhood is high, then the weights filter coefficients are reduced. Thus after filtering, the intensity of the center pixel is prevented from big changes. In other words, in BLF, the intensity values of the pixels located on edge crossings are mainly determined by the pixels in the same class. The BLF equation is given as:

$$BLF(I)_p = \frac{1}{W_p} \sum_{q \in S} G_{\sigma_s}(\|p - q\|) G_{\sigma_r}(|I_p - I_q|) I_q,$$ (3)

where σ_s is the standard deviation in the spatial domain, and σ_r is the standard deviation in the spectral domain. W is used for normalization.

2.3 Anisotropic Diffusion (AD)

Anisotropic Diffusion (AD) works by the application of the heat diffusion equation over the image as follows [5]:

$$\left\{ \begin{array}{l} \frac{\partial I}{\partial t} = div[c(|\nabla I|) \cdot \nabla I] \\ I(t=0) = I_0 \end{array} \right\}, \tag{4}$$

where I_0 is the input image, ∇ is the gradient operator, div is the divergence operator, $||$ is the magnitude operator, and $c(x)$ is the diffusion coefficient function and can be taken as one of the following:

$$c(x) = \frac{1}{1 + (x/k)^2}, \tag{5}$$

and

$$c(x) = \exp\left[-(x/k)^2 \right], \tag{6}$$

where k is the edge magnitude parameter.

Gradient magnitude is used to identify the edge areas or intensity discontinuities. At pixels where $|\nabla| >> k$, the value of $c(|\nabla|)$ becomes 0. At pixels where $|\nabla| << k$, the value of $c(|\nabla|)$ becomes 1.

The Eq. (4) can be written in discrete form as:

$$I_s^{t+\nabla t} = I_s^t + \frac{\nabla t}{|\bar{\eta}_s|} \sum_{p \in \bar{\eta}_s} c\left(\nabla I_{s,p}^t \right) \nabla I_{s,p}^t, \tag{7}$$

where I_s^t represents the discretized image, s is the location of the pixel, ∇t is the step size, $\bar{\eta}_s$ is the spatial neighborhood of s, $|\bar{\eta}_s|$ is the number of pixels in the filter window. We finally obtain:

$$\nabla I_{s,p}^t = I_p^t - I_s^t, \forall p \in \bar{\eta}_s. \tag{8}$$

2.4 Weighted Least Squares (WLS)

Given an input image f, the goal is to obtain the output image u by minimizing the energy function given in (9) using the Weighted Least Squares (WLS) method [17],

$$J(u) = \sum_p \left((u_p - f_p)^2 + \lambda \sum_{q \in N(p)} w_{p,q}(f)(u_p - u_q)^2 \right), \tag{9}$$

where $N(p)$ represents pixel p's neighbors and λ is a balancing factor. The weight equation $w_{p,q}$ is the similarity between pixels p ve q. When $J(u)$ is set to 0, the minimized u is obtained by solving a system of linear equations represented by a sparse matrix:

$$(I + \lambda A)u = f, \tag{10}$$

In [16], the authors has expressed the output image, u, as a gradient equation as follows:

$$\sum_p \left((u_p - f_p)^2 + \lambda \left(a_{x,p}(f) \left(\frac{\partial u}{\partial x} \right)_p^2 + a_{y,p}(f) \left(\frac{\partial u}{\partial y} \right)_p^2 \right) \right), \tag{11}$$

$$a_{x,p}(f) = \left(\left| \frac{\partial \ell}{\partial x} \right|^\alpha + \varepsilon \right)^{-1} a_{y,p}(f) = \left(\left| \frac{\partial \ell}{\partial y} \right|^\alpha + \varepsilon \right)^{-1}, \tag{12}$$

where ℓ is the log-luminance channel of the input image, α is the gradient sensitivity, ε is a constant to prevent division by zero at places where f is constant.

3 Experimental Results

In this section we quantitatively compare three prominent edge preserving smoothing filters; namely, AD, BLF and WLS with each other and with GLF in terms of their effects on the final detected edges. To achieve this goal, we make use of the precision/recall framework of the famous Berkeley Segmentation Dataset (BSDS) [18]. This dataset contains 200 training and 100 test images and is very popular in quantitatively comparing different boundary detection methods [19, 20]. Images in the dataset are of 481 × 321 or 321 × 481 pixel resolution and each image has between 5 and 10 human marked segmentations, which are used as the ground truth data. The edge (boundary, contour) detection results from an algorithm is automatically compared with the ground truth data using the precision/recall framework and an overall F-measure score is produced by the BSDS testbed to rate the performance an algorithm. This enables objective comparison of different edge detection algorithms.

To compare the effects of different image smoothing methods, an input image is first passed through the smoothing filter. The output of the filter is then processed

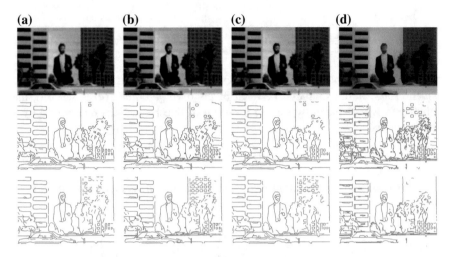

Fig. 1 Image 119082 of BSDS. Top to bottom: Smoothed image, Canny edge map having the maximum F-score, ED edge map having the maximum F-score, smoothed by **a** GLF **b** BLF **c** AD **d** WLS filters

by two different edge detectors: The famous Canny edge detector [1] and the recently proposed real-time edge segment detector, Edge Drawing (ED) [2]. Both of these detectors take a smoothed image as input and produce a binary edge map as output, which is then used in the BSDS evaluation testbed to compute an F-score for an image. The higher the F-score, the better the edge map. The edge detection algorithms are run with the same parameters for each input image. Thus, the smoothing filter that maximizes the F-measure score can be said to be the best.

Figure 1 shows a sample image from the BSDS dataset along with the smoothed images with different smoothing filters. The edge maps having the maximum F-score for each smoothed image obtained by Canny and ED are also presented in Fig. 1. Finally, the precision/recall/F-score values corresponding to each edge map are presented in Table 1.

Figure 2 shows the precision/recall curves of maximum F-score using the 100 test images in the BSDS dataset for each of the smoothing filters obtained by Canny and ED respectively, and Table 2 gives a summary of the results. Clearly, WLS yields the best performance with AD also outperforming GLF, while BLF shows mixed performance and does not live up to the expectations.

Finally, Table 3 shows the running time for each filter averaged over the 100 test images in the BSDS testbed. Clearly, GLF is two orders of magnitude faster than the other filters, which explains its wide adaption especially in applications that require real-time performance.

Table 1 Detection performance for image 119082 of the BSDS dataset

	Canny			ED		
	Precision	Recall	F-score	Precision	Recall	F-score
GLF	0.688	0.766	0.725	0.675	0.743	0.708
BLF	0.662	0.828	0.735	0. 637	0. 788	0.705
AD	0.699	0.782	0.738	0.656	0.757	0.703
WLS	0.649	0.831	0.729	0.688	0.805	0.742

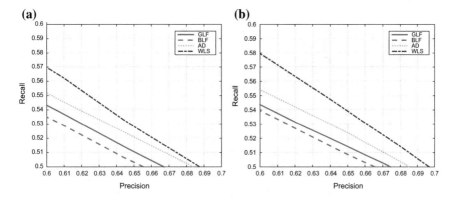

Fig. 2 **a** Canny edge detection performance **b** ED edge detection performance

Table 2 Maximum F-score obtained for each smoothing filter by Canny and ED for the BSDS 300 dataset

	Maximum Canny F-score	Maximum ED F-score
GLF	0.571	0.574
BLF	0.567	0.570
AD	0.578	0.580
WLS	0.585	0.590

Table 3 Average running time of each smoothing filter for the images in the BSDS 300 dataset

	Time (ms)
GLF	1
BLF	534
AD	395
WLS	764

4 Conclusions

Edge preserving smoothing filters try to remove noise and unwanted artifacts from an image while preserving important edge/boundary crossings. In this paper, we quantitatively compared three edge preserving smoothing filters with each other and with the widely used Gaussian Linear Filter. While BLF did not live up to the expectations, both WLS and AD were observed to produce better results compared to GLF. We conclude that edge preserving filters indeed improve the performance of the edge detectors, and further research needs to be performed to improve the existing filters or to develop new ones. Also, their running times need to be improved to make them suitable for real-time computer vision applications.

References

1. J.F. Canny, A computational approach to edge detection. IEEE Trans. Pattern Anal. Mach. Intell. **8**(6), 679–698 (1986)
2. C. Topal, C. Akinlar, Edge drawing: A combined real-time edge and segment detector. J. Visual Commun. Image Represent. **23**(6), 862–872 (2012)
3. D. Marr, E. Hildreth, Theory of edge detection. Proc. Royal Soc. London **B207**, 187–217 (1980)
4. R.M. Haralick, Digital step edges from zero crossing of second directional derivatives. IEEE Trans. Pattern Anal. Mach. Intell. **6**(1), 58–68 (1984)
5. P. Perona, J. Malik, Scale-space and edge detection using anisotropic diffusion. IEEE Trans. Pattern Anal. Mach. Intell. **12**(7), 629–639 (1990)
6. S. Konishi, A.L. Yuille, J.M. Coughlan, S.C. Zhu, Statistical edge detection: learning and evaluating edge cues. IEEE Trans. Pattern Anal. Mach. Intell. **25**(1), 57–74 (2003)
7. G. Cottet, L. Germain, Image processing through reaction combined with nonlinear diffusion. Math. Comput. **61**, 659–673 (1993)
8. J. Weickert, *Anisotropic Diffusion in Image Processing*. ECMI Series (Teubner-Verlag, 1998)
9. S. Aja-Fernández, C. Alberola-López, On the estimation of the coefficient of variation for anisotropic diffusion speckle filtering. IEEE Trans. Image Process. **15**(9), 2694–2701 (2006)
10. J. Weickert, Coherence-enhancing diffusion filtering. Int. J. Comput. Vis. **31**, 111–127 (1999)
11. Y. Yu, S. Acton, Speckle reducing an isotropic diffusion. IEEE Trans. Image Process. **11**(11), 1260–1270 (2002)
12. C. Tomasi, R. Manduchi, Bilateral filtering for gray and color images, in *IEEE International Conference on Computer Vision* (1998), pp. 839–846
13. F. Durand, J. Dorsey, Fast bilateral filtering for the display of high dynamic range images. ACM Trans. Graph. **21**(3), 257–266 (2002)
14. F. Porikli, Constant time O(1) bilateral filtering, in *IEEE Conference on Computer Vision and Pattern Recognition* (2008)
15. S. Paris, F. Durand, A fast approximation of the bilateral filter using a signal processing approach. Int. J. Comput. Vis. **81**(1), 24–52 (2009)
16. D. Lischinski, Z. Farbman, M. Uyttendaele, R. Szeliski, Interactive local adjustment of tonal values. ACM Trans. Graph. **25**(3), 646–653 (2006)
17. D. Min, S. Choi, J. Lu, B. Ham, K. Sohn, M.N. Do, Fast global image smoothing based on weighted least squares. IEEE Trans. Image Process. **23**(12), 5638–5653 (2014)
18. D. Martin, C. Fowlkes, D. Tal, J. Malik, A database of human segmented natural images and its application to evaluating segmentation algorithms and measuring ecological statistics. Proc. Eighth Int. Conf. Comput. Vis. **2**, 416–423 (2001)
19. C. Lopez-Molina, M. Galar, H. Bustince, B. De. Baets, On the impact of anisotropic diffusion on edge detection. Pattern Recogn. **47**(1), 270–281 (2014)
20. X. Hou, A. Yuille, C. Koch, A meta-theory of boundary detection benchmarks, in *Proceedings of the NIPS Workshop on Human Computation for Science and Computational Sustain Ability* (2012)

Why Sparse? Fuzzy Techniques Explain Empirical Efficiency of Sparsity-Based Data- and Image-Processing Algorithms

Fernando Cervantes, Bryan Usevitch, Leobardo Valera
and Vladik Kreinovich

Abstract In many practical applications, it turned out to be efficient to assume that the signal or an image is *sparse*, i.e., that when we decompose it into appropriate basic functions (e.g., sinusoids or wavelets), most of the coefficients in this decomposition will be zeros. At present, the empirical efficiency of sparsity-based techniques remains somewhat a mystery. In this paper, we show that fuzzy-related techniques can explain this empirical efficiency. A similar explanation can be obtained by using probabilistic techniques; this fact increases our confidence that our explanation is correct.

1 Formulation of the Problem

Sparsity-based techniques are useful. In many practical applications, it turned out to be efficient to assume that the signal or an image is sparse; see, e.g., [1–9, 12–14, 17, 18].

In precise terms, sparsity means that when we decompose the original signal $x(t)$ (or original image) into appropriate basic functions $e_1(t)$, $e_2(t)$, ... (e.g., sinusoids

F. Cervantes · B. Usevitch
Department of Electrical and Computer Engineering,
University of Texas at El Paso, 500 W. University, El Paso, TX 79968, USA
e-mail: fcervantes@miners.utep.edu

B. Usevitch
e-mail: usevitch@utep.edu

L. Valera
Computational Science Program, University of Texas
at El Paso, 500 W. University, El Paso, TX 79968, USA
e-mail: leobardovalera@gmail.com

V. Kreinovich (✉)
Department of Computer Science, University of Texas
at El Paso, 500 W. University, El Paso, TX 79968, USA
e-mail: vladik@utep.edu

© Springer International Publishing AG, part of Springer Nature 2018
L. A. Zadeh et al. (eds.), *Recent Developments and the New Direction
in Soft-Computing Foundations and Applications*, Studies in Fuzziness
and Soft Computing 361, https://doi.org/10.1007/978-3-319-75408-6_32

or wavelets), i.e., represent this signal (or image) as a linear combination $x(t) = \sum_{i=1}^{\infty} a_i \cdot e_i(t)$, then most of the coefficients a_i in this decomposition will be zeros.

Moreover, it is usually beneficial to select, among all the signals which are consistent with all the observations (and will all additional knowledge), the signal for which:

- either the number of non-zero coefficients is the smallest possible,
- or, more generally, the "weighted number" of such coefficient is the smallest possible, where the weighted number is defined as $\sum_{i: a_i \neq 0} w_i$, for some weights $w_i > 0$.

Comment. Sparsity can be viewed as a particulate case of the Occam's razor, according to which we should always select the simplest model that fits observation.

But why are sparsity-related techniques useful? At present, the empirical efficiency of sparsity-based techniques remains somewhat a mystery.

What we do in this paper. In this paper, we show that fuzzy-related techniques can explain this empirical efficiency.

We also show that a similar explanation can be obtained by using probabilistic techniques; this fact increases our confidence that our explanation is correct.

2 General Analysis of the Problem

Why do we need data processing in the first place? To better understand why different techniques are more or less empirically successful in data processing, it is important to recall why we need data processing in the first place.

One of the main goals of science and engineering is to predict the future state of the worlds, and to design gadgets and strategies that would make the future state of the world more beneficial for us.

To predict the state of the world, we need to know the current state of the world, and we need to know how this state changes in time. In general, the state of the world can be described by the numerical values of different physical quantities. In these terms, to predict the future state of the world means to predict the future values of the corresponding quantities $y_1, \ldots, y_m$.

In each practical problem, we are usually interested only in a small number of quantities. However, to predict the future values of these quantities, we often need to know the initial values of some auxiliary quantities as well. For example, when we launch a spaceship, we are interested in its location and direction when it leaves the atmosphere, and we are not directly interested in the future values of the winds on different heights. However, these winds affect the spaceship's trajectory, and, as a result, we need to know their initial values to correctly predict the desired values $y_1, \ldots, y_m$. In general, we need to know $n \gg m$ initial values $x_1, \ldots, x_n$ to make the desired prediction.

The relation between x_i and y_j is often complicated. So, to predict the values $y_1, \ldots, y_m$ based on the inputs $x_1, \ldots, x_n$, we need to apply complex computer-based algorithms, i.e., perform *data processing*.

The above description captures the main reason for data processing, but it is somewhat oversimplified, since it assumes that we know the values $x_1, \ldots, x_n$ of the original quantities. For some quantities, this is indeed true, since we can directly measure their values. However, there are many other quantities which are difficult to measure directly. For example, when we are trying to predict the state of the engine, it is desirable to know the current temperature inside; however, this temperature is difficult to measure directly. If we cannot directly measure a certain value x_i, a natural idea is to find easier-to-measure auxiliary quantities $z_1, \ldots, z_p$ which are related to x_i by a known dependence, and then use this known dependence to reconstruct x_i. The corresponding computations may be complex, so we have another reason why data processing is needed.

Before we perform data processing, we first need to know which inputs are relevant. In general, in data processing, we estimate the value of the desired quantity y_j based on the values of the known quantities $x_1, \ldots, x_n$ that describe the current state of the world.

In principle, all possible quantities $x_1, \ldots, x_N$ that describe the current state of the world could be important for predicting some future quantities. However, for each specific quantity y_j, usually, only a few of the quantities x_i are actually useful.

So, before we decide how to transform the inputs x_i into the desired output, we first need to check which inputs are actually useful. This checking is a very important stage of data processing—if we do not filter out unnecessary quantities x_i, we will waste time and resources measuring and processing these unnecessary quantities.

3 Analysis of the Problem: Let Us Use Fuzzy-Related Techniques

Description of our problem in natural-language terms. We are interested in a reconstructing a signal or image $x(t) = \sum_{i=1}^{\infty} a_i \cdot e_i(t)$ based on the measurement results and prior knowledge. In this formula, the basis functions $e_i(t)$ are known, and the coefficients a_i need to be determined. Based on measurement results and prior knowledge, we need to estimate the values a_i.

This reconstruction problem is, of course, a particular case of a general data processing problem. As we have mentioned in the previous section, a natural way to approach data processing problems in general is that:

- first, we find out which quantities are important for this particular problem, and
- then, we find the values of the important quantities.

In the above data processing problem, the quantities are the coefficients a_i. The quantity a_i is irrelevant if it does not affect the resulting signal, i.e., if $a_i = 0$. When $a_i \neq 0$, this means that this quantity affects the resulting signal $x(t) = \sum_{i=1}^{\infty} a_i \cdot e_i(t)$ and is, therefore, relevant. Thus, for our problem, the above two-stage data processing process takes the following form:

- first, we decided which values a_i are zeros and which are non-zeros, and
- then, we use an appropriate data processing algorithm to estimate the numerical values of non-zero coefficients a_i.

On the first stage, we can make several different decisions, all of which are consistent with the measurements and with the prior knowledge. For example, if in one decision, we take $a_i = 0$, then taking a_i to be very small but still different from 0 will still make this slightly modified signal consistent with all the measurement results. Out of all such possible decisions, we need to select *the most reasonable one*.

"Reasonable" is not a precise term. So, to be able to solve the problem, we need to translate this imprecise natural-language description into precise terms.

Fuzzy techniques can translate this natural-language description into a precisely formulated problem. In order to translate the above natural-language problem into precise terms, it is reasonable to use techniques specifically designed for such translations—namely, the techniques of *fuzzy logic*; see, e.g., [11, 16, 19].

In fuzzy logic, the meaning of each imprecise ("fuzzy") natural-language statement $P(x)$ about a quantity x is described by assigning, to each possible value x, the degree $\mu_P(x) \in [0, 1]$ to which we are sure that x satisfies this property P. For simple properties, we can determine these values, e.g., by simply asking the experts to mark, on a scale from 0 to 10, how much they are sure that P holds for x; if an expert marks the number 7, we take $\mu_P(x) = 7/10$.

This can be done for properties that depend on a single quantity. However, for properties like "reasonable", that depend on many values $a_1, \ldots, a_n, \ldots$, it is not feasible to ask the expert for the degrees corresponding to all possible combinations of the values a_i. In such situations, we can use the fact that from the commonsense viewpoint, a sequence $(a_1, a_2, \ldots)$ is reasonable if and only if a_1 is reasonable *and* a_2 is reasonable, *and* ... For each of the quantities a_i, we can elicit, from the expert, degree to which different values of a_i are reasonable.

Since this is all the information that we have, we need to estimate the degree to which a_1 is reasonable and a_2 is reasonable, and ..., based on the degrees to which a_1 is reasonable, to which a_2 is reasonable, etc. In other words, we know the degrees of belief $a = d(A)$ and $b = d(B)$ in statements A and B, and we need to estimate the degree of belief in the composite statement $A \& B$.

It is worth mentioning that this *estimate* cannot be always *exact*, because our degree of belief in a composite statement $A \& B$ depends not only on our degrees of belief in A and B, it also depends on the (usually unknown) dependence between A and B. Let us give an example.

- If A is "coin falls heads", and B is "coin falls tails", then for a fair coin, degrees a and b are equal: $a = b$. Here, $A \& B$ is impossible, so our degree of belief in $A \& B$ is zero: $d(A \& B) = 0$.

- However, if we take $A' = B' = A$, then $A' \& B'$ is simply equivalent to A, so we still have $a' = b' = a$ but this time $d(A' \& B') = a > 0$.

In these two cases, $d(A') = d(A)$, $d(B') = d(B)$, but $d(A \& B) \neq d(A' \& B')$.

In general, let $f_\&(a, b)$ be the estimate for $d(A \& B)$ based on the known values $a = d(A)$ and $b = d(B)$. The corresponding function $f_\&(a, b)$ must satisfy some reasonable properties: e.g.,

- since $A \& B$ means the same as $B \& A$, this operation must be commutative;
- since $(A \& B) \& C$ is equivalent to $A \& (B \& C)$, this operation must be associative, etc.

Operations with these properties are known as *"and"-operations*, or, alternatively, *t-norms*.

Let us apply an appropriate t-norm to our problem. In our case, for each variable a_i, we only need to find the degrees of belief in two situations: that $a_i = 0$ and that $a_i \neq 0$. Let us denote the degree to which it is reasonable to believe that $a_i = 0$ by $d_i^=$, and the degree to which it is reasonable to believe that $a_i \neq 0$ by $d_i^{\neq}$. Thus, we arrive at the following formulation of the first stage of data processing.

Resulting precise formulation of the first stage of data processing in precise terms. Our goal is to select a sequence $(\varepsilon_1, \varepsilon_2, \ldots)$, where each ε_i is equal either to $=$ or to $\neq$. If ε_i is $=$, this means that we have decided that $a_i = 0$, and if ε_i is $\neq$, this means that we have decided that $a_i \neq 0$.

For each such sequence $\varepsilon = (\varepsilon_1, \varepsilon_2, \ldots)$, we can determine the degree $d(\varepsilon)$ to which this sequence is reasonable, by applying the selected t-norm $f_\&(a, b)$ to the degrees $d_i^{\varepsilon_i}$ to which we belief that each choice ε_i is reasonable:

$$d(\varepsilon) = f_\&(d_1^{\varepsilon_1}, d_2^{\varepsilon_2}, \ldots).$$

Out of all sequences ε which are consistent with the measurements and with the prior knowledge, we must select the one for which this degree of belief is the largest possible.

An additional fact that we can use. If we have no information about the signal, i.e., in other words, if there is no evidence that there is a non-zero signal, then the most reasonable choice is to select $x(t) = 0$, i.e., to select a signal for which

$$a_1 = a_2 = \cdots = 0.$$

In other words, if we do not have any way to impose restrictions on the sequence ε, then the most reasonable should be the sequence $(=, =, \ldots)$.

Similarly, the worst reasonable is the sequence in which we take all the values into account, i.e., the sequence $(\neq, \ldots, \neq)$.

A comment about t-norms. In principle, there are many possible t-norms. However, it is known (see, e.g., [15]) that an arbitrary continuous t-norm can be approximated, with an arbitrary accuracy, by an *Archimedean* t-norm, i.e., by a t-norm of the type $f_\&(a, b) = f^{-1}(f(a) \cdot f(b))$, for some continuous strictly increasing function $f(x)$. Thus, without losing generality, we can assume that the actual t-norm is Archimedean.

Now, we are ready to formulate and solve the corresponding problem.

4 Definitions and the Main Result: Fuzzy-Related Techniques Explain Sparsity

Definition 1

- By a *t-norm*, we means a function $f_\& : [0, 1] \times [0, 1] \to [0, 1]$ of the form $f_\&(a, b) = f^{-1}(f(a) \cdot f(b))$, where $f : [0, 1] \to [0, 1]$ is a continuous strictly increasing function for which $f(0) = 0$ and $f(1) = 1$.
- By a *sequence*, we mean a sequence $\varepsilon = (\varepsilon_1, \ldots, \varepsilon_N)$, where each symbol ε_i is equal either to $=$ or to $\neq$.
- Let $d^= = (d_1^=, \ldots, d_N^=)$ and $d^{\neq} = (d_1^{\neq}, \ldots, d_N^{\neq})$ be sequences of real numbers from the interval $[0, 1]$. For each sequence ε, we define its *degree of reasonableness* as $d(\varepsilon) \stackrel{\text{def}}{=} f_\&(d_1^{\varepsilon_1}, \ldots, d_N^{\varepsilon_N})$.
- We say that the sequences $d^=$ and $d^{\neq}$ *properly describe reasonableness* if the following two conditions are satisfied:
  - the sequence $\varepsilon_= \stackrel{\text{def}}{=} (=, \ldots, =)$ is more reasonable than all others, i.e., $d(\varepsilon_=) > d(\varepsilon)$ for all $\varepsilon \neq \varepsilon_=$, and
  - the sequence $\varepsilon_{\neq} \stackrel{\text{def}}{=} (\neq, \ldots, \neq)$ is less reasonable than all others, i.e., $d(\varepsilon_{\neq}) < d(\varepsilon)$ for all $\varepsilon \neq \varepsilon_{\neq}$.
- For each set S of sequences, we say that a sequence $\varepsilon \in S$ is the most reasonable if its degrees of reasonableness is the largest possible, i.e., if $d(\varepsilon) = \max_{\varepsilon' \in S} d(\varepsilon')$.

Proposition 1 *Let us assume that the sequences $d^=$ and $d^{\neq}$ properly describe reasonableness. Then, there exist weights $w_i > 0$ for which, within each set S, a sequence $\varepsilon \in S$ is the most reasonable if and only if for this sequence, the sum $\sum\limits_{i:\varepsilon_i = \neq} w_i$ is the smallest possible.*

Discussion. In other words, a sequence is the most reasonable if and only if the sum $\sum\limits_{i:a_i \neq p} w_i$ attains the smallest possible value. Thus, fuzzy-based techniques indeed naturally lead to the sparsity condition.

Proof By definition of the t-norm, we have

$$d(\varepsilon) = f_\&(d_1^{\varepsilon_1}, \ldots, d_N^{\varepsilon_N}) = f^{-1}(f(d_1^{\varepsilon_1}) \cdot \ldots \cdot f(d_N^{\varepsilon_N})),$$

i.e.,

$$d(\varepsilon) = f_{\&}(d_1^{\varepsilon_1}, \ldots, d_N^{\varepsilon_N}) = f^{-1}(e_1^{\varepsilon_1} \cdot \ldots \cdot e_N^{\varepsilon_N}),$$

where we denoted $e_i^{\varepsilon_i} \overset{\text{def}}{=} f(d_i^{\varepsilon_i})$.

Since the continuous function $f(x)$ is strictly increasing, its inverse $f^{-1}(x)$ is also strictly increasing. Thus, maximizing $d(\varepsilon)$ is equivalent to maximizing the function $e(\varepsilon) \overset{\text{def}}{=} f(d(\varepsilon))$. This function has the form

$$e(\varepsilon) = f(d(\varepsilon)) = f(f^{-1}(e_1^{\varepsilon_1} \cdot \ldots \cdot e_N^{\varepsilon_N})),$$

i.e., the form

$$e(\varepsilon) = e_1^{\varepsilon_1} \cdot \ldots \cdot e_N^{\varepsilon_N}.$$

From the condition that the sequences $d^=$ and $d^{\neq}$ properly describe reasonableness, it follows, in particular, that for each i, we have $d(\varepsilon_=) > d(\varepsilon_=^{(i)})$, where

$$\varepsilon_=^{(i)} \overset{\text{def}}{=} (=, \ldots, =, \neq \text{ (on } i\text{th place)}, =, \ldots, =).$$

This inequality is equivalent to $e(\varepsilon_=) > e(\varepsilon_=^{(i)})$.

Since the values $e(\varepsilon)$ are simply the products, we thus conclude that

$$\prod_{j=1}^{N} e_j^= > \left(\prod_{j \neq i} e_j^= \right) \cdot e_i^{\neq}.$$

The values $e_j^=$ corresponding to $j \neq i$ cannot be equal to 0, since otherwise, both products would be equal to 0s. Thus, these values are non-zeros. Dividing both sides of the inequality by all these values, we conclude that $e_i^= > e_i^{\neq}$.

Similarly, from the condition that the sequences $d^=$ and $d^{\neq}$ properly describe reasonableness, it also follows, in particular, that for each i, we have $d(\varepsilon_{\neq}) < d(\varepsilon_{\neq}^{(i)})$, where

$$\varepsilon_{\neq}^{(i)} \overset{\text{def}}{=} (\neq, \ldots, \neq, = \text{ (on } i\text{th place)}, \neq, \ldots, \neq).$$

This inequality is equivalent to $e(\varepsilon_{\neq}) > e(\varepsilon_{\neq}^{(i)})$.

Since the values $e(\varepsilon)$ are simply the products, we thus conclude that

$$\prod_{j=1}^{N} e_j^{\neq} < \left(\prod_{j \neq i} e_j^{\neq} \right) \cdot e_i^=.$$

The values $e_j^{\neq}$ corresponding to $j \neq i$ cannot be equal to 0, since otherwise, both products would be equal to 0s.

Thus, for all i, we have $e_i^= > e_i^{\neq} > 0$.

Now, in general, maximizing the product $e(\varepsilon) = \prod_{i=1}^{N} d_i^{\varepsilon_i}$ is equivalent to maximizing the same product divided by a constant $c \stackrel{\text{def}}{=} \prod_{i=1}^{N} d_i^{\neq}$. The ratio $\dfrac{e(\varepsilon)}{c}$ can be equivalently reformulated as $\dfrac{e(\varepsilon)}{c} = \prod_{i:\varepsilon_i \neq} \dfrac{e_i^=}{e_i^{\neq}}$.

Since logarithm is a strictly increasing function, maximizing this product is, in its turn, equivalent to maximizing its logarithm, i.e., the value

$$L(\varepsilon) \stackrel{\text{def}}{=} \ln\left(\frac{e(\varepsilon)}{c}\right) = \sum_{i:\varepsilon_i \neq} w_i,$$

where we denoted $w_i \stackrel{\text{def}}{=} \ln\left(\dfrac{e_i^=}{e_i^{\neq}}\right)$. Since $e_i^= > e_i^{\neq} > 0$, we have $\dfrac{e_i^=}{e_i^{\neq}} > 1$ and thus, $w_i > 0$. The proposition is proven.

5 A Similar Derivation Can Be Obtained in the Probabilistic Case

Towards a probabilistic reformulation of the problem. In the probabilistic approach, reasonableness can be described by assigning a prior probability $p(\varepsilon)$ to each possible sequences ε. In this case, out of each set of sequences, we should select the most probable one, i.e., the one with the largest value of the prior probability.

Let $p_i^=$ be the prior probability that $a_i = 0$, and let $p_i^{\neq} = 1 - p_i^=$ be the probability that $a_i \neq 0$. A priori we do not know the relation between the values ε_i and ε_j corresponding to different coefficients $i \neq j$, so it makes sense to assume that the corresponding random variables ε_i and ε_j are independent.

This assumption is in perfect agreement with the maximum entropy idea (also known as the Laplace's indeterminacy principle), according to which, out of all probability distributions which are consistent with our observations, we should select the one for which the entropy $- \sum p_i \cdot \ln(p_i)$ is the largest possible; see, e.g., [10]. Indeed, if we only know marginal distributions, then the maximum entropy idea implies that, according to the joint distribution, all the random variables are independent.

Under this assumption, $p(\varepsilon) = \prod_{i=1}^{N} p_i^{\varepsilon_i}$. Thus, we arrive at the following definition.

Definition 2

- Let $p^= = (p_1^=, \ldots, p_N^=)$ be a sequence of real numbers from the interval $[0, 1]$, and let $p_i^{\neq} \stackrel{\text{def}}{=} 1 - p_i^=$. For each sequence ε, we define its *prior probability* as

$$p(\varepsilon) \stackrel{\text{def}}{=} \prod_{i=1}^{N} p_i^{\varepsilon_i}.$$

- We say that the sequence $p^=$ *properly describes reasonableness* if the following two conditions are satisfied:

  - the sequence $\varepsilon_= \stackrel{\text{def}}{=} (=, \ldots, =)$ is more probable than all others, i.e., $p(\varepsilon_=) > p(\varepsilon)$ for all $\varepsilon \neq \varepsilon_=$, and
  - the sequence $\varepsilon_{\neq} \stackrel{\text{def}}{=} (\neq, \ldots, \neq)$ is less probable than all others, i.e., $p(\varepsilon_{\neq}) < p(\varepsilon)$ for all $\varepsilon \neq \varepsilon_{\neq}$.

- For each set S of sequences, we say that a sequence $\varepsilon \in S$ *is the most probable* if its prior probability is the largest possible, i.e., if $p(\varepsilon) = \max_{\varepsilon' \in S} p(\varepsilon')$.

Proposition 2 *Let us assume that the sequence $p^=$ properly describes reasonableness. Then, there exist weights $w_i > 0$ for which, within each set S, a sequence $\varepsilon \in S$ is the most probable if and only if for this sequence, the sum $\sum\limits_{i:\varepsilon_i=\neq} w_i$ is the smallest possible.*

Discussion. In other words, probabilistic techniques also lead to the sparsity condition.

Proof is similar to the Proof of Proposition 1.

Comments.

- The fact that the probabilistic approach leads to the same conclusion as the fuzzy approach makes us more confident that our justification of sparsity is valid.
- The comparison of the above two derivations shows an important advantage of fuzzy-based approach in situations like this, when we have a large amount of uncertainty:

  - the probability-based result is based on the assumption of independence, while
  - the fuzzy-based result can allow different types of dependence—as described by different t-norms.

Remaining open questions. In this paper, we showed that fuzzy techniques help explain empirical efficiency of sparsity-based data- and image-processing algorithms. This fact makes us hopeful that similar fuzzy-based ideas can help explain not just the general idea behind such algorithms, but also empirical recommendations for selecting specific parameters of sparsity-based algorithms.

Acknowledgements This work was supported in part by the National Science Foundation grants HRD-0734825 and HRD-1242122 (Cyber-ShARE Center of Excellence) and DUE-0926721, and by an award "UTEP and Prudential Actuarial Science Academy and Pipeline Initiative" from Prudential Foundation.

The authors are thankful to the anonymous referees for valuable suggestions.

References

1. B. Amizic, L. Spinoulas, R. Molina, A.K. Katsaggelos, Compressive blind image deconvolution. IEEE Trans. Image Process. **22**(10), 3994–4006 (2013)
2. E.J. Candès, J. Romberg, T. Tao, Stable signal recovery from incomplete and inaccurate measurements. Comm. Pure Appl. Math. **59**, 1207–1223 (2006)
3. E.J. Candès, J. Romberg, T. Tao, Robust uncertainty principles: exact signal reconstruction from highly incomplete frequency information. IEEE Trans. Inf. Theory **52**(2), 489–509 (2006)
4. E.J. Candès, T. Tao, Decoding by linear programming. IEEE Trans. Inf. Theory **51**(12), 4203–4215 (2005)
5. E.J. Candès, M.B. Wakin, An introduction to compressive sampling. IEEE Signal Process. Mag. **25**(2), 21–30 (2008)
6. D.L. Donoho, Compressed sensing. IEEE Trans. Inf. Theory **52**(4), 1289–1306 (2005)
7. M.F. Duarte, M.A. Davenport, D. Takhar, J.N. Laska, T. Sun, K.F. Kelly, R.G. Baraniuk, Single-pixel imaging via compressive sampling. IEEE Signal Process. Mag. **25**(2), 83–91 (2008)
8. T. Edeler, K. Ohliger, S. Hussmann, A. Mertins, Super-resolution model for a compressed-sensing measurement setup. IEEE Trans. Instrum. Meas. **61**(5), 1140–1148 (2012)
9. M. Elad, *Sparse and Redundant Representations* (Springer, 2010)
10. E.T. Jaynes, G.L. Bretthorst, *Probability Theory: The Logic of Science* (Cambridge University Press, Cambridge, UK, 2003)
11. G. Klir, B. Yuan, *Fuzzy Sets and Fuzzy Logic* (Prentice Hall, Upper Saddle River, New Jersey, 1995)
12. J. Ma, F.-X. Le Dimet, Deblurring from highly incomplete measurements for remote sensing. IEEE Trans. Geosci. Remote Sens. **47**(3), 792–802 (2009)
13. L. McMackin, M.A. Herman, B. Chatterjee, M. Weldon, A high-resolution SWIR camera via compressed sensing. Proc. SPIE **8353**(1), 835303 (2012)
14. B.K. Natarajan, Sparse approximate solutions to linear systems. SIAM J. Comput. **24**, 227–234 (1995)
15. H.T. Nguyen, V. Kreinovich, P. Wojciechowski, Strict archimedean t-norms and t-conorms as universal approximators. Int. J. Approximate Reasoning **18**(3–4), 239–249 (1998)
16. H.T. Nguyen, E.A. Walker, *A First Course in Fuzzy Logic* (Chapman and Hall/CRC, Boca Raton, Florida, 2006)
17. Y. Tsaig, D. Donoho, Compressed sensing. IEEE Trans. Inf. Theory **52**(4), 1289–1306 (2006)
18. L. Xiao, J. Shao, L. Huang, Z. Wei, Compounded regularization and fast algorithm for compressive sensing deconvolution, in *Proceedings of the 6th International Conference on Image Graphics* (2011), pp. 616–621
19. L.A. Zadeh, Fuzzy sets. Inf. Control **8**, 338–353 (1965)

Part X
Fuzziness in Education

Bilingual Students Benefit from Using Both Languages

Julian Viera, Olga Kosheleva and Shahnaz N. Shahbazova

Abstract When using an individualized learning system ALEKS to study mathematics, bilingual students can use both English-language and Spanish-language modules. When we started our study, we expected that those Spanish-language students whose knowledge of English is still not perfect would first use mostly Spanish-language modules, and that their use of English-language modules will increase as their English skills increase. Instead, what we found is that even students who are not very skilled in English use both Spanish-language and English-language modules. This raises a natural question: why, in spite of the presence of well-designed well-tested easy to Spanish-language models, the students benefit from also using English-language modules—which for them are not so easy to access (they use Google translator). In this paper, we use fuzzy logic to provide a possible theoretical explanation for this surprising behavior.

1 Formulation of the Problem

How the study started. Many schools and universities use the ALEKS leaning system to help students learn math and other subjects; see, e.g., [2, 6–8]. According to the official ALEKS website [2], "ALEKS is an adaptive, artificially-intelligent learning system that provides students with an individualized learning experience tailored to their unique strengths and weaknesses."

J. Viera · O. Kosheleva (✉)
Department of Teacher Education, University of Texas at El Paso, 500 W. University,
El Paso, TX 79968, USA
e-mail: olgak@utep.edu

J. Viera
e-mail: jviera1@utep.edu

S. N. Shahbazova
Azerbaijan Technical University, Baku, Azerbaijan
e-mail: shahbazova@gmail.com

© Springer International Publishing AG, part of Springer Nature 2018
L. A. Zadeh et al. (eds.), *Recent Developments and the New Direction
in Soft-Computing Foundations and Applications*, Studies in Fuzziness
and Soft Computing 361, https://doi.org/10.1007/978-3-319-75408-6_33

At our University of Texas at El Paso, most students are bilingual. Many of them are *English Learners* in the sense that their native language is Spanish, and:

- while they have already mastered enough English to be able to attend English-language classes,
- they are still taking English classes to further improve their level of understanding English.

For such students, ALEKS is a good way to study, since this learning system has both English-language and Spanish-language modules available to students.

ALEKS is a system in progress. Many researchers and practitioners are studying the students' interaction with this system, and the results of these studies are used to improve and update this system. Most of the current studies are about students who use ALEKS in only one language.

(To the best of our knowledge, so far, no one has studied how bilingual students use ALEKS in both languages.)

What we expected. We expected that students at the beginning of their English studies use mostly Spanish-language ALEKS modules, and that:

- as their English skills improve,
- they will start using English-language modules more.

What we found, to our surprise. Surprisingly, this is not what we found. What we found is that even the students with the least knowledge of English,

- instead of using just Spanish-language modules,
- use *both* English-language and Spanish-language modules.

Using English-language modules is not easy on them: they use Google translation to better understand these modules, but they still go to all these trouble because, in their experience, the use of both modules is beneficial to them.

Resulting question, and what we do in this paper. Our findings lead to a natural question: why? Why, in spite of the existence of the well-designed actively used English-language modules, the students benefit from also accessing Spanish-language modules?

In this paper, we use fuzzy logic ideas (see, e.g., [12, 15, 19]) to provide a possible theoretical explanation for this empirical phenomenon.

2 Our Explanation

Main idea behind our explanation: texts are (somewhat) fuzzy. If the texts were absolutely mathematically precise, there would be no need to look into the same text in two different languages—because in this case, the Spanish-language text carries the exact same information as the English-language text.

Thus, the very fact that using both language help indicates that some of the information that the students get from the text is not absolutely precise, it is—to some extent—fuzzy.

When statements are fuzzy, their translation into a different language can somewhat change the meaning. Every time we translate a (somewhat) fuzzy statement from one language to another, the meaning can slightly change—because:

- we are trying to describe the same meaning by using words from different languages, and
- meanings tend to change (usually slightly) when we move from language to another, from one culture to another.

Examples are easy to find: e.g.,

- what is a quiet conversation to an Italian may not be perceived as a quite one by an average American, and, vice versa,
- what a stereotypical WASP American may perceive as a passionate political speech may be perceived a very stiff and un-emotional performance by an Italian spectator.

A positive twist on the same observation. Up to now, the imprecise character of translation was presented as a negative fact: for example, a person who does not speak Italian cannot fully appreciate Dante or Petrarca, since every translation changes the meaning slightly.

However, this same observation can be given a positive twist in situations—like the educational one that we described earlier—when our objective is *not* to understand a text written in a specific language, but rather understand a *material*.

In this case, every time this material is described in a new language, this translation from meaning to language inevitably leads to a slight change in the meaning. But the good news is that since different languages are different, we get slightly different changes. And so, by combining different changed meanings, we can hopefully get a better understanding of the original meaning.

Example. For example, the Russian names for colors are often more exact—there are special words for light blue and other shades of blue, all corresponding to the same English word "blue". On the other hand, English has more words for different types of cars. So, a person interested in a car may benefit from reading its description in both languages: that way, this person will:

- get a better understanding of the car's color from a Russian-language description and
- get a better understanding of the car's type from the English-language description.

Why combining several changed meanings, we can better understand the original text: a simple explanation. To understand this phenomenon, let us consider a simple example when the meaning is described by a single number d. This number can be, e.g., the fuzzy degree of certainty in a corresponding statement.

When we describe the original meaning in a language i, we use a word or a phrase from this language to reflect this meaning. In every language, there is only a limited number of words and phrases describing the corresponding notion, so inevitably, the selected word or phrase will not describe the original meaning exactly. As a result, the fuzzy degree d_i of the selected word will be, in general, somewhat different from the original intended degree: $d_i \neq d$.

For different languages, we have different degrees $d_1 \neq d$, $d_2 \neq d$, etc. In general, the difference $d_i - d$ is caused by many different factors. Such situations, when many small independent factors are in place, is ubiquitous. Such situations are covered by the *Central Limit Theorem* (see, e.g., [17]), according to which in such cases, the probability distribution of the corresponding differences $d_i - d$ is close to Gaussian.

There is no reason to believe that translation is biased, so it is reasonable to assume that the mean value of each difference $d_i - d$ is zero. Similarly, since there is no reason to believe that one language is more accurate than another one, it is reasonable to assume that for all the languages, the standard deviation of the difference $d_i - d$ is approximately the same. Let us denote this common standard deviation by σ.

Also, there is no reason to believe that two distortions caused by different languages are correlated—unless these languages are very similar. Thus, it makes sense to assume that the differences $d_i - d$ corresponding to different languages i are statistically independent.

So, when we have several translations, we have several estimates $d_i \approx d$ for the desired meaning. The corresponding differences $d_i - d$ are independent normally distributed random variables with 0 mean and the same standard deviation σ. We would like to use all available estimates $d_1, \ldots, d_n$ to come up with—ideally—a more accurate estimate for the degree d.

Such a situation is very common in statistics, and in statistics, there are known methods for coming up with such a better estimate. Namely, for each i, normal distribution means that if the actual degree is d, then the probability that the observed estimate is d_i (to be more precise, probability density) can be computed as

$$\rho(d_i) = \frac{1}{\sqrt{2\pi} \cdot \sigma} \cdot \exp\left(-\frac{(d_i - d)^2}{2\sigma^2}\right). \tag{1}$$

Since different deviations $d_i - d$ are independent, the overall probability that, given d, we observe the values $d_1, \ldots, d_n$, can be computed as the product of the corresponding probabilities (1), i.e., as

$$\rho(d_1, \ldots, d_n) = \prod_{i=1}^{n} \rho(d_i) = \prod_{i=1}^{n} \frac{1}{\sqrt{2\pi} \cdot \sigma} \cdot \exp\left(-\frac{(d_i - d)^2}{2\sigma^2}\right). \tag{2}$$

Of all possible values d, it is reasonable to select a value d for which this probability is the largest—this natural idea is known as the *Maximum Likelihood method* [17].

For the formula (2), maximization of this expression can be simplified if we take into account that $\ln(x)$ is a monotonic function and thus, maximizing the probability

$\rho(d_1, \ldots, d_n)$ is equivalent to maximizing its logarithm

$$L \overset{\text{def}}{=} \ln(\rho(d_1, \ldots, d_n)).$$

For the expression (2), by using the fact that logarithm of the product is equal to the sum of the logarithms, we indeed conclude that the logarithm L has a much simpler form than the original probability: namely,

$$L = \text{const} - \sum_{i=1}^{n} \frac{(d_i - d)^2}{2\sigma^2},$$

where the constant does not depend on d at all and is, this, irrelevant in maximization. Maximizing this sum is equivalent to minimizing the expression

$$J \overset{\text{def}}{=} -\sigma^2(L - \text{const}) = \sum_{i=1}^{n}(d_i - d)^2.$$

Differentiating this expression with respect to d and equating the derivatives to 0, we conclude that the optimal estimate has the form

$$d = \frac{d_1 + \cdots + d_n}{n}. \tag{3}$$

This average is indeed actively used in statistical data processing [17]. It is also in perfect accordance with the fact that d is the mean of d_i, and the mean, by definition, is the limit of sample means (3). Thus, for a given sample, a sample mean is a good approximation to the actual mean.

Good news is that, as one can easily check, the standard deviation of the mean d is equal to $\dfrac{\sigma}{\sqrt{n}}$.

So, if we use two languages—as the students in the above study do—we decrease the inaccuracy of our understanding from σ to

$$\frac{\sigma}{\sqrt{2}} \approx 0.7 \cdot \sigma,$$

i.e., by 30%. This clearly explains why using the same text in both languages leads to a better understanding that using one of these languages.

From the simple cases to more complex cases. The above explanation only covers the simplest cases, when we have a single value d that we want to estimate. But more complex situations can be described as recovering several such values. For example, a complex (imprecise) notion can be described, in fuzzy logic, by its membership function—i.e., in effect, by several numbers that describe,

- for different possible situations s,
- the degree $d(s)$ to which this particular situation reflects this notion.

By invoking descriptions $d_i(s)$ from several languages, we can get, for each situation s, a more accurate description of the desired notion, as

$$d(s) \approx \frac{d_1(s) + \cdots + d_n(s)}{n}.$$

Thus, in more complex situations, the use of several languages is beneficial as well.

Conclusion. Thus, the above seemingly counterintuitive empirical fact—that students benefit from looking in descriptions in two languages—can be naturally explained if we take into account the fuzzy (imprecise) character of the knowledge.

3 Discussion

General consequence for bilingual students. Several studies that shown that bilingual students have, an average, an intellectual advantage over monolingual ones; see, e.g., [1, 3–5, 9–11, 13, 14, 16]. Until recently, this empirical fact was explained by the fact that for bilingual students, their experience of constantly switching from one language to another makes them, in general, more skilled in switching form one context to another—thus enhancing their intellectual abilities.

However, a recent research [4] shows that the intellectual advantages of bilingual students go beyond switching. The above argument explains where this additional advantage comes from: namely, the possibility to describe the imprecise situation in both languages enables the students to gain additional information about the situation.

Other relations to problem solving. The advantages of using descriptions of two different languages when learning the material are similar to the advantages of looking into several aspects of a problem in problem solving; see, e.g., [18].

Acknowledgements The authors are thankful to Mourat Tchoshanov from the University of Texas at El Paso and to Mark Leikin from the University of Haifa, Israel, for valuable discussions.

The authors are also grateful to the anonymous referees for useful suggestions.

References

1. O.O. Adesope et al., A systematic review and meta-analysis of the cognitive correlates of bilingualism. Rev. Educ. Res. **80**, 207–245 (2010)
2. *Assessment and LEarning in Knowledge Spaces (ALEKS)* (McGraw Hill Education, 2016), https://www.aleks.com/

3. S. Ben-Zeev, The influence of bilingualism on cognitive strategy and cognitive development. Child Dev. **48**, 1009–1018 (1977)
4. E. Bialystok, F.I.M. Craik, G. Luk, Bilingualism: consequences for mind and brain., Trends Cogn. Sci. **16**(4), 240–250 (2012)
5. E. Bialystok, S. Majumder, The relationship between bilingualism and the development of cognitive processes in problem-solving. Appl. Psycholinguist. **19**, 69–85 (1998)
6. J.-P. Doignon, J.-C. Falmagne, *Learning Spaces* (Springer, Berlin, New York, 2011)
7. J.-C. Falmagne, E. Cosyn, J.-P. Doignon, N. Thiéry, The assessment of knowledge, in theory and in practice, by ed. R. Missaoui, J. Schmidt, *Formal Concept Analysis: Proceedings of the 4th International Conference on Formal Concept Analysis ICFCA'2006, Dresden, Germany, 13–17 February 2006* (Springer, Berlin, New York, 2006), pp. 61–79
8. J.-C. Falmagne, M. Koppen, M. Villano, J.-P. Doignon, L. Johannesen, Introduction to knowledge spaces: how to build, test, and search them. Psychol. Rev. **97**, 201–224 (1990)
9. W.S. Francis, Analogical transfer of problem solutions within and between languages in Spanish-English bilinguals. J. Mem. Lang. **40**, 301–329 (1999)
10. K. Hakuta, *Mirror of Language: The Debate on Bilingualism* (Basic Books, New York, 1986)
11. A. Ianco-Worrall, Bilingualism and cognitive development. Child Dev. **43**, 1390–1400 (1972)
12. G. Klir, B. Yuan, *Fuzzy Sets and Fuzzy Logic* (Prentice Hall, Upper Saddle River, New Jersey, 1995)
13. A.M. Kovács, J. Mehler, Cognitive gains in 7-month-old bilingual infants. Proceedings of the National Academy of Sciences of the USA **106**(16), 6556–6560 (2009)
14. M. Leikin, The effect of bilingualism on creativity: developmental and educational perspectives. Int. J. Bilingualism **17**(4), 431–447 (2013)
15. H.T. Nguyen, E.A. Walker, *A First Course in Fuzzy Logic* (Chapman and Hall/CRC, Boca Raton, Florida, 2006)
16. E. Peal, W. Lambert, The relation of bilingualism to intelligence. Psychol. Monogr. **76**, 1–23 (1962)
17. D.J. Sheskin, *Handbook of Parametric and Nonparametric Statistical Procedures* (Chapman and Hall/CRC, Boca Raton, Florida, 2011)
18. M. Tchoshanov, O. Kosheleva, V. Kreinovich, On the importance of duality and multi-ality in mathematics education, in *Proceedings of the 5th International Conference Mathematics Education: Theory and Practice MATHEDU'2015, Kazan, Russia, 27–28 November 2015*, pp. 8–13
19. L.A. Zadeh, Fuzzy sets. Inf. Control **8**, 338–353 (1965)

Decomposable Graphical Models on Learning, Fusion and Revision

Fabian Schmidt, Jörg Gebhardt and Rudolf Kruse

Abstract Industrial applications often face elaborated problems. In order to solve them properly a great deal of complexity and data diversity has to be managed. In this paper we present a planning system that is used globally by the Volkswagen Group. We introduce the specific challenges that this industrial application faces, namely a high complexity paired with diverse heterogeneous data sources, and describe how the problem has been modelled and solved. We further introduce the core technology we used, the revision of Markov networks. We further motivate the need to handle planning inconsistencies and present our framework consisting of six main components: Prevention, Detection, Analysis, Explanation, Manual Resolution, and Automatic Elimination.

1 Introduction

Modern industrial applications face complex problems involving vast heterogeneous knowledge bases and highly complex domains. In this work we present the challenges and solutions to an item planning system of a car manufacturer. One important aspect of creating solutions for industrial clients is the maintenance and fusion of knowledge in order to create useful expert systems. One possible result of data fusion are probability distributions. In order to store those properly Markov networks are a good option. They decompose high dimensional probability spaces into a number of smaller low dimensional probability distributions. They belong to a group of techniques called probabilistic graphical models [1–5]. Other common

F. Schmidt (✉) · J. Gebhardt
ISC Gebhardt, Celle, Germany
e-mail: schmidt@isc-gebhardt.de

J. Gebhardt
e-mail: gebhardt@isc-gebhardt.de

R. Kruse
Otto-von-Guericke University, Magdeburg, Germany
e-mail: rudolf.kruse@ovgu.de

© Springer International Publishing AG, part of Springer Nature 2018   439
L. A. Zadeh et al. (eds.), *Recent Developments and the New Direction in Soft-Computing Foundations and Applications*, Studies in Fuzziness and Soft Computing 361, https://doi.org/10.1007/978-3-319-75408-6_34

types of graphical models are possibilistic graphical models [6–9] and relational probabilistic graphical models [10, 11]. Not only fusing and storing knowledge is important. Modern knowledge based systems need the ability to react to changes in beliefs quickly and frequently. Therefore, methods have been developed to properly adapt knowledge to new beliefs. One important aspect of proper adaptation is formulated in the principle of minimal change [12], which states that in order to incorporate given new beliefs, only absolutely necessary changes can be made in a knowledge base. This means, after the incorporation of the new beliefs, the knowledge base should be as close to the original one as possible, in an information theoretic sense. The revision operation has been introduced as a belief change operation that applies new beliefs respecting this principle [13]. From the perspective of knowledge based systems, further properties a revision operation should satisfy have been formulated as postulates in [14–16]. How to approach revision algorithmically has been outlined in [17], and computational considerations have been made in [18]. Our work focuses on the revision of probability distributions as it has been introduced in [19]. In this context the revision operation has been successfully implemented for Markov networks [20] using iterative proportional fitting [21, 22]. This method is well known in the area of statistics and shows beneficial properties for our context.

The growing complexity and interconnectedness of knowledge bases and an increasing number of new beliefs lead almost inevitably to inconsistencies in the formulation of revision problems. In almost any type of knowledge based systems, inconsistencies render the underlying knowledge useless and should consequently be addressed. The handling of inconsistencies is a multi-facet problem and different aspects of it have been introduced in [23]. Furthermore, two types of inconsistencies and a revision control algorithm have been described in [24].

In this work we give a brief overview over our industrial application, its challenges and our solutions using probabilistic knowledge revision. In Sect. 2 of this paper, we present the industrial example. Section 3 describes the main problems we are facing in that application. Section 4 then presents the model we use to solve that problem. Section 5 introduces the revision operation. Section 6 discusses revision inconsistencies. In Sect. 7, we then present a framework we developed to address inconsistencies and finally in Sect. 8, we conclude this work.

2 Application

Our research is conducted in cooperation with the Volkswagen Group and their system for estimating part demands and managing capacities for short- and medium-term forecasts, called EPL (EigenschaftsPLanung: item planning). EPL combines several heterogeneous input sources such as rules describing buildable vehicle specifications, production history reflecting customer preferences, and market forecasts leading to stipulations of modified item rates, and capacity restrictions that are modelled as revision assignments. Those sources are fused into

Markov Networks and the revision operation is then used to estimate the part demands. We will explain the modelling in more detail in the next sections. For more information we refer to previous works on the topic [19, 25]. EPL is currently the biggest industrial application for Markov networks. Using EPL the demands for more than a hundred different model groups from multiple car manufacturers of the Volkswagen Group are estimated every week. In case of the VW Golf being Volkswagens most popular car class there are about 200 item families with typically 4–8 (but up to 150) values each, that together fully describe a car to be produced, and many of these families can directly be chosen by the customer.

3 Problem

The Volkswagen Group favours a marketing policy that provides their customers with a maximum degree of freedom in choosing individual specifications of vehicles. As mentioned before, for the VW Golf, being Volkswagens most popular car class, there are about 200 item families with typically 4–8 (but up to 150) values each, that together fully describe a car to be produced, and many of these families can directly be chosen by the customer. Although of course not all item combinations are possible, the customers utilize the given variety, as in spite of the vast number of produced cars only a diminishing fraction thereof are completely identical.

This kind of market strategy requires a planning system that not only handles the complexity well, but also accommodates a number of very heterogeneous data sources that have to be combined in order to create a model that is able facilitate the planning operation. Figure 1 shows an overview over the kinds of information that have to be fused in order to generate a proper model for Volkswagens planning problem.

3.1 Technical Rules

In the example of EPL the model had to incorporate technical and logistical data about cars that is stored as rules describing how valid cars can be assembled. The rules describe which items are compatible as well as specific requirements for certain markets, since regulations differ from country to country. Those rules are logical in nature. One simple example of such a rule could be: IF *Engine* $= m_4$ AND *Heating* $= h_1$ THEN *Generator* $\in \{g_2, g_5, g_7\}$. This rule would describe the fact, that if a costumer would choose the engine m_4 in combination with the heating system h_1, then the car can only have one of the three generators g_2, g_5 or g_7, other generators are not possible. Rules used in our application can have more than 15 dimension. Further, there are roughly ten thousand rules used in order to describe all possible variations of one specific model group.

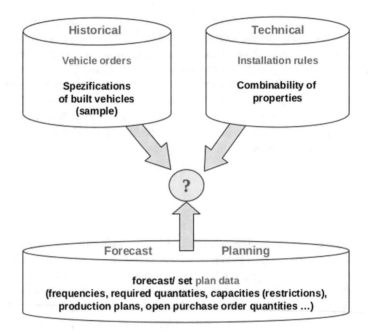

Fig. 1 Fusion of different knowledge types

3.2 Historical Data

Another type of knowledge represents the cars ordered in the past. In this knowledge base, sets of complete vehicle specifications of ordered cars are stored. Each car is represented by a high dimensional tuple where every variable is instantiated. It has to be mentioned here that the variety of those tuples is rather large. Under normal circumstances it is fairly rare for two cars to have the exact same specification. This is due to the fact, that costumers are allowed to customise their car to a great extend.

3.3 Planning Data

The last type of knowledge used in our application is actual planning data. This is represented by lower dimensional probability assignments. Those can represent frequencies, desired quantities or production capacities and other planning related estimations.

4 Model

For Volkswagens system EPL the problem was modelled and solved in a three step process.

(1) Obtaining structural information from the logic rules
(2) Enriching the structural data with quantitative information
(3) Modify the enriched data structure to accommodate the planning data.

The first step is to transform the rules describing valid cars and item combinations into a relational network. This is a rather time consuming effort, however the resulting model (Fig. 2) makes further operations easier than using the rule database directly.

In order to use the resulting relational network to perform a planning operation, in a second step the historical data is sampled and used to transform the relational network into a probability distribution. In the case of EPL, Markov networks have been chosen to represent the probability distribution that stores the combination of technical rules and historical customer data. Figure 3 shows an anonymised example of a real world Markov network used to plan a VW Golf.

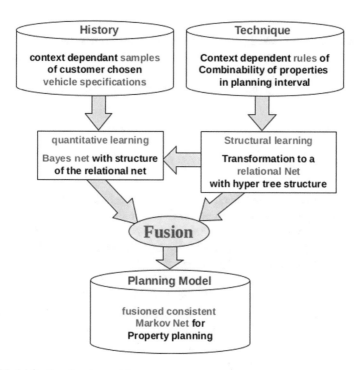

Fig. 2 Model for the planning problem

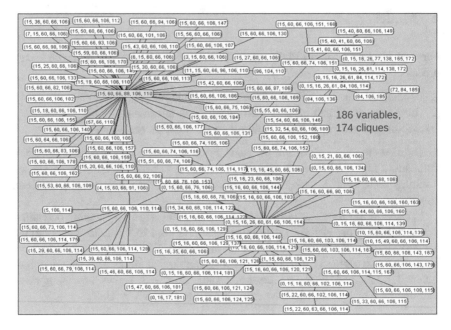

Fig. 3 Markov network used to plan a VW Golf

The last step is the actual planning operation. Here an operation called revision is used in order to incorporate the forecast data and modify the Markov network according to the principle of minimal change.

Figure 2 shows a schematic representation of the process as it was modelled for VW.

5 Revision

The goal of (probabilistic) revision is to compute a posterior probability distribution which satisfies given new distribution conditions, only accepting a minimal change of the quantitative interaction structures of the underlying prior distribution.

More formally, in our setting, a revision operation (see [24, 25]) operates on a joint probability distribution $P(V)$ on a set $V = \{X_1, \ldots, X_n\}$ of variables with finite domains $\Omega(X_i)$, $i = 1, \ldots, n$. The purpose of the operation is to adapt $P(V)$ to new sets of beliefs. The beliefs are formulated in a socalled revision structure. $\Sigma = (\sigma_s)_{s=1}^{S}$. This structure consists of revision assignments σ_s, each of which is referred to a (conditional) assignment scheme $(R_s \mid K_s)$ with a context scheme K_s, $K_s \subseteq V$, and a revision scheme R_s, where $\emptyset\, 6 = R_s \subseteq V$ and $K_s \cap R_s = \emptyset$. The pair $(P(V), \Sigma)$ is called revision problem (Fig. 4).

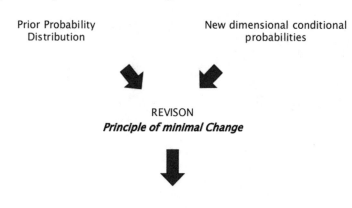

Fig. 4 Overview over probabilistic revision

For example, in the revision assignment $(GPS = nav1 \mid Country = France)$: $= 0.2$, which sets the probability for the GPS system $nav1$ in the country $France$ to 0.2, the context scheme K_s is $\{Country\}$ and the revision scheme R_s is $\{GPS\}$.

The result of the revision, and solution to the revision problem, is a probability distribution $P_{\Sigma}(V)$ which

- satisfies the revision assignments (the postulated new probabilities)
- preserves the probabilistic interaction structure as far as possible.

By preserving the interaction structure we mean that, except from the modifications induced by the revision assignments in Σ, all probabilistic dependencies of $P(V)$ are to be invariant. This requirement ensures that modifications are made according to the principle of minimal change.

It can be proven (see, e.g. [19]) that in case of existence, the solution of the revision problem $(P(V), \Sigma)$ is uniquely defined. This solution can be determined using iterative proportional fitting [2, 21]. Starting with the initial probability distribution, this process adapts the initial probability distribution iteratively, one revision assignment at the time, and converges to a limit distribution that solves the revision problem, given there are no inconsistencies.

6 Inconsistencies

Due to complex industrial data and revision problems, inconsistencies are practically not avoidable, see Fig. 5.

Inconsistencies in the context of revising probability distributions have been analysed in [24], and two types of inconsistencies of revision problems have been

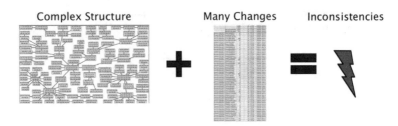

Fig. 5 Inconsistencies are practically not avoidable

distinguished, which are *inner inconsistencies and outer inconsistencies,* respectively.

Inner consistency of a revision structure Σ is given, if and only if a probability distribution exists that satisfies the revision assignments of Σ; otherwise we refer to inner inconsistencies of Σ.

In Fig. 6, a simple example is shown where the given revision assignments contradict each other and hence do not form a single probability distribution. The filled entries in the left table represent the revision assignments. In the right table consequences for the rest of the distribution are shown and one conflict is highlighted.

Given that there is a probability distribution that satisfies Σ, it is still possible that due to the zero probabilities of $P(V)$ the revision problem $(P(V), \Sigma)$ is not solvable. This is the case when one of those zero values would need to be modified in order to satisfy the revision assignments. Such a modification of the interaction structure of $P(V)$ is not permitted during a revision operation. Therefore, a second type of inconsistency is defined as follows:

Given that Σ has the property of inner consistency, the revision problem $(P(V), \Sigma)$ shows the property of outer inconsistency, if and only if there is no solution to this revision problem.

Figure 7 illustrates an outer inconsistency. In the left table again the numbers represent revision assignments. This time there are additional circles representing zero values that cannot be changed during the revision operation. As before, the right table shows consequences for the remaining table entries as well as an inconsistency.

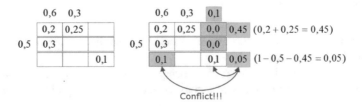

Fig. 6 Inner inconsistency

Fig. 7 Outer inconsistency

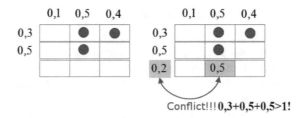

Conflict!!! 0,3+0,5+0,5>1!

7 Handling Inconsistencies

Our most recent research is geared towards the systematic handling of inconsistencies. We proposed a framework in [23]. It is the result of structuring, systematising and complementing the mostly manual efforts done in the EPL system. The following six important components have been identified:

- Detection
- Automatic Elimination
- Manual Elimination
- Analysis
- Explanation/Presentation
- Prevention.

In the following sections the purpose for each of those components is explained and requirements are listed if applicable.

7.1 Detection

As described in the previous section, a revision problem that contains inconsistencies is not solvable. For that reason it is important to detect potential inconsistencies and remove them in order to ensure a revision operation provides a usable solution.

The main task of this component is to determine whether a given revision problem $(P(V), \Sigma)$ contains inconsistencies. In principle there are two possible approaches to achieve that goal. The first approach is to take the given revision problem and run different consistency checks. The second idea is based on the definition of revision inconsistencies. By definition, the revision operation converges towards one limit distribution if and only if no inconsistency is present [25]. Consequently, if the revision operation is performed and does not converge towards a single unique distribution, the revision problem $(P(V), \Sigma)$ is not solvable, and hence contains inconsistencies. However, in industrial applications it is not feasible to wait indefinitely in order to decide whether a revision problem converges towards a single solution. Therefore, this component is concerned with techniques to test a revision problem for inconsistencies or detect inconsistencies during the revision operation in a reasonable amount of time.

7.2 Automatic Elimination

Inconsistencies are practically unavoidable in real world problems and usually a solution is expected, even if the revision problem is not actually solvable. Therefore, a mechanism is needed to eliminate inconsistencies automatically during the revision operation.

Such a mechanism should ensure that the revision operation always returns a solution. Since, the original revision problem is not solvable if inconsistencies are present, a suitable method should ensure it creates a modified solvable revision problem that is as close as possible to the original unsolvable problem. Some reasonable requirements for such a method are:

- After the resolution the (modified) revision structure Σ_{rev} has the property of inner consistency.
- Σ_{rev} is also consistent to the underlying interaction structure so that outer consistency is given.
- The revision structure after adaptation Σ_{rev} should be as close to the originally specified revision structure Σ as possible in an information theoretic sense.
- There should be no unnecessary modifications to the revision problem.
- The process should be deterministic and the modifications it applies comprehensible by domain experts.

7.3 Manual Elimination

In general, manual elimination of inconsistencies using the knowledge of a domain expert is preferable to automatic resolution using some sort of heuristic method. By using their domain knowledge, experts are usually able to resolve inconsistencies correctly with respect to that specific domain. Since it is rather difficult to model and apply the domain knowledge of experts, our focus so far has been on providing information to domain experts so they can make an educated decision. Furthermore, inconsistencies are often not obvious to domain experts. For that reason the following components address aspects that are important to automatically create descriptions for given inconsistencies.

7.4 Analysis

In order to create useful descriptions, it is necessary to capture the core of inconsistency. We identified three important aspects in [26] that should be addressed in an explanation, namely:

- a set of revision assignments (ideally minimal) that are inconsistent
- the relevant parts of the interaction structure
- the relation between the revision assignments and the interaction structure.

The idea of searching for minimal inconsistent sets of facts has been discussed in literature for different types of knowledge bases. For knowledge that is described with logical expressions algorithms can be found in [27, 28]. An algorithm to find a minimal set of inconsistent revision assignments was presented in [23].

In the context of probability distribution(s) finding a minimal explaining set of revision assignments is often times insufficient to explain an inconsistency enough in order for domain experts to understand it. In real world problems, probability spaces usually have a high number of dimensions. In our industrial application for example, the probability distributions have between 200 and 250 dimensions. Since, humans have usually problems comprehending that many dimensions, it is necessary to identify a lower dimensional subspace that is affected by the inconsistency.

Further, it is desirable to properly explain how exactly the lower dimensional subspace is affected by the inconsistent revision assignments.

Consequently, the analysis component is concerned with methods for the analysis of inconsistent revision problems in order to gather information related to the three mentioned aspects. The collected information can then be used to create useful explanations.

7.5   Explanation

Since domain experts are usually not trained in understanding probabilistic models, presenting the raw analysis results does not necessarily help them to properly understand a given inconsistency. In our application usually data analysis experts are manually post processing the results and create a comprehensible explanation from them. Further, practical experience with human experts has shown that it is beneficial to reduce the provided information to a minimum that still explains the problem. Human experts are better able to understand a problem if the explanation is less complicated. Since the manual creation of explanations from raw analysis results is a time consuming and costly effort, this component is concerned with processing and combining of the results obtained by the previous analysis step automatically in order to create explanations that are useful to human domain expert. Additionally, this component addresses the presentation of an explanation as well as methods to reduce the complexity of those explanations.

7.6   Prevention

The analysis and elimination of inconsistencies is a time consuming and expensive task. Further, the higher the number of inconsistencies in a given revision problem, the more time consuming the revision operation becomes. In addition to the higher time consumption also the quality of the results usually declines.

Fig. 8 Framework for
handling revision
inconsistencies

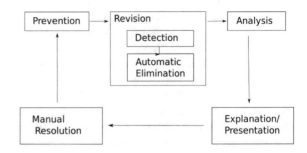

For that reason this component is concerned with methods and ideas that can help to keep the actually number of inconsistencies low during the formulation of revision problems.

7.7 Organising the Six Components

Those six components form an interdependent cycle. Figure 8 shows how the components interact, as well as the work flow that combines them properly.

The best place to start in this cycle is to deploy methods that prevent inconsistencies during the formulation of the revision problem. Since, inconsistencies are practically not avoidable even given some methods of prevention, detection and automatic elimination methods should be included in the revision process. After the revision operation has provided a result, domain experts are usually interested in the cause of specific modifications of the revision problem and hence need explanations. Therefore, the next step after the revision operation, is the analysis of a given revision inconsistency, followed by the creation of an explanation. After the domain expert is presented with the explanation of the inconsistency (s)he can then resolve the inconsistency in the correct way manually, if they are not satisfied with the automatic resolution. From here the circle restarts with the reformulation of the revision problem.

While displayed separately in the diagram some of tasks corresponding to each component may be performed together in an application. Especially the detection and automatic resolution are often performed in conjunction, since they ensure that the revision operation produces a usable result.

8 Conclusion

Problems that have to be solved in real world industrial applications are very complex. Many different types of knowledge often have to be combined in order to achieve reliable and usable results. Decomposable models whether relational, possibilistic, or Markovian or a combination of those are able to handle hundreds of variables, high-dimensional dependencies, and meta information as well as uncertainty in highly dynamic environments.

Using a powerful framework in combination with global and local learning methods, belief revision, constraint propagation, information visualization, inconsistency management, etc. we are able to address the needs of industrial requirements.

Even though complex problems can be handled using the described methods there are still many open research topics especially concerning information visualization and the proper management of inconsistencies.

References

1. R. Kruse et al., *Computational Intelligence, A Methodological Introduction*, 2nd edn. (Springer, London, 2016)
2. J. Whittaker, *Graphical Models in Applied Multivariate Statistics* (Wiley, Chichester, 1990)
3. J. Pearl, *Probabilistic Reasoning in Intelligent Systems: Networks of Plausible Inference* (Morgan Kaufmann, 1991)
4. D. Koller, N. Friedman, *Probabilistic Graphical Models: Principles and Techniques* (MIT Press, Cambridge, Mass, 2009)
5. C. Borgelt, M. Steinbrecher, R. Kruse, *Graphical Models: Representations for Learning, Reasoning and Data Mining*, 2nd edn. (Wiley, Wiley Series in Computational Statistics, 2009)
6. J. Gebhardt, R. Kruse, Background to and perspectives of possibilistic graphical models, in *Applications of Uncertainty Formalisms*, ed. by A. Hunter, S. Parsons (Springer, Berlin, Heidelberg, 1998), pp. 397–415
7. C. Borgelt, J. Gebhardt, R. Kruse, Possibilistic graphical models, in *Computational Intelligence in Data Mining, ISSEK'98 (Udine, Italy)*, ed. by G.D. Riccia, R. Kruse, H.-J. Lenz (Springer, Wien, 2000), pp. 51–68
8. J. Gebhardt, R. Kruse, Learning possibilistic networks from data, in *Learning from Data, Artificial Intelligence and Statistics 5*, ed. by D. Fisher, H. Lenz. Lecture Notes in Statistics, vol. 112 (Springer, New York, 1996), pp. 143–153
9. B. Amor, Possibilistic graphical models: from reasoning to decision making, in *Fuzzy Logic and Applications: 10th International Workshop, WILF, 2013, Genoa, Italy, November 19–22, 2013. Proceedings*, ed. by F. Masulli, G. Pasi, R. Yager (Springer International Publishing, Cham, 2013), pp. 86–99
10. L.E. Sucar, *Relational Probabilistic Graphical Model*. (Springer, London, 2015), pp. 219–235
11. L. Getoor, B. Taskar (eds.), *Introduction to Statistical Relational Learning* (MIT Press, Cambridge, MA, 2007)
12. P. Gardenfors, *Knowledge in Flux: Modeling the Dynamics of Epistemic States* (MIT Press, Cambridge, Mass, 1988)
13. D. Gabbay, P. Smets (eds.), *Handbook of Defeasable Reasoning and Uncertainty Management Systems: Belief Change*, vol. 3 (Kluwer Academic Press, Dordrecht, Netherlands, 1998)
14. C.E. Alchourron, P.G. Ardenfors, D. Makinson, On the logic of theory change: partial meet contraction and revision functions. J. Symbol. Logic **50**(02), 510–530 (1985)
15. A. Darwiche, On the logic of iterated belief revision. Artif. Intell. **89**(1–2), 1–29 (1997)

16. H. Katsuno, A.O. Mendelzon, Propositional knowledge base revision and minimal change. Artif. Intell. **52**(3), 263–294 (1991)
17. D. Gabbay, Controlled revision—an algorithmic approach for belief revision. J. Logic Comput. **13**(1), 3–22 (2003)
18. B. Nebel, Base revision operations and schemes: representation, semantics and complexity, in *Proceedings of the Eleventh European Conference on Artificial Intelligence (ECAI94)* (Wiley, Amsterdam, The Netherlands, 1994), pp. 341–345
19. J. Gebhardt, H. Detmer, A.L. Madsen, Predicting parts demand in the automotive industry—an application of probabilistic graphical models, in *Proceedings 19th International Joint Conference on Uncertainty in Artificial Intelligence* (Acapulco, 2003)
20. J. Gebhardt, A. Klose, H. Detmer, F. Ruegheimer, R. Kruse, Graphical models for industrial planning on complex domains, in *Decision Theory and Multi-Agent Planning, ser. CISM Courses and Lectures*, ed. by G. Della, D. Riccia, Dubois, R. Kruse, H.-J. Lenz, vol. 482 (Springer, 2006), pp. 131–143
21. Y.W. Teh, M. Welling, On improving the efficiency of the iterative proportional fitting procedure, in *Proceedings of the 9th international Workshop on Artificial Intelligence and Statistics in Key West, Florida*, ed. by C.M. Bishop, B.J. Frey (2003), pp. 1–8
22. F. Pukelsheim, B. Simeone, On the iterative proportional fitting procedure: structure of accumulation points and L1-error analysis, in *Structure*, vol. 05 (2009), p. 28
23. F. Schmidt, J. Wendler, J. Gebhardt, R. Kruse, Handling inconsistencies in the revision of probability distributions, in *Hybrid Artificial Intelligent Systems: 8th International Conference, HAIS 2013, Salamanca, Spain, September 11–13, 2013. Proceedings*, ed. by J.-S. Pan, M.M. Polycarpou, M. Wozniak, L.F. de Carvalho, H. Quinti´an, E. Corchado (Springer, Berlin, Heidelberg, 2013), pp. 598–607
24. J. Gebhardt, A. Klose, J. Wendler, Markov network revision: on the handling of inconsistencies, in *Computational Intelligence in Intelligent Data Analysis, ser. Studies in Computational Intelligence*, ed. by C. Moewes, A. Nurnberger, vol. 445 (Springer, Berlin, Heidelberg, 2012), pp. 153–165
25. J. Gebhardt, C. Borgelt, R. Kruse, H. Detmer, Knowledge revision in Markov networks. Math. Soft Comput. **11**(2–3), 93–107 (2004)
26. F. Schmidt, J. Gebhardt, R. Kruse, Handling revision inconsistencies: creating useful explanations, in *HICSS-48, Proceedings, 5–8 January 2015, Koloa, Kauai, HI, USA*, ed. by T.X. Bui, R.H. Sprague (IEEE Computer Society, 2015), pp. 1–8
27. A. Hunter, S. Konieczny et al., Measuring inconsistency through minimal inconsistent sets, in *Proceedings of the Eleventh International Conference on Principles of Knowledge Representation and Reasoning* (Sydney, Australia, 2008), pp. 358–366
28. K. Mu, W. Liu, Z. Jin, A general framework for measuring inconsistency through minimal inconsistent sets. Knowl. Inf. Syst. **27**(1), 85–114 (2011)

Optimal Academic Ranking of Students in a Fuzzy Environment: A Case Study

Satish S. Salunkhe, Yashwant Joshi and Ashok Deshpande

Abstract Traditionally, academic ranking of students' performance is based on test score which can be interpreted in linguistic terms such as 'very good', 'good', 'poor', 'very poor' with varying degree of certainty attached to each description. There could be several students in a school having 'very poor' performance with varying degree of certainty. The authorities would certainly like to improve students' academic performance based on their ranking. The case study relates to the combination of Zadeh-Deshpande formalism with Bellman-Zadeh method to arrive at an optimal ranking of especially 'very poor' students based on well-defined performance shaping factors.

Keywords Students ranking · Academic performance · Goal
Constraints · Decision · Zadeh-Deshpande formalism · Bellman-Zadeh fuzzy
decision-making model

1 Introduction

Examination process tries to ensure students' abilities in any area of the academic program. Test scores, Grade point average (GPA) are widely used indicators of academic performance in educational systems to rank students [1]. Test Scores often reflect limited measures of some aspect of student proficiency. Many factors could act as barriers to students attaining and maintaining a high GPA that reflects their overall

S. S. Salunkhe (✉)
Terna Engineering College, Navi Mumbai, Maharashtra, India
e-mail: satishssalunkhe@gmail.com

Y. Joshi
SGGSIE&T, SRTM University, Nanded, Maharashtra, India
e-mail: yvjoshi@sggs.ac.in

A. Deshpande
BISC -SIG-EMS, Berkeley, CA 94720-5800, USA
e-mail: ashok_deshpande@hotmail.com

© Springer International Publishing AG, part of Springer Nature 2018   453
L. A. Zadeh et al. (eds.), *Recent Developments and the New Direction
in Soft-Computing Foundations and Applications*, Studies in Fuzziness
and Soft Computing 361, https://doi.org/10.1007/978-3-319-75408-6_35

academic performance. Constraints (or factors) which can affect students' academic performance have been investigated in many studies in recent years [2–13].

The constraints analysed in academic performance of students relates to, motivation, personality, and psychosocial factors [14]. Students ranking using only achievement tests is, in our view is inadequate to assess the real performance as some other constraints as additional information might affect students' performance. These are *problem-solving skill, retention rate, the level of motivation, intellectual curiosity, test anxiety,* etc. Ranking of students' academic achievement with this additional information will be more meaningful for policy makers and educators to measure the real performance to distinguish differences among students [15]. Therefore, ranking students using cognitive as well as affective factors to define performance measure may be a realistic approach. Decision-making in a fuzzy environment [16] is one of the simplest kinds of an algorithm for optimal ranking strategy.

The objective of the study is to develop formalism for optimal ranking of students based on factors affecting academic performance in a Fuzzy Environment. Traditionally, only test marks are considered to rank the performance of a student. Therefore, in *Zadeh-Deshpande* (ZD) formalism only test-marks considered [17]. The academic performance of students as linguistic descriptions with varying *degree of certainty* (DC) attached to the report obtained from ZD formalism is the input to *Bellman-Zadeh* (BZ) formalism as a *goal* (G).

The paper is organized as follows: Sect. 2 relates to mathematical preliminaries of Bellman-Zadeh method while Sect. 3 presents the proposed methodology. The case study for ranking students considering the factors affecting students' academic performance using Bellman-Zadeh formalism is covered in Sect. 4. Results and Discussion of the case study are discussed in Sect. 5. A concluding remark is an integral part of Sect. 6.

2   Mathematical Preliminaries

Zadeh-Deshpande (ZD) formalism is shown in Fig. 1.

Brief Commentary on Bellman-Zadeh Approach

Bellman and Zadeh [16] proposed a fuzzy decision-making model in which decision are obtained by aggregate operations on goals and constraints which are expressed as fuzzy sets. The principal components of a decision process in this model are

(a) a set A of possible actions;

(b) a set of goals G_i $(i \in N_n)$, each of which is expressed concerning a fuzzy set defined on A;

(c) a set of constraints C_j $(j \in N_m)$, each of which is also represented by a fuzzy set defined on A.

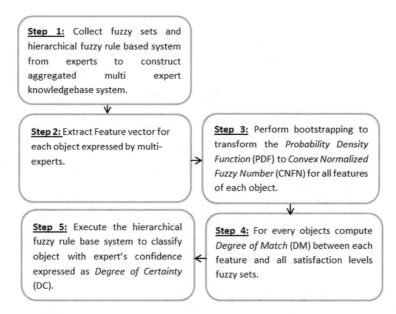

Step 1: Collect fuzzy sets and hierarchical fuzzy rule based system from experts to construct aggregated multi expert knowledgebase system.

Step 2: Extract Feature vector for each object expressed by multi-experts.

Step 3: Perform bootstrapping to transform the *Probability Density Function* (PDF) to *Convex Normalized Fuzzy Number* (CNFN) for all features of each object.

Step 5: Execute the hierarchical fuzzy rule base system to classify object with expert's confidence expressed as *Degree of Certainty* (DC).

Step 4: For every objects compute *Degree of Match* (DM) between each feature and all satisfaction levels fuzzy sets.

Fig. 1 Zadeh-Deshpande (ZD) formalism to classify objects using experts knowledgebase

Definition Assume that we are given a fuzzy goal G and fuzzy constraints $C_1, \ldots, C_n$ in a space of alternatives X. Then aggregation on G and C forms a decision D which is a fuzzy set resulting from the intersection of G and C. In symbols,

$$D = G \cap C \tag{1}$$

and correspondingly $\mu_D = \mu_G \wedge \mu_C$.

The relation between G, C and D is depicted in Fig. 2. Normalization is a major step which provides a common denominator for fuzzy goals and fuzzy constraints thereby making it possible to treat them alike. It is done as membership grade of individual constraint divided by sum of the same constraint across all alternatives or sites. Same is done for the fuzzy goals.

This concept explains why it is perfectly justified to regard fuzzy concepts of goals and constraints rather than performance function as one of the major component for decision making in a fuzzy environment. Normalization makes it possible to treat the fuzzy 'goals' and 'constraints' identically in the formulation of a decision.

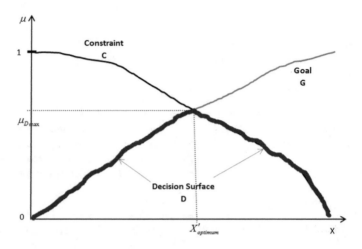

Fig. 2 Fuzzy decision model of Bellman-Zadeh formalism

3 The Proposed Study Methodology

The stepwise brief description of the activities is described below:

Step 1 Information of all constraints C_i are collected from academic records. Output of students test performance (Goal G) with expert's degree of certainty (DC) are obtained from ZD formalism for each student S_k, where $1 \le i \le m$, and $1 \le k \le n$.

Step 2 For each cluster of students P_j, normalize and transform goal G and all constraints C_i to either monotonic increasing or decreasing fuzzy sets depending on maximization or minimization objective function to obtain membership values of G and C_i for every student S_k.

Step 3 For each student S_{kj} of the cluster P_j, get the resultant decision as the intersection of the given goals $<G_{1j}, G_{2j}, \ldots, G_{tj}>$ and the given constraints $<C_{1j}, C_{2j}, \ldots, C_{mj}>$.

$$\mu_{D_{kj}} = \mu_{G_{1j}} \wedge \cdots \wedge \mu_{G_{tj}} \wedge \mu_{C_{1j}} \wedge \cdots \wedge \mu_{C_{mj}} \tag{2}$$

$$\mu_{D_{kj}} = \min\left(\mu_{G_{1j}}, \ldots, \mu_{G_{tj}}, \mu_{C_{1j}}, \cdots \wedge \mu_{C_{mj}}\right) \tag{3}$$

Step 4 Compute the optimal ranking of students (highest to lowest) in each cluster P_j for decision $\mu_{D_{kj}}$ obtained in step 3 in descending order. Select the student with maximum decision value as highest rank.

$$Highest\_Rank = \max\left(\mu_{D_{1j}}, \mu_{D_{2j}}, \ldots, \mu_{D_{kj}}\right) \tag{4}$$

Repeat this process to obtain next highest rank student by eliminating ranked students from the cluster. Repeat step 1 to 4 for all groups of students.

Step 5 Results obtained from Step 4 using optimal ranking strategy can be utilized by policy makers to take necessary corrective actions and decisions.

4 Case Study

The case study relates to optimal ranking of students' using test ranking results obtained from Zadeh-Deshpande (ZD) formalism [17] by considering all factors influencing academic performance. The examination answer script samples in subject 'Marathi' language were obtained from 237 secondary school students from three distinct institutions in Mumbai, India during the academic year 2013–2014. Each student wrote a total of 10–12 page solution for 12 subjective questions. Answers were evaluated using ten point rubric from the Secondary School (SSC) Board, Maharashtra. Twenty subject matter experts (teachers) from different schools were identified for the answer scripts evaluation. Total ten factors (i.e. constraints) influencing students' academic performance are measured for every student with evidence and survey carried out periodically throughout the school course by experts, while data collection for the test performance was conducted only on a sole day. The survey instrument consisted of a single page, back-to-back, with 30 items and questions to obtain descriptive data. Students test performance was expressed in a linguistic term like *'Very Poor'*, *'Poor'*, *'Average'*, *'Good'*, *'Very Good'* associated with evaluators *degree of certainty* (DC) using ZD formalism.

The factors (i.e. constraints) employed in Bellman-Zadeh method are *Attendance (C1), Previous Test marks (C2), Discipline (C3), Problem Solving Skills (C4), Motivation (C5), Retention (C6), Anxiety (C7), Number of punishments and scolding received (C8), Participation in extracurricular activities (C9),* and *Accuracy (C10).* These factors are normalized and transformed to either monotonic increasing or decreasing fuzzy sets. *Anxiety (C7)* and *Number of punishments and scolding received (C8)* have negative influence while rest of the factors have the positive impact on students' academic achievement. *Attendance, Previous Test marks, Number of punishments and scolding received, Participation in extracurricular activities, Accuracy* were recorded in quantitative form periodically in an academic term. *Discipline, Problem Solving Skills, Motivation, Retention, Anxiety* were measured using a 5-point Likert scale where 1—strongly agree, 2—agree, 3—neutral, 4—disagree, and 5—strongly disagree. The *degree of certainty* (DC) obtained from ZD formalism is input for BZ formalism as a goal (G).

5   Results and Discussion

Table 1 shows the students having *'very poor'* grades classified using ZD formalism and the factor affecting their scholarly performance. In BZ method, the Goal is the *'degree of certainty'* (DC) while constraints are *'C1'* to *'C10'*. Furthermore, Goal and Constraints are normalized into fuzzy sets using the standard procedure of normalization. The proposed methodology is applied, and student's optimal ranking is shown in Table 2.

For example student id-129 have an aggregation of goal and constraints to form fuzzy decision value 0.33. $\mu_{D_{129}} = \min(0.85, 0.74, 0.91, 0.62, 0.33, 0.8, 0.33, 0.5, 1, 0.78, 0.6, 1)$ $\mu_{D_{129}} = 0.33$ Student id-129 with maximum decision value is ranked top among the *'very poor'* category students.

Ranking of student id-87 with degree of certainty (DC) 0.565 with Membership Function (Goal) = 0, remains at the last position even with BZ method. The students with boundary marks have *'very poor'* academic performance as the DC is 0.565. In fact, in partial belief, the student is more likely in *'poor'* category than *'very poor'*. It is the reason why all other constraints, though with high membership value, does not alter his ranking and student id-87 remain last in *'very poor'* cluster. We can say that 0.565 is the cut-off value of membership function (MF) to decide the category of a student. He is belonging to *'very poor'* category, marginally and therefore ranks last. In summary, in such a case, it is not always the other constraints are significant—even if their membership value is high, but the test marks governs the category. If you have all other constraints having high membership value and do not score marks required to be in that category, then your large membership function-valued constraints will not take you to the higher rank. The purpose of the examination is to score high marks and not always score high MF in constraints. Table 3 is the outcome of the study. Ranking of *'very poor'* category students based on a single criterion of *'test score'* is computed using Zadeh-Deshpande formalism. Considering relevant *constraints* and using Bellman-Zadeh method, an optimal ranking of students has been worked out which might serve as a guideline to initiate suitable corrective actions to improve the performance of *'very poor'* category students. For example student id-93 with *'very poor'* category ranked 7th using ZD formalism while is rank 1st in BZ method and is considered as Optimal ranked. The management is expected to initiate corrective action to improve the performance of student id-87 more rigorously. The software implementation of this approach has been done in MATLAB R2008a and Microsoft Access 2010.

Table 1 Degree of certainty from ZD and parameters affecting students academic performance in 'Very Poor' category

Stud_ID	Category	Confidence	DC	C1	C2	C3	C4	C5	C6	C7	C8	C9	C10
93	Very poor	Very high	0.9539	98	9.536	8	0	3	2	6	9	10	6
203	Very poor	Very high	0.9528	100	23.38	9	0	5	1	7	5	50	7
130	Very poor	Very high	0.9122	83	23.5	7	5	3	0	9	6	40	8
23	Very poor	High	0.8872	80	27.07	6	8	2	2	10	4	20	3
137	Very poor	High	0.8676	93	11.22	5	10	4	1	9	3	15	5
91	Very poor	High	0.8594	78	27.19	4	20	3	3	8	0	25	4
129	Very poor	High	0.853	98	25.37	3	40	3	4	5	2	30	8
133	Very poor	High	0.8487	100	20.91	1	50	4	1	3	3	10	5
85	Very poor	High	0.7994	100	37.08	5	34	2	3	5	1	35	4
237	Very poor	High	0.7895	93	36.04	6	25	1	6	7	5	40	3
84	Very poor	High	0.7583	85	16.08	7	10	2	8	6	2	45	6
134	Very poor	High	0.745	80	18.77	5	5	3	3	0	8	10	4
31	Very poor	High	0.7415	94	28.79	7	23	5	2	6	3	15	5
4	Very poor	High	0.7382	92	35.51	8	0	3	2	4	4	20	6
83	Very poor	High	0.7028	99	17.2	0	10	2	6	3	2	10	5
5	Very poor	High	0.6606	100	14.45	2	25	4	2	6	0	5	3
98	Very poor	High	0.6438	98	5.985	7	5	7	3	9	1	0	7
20	Very poor	Fair	0.5924	89	34.7	6	10	3	1	8	3	10	8
87	Very poor	Fair	0.565	88	23.59	4	25	2	4	6	4	15	5

Table 2 Optimal ranking of students using BZ technique

Stud_ID	Category	Confidance	DC	Goal	C1	C2	C3	C4	C5	C6	C7	C8	C9	C10	Decision
129	Very poor	High	0.853	0.74	0.91	0.62	0.33	0.8	0.33	0.5	1	0.78	0.6	1	0.33333
31	Very poor	High	0.7415	0.45	0.73	0.73	0.78	0.46	0.67	0.25	0	0.67	0.3	0.4	0.25
85	Very poor	High	0.7994	0.6	1	1	0.56	0.68	0.17	0.38	1	0.89	0.7	0.2	0.16667
84	Very poor	High	0.7583	0.5	0.32	0.32	0.78	0.2	0.17	1	0	0.78	0.9	0.6	0.16667
133	Very poor	High	0.8487	0.73	1	0.48	0.11	0.2	0.5	0.13	1	0.67	0.2	0.4	0.11111
137	Very poor	High	0.8676	0.78	0.68	0.17	0.56	0.2	0.5	0.13	0	0.67	0.3	0.4	0.1
134	Very poor	High	0.745	0.46	0.09	0.41	0.56	0.1	0.33	0.38	1	0.11	0.2	0.2	0.09091
20	Very poor	Fair	0.5924	0.07	0.5	0.92	0.67	0.2	0.33	0.13	0	0.67	0.2	1	0.0704
93	Very poor	Very high	0.9539	1	0.91	0.11	0.89	0	0.33	0.25	0	0	0.2	0.6	0
203	Very poor	Very high	0.9528	1	1	0.56	1	0	0.67	0.13	0	0.44	1	0.8	0
130	Very poor	Very high	0.9122	0.89	0.23	0.56	0.78	0.1	0.33	0	0	0.33	0.8	1	0
23	Very poor	High	0.8872	0.83	0.09	0.68	0.67	0.16	0.17	0.25	0	0.56	0.4	0	0
91	Very poor	High	0.8594	0.76	0	0.68	0.44	0.4	0.33	0.38	0	1	0.5	0.2	0
237	Very poor	High	0.7895	0.58	0.68	0.97	0.67	0.5	0	0.75	0	0.44	0.8	0	0
4	Very Poor	High	0.7382	0.45	0.64	0.95	0.89	0	0.33	0.25	1	0.56	0.4	0.6	0
83	Very Poor	High	0.7028	0.35	0.95	0.36	0	0.2	0.17	0.75	1	0.78	0.2	0.4	0
5	Very Poor	High	0.6606	0.25	1	0.27	0.22	0.5	0.5	0.25	0	1	0.1	0	0
98	Very Poor	High	0.6438	0.2	0.91	0	0.78	0.1	1	0.38	0	0.89	0	0.8	0
87	Very Poor	Fair	0.565	0	0.45	0.57	0.44	0.5	0.17	0.5	0	0.56	0.3	0.4	0

Table 3 Optimal ranking of students using BZ technique

Stud_ID	Ranking of students using ZD technique	Optimal ranking of students using BZ technique
93	7	1
203	13	2
130	9	3
23	11	4
137	8	5
91	5	6
129	12	7
133	18	8
85	1	9
237	2	10
84	3	11
134	4	12
31	6	13
4	10	14
83	14	15
5	15	16
98	16	17
20	17	18
87	19	19

6 Concluding Remarks

Multi-constraint optimization based on fuzzy membership is different than traditional optimization techniques. The authors have successfully demonstrated in deciding the optimal ranking of students' performance based on relevant performance shaping factors. Much more needs to be done.

References

1. D. Koretz, Limitations on the use of achievement tests. J. Hum. Res. 1–42, July 2000
2. L. Cheung, A. Kan, Evaluation of factors related to student performance in a distance-learning business communication course. J. Educ. Bus. **77**(5), 257–263 (2002)
3. T.R. Ford, Social factors affecting academic performance: further evidence, in *The School Review*, vol. 65, no. 4 (The University of Chicago Press, 1957), pp. 415–422
4. H.G. Gough, What determines the academic achievement of high school students? J. Educ. Res. **XLVI**, 321–331 (1953)
5. F. Gull, S. Fong, Predicting success for introductory accounting students; some further Hong Kong evidence. Acc. Educ. Int. J. **2**(1), 33–42 (1993)

6. Y. Guney, Exogenous and endogenous factors impacting student performance in undergraduate accounting modules. Acc. Educ. Int. J. **18**(1), 51–73 (2009)
7. R.A. Heimann, Q.F. Schenk, Relations of social-class and sex differences to high-school achievement, in *School Review*, vol. LXII (Apr 1954), pp. 213–221
8. A.B. Hollingshead, *Elmtown's Youth* (Wiley, New York, 1949)
9. J.J. Kurtz, E.J. Swenson, Factors related to over-achievement and under-achievement in school, in *School Review*, 472–480, Nov 1951
10. J.P. Noble, W.L. Roberts, R.L. Sawyer, Student achievement, behavior, perceptions, and other factors affecting ACT scores (ACTR, 2006)
11. S. Rasul, Q. Bukhsh, A study of factors affecting students' performance in an examination at the university level, WCES-2011. Procedia Soc. Behav. Sci. 2042–2047 (2011)
12. A. Raychaudhuri, M. Debnath, S. Sen, B.G. Majumder, Factors affecting students' academic performance: a case study in Agartala municipal council area. Bangladesh e-J. Sociol. **7**(2), 34–41 (2010)
13. R. Robinowitz, Attributes of pupils achieving beyond their level of expectancy. J. Pers. **XXIV**, 308–317 (1956)
14. J.G. Rdia, et al., Factors related to the academic performance of students in the statistics course in psychology, in *Quality and Quantity*, vol. 40 (Springer, 2006), pp. 661–674
15. S.S. Sansgiry, M. Bhosle, K. Sail, Factors that affect academic performance among pharmacy students. Am. J. Pharm. Educ. **70**(5), 1–9 (2006)
16. R.E. Bellman, L.A. Zadeh, Decision-making in a fuzzy environment. Manage. Sci. **17**(4), 141–164 (1970)
17. S.S. Salunkhe, Y. Joshi, A. Deshpande, Degree of certainty in students' academic performance evaluation using new fuzzy inference system. J. Intell. Syst. (2017) (accepted for publication)

Part XI
Applications

Beyond Traditional Applications of Fuzzy Techniques: Main Idea and Case Studies

Vladik Kreinovich, Olga Kosheleva and Thongchai Dumrongpokaphan

Abstract Fuzzy logic techniques were originally designed to translate expert knowledge—which is often formulated by using imprecise ("fuzzy") from natural language (like "small")—into precise computer-understandable models and control strategies. Such a translation is still the main use of fuzzy techniques. Lately, it turned out that fuzzy methods can help in another class of applied problems: namely, in situations when there are semi-heuristic techniques for solving the corresponding problems, i.e., techniques for which there is no convincing theoretical justification. Because of the lack of a theoretical justification, users are reluctant to use these techniques, since their previous empirical success does not guarantee that these techniques will work well on new problems. In this paper, we show that in many such situations, the desired theoretical justification can be obtained if, in addition to known (crisp) requirements on the desired solution, we also take into account requirements formulated by experts in natural-language terms. Naturally, we use fuzzy techniques to translate these imprecise requirements into precise terms.

1 Introduction

Fuzzy logic techniques (see, e.g., [8, 11, 13]) were originally designed to translate expert knowledge—which is often formulated by using imprecise ("fuzzy") from natural language (like "small")—into precise computer-understandable models and control strategies. Such a translation is still the main use of fuzzy techniques.

V. Kreinovich (✉) · O. Kosheleva
University of Texas at El Paso, 500 W. University, El Paso, TX 79968, USA
e-mail: vladik@utep.edu

O. Kosheleva
e-mail: olgak@utep.edu

T. Dumrongpokaphan
Faculty of Science, Department of Mathematics,
Chiang Mai University, Chiang Mai, Thailand
e-mail: tcd43@hotmail.com

© Springer International Publishing AG, part of Springer Nature 2018
L. A. Zadeh et al. (eds.), *Recent Developments and the New Direction in Soft-Computing Foundations and Applications*, Studies in Fuzziness and Soft Computing 361, https://doi.org/10.1007/978-3-319-75408-6_36

465

For example, we want to control a complex plant for which no good control technique is known, but for which there are experts how can control this plant reasonably well. So, we elicit rules from the experts, and then we use fuzzy techniques to translate these rules into a control strategy.

Lately, it turned out that fuzzy methods can help in another class of applied problems: namely, in situations when there are semi-heuristic techniques for solving the corresponding problems, i.e., techniques for which there is no convincing theoretical justification. Because of the lack of a theoretical justification, users are reluctant to use these techniques, since their previous empirical success does not guarantee that these techniques will work well on new problems.

Also, these techniques are usually not perfect, and without an underlying theory, it is not clear how to improve their performance. For example, linear models can be viewed as first approximation to Taylor series, so a natural next approximation is to use quadratic models. However, e.g., for ℓ^p-models, when they do not work well, it is not immediately clear what is a reasonable next approximation.

In this paper, we show that in many such situations, the desired theoretical justification can be obtained if, in addition to known (crisp) requirements on the desired solution, we also take into account requirements formulated by experts in natural-language terms. Naturally, we use fuzzy techniques to translate these imprecise requirements into precise terms. To make the resulting justification convincing, we need to make sure that this justification works not only for one specific choice of fuzzy techniques (i.e., membership function, "and"- and "or"-operations, etc.), but for all combinations of such techniques which are consistent with the corresponding practical problem.

As examples, we provide a reasonably detailed justification of:

- sparsity techniques in data and image processing—a very successful hot-topic technique whose success is often largely a mystery; and
- ℓ^p-regularization techniques in solving inverse problems—an empirically successful alternative to Tikhonov regularization appropriate for situations when the desired signal or image is not smooth.

2   Fuzzy Logic: From Traditional to New Applications

Traditional use of fuzzy logic. Expert knowledge is often formulated by using imprecise ("fuzzy") from natural language (like "small"). Fuzzy logic techniques was originally invented to translate such knowledge into precise terms. Such a translation is still the main use of fuzzy techniques.

Example A typical example is that we want to control a complex plant for which:

- no good control technique is known, but
- there are experts how can control this plant reasonably well.

So, we elicit rules from the experts. Then, we use fuzzy techniques to translate these rules into a control strategy.

Other situations in which we need help. Lately, it turned out that fuzzy techniques can help in another class of applied problems: in situations when

- there are semi-heuristic techniques for solving the corresponding problems, i.e.,
- techniques for which there is no convincing theoretical justification.

These techniques lack theoretical justification. Their previous empirical success does not guarantee that these techniques will work well on new problems. Thus, users are reluctant to use these techniques.

An additional problem of semi-heuristic techniques is that they are often not perfect. Without an underlying theory, it is not clear how to improve their performance.

For example, linear models can be viewed as first approximation to Taylor series. So, a natural next approximation is to use quadratic models. However, e.g., for ℓ^p-models (described later), when they do not work well, it is not immediately clear what is a reasonable next approximation.

What we show in this paper. We show that in many such situations, the desired theoretical justification can be obtained if:

- in addition to known (crisp) requirements on the desired solution,
- we also take into account requirements formulated by experts in natural-language terms.

Naturally, we use fuzzy techniques to translate these imprecise requirements into precise terms.

To make the resulting justification convincing, we need to make sure that this justification works not only for one specific choice of fuzzy techniques (membership function, t-norm, etc.), but for all techniques which are consistent with the practical problem.

Case studies. As examples, we provide a reasonably detailed justification:

- of sparsity techniques in data and image processing—a very successful hot-topic technique whose success is often largely a mystery; and
- of ℓ^p-regularization techniques in solving inverse problems, an empirically successful alternative to smooth approaches which is appropriate for situations when the desired signal or image is not smooth.

Comment. A detailed description of the corresponding case studies can be found in [3–6].

3   Why Sparse? Fuzzy Techniques Explain Empirical Efficiency of Sparsity-Based Data- and Image-Processing Algorithms

Sparsity is useful, but why? In many practical applications, it turned out to be efficient to assume that the signal or an image is *sparse* (see, e.g., [7]):

- when we decompose the original signal $x(t)$ (or image) into appropriate basic functions $e_i(t)$:

$$x(t) = \sum_{i=1}^{\infty} a_i \cdot e_i(t),$$

- then most of the coefficients a_i in this decomposition will be zeros.

It is often beneficial to select, among all the signals consistent with the observations, the signal for which the number of non-zero coefficients—sometimes taken with weights—is the smallest possible:

$$\#\{i \,:\, a_i \neq 0\} \to \min \text{ or } \sum_{i:a_i \neq 0} w_i \to \min.$$

At present, the empirical efficiency of sparsity-based techniques remains somewhat a mystery.

Before we perform data processing, we first need to know which inputs are relevant. In general, in data processing, we estimate the value of the desired quantity y_j based on the values of the known quantities $x_1, \ldots, x_n$ that describe the current state of the world.

In principle, all possible quantities $x_1, \ldots, x_n$ could be important for predicting some future quantities. However, for each specific quantity y_j, usually, only a few of the quantities x_i are actually useful. So, we first need to check which inputs are actually useful.

This checking is an important stage of data processing: else we waste time processing unnecessary quantities.

Analysis of the problem. We are interested in a reconstructing a signal or image $x(t) = \sum_{i=1}^{\infty} a_i \cdot e_i(t)$ based on:

- the measurement results and
- prior knowledge.

First, we find out which quantities a_i are relevant. The quantity a_i is irrelevant if it does not affect the resulting signal, i.e., if $a_i = 0$. So, first, we decide which values a_i are zeros and which are non-zeros.

Out of all such possible decisions, we need to select *the most reasonable one*. The problem is that "reasonable" is not a precise term.

Let us use fuzzy logic. The problem is that we want the most reasonable decision, but "reasonable" is not a precise term. So, to be able to solve the problem, we need to translate this imprecise description into precise terms. Let's use fuzzy techniques which were specifically designed for such translations.

In fuzzy logic, we assign, to each statement S, our degree of confidence d in S. For example, we ask experts to mark, on a scale from 0 to 10, how confident they are in S. If an expert marks the number 7, we take $d = 7/10$. There are many other ways to assign these degrees.

Thus, for each i, we can learn to what extent $a_i = 0$ or $a_i \neq 0$ are reasonable.

Need for an "and"-operation. We want to estimate, for each tuple of signs, to which extent this tuple is reasonable. There are 2^n such tuples, so for large n, it is not feasible to directly ask the expert about all these tuples.

In such situations, we need to estimate the degree to which a_1 is reasonable *and* a_2 is reasonable ... based on individual degrees to which a_i are reasonable. In other words, we need to be able to solve the following problem:

- we know the degrees of belief $a = d(A)$ and $b = d(B)$ in statements A and B, and
- we need to estimate the degree of belief in the composite statement $A \& B$, as $f_\&(a, b)$.

The "and"-estimate is not always exact: an example. It is important to emphasize that the resulting estimate cannot be exact. Let us give two examples.

In the first example, A is "coin falls heads", B is "coin falls tails". For a fair coin, degrees a and b are equal: $a = b$. Here, $A \& B$ is impossible, so our degree of belief in $A \& B$ is zero: $d(A \& B) = 0$.

Let us now consider the second example. If we take $A' = B' = A$, then $A' \& B'$ is simply equivalent to A. So we still have $a' = b' = a$ but this time $d(A' \& B') = a > 0$.

In these two examples, we have $d(A') = d(A) = a$ and $d(B') = d(B) = b$, but $d(A \& B) \neq d(A' \& B')$.

Which "and"-operation (t-norm) should we choose. The corresponding function $f_\&(a, b)$ must satisfy some reasonable properties.

For example, since $A \& B$ means the same as $B \& A$, this operation must be commutative. Since $(A \& B) \& C$ is equivalent to $A \& (B \& C)$, this operation must be associative, etc.

It is known that each such operation can be approximated, with any given accuracy, by an *Archimedean* t-norm of the type $f_\&(a, b) = f^{-1}(f(a) \cdot f(b))$, for some strictly increasing function $f(x)$; see, e.g., [10].

Thus, without losing generality, we can assume that the actual t-norm is Archimedean.

Let us use fuzzy logic. Let $d_i^= \stackrel{\text{def}}{=} d(a_i = 0)$ and $d_i^{\neq} \stackrel{\text{def}}{=} d(a_i \neq 0)$. So, for each sequence $(\varepsilon_1, \varepsilon_2, \ldots)$, where ε_i is $=$ or $\neq$, we estimate the degree that this sequence is reasonable as:

$$d(\varepsilon) = f_\&(d_1^{\varepsilon_1}, d_2^{\varepsilon_2}, \ldots).$$

Out of all sequences ε which are consistent with the measurements and with the prior knowledge, we must select the one for which this degree of belief is the largest possible.

If we have no information about the signal, then the most reasonable choice is $x(t) = 0$, i.e.,

$$a_1 = a_2 = \cdots = 0 \text{ and } \varepsilon = (=, =, \cdots).$$

Similarly, the least reasonable is the sequence in which we take all the values into account, i.e., $\varepsilon = (\neq, \cdots, \neq)$.

Thus, we arrive at the following definitions.

Definition 1

- By a *t-norm*, we mean $f_\&(a, b) = f^{-1}(f(a) \cdot f(b))$, where $f : [0, 1] \to [0, 1]$ is continuous, strictly increasing, $f(0) = 0$, and $f(1) = 1$.
- By a *sequence*, we mean a sequence $\varepsilon = (\varepsilon_1, \ldots, \varepsilon_N)$, where each symbol ε_i is equal either to $=$ or to $\neq$.
- Let $d^= = (d_1^=, \ldots, d_N^=)$ and $d^{\neq} = (d_1^{\neq}, \ldots, d_N^{\neq})$ be sequences of real numbers from the interval $[0, 1]$.
- For each sequence ε, we define its *degree of reasonableness* as

$$d(\varepsilon) \overset{\text{def}}{=} f_\&(d_1^{\varepsilon_1}, \ldots, d_N^{\varepsilon_N}).$$

- We say that the sequences $d^=$ and $d^{\neq}$ *properly describe reasonableness* if the following two conditions hold:

  – for $\varepsilon_= \overset{\text{def}}{=} (=, \cdots, =)$, $d(\varepsilon_=) > d(\varepsilon)$ for all $\varepsilon \neq \varepsilon_=$,
  – for $\varepsilon_{\neq} \overset{\text{def}}{=} (\neq, \cdots, \neq)$, $d(\varepsilon_{\neq}) < d(\varepsilon)$ for all $\varepsilon \neq \varepsilon_{\neq}$.

- For each set S of sequences, we say that a sequence $\varepsilon \in S$ is the *most reasonable* if $d(\varepsilon) = \max\limits_{\varepsilon' \in S} d(\varepsilon')$.

Now, we can formulate the main result of this section.

Proposition 1 *Let us assume that the sequences $d^=$ and $d^{\neq}$ properly describe reasonableness. Then, there exist weights $w_i > 0$ for which, for each set S, the following two conditions are equivalent:*

- *the sequence $\varepsilon \in S$ is the most reasonable,*
- *the sum $\sum\limits_{i: \varepsilon_i = \neq} w_i = \sum\limits_{i: a_i \neq 0} w_i$ is the smallest possible.*

Discussion. Thus, fuzzy-based techniques indeed naturally lead to the sparsity condition.

Proof of Proposition 1 By definition of the t-norm, we have

$$d(\varepsilon) = f_\&(d_1^{\varepsilon_1}, \ldots, d_N^{\varepsilon_N}) = f^{-1}(f(d_1^{\varepsilon_1}) \cdot \ldots \cdot f(d_N^{\varepsilon_N})).$$

So, $d(\varepsilon) = f_\&(d_1^{\varepsilon_1}, \ldots, d_N^{\varepsilon_N}) = f^{-1}(e_1^{\varepsilon_1} \cdot \ldots \cdot e_N^{\varepsilon_N})$, where we denoted $e_i^{\varepsilon_i} \stackrel{\text{def}}{=} f(d_i^{\varepsilon_i})$.

Since the function $f(x)$ is increasing, maximizing $d(\varepsilon)$ is equivalent to maximizing $e(\varepsilon) \stackrel{\text{def}}{=} f(d(\varepsilon)) = e_1^{\varepsilon_1} \cdot \ldots \cdot e_N^{\varepsilon_N}$.

We required that the sequences $d^=$ and $d^{\neq}$ properly describe reasonableness. Thus, for each i, we have $d(\varepsilon_=) > d(\varepsilon_=^{(i)})$, where

$$\varepsilon_=^{(i)} \stackrel{\text{def}}{=} (=, \cdots, =, \neq \text{ (on } i\text{-th place)}, =, \cdots, =).$$

This inequality is equivalent to $e(\varepsilon_=) > e(\varepsilon_=^{(i)})$. Since the values $e(\varepsilon)$ are simply the products, we thus conclude that $e_i^= > e_i^{\neq}$.

Maximizing $e(\varepsilon) = \prod_{i=1}^{N} e_i^{\varepsilon_i}$ is equivalent to maximizing $\dfrac{e(\varepsilon)}{c}$, for a constant $c \stackrel{\text{def}}{=}$

$\prod_{i=1}^{N} e_i^=$. The ratio $\dfrac{e(\varepsilon)}{c}$ can be reformulated as $\dfrac{e(\varepsilon)}{c} = \prod_{i:\varepsilon_i=\neq} \dfrac{e_i^{\neq}}{e_i^=}$.

Since $\ln(x)$ is an increasing function, maximizing this product is equivalent to minimizing minus logarithm of this product:

$$L(\varepsilon) \stackrel{\text{def}}{=} -\ln\left(\frac{e(\varepsilon)}{c}\right) = \sum_{i:\varepsilon_i=\neq} w_i, \text{ where } w_i \stackrel{\text{def}}{=} -\ln\left(\frac{e_i^{\neq}}{e_i^=}\right).$$

Since $e_i^= > e_i^{\neq} > 0$, we have $\dfrac{e_i^{\neq}}{e_i^=} < 1$ and thus, $w_i > 0$.

The proposition is proven.

A similar derivation can be obtained in the probabilistic case. Alternatively, reasonableness can be described by assigning a *probability* $p(\varepsilon)$ to each possible sequence ε.

Let $p_i^=$ be the probability that $a_i = 0$, and let $p_i^{\neq} = 1 - p_i^=$ be the probability that $a_i \neq 0$. We do not know the relation between the values ε_i and ε_j corresponding to different coefficients $i \neq j$. So, it makes sense to assume that the corresponding random variables ε_i and ε_j are independent, thus

$$p(\varepsilon) = \prod_{i=1}^{N} p_i^{\varepsilon_i}.$$

So, we arrive at the following definition.

Definition 2

- Let $p^= = (p_1^=, \ldots, p_N^=)$ be a sequence of real numbers from the interval $[0, 1]$, and let $p_i^{\neq} \stackrel{\text{def}}{=} 1 - p_i^=$.

- For each sequence ε, its *probability* is $p(\varepsilon) \stackrel{\text{def}}{=} \prod_{i=1}^{N} p_i^{\varepsilon_i}$.

- We say that the sequence $p^=$ *properly describes reasonableness* if the following two conditions are satisfied:

  - the sequence $\varepsilon_= \stackrel{\text{def}}{=} (=, \ldots, =)$ is more probable than all others, i.e., $p(\varepsilon_=) > p(\varepsilon)$ for all $\varepsilon \neq \varepsilon_=$,
  - the sequence $\varepsilon_{\neq} \stackrel{\text{def}}{=} (\neq, \ldots, \neq)$ is less probable than all others, i.e., $p(\varepsilon_{\neq}) < p(\varepsilon)$ for all $\varepsilon \neq \varepsilon_{\neq}$.

- For each set S of sequences, we say that a sequence $\varepsilon \in S$ *is the most probable* if $p(\varepsilon) = \max_{\varepsilon' \in S} p(\varepsilon')$.

Proposition 2 *Let us assume that the sequence $p^=$ properly describes reasonableness. Then, there exist weights $w_i > 0$ for which, for each set S, the following two conditions are equivalent to each other:*

- *the sequence $\varepsilon \in S$ is the most probable,*
- *the sum $\sum_{i: \varepsilon_i = \neq} w_i$ is the smallest possible.*

Proof of Proposition 2 The proof of this proposition is similar to the proof of Proposition 1.

Discussion. In other words, probabilistic techniques also lead to the sparsity condition.

Fuzzy approach versus probabilistic approach. The fact that the probabilistic approach leads to the same conclusion as the fuzzy approach makes us more confident that our justification of sparsity is valid.

It should be mentioned, however, that the probability-based result is based on the assumption of independence, while the fuzzy-based result can allow different types of dependence—as described by different t-norms. This is an important advantage of the fuzzy-based approach.

4 Why ℓ_p-Methods in Signal and Image Processing: A Fuzzy-Based Explanation

Need for beblurring. The second case study deals with signal and image processing.

Cameras and other image-capturing devices are getting better and better every day. However, none of them is perfect, there is always some blur, that comes from the fact that:

- while we would like to capture the intensity $I(x, y)$ at each spatial location (x, y),
- the signal $s(x, y)$ is influenced also by the intensities $I(x', y')$ at nearby locations (x', y'):

$$s(x, y) = \int w(x, y, x', y') \cdot I(x', y') \, dx' \, dy'.$$

When we take a photo of a friend, this blur is barely visible—and does not constitute a serious problem. However, when a spaceship takes a photo of a distant planet, the blur is very visible—so deblurring is needed.

In general, signal and image reconstruction are ill-posed problems. The image reconstruction problem is *ill-posed* in the sense that large changes in $I(x, y)$ can lead to very small changes in $s(x, y)$.

Indeed, the measured value $s(x, y)$ is an average intensity over some small region. Averaging eliminates high-frequency components. Thus, for

$$I^*(x, y) = I(x, y) + c \cdot \sin(\omega_x \cdot x + \omega_y \cdot y),$$

the signal is practically the same: $s^*(x, y) \approx s(x, y)$. However, the original images, for large c, may be very different.

Need for regularization. To reconstruct the image reasonably uniquely, we must impose additional conditions on the original image. This imposition is known as *regularization*.

Often, a signal or an image is smooth (differentiable). Then, a natural idea is to require that the vector $d = (d_1, d_2, \ldots)$ formed by the derivatives is close to 0:

$$\rho(d, 0) \le C \Leftrightarrow \sum_{i=1}^{n} d_i^2 \le c \overset{\text{def}}{=} C^2.$$

For continuous signals, sum turns into an integral:

$$\int (\dot{x}(t))^2 \, dt \le c \text{ or } \int \left(\left(\frac{\partial I}{\partial x} \right)^2 + \left(\frac{\partial I}{\partial y} \right)^2 \right) dx \, dy \le c.$$

Tikhonov regularization. Out of all smooth signals or images, we want to find the best fit with observation: $J \overset{\text{def}}{=} \sum_i e_i^2 \to \min$, where e_i is the difference between the actual and the reconstructed values. Thus, we need to minimize J under the constraint

$$\int (\dot{x}(t))^2 \, dt \le c \text{ and } \int \left(\left(\frac{\partial I}{\partial x} \right)^2 + \left(\frac{\partial I}{\partial y} \right)^2 \right) dx \, dy \le c.$$

The Lagrange multiplier method reduced this constraint optimization problem to the unconstrained one:

$$J + \lambda \cdot \int \left(\left(\frac{\partial I}{\partial x} \right)^2 + \left(\frac{\partial I}{\partial y} \right)^2 \right) dx \, dy \to \min_{I(x,y)}.$$

This idea is known as *Tikhonov regularization*; see, e.g., [12].

From continuous to discrete images. In practice, we only observe an image with a certain spatial resolution. So we can only reconstruct the values $I_{ij} = I(x_i, y_j)$ on a certain grid $x_i = x_0 + i \cdot \Delta x$ and $y_j = y_0 + j \cdot \Delta y$.

In this discrete case, instead of the derivatives, we have differences:

$$J + \lambda \cdot \sum_i \sum_j ((\Delta_x I_{ij})^2 + (\Delta_y I_{ij})^2) \to \min_{I_{ij}},$$

where $\Delta_x I_{ij} \overset{\text{def}}{=} I_{ij} - I_{i-1,j}$, and $\Delta_y I_{ij} \overset{\text{def}}{=} I_{ij} - I_{i,j-1}$.

Limitations of Tikhonov regularization and ℓ^p-method. Tikhonov regularization is based on the assumption that the signal or the image is smooth. In real life, images are, in general, not smooth. For example, many of them exhibit a fractal behavior; see, e.g., [9].

In such non-smooth situations, Tikhonov regularization does not work so well. To take into account non-smoothness, researchers have proposed to modify the Tikhonov regularization:

- instead of the squares of the derivatives,
- use the *p*-th powers for some $p \ne 2$:

$$J + \lambda \cdot \sum_i \sum_j (|\Delta_x I_{ij}|^p + |\Delta_y I_{ij}|^p) \to \min_{I_{ij}}.$$

This works much better than Tikhonov regularization; see, e.g., [2].

Remaining problem. A big problem is that the ℓ^p-methods are heuristic. For example, there is no convincing explanation of why necessarily we replace the square with a *p*-th power and not with some other function.

What we show. In this section, we show that a natural formalization of the corresponding intuitive ideas indeed leads to ℓ^p-methods.

To formalize the intuitive ideas behind image reconstruction, we use *fuzzy techniques*, techniques that were designed to transform imprecise intuitive ideas into exact formulas.

Let us apply fuzzy techniques. We are trying to formalize the statement that the image is continuous. This means that the differences $\Delta x_k \overset{\text{def}}{=} \Delta_x I_{ij}$ and $\Delta_y I_{ij}$ between image intensities at nearby points are small.

Let $\mu(x)$ denote the degree to which x is small, and $f_\&(a, b)$ denote the "and"-operation. Then, the degree d to which Δx_1 is small *and* Δx_2 is small, etc., is:

$$d = f_\&(\mu(\Delta x_1), \mu(\Delta x_2), \mu(\Delta x_3), \ldots).$$

We have already mentioned, in the previous section, that each "and"-operation can be approximated, for any $\varepsilon > 0$, by an *Archimedean* one:

$$f_\&(a, b) = f^{-1}(f(a)) \cdot f(b)).$$

Thus, without losing generality, we can safely assume that the actual "and"-operation is Archimedean.

Analysis of the problem. We want to select an image with the largest degree d of satisfying the above condition:

$$d = f^{-1}(f(\mu(\Delta x_1)) \cdot f(\mu(\Delta x_2)) \cdot f(\mu(\Delta x_3)) \cdot \ldots) \to \max.$$

Since the function $f(x)$ is increasing, maximizing d is equivalent to maximizing

$$f(d) = f(\mu(\Delta x_1)) \cdot f(\mu(\Delta x_2)) \cdot f(\mu(\Delta x_3)) \cdot \ldots$$

Maximizing this product is equivalent to minimizing its negative logarithm

$$L \stackrel{\text{def}}{=} -\ln(d) = \sum_k g(\Delta x_k), \quad \text{where } g(x) \stackrel{\text{def}}{=} -\ln(f(\mu(x))).$$

In these terms, selecting a membership function is equivalent to selecting the related function $g(x)$.

Which function $g(x)$ should we select: idea. The value $\Delta x_i = 0$ is absolutely small, so we should have $\mu(0) = 1$ and $g(0) = -\ln(1) = 0$.

The numerical value of a difference Δx_i depends on the choice of a measuring unit. If we choose a measuring unit which is a times smaller, then $\Delta x_i \to a \cdot \Delta x_i$. It is reasonable to request that the requirement $\sum_k g(\Delta x_k) \to \min$ not change if we change a measuring unit. For example, if $g(z_1) + g(z_2) = g(z_1') + g(z_2')$, then

$$g(a \cdot z_1) + g(a \cdot z_2) = g(a \cdot z_1') + g(a \cdot z_2').$$

Which functions $g(z)$ satisfy this property?

Definition 3 A function $g(z)$ is called *scale-invariant* if it satisfies the following two conditions:

- $g(0) = 0$ and
- for all z_1, z_2, z_1', z_2', and a, $g(z_1) + g(z_2) = g(z_1') + g(z_2')$ implies

$$g(a \cdot z_1) + g(a \cdot z_2) = g(a \cdot z_1') + g(a \cdot z_2').$$

Proposition 3 *A function $g(z)$ is scale-invariant if and only if it has the form $g(a) = c \cdot a^p$, for some c and $p > 0$.*

Discussion. Minimizing $\sum_k g(\Delta x_k)$ is equivalent to minimizing the sum $\sum_k |\Delta x_k|^p$. Minimizing the sum $\sum_k |\Delta x_k|^p$ under condition $J \leq c$ is equivalent to minimizing the expression $J + \lambda \cdot \sum_k |\Delta x_k|^p$. Thus, fuzzy techniques indeed justify the ℓ^p-method.

Proof of Proposition 3 We are looking for a function $g(x)$ for which $g(z_1) + g(z_2) = g(z_1') + g(z_2')$, then $g(a \cdot z_1) + g(a \cdot z_2) = g(a \cdot z_1') + g(a \cdot z_2')$.

Let us consider the case when $z_1' = z_1 + \Delta z$ for a small Δz, and

$$z_2' = z_2 + k \cdot \Delta z + o(\Delta z)$$

for an appropriate k. Here, $g(z_1 + \Delta z) = g(z_1) + g'(z_1) \cdot \Delta z + o(\Delta z)$, so $g'(z_1) + g'(z_2) \cdot k = 0$ and $k = -\dfrac{g'(z_1)}{g'(z_2)}$.

The condition $g(a \cdot z_1) + g(a \cdot z_2) = g(a \cdot z_1') + g(a \cdot z_2')$ similarly takes the form $g'(a \cdot z_1) + g'(a \cdot z_2) \cdot k = 0$, so

$$g'(a \cdot z_1) - g'(a \cdot z_2) \cdot \frac{g'(z_1)}{g'(z_2)} = 0.$$

Thus, $\dfrac{g'(a \cdot z_1)}{g'(z_1)} = \dfrac{g'(a \cdot z_2)}{g'(z_2)}$ for all a, z_1, and z_2.

This means that the ratio $\dfrac{g'(a \cdot z_1)}{g'(z_1)}$ does not depend on z_i: $\dfrac{g'(a \cdot z_1)}{g'(z_1)} = F(a)$ for some $F(a)$.

For $a = a_1 \cdot a_2$, we have

$$F(a) = \frac{g'(a \cdot z_1)}{g'(z_1)} = \frac{g'(a_1 \cdot a_2 \cdot z_1)}{g'(z_1)} = \frac{g'(a_1 \cdot (a_2 \cdot z_1))}{g'(a_2 \cdot z_1)} \cdot \frac{g'(a_2 \cdot z_1)}{g'(z_1)} = F(a_1) \cdot F(a_2).$$

So, $F(a_1 \cdot a_2) = F(a_1) \cdot F(a_2)$. Continuous solutions of this functional equations are well known (see, e.g., [1]), so we conclude that $F(a) = a^q$ for some real number q. For this function $F(a)$, the equality $\dfrac{g'(a \cdot z_1)}{g'(z_1)} = F(a)$ becomes $g'(a \cdot z_1) = g'(z_1) \cdot a^q$.

In particular, for $z_1 = 1$, we get $g'(a) = C \cdot a^q$, where $C \overset{\text{def}}{=} g'(1)$.

In general, we could have $q = -1$ or $q \neq -1$. For $q = -1$, we get $g(a) = C \cdot \ln(a) + \text{const}$, which contradicts to $g(0) = 0$. Thus, this case is impossible, and $q \neq -1$. Integrating, for $q \neq -1$, we get $g(a) = \dfrac{C}{q+1} \cdot a^{q+1} + \text{const}$. The condition $g(0) = 0$ implies that $\text{const} = 0$.

Thus, the proposition is proven, for $p = q + 1$.

5 How to Improve the Existing Semi-Heuristic Technique

What we do in this section. Until now, we have discussed how to justify the existing semi-heuristic techniques. However, often, these techniques are not perfect, so it is desirable to improve them. Let us describe an example of how this can be done.

Blind image deconvolution: formulation of the problem. In general, the measurement results y_k differ from the actual values x_k dues to additive noise and blurring:

$$y_k = \sum_i h_i \cdot x_{k-i} + n_k.$$

From the mathematical viewpoint, y is a *convolution* of h and x: $y = h \star x$.

Similarly, the observed image $y(i,j)$ differs from the ideal one $x(i,j)$ due to noise and blurring:

$$y(i,j) = \sum_{i'} \sum_{j'} h(i - i', j - j') \cdot x(i',j') + n(i,j).$$

It is desirable to reconstruct the original signal or image, i.e., to perform *deconvolution*.

Ideal no-noise case. In the ideal case, when noise $n(i,j)$ can be ignored, we can find $x(i,j)$ by solving a system of linear equations:

$$y(i,j) = \sum_{i'} \sum_{j'} h(i - i', j - j') \cdot x(i',j').$$

However, already for 256×256 images, the matrix h is of size $65{,}536 \times 65{,}536$, with billions entries. Direct solution of such systems is not feasible.

A more efficient idea is to use Fourier transforms, since $y = h \star x$ implies $Y(\omega) = H(\omega) \cdot X(\omega)$; hence:

- we compute $Y(\omega) = \mathscr{F}(y)$;
- we compute $X(\omega) = \dfrac{Y(\omega)}{H(\omega)}$, and
- finally, we compute $x = \mathscr{F}^{-1}(X(\omega))$.

Deconvolution in the presence of noise with known characteristics. Suppose that signal and noise are independent, and we know the power spectral densities

$$S_I(\omega) = \lim_{T \to \infty} E\left[\frac{1}{T} \cdot |X_T(\omega)|^2\right], \quad S_N(\omega) = \lim_{T \to \infty} E\left[\frac{1}{T} \cdot |N_T(\omega)|^2\right].$$

Then, we minimize the expected mean square difference

$$d \stackrel{\text{def}}{=} \lim_{T \to \infty} \frac{1}{T} \cdot E\left[\int_{-T/2}^{T/2} (\widehat{x}(t) - x(t))^2 \, dt\right].$$

Minimizing d leads to the known Wiener filter formula

$$\widehat{X}(\omega_1, \omega_2) = \frac{H^*(\omega_1, \omega_2)}{|H(\omega_1, \omega_2)|^2 + \dfrac{S_N(\omega_1, \omega_2)}{S_I(\omega_1, \omega_2)}} \cdot Y(\omega_1, \omega_2).$$

Blind image deconvolution in the presence of prior knowledge. Wiener filter techniques assume that we know the blurring function h. In practice, we often only have partial information about h. Such situations are known as *blind deconvolution*.

Sometimes, we know a joint probability distribution $p(\Omega, x, h, y)$ corresponding to some parameters Ω:

$$p(\Omega, x, h, y) = p(\Omega) \cdot p(x|\Omega) \cdot p(h|\Omega) \cdot p(y|x, h, \Omega).$$

In this case, we can find

$$\widehat{\Omega} = \arg \max_{\Omega} p(\Omega|y) = \int \int_{x,h} p(\Omega, x, h, y) \, dx \, dh \text{ and}$$

$$(\widehat{x}, \widehat{h}) = \arg \max_{x,h} p(x, h|\widehat{\Omega}, y).$$

Blind image deconvolution in the absence of prior knowledge: sparsity-based techniques. In many practical situations, we do not have prior knowledge about the blurring function h. Often, what helps is *sparsity* assumption: that in the expansion $x(t) = \sum_i a_i \cdot e_i(x)$, most a_i are zero. In this case, it makes sense to look for a solution with the smallest number of non-zero coefficients:

$$\|a\|_0 \stackrel{\text{def}}{=} \#\{i : a_i \neq 0\}.$$

The function $\|a\|_0$ is not convex and thus, difficult to optimize. It is therefore replaced by a close *convex* objective function $\|a\|_1 \overset{\text{def}}{=} \sum_i |a_i|$.

State-of-the-art technique for sparsity-based blind deconvolution. Sparsity is the main idea behind the algorithm described in [2] that minimizes

$$\frac{\beta}{2} \cdot \|y - Wa\|_2^2 + \frac{\eta}{2} \cdot \|Wa - Hx\|_2^2 + \tau \cdot \|a\|_1 + \alpha \cdot R_1(x) + \gamma \cdot R_2(h).$$

Here, $R_1(x) = \sum_{d \in D} 2^{1-o(d)} \sum_i |\Delta_i^p(x)|^p$, where $\Delta_i^p(x)$ is the difference operator, and $R_2(h) = \|Ch\|^2$, where C is the discrete Laplace operator.

The ℓ^p-sum $\sum_i |v_i(x)|^p$ is optimized as $\sum_i \dfrac{(v_i(x^{(k)}))^2}{v_i^{2-p}}$, where $v_i = v_i(x^{(k-1)})$ for x

from the previous iteration.

This method results in the best blind image deconvolution.

Need for improvement. The current technique is based on minimizing the sum $|\Delta_x I|^p + |\Delta_y I|^p$. This is a discrete analog of the term $\left|\dfrac{\partial I}{\partial x}\right|^p + \left|\dfrac{\partial I}{\partial y}\right|^p$.

For $p = 2$, this is the square of the length of the gradient vector and is, thus, rotation-invariant. However, for $p \neq 2$, the above expression is not rotation-invariant. Thus, even if it works for some image, it may not work well if we rotate this image.

To improve the quality of image deconvolution, it is thus desirable to make the method rotation-invariant. We show that this indeed improves the quality of deconvolution.

Rotation-invariant modification: description and results. We want to replace the expression $\left|\dfrac{\partial I}{\partial x}\right|^p + \left|\dfrac{\partial I}{\partial y}\right|^p$ with a rotation-invariant function of the gradient.

The only rotation-invariant characteristic of a vector a is its length $\|a\| = \sqrt{\sum_i a_i^2}$.

Thus, we replace the above expression with

$$\left(\left|\frac{\partial I}{\partial x}\right|^2 + \left|\frac{\partial I}{\partial y}\right|^2 \right)^{p/2}.$$

Its discrete analog is $((\Delta_x I)^2 + (\Delta_y I)^2)^{p/2}$.

This modification indeed leads to a statistically significant improvement in reconstruction accuracy $\|\hat{x} - x\|_2$.

Specifically, to compare the new methods with the original method from [2], we applied each of the two algorithms 30 times, and for each application, we computed the reconstruction accuracy. To make the results of the comparison more robust, for each of the algorithms, we eliminated the smallest and the largest value of this distance, and got a list of 28 values. For the original algorithm, the average of these values is 1195.21. For the new method, the average is 1191.01, which is smaller than the average distance corresponding to the original algorithm. To check whether this difference is statistically significance, we applied the t-test for two independent means. The t-test checks whether the null hypothesis—that both samples comes from the populations with same mean—can be rejected. For the two samples, computations lead to rejection with $p = 0.002$. This is much smaller than the p-values 0.01 and 0.05 normally used for rejecting the null hypothesis. So, we can conclude that the null hypothesis can be rejected, and that, therefore, the modified algorithm is indeed statistically significantly better than the original one (see [3] for details).

How can we go beyond ℓ^p-methods? While ℓ^p-methods are efficient, they are not always perfect. A reasonable idea is to try to improve the quality of signal and image reconstruction by using functions $g(z)$ more general than $g(z) = C \cdot |z|^p$. For example, instead of considering only functions from this 1-parametric family, we can consider a more general 2-parametric family of functions

$$g(z) = C \cdot |z|^p + C_1 \cdot g_1(z).$$

Which function $g_1(z)$ should we use?

In [6], we used the same ideas of scale-invariance—that are used above to justify ℓ^p-techniques—to show that the best choice is to use functions $g_1(z) = |z|^p \cdot \ln(z)$ or $g_1(z) = |z|^{p_1}$ for some p_1. The same approach also helps to decide which functions to use if we consider 3- and more-parametric families instead of 2-parametric ones [6].

Acknowledgements We are very thankful to the organizers of the 2016 World Conference on Soft Computing for their support, and to all the conference participants for their valuable suggestions. This work was also supported in part by the National Science Foundation grants HRD-0734825 and HRD-1242122 (Cyber-ShARE Center of Excellence) and DUE-0926721, by an award from Prudential Foundation, and by Chiang Mai University, Thailand.

References

1. J. Aczél, J. Dhombres, *Functional Equations in Several Variables* (Cambridge University Press, 2008)
2. B. Amizic, L. Spinoulas, R. Molina, A.K. Katsaggelos, Compressive blind image deconvolution. IEEE Trans. Image Process. **22**(10), 3994–4006 (2013)
3. F. Cervantes, B. Usevitch, V. Kreinovich, Rotation-invariance can further improve state-of-the-art blind deconvolution techniques, in *Proceedings of International IEEE Conference on Systems, Man, and Cybernetics SMC'2016* (Budapest, Hungary, 9–12 Oct 2016)
4. F. Cervantes, B. Usevitch, V. Kreinovich, Why ℓ^p-methods in signal and image processing: a fuzzy-based explanation, in *Proceedings of the 2016 Annual Conference of the North American Fuzzy Information Processing Society NAFIPS'2016* (El Paso, Texas, Oct 31–Nov 4 2016)
5. F. Cervantes, B. Usevitch, L. Valera, V. Kreinovich, Why sparse? fuzzy techniques explain empirical efficiency of sparsity-based data- and image-processing algorithms, in *Proceedings of the 2016 World Conference on Soft Computing* (Berkeley, California, 22–25 May 2016), pp. 165–169

6. F. Cervantes, B. Usevitch, L. Valera, V. Kreinovich, O. Kosheleva, Fuzzy techniques provide a theoretical explanation for the heuristic ℓ^p-regularization of signals and images, in *Proceedings of the 2016 IEEE International Conference on Fuzzy Systems FUZZ-IEEE'2016* (Vancouver, Canada, 24–29 July 2016)
7. M. Elad, *Sparse and Redundant Representations* (Springer, 2010)
8. G. Klir, B. Yuan, *Fuzzy Sets and Fuzzy Logic* (Prentice Hall, Upper Saddle River, New Jersey, 1995)
9. B. Mandelbrot, *The Fractal Geometry of Nature* (Freeman, San Francisco, California, 1983)
10. H.T. Nguyen, V. Kreinovich, P. Wojciechowski, Strict Archimedean t-Norms and t-Conorms as universal approximators. Int. J. Approx. Reason. **18**(3–4), 239–249 (1998)
11. H.T. Nguyen, E.A. Walker, *A First Course in Fuzzy Logic* (Chapman and Hall/CRC, Boca Raton, Florida, 2006)
12. A.N. Tikhonov, V.Y. Arsenin, *Solutions of Ill-Posed Problems* (V. H. Winston & Sons, Washington, DC, 1977)
13. L.A. Zadeh, Fuzzy sets. Inf. Control **8**, 338–353 (1965)

A Survey of the Applications of Fuzzy Methods in Recommender Systems

B. Sziová, A. Tormási, P. Földesi and L. T. Kóczy

Abstract In the past half century of fuzzy systems they were used to solve a wide range of complex problems, and the field of recommendation is no exception. The mathematical properties and the ability to efficiently process uncertain data enable fuzzy systems to face the common challenges in recommender systems. The main contribution of this paper is to give a comprehensive literature overview of various fuzzy based approaches to the solving of common problems and tasks in recommendation systems. As a conclusion possible new areas of research are discussed.

1 Introduction

Nowadays recommender systems [1, 2] are widely used in various aspects of life, for example, recommending books for consumers to buy, or on-demand videos which are in particular users' interests with high possibility. The Diffusion Group forecasted that in 10 years 75% of all TV viewings will be driven by recommendation-driven guides.

The recommendation task requires the understanding of the consumers' preferences regarding the items from various aspects, which is based on the users'

B. Sziová (✉) · A. Tormási · L. T. Kóczy
Department of Information Technology, Széchenyi István University, Győr, Hungary
e-mail: szi.brigitta@sze.hu

A. Tormási
e-mail: tormasi@sze.hu

L. T. Kóczy
e-mail: koczy@sze.hu; koczy@tmit.bme.hu

P. Földesi
Department of Logistics and Forwarding, Széchenyi István University, Győr, Hungary
e-mail: foldesi@sze.hu

L. T. Kóczy
Department of Telecommunications and Media Informatics, Budapest University of
Technology and Economics, Budapest, Hungary

© Springer International Publishing AG, part of Springer Nature 2018
L. A. Zadeh et al. (eds.), *Recent Developments and the New Direction
in Soft-Computing Foundations and Applications*, Studies in Fuzziness
and Soft Computing 361, https://doi.org/10.1007/978-3-319-75408-6_37

subjective judgment. The collection of explicit feedback data on customer preferences and item properties are often difficult and limited [3–5]. It follows from this that handling uncertainty in these systems is a key for success. The concept and mathematical properties of fuzzy systems [6–13] enables them to efficiently solve complex tasks in various fields, thus take over main difficulties in recommendation problem.

In this paper 37 articles on the application of fuzzy methods in recommender systems are discussed and reviewed from the period 1999 and 2015. The authors attempt covering a wide range of fuzzy techniques used in the field to support researchers by giving an overview and point out possible new research areas in the application of computational intelligence methods in recommendation related problems.

After the Introduction in Sect. 2 some of the challenges in recommender systems are outlined with common (non-fuzzy) solutions found in the literature. In Sect. 3 recommender systems, which employ fuzzy techniques are briefly discussed and reviewed from the aspect of the applied fuzzy methods. Conclusions of the literature review are made and some possible directions of research in the fuzzy based recommender systems are pointed out in Sect. 4.

2 Challenges and Common Solutions in Recommender Systems

The most essential challenge of recommender systems is to create user profiles from available data. The generation of a proper user profile has a high significance for the success of a recommendation. Another challenge on creating a detailed user profile is the fact that the preferences may change over time. Creating models for new items is also difficult due to poor quality and low availability of data. For such items mostly only Meta data are available. The main issue in practice is that they contain only basic information or even no data at all, which could be still misclassified causing low quality recommendations. Meta data is rarely well structured, which also render it more difficult to create the profiles.

User interactions are also used to extract behavioral patterns, which cannot be explained by Meta data. Collaborative filtering (CF) methods [14] are used to generate recommendations based on what kinds of items were consumed by similar users. Another type of CF methods (item-based) is comparing items and recommends the ones with the highest similarity. Main disadvantage of the recommender systems employing CF is that they are unable to recommend when neither explicit nor implicit feedbacks are present for a user or an item, which is called the cold-start problem [15, 16].

To overcome the cold-start problem, content-based filtering (CBF) methods [15–18] were introduced. In this case only basic data are available and used by the recommender systems. In such systems Meta data is essential to enhance the user and item models. A main disadvantage of CBF based solutions is the lack of diversity in the recommendations. Such methods are usually limited to recommended items, which are closely related to the known user preferences and do not discover other types of user interests.

Hybrid filtering methods [19, 20] are combining CF and CBF methods in order to exploit the advantages of both. There are several possibilities to create hybrid recommender systems, one example is to use CF and CBF methods independently in the same system and combining their results, and another example is to create a composite model, which uses the concepts of both CF and CBF methods. These systems are able to solve the cold-start problem and extract consumption patterns at a time, and also able to detect and handle misclassified Meta data.

There is a tendency that both end users and businesses are showing a growing interest in the explanation of provided recommendations. The explanations of recommendations [21–27] are integral parts of the modeling algorithms. The aim of the explanation process is to show the inner workings of the recommender algorithms in a form, which is interpretable for humans. The complexity of such methods varies and depends on the methods used by the recommender system.

3   Review of Fuzzy Methods in Recommender Systems

3.1   Fuzzy Sets, Fuzzy Numbers and Linguistic Variables

R. R. Yager in [28] discussed the possibilities of using fuzzy (sub)sets to describe objects and user preferences in recommender systems. He also detailed the mathematical background and suggested the use of linguistic expressions to express user preferences. The studied approaches were called reclusive methods, which differs from collaborative filtering, since the recommendations were not based on the preferences of similar users.

J. Carbo and J. M. Molina proposed a new agent-based CF approach in [29], which used fuzzy sets, more specifically fuzzy labels to describe users' preferences and recommendations. The proposed method improved the hit-rate and false-alarm rate compared to other systems on a data set with movies.

J. Lu proposed a personalized recommender system to supports students to choose suitable learning materials in [30]. He applied fuzzy sets, more specifically triangular fuzzy numbers to describe linguistic terms and to handle the uncertainty in criteria values.

Y. Cao, Y. Li and X. Liao proposed a fuzzy-based recommender system for situations, where users' previous preferences are neither present nor useful for the current purchase [31]. They used triangular fuzzy numbers and linguistic terms to describe the items' features and customers' needs. The system was evaluated on consumer electronics data; the presented results were promising with 83.82% precision.

M. Y. H. Al-Shamri and K. K. Bharadwaj in [32] proposed a new recommender system, which used fuzzy sets to describe items' features and users' preferences (FRS). They also proposed an extension of the system, in which genetic algorithm was used in order to determine the values of feature weights (FGRS). These values describe how much importance of features has from the users' perspective. The results of experiments showed that the time complexity of the proposed methods was lower compared to a recommender system, which used Pearson correlation coefficient to determine the distance between users (PRS). The introduced FRS and FGRS methods also outperformed and had a greater coverage compared to PRS.

G. Castellano, A. M. Fanelli, P. Plantamura and M. A. Torsello proposed a recommender system based on neuro-fuzzy strategy in [33], which used linguistic labels for fuzzy sets to describe the input parameters and the recommendation results.

L. M. de Campos, J. M. Fernández-Luna and J. F. Huete proposed a Bayesian network based CF method in [34]. They investigated three models: (1) the CIFO model in which they used crisp input and fuzzy output; (2) FICO, where the inputs are fuzzy, while the outputs are crisp values; and (3) FIFO, where both the inputs and the outputs are also fuzzy values. Fuzzy labels (described by fuzzy sets) were used to collect and describe user ratings. Their experimental results on movie data showed that it is better to use fuzzy ratings, but when a crisp rating scheme is used, then it might be better to use fuzzy definitions as output.

A. Zenebe and A. F. Norcio defined a CBF recommender method using fuzzy set theory in [35]. In the proposed system the items' features were described with fuzzy sets, while users' preferences were represented by fuzzy numbers. Their results showed that the fuzzy based method is better in terms of precision, model size and recommendation size.

M. Maatallah and H. Seridi used fuzzy linguistic variables to describe the input and output parameters in their rule based recommender system [36].

V. Ramkumar, S. Rajasekar and S. Swamynathan proposed an item scoring technique for recommender systems [37]. The proposed method uses fuzzy logic to describe the spam levels of users' review and to determine the items' score accordingly. Their experimental results showed that the new method had a significant increase in the precision factor.

L. Terán and A. Meier proposed a recommender system for eElections [38]. They used a so called fuzzy interface, which applied fuzzy sets describing linguistic values to collect user preferences. The paper also presented fuzzy profiles to describe users' and items' (in this case candidates') properties.

A. Zenebe, L. Zhou and A. F. Norcio proposed a fuzzy set-based framework for user preference discovery [39]. User preferences and item features were described with fuzzy sets in the proposed model.

J. J. Castro-Schez, R. M. D. Vallejo and L. M. López-López proposed a recommender system, which used fuzzy sets to describe products and linguistic labels to identify groups [40].

L. C. Cheng and H. A. Wang proposed a novel CF framework based on fuzzy set theory [41]. The user preferences were described by linguistic terms and represented as fuzzy numbers. The results of their experiments showed that it outperformed traditional CF methods and could handle the cold-start problem, when new user or item is presented.

Á. García-Crespo, J. L. López-Cuadrado, I. González-Carrasco, R. Colomo-Palacios and B. Ruiz-Mezcua presented a recommender system for investments in [42]. The users' behavior and items' properties were described with fuzzy labels represented by fuzzy sets.

V. Kant and K. K. Bharadwaj proposed a fuzzy CF, a fuzzy CBF and a hybrid Fuzzy-CF-CBF system [43]. The Fuzzy-CF model used fuzzy sets to describe users' demographical data and the user ratings of items; the Fuzzy-CBF used fuzzy sets to represent items. The results of their experiments (comparing the three models and a CF-CBF method) showed that the proposed hybrid Fuzzy-CF-CBF consistently outperformed the other approaches. It is important to highlight that the CF-CBF model was performing worse than the other three methods in all cases during the experiment.

J. P. Lucas, A. Laurent, M. N. Moreno and M. Teisseire proposed a hybrid CF-CBF recommender system, which used fuzzy logic to represent user and item properties in their method [44].

D. Wu, G. Zhang and J. Lu used fuzzy sets as linguistic terms to describe item parameters and user preferences in [45].

Z. Zhang et al. proposed a hybrid user- and item-based CF approach for recommender systems in [46]. The presented method used fuzzy numbers as linguistic variables to describe user preferences. In their experiment on telecom services achieved excellent performance compared to other hybrid filtering methods.

D. Anand and B. S. Mampilli used fuzzy sets to describe user preferences, item properties and linguistic terms for item tags in their profiling method to improve recommendations [47]. In their experiment fuzzy collaborative filtering approaches outperformed the traditional methods for sparse data, but required the data to be slightly dense.

L.-C. Cheng and H.-A. Wang proposed a novel CF recommender system, which applied fuzzy numbers as linguistic terms to describe user preferences [48]. The experiments showed that the recommender systems using fuzzy methods to represent preferences could achieve better results compared to the ones with crisp methods.

W. Liu and L. Gao proposed a fuzzy-based recommender system to recommend academic papers [49]. The items in the system were represented with fuzzy sets.

M. Nilashi, O. bin Ibrahim and N. Ithnin proposed a hybrid approaches for multi-criteria CF recommender system [50]. The authors used fuzzy sets as linguistic variables to describe user preferences.

L. H. Son proposed a hybrid user-based collaborative filtering model in [51]. The model uses fuzzy linguistic labels to describe user properties and preferences.

G. Posfai, G. Magyar and L. T. Kóczy proposed novel information diffusion based social recommender system, which used fuzzy sets to describe user preferences in [52].

3.2 Fuzzy Similarities and Fuzzy Distances

R. R. Yager also discussed the methodology to determine similarities between objects described with fuzzy logic in [28].

J. Carbo and J. M. Molina used similarities to determine the level of success/failure of last prediction in [29].

J. Lu recommended to use fuzzy distance and fuzzy matching rules to measure the similarity between student requirements and learning materials [30].

Y. Cao, Y. Li and X. Liao used Euclidean fuzzy near compactness to measure the similarities between fuzzy numbers describing users' needs and items' features [31].

M. Y. H. Al-Shamri and K. K. Bharadwaj introduced a fuzzy distance function to match different users with many features used by the proposed FRS and FGRS [32].

L. M. de Campos, J. M. Fernández-Luna and J. F. Huete used a geometric distance model [34] to express the similarity of fuzzy sets, where smaller distance represents greater similarity between them. They achieved the best accuracy when they used fuzzy similarity measure to compare the a posteriori probability values in the collaborative node with the set of vague ratings.

A. Zenebe and A. F. Norcio used and compared various (crisp set-theoretic, fuzzy set-theoretic, cosine, proximity-based and correlation-like) similarity measures to determine the similarity between users and between items [35]. Their results showed that the different similarity measures have significant and different impact on the accuracy of recommendations.

A. Zenebe, L. Zhou and A. F. Norcio used fuzzy theoretic cosine measure was used to determine the similarity between users' preferences and items' descriptions [39]. The proposed method showed superior performance.

L. C. Cheng and H. A. Wang used Euclidean fuzzy near compactness to determine the distance/similarity between fuzzy numbers in [41].

V. Kant and K. K. Bharadwaj used local fuzzy distance (LFD) to calculate the similarity between users for the Fuzzy-CF system and the similarity between the items for the Fuzzy-CBF method [43].

Fuzzy tree similarity was used by D. Wu, G. Zhang and J. Lu to determine the similarity between user profiles to recommend an item in [45].

Z. Zhang et al. used Pearson correlation as similarity measure to calculate the similarity between two items and between two users [46].

L.-C. Cheng and H.-A. Wang used Euclidean fuzzy near compactness in their system to measure the similarity between the user preferences [48].

M. Nilashi, O. bin Ibrahim and N. Ithnin compared the proposed model's performance with Pearson correlation, fuzzy-based distance similarity, average similarity and Euclidean distance to determine the similarity between users' preferences [50].

L. H. Son used Pearson coefficient to measure similarity between users in [51]. The proposed method outperformed other methods in accuracy, however it had a greater computational time, but it was still acceptable.

3.3 Fuzzy Relations, Fuzzy Rules and Inference

The method proposed by G. Castellano, A. M. Fanelli, P. Plantamura and M. A. Torsello used fuzzy rules to describe the relation between users and items [33]. They used a neural network in order to identify the structure and the parameters of the fuzzy rules used in the system. The proposed neuro-fuzzy method was tested on real and synthetic Web usage data to recommend URLs to users based on their activities in sessions.

M. Maatallah and H. Seridi proposed a recommendation technique based on fuzzy logic that combines a collaborative filtering and taxonomic based filtering in [36], which used Mamdani method to evaluate the rule base and determine the recommendations.

J. J. Castro-Schez, R. M. D. Vallejo and L. M. López-López used fuzzy association rules to model the relationship between the described variables [40]. The rule base was determined from training data.

Á. García-Crespo, J. L. López-Cuadrado, I. González-Carrasco, R. Colomo-Palacios and B. Ruiz-Mezcua in their system used Mamdani-type fuzzy rules to determine the relation between the users' demographical data and their preferences in portfolio investment [42]. The proposed system was evaluated with the collaboration of experts (investment advisors) and the results showed that the model had a good performance.

Fuzzy association rules were used to determine the recommendations (associative classification) by J. P. Lucas, A. Laurent, M. N. Moreno and M. Teisseire in their hybrid recommender system [44]. The results in their experiments showed that these methods provide a fast and comprehensible learning model and generates a low number of false positive results. The drawback of the systems was that the false negative results were generated more frequently.

The knowledge was represented by fuzzy rules and first-order Sugeno method was used for inferences in [50]. The results of their experiments showed that the proposed methods significantly increased the accuracy. The drawback of the proposed model is the lack of using incremental learning.

P. Perny and J.-D. Zucker introduced a new collaborative decision making process in [53], which used fuzzy preference and fuzzy similarity relations between users expressing both positive and negative preferences.

O. Nasraoui and C. Petenes used fuzzy approximate reasoning for recommendation in [54]. They used the connection between fuzzy relations and fuzzy rules. The main advantage of the proposed system was that it used prediscovered (offline) profile data for the recommendation, which enables the method to provide real-time results.

C. Cornelis, X. Guo, J. Lu and G. Zhang proposed a new hybrid CF and CBF method in [55]. The proposed model used fuzzy relations to describe relationship between users, between items and between users to items. The method uses both positive and negative preferences for user-item pairs. The introduced hybrid system was able to work even with in sparse or non-existent rating data.

C. Cornelis, J. Lu, X. Guo and G. Zhang proposed a conceptual hybrid CF-CBF recommender method, which models as well as user and item similarities with fuzzy relations in [56].

L. G. Pérez, M. Barranco and L. Martínez proposed a knowledge-based recommender system, which used fuzzy preference relations to describe user preferences in [57]. The advantage of the proposed model is that it requires minimal information from the users to generate their profiles, while still provides accurate recommendations.

C. Porcel and E. Herrera-Viedma proposed a recommender system for university digital libraries in [58], which used fuzzy preference relations to represent user preferences. The experimental results were compared to recommendations of librarians and showed good user satisfaction.

M. Nilashi, O. bin Ibrahim and N. Ithnin presented a new multi-criteria CF method using neuro-fuzzy inference system in [59]. The user preferences were represented by fuzzy rules. The properties of rules were determined by neural network.

3.4 Fuzzy Clusters

G. Castellano, A. M. Fanelli, P. Plantamura and M. A. Torsello used fuzzy clustering to determine the user profile clusters [33]. Their experiments showed that fuzzy clustering is an effective tool to extract the users' profiles. This advantage was a result of that the users could belong to more than one (overlapping) clusters, which is more realistic compared to classic clustering methods, in which a user belongs to a single category.

M. Maatallah and H. Seridi applied fuzzy clustering [36] to determine user groups and to track the changes in the preferences of a particular user. The results of their experiments confirmed that assigning a user to multiple clusters at the same has positive effect on the system.

The recommendation engine presented in [38] used fuzzy c-means clustering to determine which fuzzy profiles are belonging to particular categories.

C. Birtolo, D. Ronca and R. Armenise proposed an item-based CF recommender system, which applies fuzzy c-means clustering (IFCCF) [60]. The method was tested on movies and jokes data and the results were compared to collaborative filtering techniques, one which uses k-means clustering (KMCF) and a memory-based approach (MBCF). It was shown in the experiment that the IFCCF model increased the accuracy.

W. Liu and L. Gao used a fuzzy clustering method, which integrated information entropy theory to determine the classes of disciplines and keywords and to prepare for constructing the nodes of the used fuzzy cognitive maps [49].

J. Kim and E. Lee proposed a recommender model [61] based on a new method with fuzzy clustering to merging/eliminating small clusters. The clustering method is used to classify user into categories.

B. Suryavanshi, N. Shiri and S. Mudur proposed a new hybrid CF technique in [62], which used relational fuzzy subtractive clustering for extracting user profiles. The main advantages of the proposed clustering method were that it requires low computational time and the achieved accuracy was comparable to memory-based CF's.

S. Nadi, M. Saraee and M. Davarpanah-Jazi introduced a novel recommendation model in [63], which used fuzzy c-means clustering to determine the groups of users and items. An item is recommended to user if the item belongs to a cluster, which is related to the user's cluster. The model is effective in identifying user preferences, and provides dynamic user clustering. The main disadvantage of the model is that the separate clustering processes of users and items are time consuming.

C. Birtolo and D. Ronca proposed a framework for item-based fuzzy clustering CF (IFCCF) and trust-aware clustering CF (TRACCF) in [64]. The IFCCF used fuzzy c-means, while TRACCF used K-means to determine user and item clusters. The experiment results showed that the TRACCF model could increase the diversity (coverage), while it maintained quality and increased the accuracy of recommendations.

3.5   Fuzzy Trees and Fuzzy Cognitive Maps

D. Wu, G. Zhang and J. Lu proposed a system, which uses fuzzy trees to describe the items' parameters [45]. This special form of representation was explained by the special properties of the items such as telecom services. The proposed method was able to handle both fuzzy and crisp values.

Ref. [49] was the only method, which used fuzzy cognitive maps for the recommendation process and can implement effective recommendation with little or incomplete user information.

4   Conclusions

It is clear from the previous section that fuzzy sets, numbers and linguistic variables are widely used to describe user/item parameters and are able to increase the recommendations. Fuzzy clustering techniques are also popular and useful to describe the groups of users and items. Fuzzy relations and rules are mainly used to describe user-item relationships and able to handle uncertain and sparse data. The mathematical properties of fuzzy methods like rules and linguistic variables could be also useful to overcome difficulties in both understandability and computational complexity of recommendation explanations.

The investigated methods had great benefits, however it is important to highlight the fact that most of the recommender systems use fuzzy techniques to improve traditional methods and only a few models use higher level fuzzy approaches, for example, fuzzy cognitive maps and fuzzy signatures. The mathematical properties of such methods could increase the accuracy of recommendation and overcome several issues in recommender systems and should be target of future research in the field.

Acknowledgements This paper was partially supported by the GOP-1.1.1-11-2012-0172 and the National Research, Development and Innovation Office (NKFIH) K105529, K108405.

References

1. P. Resnick, H.R. Varian, Recommender systems. Commun. ACM **40**(3), 56–58 (1997)
2. H. Kautz, B. Selman, Creating models of real-world communities with referralweb, in *Working Notes of the Workshop on Recommender Systems, Held in Conjunction with AAAI-98*, Madison, WI, vol. 82, pp. 58–59 (1998)
3. G. Jawaheer, M. Szomszor, P. Kostkova, Comparison of implicit and explicit feedback from an online music recommendation service, in *Proceedings of the 1st International Workshop on Information Heterogeneity and Fusion in Recommender Systems (HetRec '10)* (ACM, USA, 2010), pp. 47–51
4. D. Parra, A. Karatzoglou, X. Amatriain, I. Yavuz, Implicit feedback recommendation via implicit-to-explicit ordinal logistic regression mapping, in *Proceedings of the CARS-2011*, USA, October 2011, p. 5
5. X. Amatriain, J. Pujol, N. Oliver, *I Like It... I Like It Not: Evaluating User Ratings Noise in Recommender Systems, User Modeling, Adaptation, and Personalization* (Springer, Berlin, 2009), pp. 247–258
6. L.A. Zadeh, Fuzzy Sets. Inf. Control **8**(3), 338–353 (1965)
7. L.A. Zadeh, Outline of a new approach to the analysis of complex systems and decision processes. IEEE Trans. Syst. Man Cybern. **SMC-3**(1), 28–44 (1973)
8. L.A. Zadeh, The concept of a linguistic variable and its application to approximate reasoning —I. Inf. Sci. **8**(3), 199–249 (1975)
9. L.A. Zadeh, The concept of a linguistic variable and its application to approximate reasoning —II. Inf. Sci. **8**(4), 301–357 (1975)
10. L.A. Zadeh, The concept of a linguistic variable and its application to approximate reasoning —III. Inf. Sci. **9**(1), 43–80 (1975)
11. L.A. Zadeh, Fuzzy logic and approximate reasoning. Synthese **30**(3–4), 407–428 (1975)
12. E.H. Mamdani, S. Assilian, An experiment in linguistic synthesis with a fuzzy logic controller. Int. J. Man Mach. Stud. **7**(1), 1–13 (1975)
13. T. Takagi, M. Sugeno, Fuzzy identification of systems and its applications to modeling and control. IEEE Trans. Syst. Man Cybern. **SMC-15**(1), 116–132 (1985)
14. Y. Hu, Y. Koren, C. Volinsky, Collaborative filtering for implicit feedback datasets, in *Eighth IEEE International Conference on Data Mining, ICDM '08*, Pisa, pp. 263–272 (2008)

15. G. Shaw, Y. Xu, S. Geva, Using association rules to solve the cold-start problem in recommender systems, in *Advances in Knowledge Discovery and Data Mining* (Springer, Berlin, Heidelberg, 2010), pp. 340–347
16. H. Sobhanam, A.K. Mariappan, Addressing cold start problem in recommender systems using association rules and clustering technique, in *2013 International Conference on Computer Communication and Informatics (ICCCI)*, Coimbatore, pp. 1–5 (2013)
17. R. Van Meteren, M. Van Someren, Using content-based filtering for recommendation, in *Proceedings of the Machine Learning in the New Information Age: MLnet/ECML2000 Workshop* (2000), pp. 47–56
18. A.M. Rashid, G. Karypis, J. Riedl, Learning preferences of new users in recommender systems: an information theoretic approach. ACM SIGKDD Explor. News l **10**, 90 (2008)
19. R. Burke, Hybrid recommender systems: survey and experiments. User Model. User-Adap. Inter. **12**(4), 331–370 (2002)
20. J. Basilico, T. Hofmann, Unifying collaborative and content-based filtering, in *Proceedings of the Twenty-First International Conference on Machine Learning* (ACM, 2004), p. 9
21. D. Billsus, M.J. Pazzani, A personal news agent that talks, learns and explains, in *Proceedings of the Third Annual Conference on Autonomous Agents* (ACM, 1999), pp. 268–275
22. J.L. Herlocker, J.A. Konstan, J. Riedl, Explaining collaborative filtering recommendations, in *Proceedings of the 2000 ACM Conference on Computer Supported Cooperative Work* (New York, USA, 2000), pp. 241–250
23. M. Bilgic, R.J. Mooney, Explaining recommendations: satisfaction vs. promotion, in *Beyond Personalization Workshop, IUI*, vol. 5 (2005), pp. 1–8
24. D. Mcsherry, Explanation in recommender systems. Artif. Intell. Rev. **24**(2), 179–197 (2005)
25. F. Sormo, J. Cassens, A. Aamodt, Explanation in case-based reasoning–perspectives and goals. Artif. Intell. Rev. **24**(2), 109–143 (2005)
26. P. Pu, L. Chen, Trust building with explanation interfaces, in *Proceedings of the 11th International Conference on Intelligent User Interfaces* (ACM, New York, USA, 2006), pp. 93–100
27. N. Tintarev, J. Masthoff, A survey of explanations in recommender systems, in *IEEE 23rd International Conference on Data Engineering Workshop, 2007*, Istanbul, pp. 801–810 (2007)
28. R.R. Yager, Fuzzy logic methods in recommender systems. Fuzzy Sets Syst. **136**(2), 133–149 (2003)
29. J. Carbo, J.M. Molina, Agent-based collaborative filtering based on fuzzy recommendations. Int. J. Web Eng. Technol. **1**(4), 414–426 (2004)
30. L. Jie, Personalized e-learning material recommender system, in *International Conference on Information Technology for Application* (2004), pp. 374–379
31. Y. Cao, Y. Li, X. Liao, Applying fuzzy logic to recommend consumer electronics, in *Distributed Computing and Internet Technology* (Springer, Berlin Heidelberg, 2005), pp. 278–289
32. M.Y.H. Al-Shamri, K.K. Bharadwaj, Fuzzy-genetic approach to recommender systems based on a novel hybrid user model. Expert Syst. Appl. **35**(3), 1386–1399 (2008)
33. G. Castellano, A. M. Fanelli, P. Plantamura, and M. A. Torsello, A neuro-fuzzy strategy for web personalization, in *Proceedings of the Twenty-Third AAAI Conference on Artificial Intelligence*, pp. 1784–1785 (2008)
34. L.M. de Campos, J.M. Fernández-Luna, J.F. Huete, A collaborative recommender system based on probabilistic inference from fuzzy observations. Fuzzy Sets Syst. **159**(12), 1554–1576 (2008)
35. A. Zenebe, A.F. Norcio, Representation, similarity measures and aggregation methods using fuzzy sets for content-based recommender systems. Fuzzy Sets Syst. **160**(1), 76–94 (2009)
36. M. Maatallah, H. Seridi, A fuzzy hybrid recommender system, in *2010 International Conference on Machine and Web Intelligence* (2010), pp. 258–263
37. V. Ramkumar, S. Rajasekar, S. Swamynathan, Scoring products from reviews through application of fuzzy techniques. Expert Syst. Appl. **37**(10), 6862–6867 (2010)

38. L. Terán, A. Meier, *A Fuzzy Recommender System for eElections, in Electronic Government and the Information Systems Perspective* (Springer, Berlin Heidelberg, 2010), pp. 62–76
39. A. Zenebe, L. Zhou, A.F. Norcio, User preferences discovery using fuzzy models. Fuzzy Sets Syst. **161**(23), 3044–3063 (2010)
40. J.J. Castro-Schez, R.M.D. Vallejo, L.M. López-López, A highly adaptive recommender system based on fuzzy logic for B2C e-commerce portals. Expert Syst. Appl. **38**(3), 2441–2454 (2011)
41. L. C. Cheng, H. A. Wang, A novel fuzzy recommendation system integrated the experts' opinion, in *2011 IEEE International Conference on Fuzzy Systems (FUZZ)*, Taipei, pp. 2060–2065 (2011)
42. Á. García-Crespo, J.L. López-Cuadrado, I. González-Carrasco, R. Colomo-Palacios, B. Ruiz-Mezcua, SINVLIO: Using semantics and fuzzy logic to provide individual investment portfolio recommendations. Knowl. Based Syst. **27**, 103–118 (2012)
43. V. Kant, K.K. Bharadwaj, Enhancing recommendation quality of content-based filtering through collaborative predictions and fuzzy similarity measures. Proced. Eng. **38**, 939–944 (2012)
44. J.P. Lucas, A. Laurent, MN. Moreno, M. Teisseire, A fuzzy associative classification approach for recommender systems. Int. J. Uncertain. Fuzziness Knowl. Based Syst. World Scientific, **20**(4), 579–617 (2012)
45. D. Wu, G. Zhang, J. Lu, A fuzzy tree similarity measure and its application in telecom product recommendation, in *2013 IEEE International Conference on Systems, Man, and Cybernetics (SMC)*, Manchester, pp. 3483–3488 (2013)
46. Z. Zhang, H. Lin, K. Liu, D. Wu, G. Zhang, J. Lu, A hybrid fuzzy-based personalized recommender system for telecom products/services. Inf. Sci. **235**, 117–129 (2013)
47. D. Anand, B.S. Mampilli, Folksonomy-based fuzzy user profiling for improved recommendations. Expert Syst. Appl. **41**(5), 2424–2436 (2014)
48. L.-C. Cheng, H.-A. Wang, A fuzzy recommender system based on the integration of subjective preferences and objective information. Appl. Soft Comput. **18**, 290–301 (2014)
49. W. Liu, L. Gao, Recommendation system based on fuzzy cognitive map. J. Multimed. **9**(7), 970–976 (2014)
50. M. Nilashi, O. bin Ibrahim, N. Ithnin, Hybrid recommendation approaches for multi-criteria collaborative filtering. Expert Syst. Appl. **41**(8), 3879–3900 (2014)
51. L.H. Son, HU-FCF: a hybrid user-based fuzzy collaborative filtering method in recommender systems. Expert Syst. Appl. Int. J. **41**(15), 6861–6870 (2014)
52. G. Posfai, G. Magyar, L.T. Kóczy, IDF-social: an information diffusion-based fuzzy model for social recommender systems, in *Proceedings of the Congress on Information Technology, Computational and Experimental Physics (CITCEP 2015)* (2015), pp. 106–112
53. P. Perny, J.-D. Zucker, Collaborative filtering methods based on fuzzy preference relations, in *Proceedings of EUROFUSE-SIC 99* (1999), pp. 279–285
54. O. Nasraoui, C. Petenes, An intelligent web recommendation engine based on fuzzy approximate reasoning, in *The 12th IEEE International Conference on Fuzzy Systems, 2003, FUZZ'03* (IEEE, 2003), pp. 1116–1121
55. C. Cornelis, X. Guo, J. Lu, G. Zhang, A fuzzy relational approach to event recommendation. IICAI **5**, 2231–2242 (2005)
56. C. Cornelis, J. Lu, X. Guo, G. Zhang, One-and-only item recommendation with fuzzy logic techniques. Inf. Sci. **177**(22), 4906–4921 (2007)
57. L.G. Pérez, M. Barranco, L. Martínez, Building user profiles for recommender systems from incomplete preference relations, in *IEEE International on Fuzzy Systems Conference, 2007, FUZZ-IEEE 2007* (2007), pp. 1–6
58. C. Porcel, E. Herrera-Viedma, Dealing with incomplete information in a fuzzy linguistic recommender system to disseminate information in university digital libraries. Knowl.-Based Syst. **23**(1), 32–39 (2010)

59. M. Nilashi, O. bin Ibrahim, N. Ithnin, Multi-criteria collaborative filtering with high accuracy using higher order singular value decomposition and neuro-fuzzy system. Knowl.-Based Syst. **60**, 82–101 (2014)
60. C. Birtolo, D. Ronca, R. Armenise, Improving accuracy of recommendation system by means of item-based fuzzy clustering collaborative filtering, in *2011 11th International Conference on Intelligent Systems Design and Applications (ISDA)* (IEEE, 2011), pp. 100–106
61. J. Kim, E. Lee, XFC—XML based on fuzzy Clustering—method for personalized user profile based on recommendation system, in *IEEE Conference on Cybernetics and Intelligent Systems, 2004* (Singapore, 2004), pp. 1202–1206
62. B. Suryavanshi, N. Shiri, S. Mudur, A fuzzy hybrid collaborative filtering technique for web personalization, in *Proceedings of 3rd International Workshop on Intelligent Techniques for Web Personalization (ITWP 2005), 19th International Joint Conference on Artificial Intelligence (IJCAI 2005)* (2005), pp. 1–8
63. S. Nadi, M. Saraee, M. Davarpanah-Jazi, A fuzzy recommender system for dynamic prediction of user's behavior, in *2010 International Conference on Internet Technology and Secured Transactions (ICITST)* (London, 2010), pp. 1–5
64. C. Birtolo, D. Ronca, Advances in clustering collaborative filtering by means of fuzzy C-means and trust. Expert Syst. Appl. **40**(17), 6997–7009 (2013)

Fuzzy Physiologically Based Pharmacokinetic (PBPK) Model of Chloroform in Swimming Pools

R. A. Dyck, R. Sadiq and M. J. Rodriguez

Abstract Chloroform is one of the most prevalent disinfection byproducts (DBPs) formed in swimming pools through reactions between disinfectants and organic contaminants. Chloroform and related DBPs have been a subject of research in exposure and human health risk assessments over the last several decades. Physiologically based pharmacokinetic (PBPK) models are one tool that is being used increasingly by researchers to evaluate the health impacts of swimming pool exposures. These models simulate the absorption, distribution, metabolism and excretion of chemicals in the human body to assess doses to sensitive organs. As with any model, uncertainties arise from variability and imprecision in inputs. Among the most uncertain model parameters are the partition coefficients which describe uptake and distribution of chemical to different tissues of the body. In this paper, a fuzzy based model is presented for improving the description and incorporation of uncertain parameters into the model. The fuzzy PBPK model compares well with the deterministic model and measured concentrations while providing more information about uncertainty.

1 Introduction

Chlorination of water in swimming pools can result in unwanted disinfection byproducts (DBPs). While the benefits of swimming have been shown to outweigh health risks [1], research suggests a link between DBP exposure and a variety of adverse health effects such bladder cancer, asthma and fetal-growth related pregnancy

R. A. Dyck (✉) · R. Sadiq
School of Engineering, University of British Columbia Okanagan, Kelowna, BC, Canada
e-mail: roberta.dyck@alumni.ubc.ca

R. Sadiq
e-mail: rehan.sadiq@ubc.ca

M. J. Rodriguez
École supérieure d'aménagement du territoire, Université Laval, Québec, QC, Canada
e-mail: manuel.rodriguez@esad.ulaval.ca

© Springer International Publishing AG, part of Springer Nature 2018
L. A. Zadeh et al. (eds.), *Recent Developments and the New Direction in Soft-Computing Foundations and Applications*, Studies in Fuzziness and Soft Computing 361, https://doi.org/10.1007/978-3-319-75408-6_38

outcomes [2–8]. In an effort to provide support for risk management and guideline development, risk assessors have developed physiologically based pharmacokinetic (PBPK) models to estimate the dose of DBPs delivered to target organs [9–11].

The purpose of this paper is to illustrate the benefit of using fuzzy numbers to represent uncertain PBPK model parameters to improve understanding of the transport and fate in the human body of chloroform (one of the most prevalent and studied DBPs). The ultimate goal of this research is to reduce human health risk through engineering interventions in pool design and management and the development of limits for allowable concentrations of DBPs in indoor swimming pools.

2   Background

2.1   Disinfection Byproducts in Indoor Swimming Pools

Chlorination of drinking water has been fundamental in protecting public health for over a hundred years [12, 13]. Disinfection of water in swimming pools has been protecting swimmers from communicable disease for almost as long [14]. In 1974, researchers identified compounds unintentionally generated during reactions between chlorine and organic matter in water [15]. These DBPs were soon suggested to be linked to health effects such as cancer [16]. While not conclusive, studies have reported possible connections to several types of cancer [2, 3] as well as reproductive effects [4], respiratory irritation and asthma [5–8].

DBPs were first reported in swimming pools in 1980 [17]. Over 100 different DBPs have been detected in swimming pools [18]. Trihalomethanes (THMs) are the most prevalent [19]. Health Canada sets limits for total THMs in drinking water [20]; however, no comparable guidelines are used in Canada limiting the amount of THMs in swimming pools.

Swimming is the fourth most popular leisure activity in Canada, after walking, gardening and home exercise [21]. The World Health Organization [22] recognizes the benefits of swimming to health and well-being through increased social interaction, relaxation and exercise. A recent study has confirmed that the health benefits of swimming outweigh the risk of DBP exposure [1]. However, given the popularity of swimming, a thorough understanding of exposure and associated risks will contribute to improved risk management strategies for regulators, pool designers and pool managers.

2.2   PBPK Modeling for Human Health Risk Assessment

Assessment of human health risk for environmental exposures consists of four main steps: (a) problem formulation (identification of contaminants, receptors and pathways), (b) exposure assessment (based on detailed information about the contaminants, receptors and the environment), (c) toxicity assessment (traditionally the dose-response relationship), and (d) risk characterization (integration of the exposure and toxicity assessments to assess the significance of the risk) [23].

Because many DBPs, including THMs, are volatile, when they are present in the water of a swimming facility, they are also present in the air. This leads to three potential routes of exposure to DBPs: inhalation of vapors, accidental ingestion of water and absorption through the skin. Inhalation and dermal absorption are considered the primary uptake pathways for volatile DBPs in chlorinated pools, including THMs [24–26].

The dose-response relationships used in toxicity assessment are often based on high-dose animal studies (relative to environmental concentrations) leading to uncertainty through the use of extrapolation factors [27]. Increasing knowledge of biological systems has resulted in better estimates of relationships between toxic response and external, internal and target–organ doses [27]. Physiologically based pharmacokinetic (PBPK) or physiologically based toxicokinetic (PBTK)[1] models allow risk assessors to use existing physiological, biochemical and physicochemical data about absorption, distribution, metabolism and excretion within the human (or test animal) body to relate internal doses to toxic responses thereby reducing the uncertainty [27].

In a PBPK model, the virtual body is divided into tissue compartments connected by circulation of venous and arterial blood. PBPK models use concentrations and partitioning properties to express the fate and transport of chemicals. Differential equations representing mass balances between input and output amounts of chemical in each compartment are integrated to estimate concentrations versus time in each compartment. Models are evaluated by comparing output to measurements in biological media such as urine, blood and alveolar air (exhaled). Estimates of dose to target organs can then be used to connect environmental concentrations with internal doses expected to cause an adverse effect. Physiological model parameters (e.g. tissue volumes and blood flows) and chemical specific model parameters (e.g. partition coefficients and molecular weight) are derived from published literature or determined experimentally.

2.3 Uncertainty in PBPK Modeling

Partition coefficients for transfer of chemicals from the environment to the body (e.g. blood:air) as well as partition coefficients from the blood to the tissues (e.g. blood:liver) have been identified as significant sources of uncertainty [28, 29]. Probabilistic methods of addressing uncertainty have been in use since early in the

---

[1]While models that evaluate environmental exposures are technically considered "toxico" kinetic models, much of the literature refers to PBPK because the pharmaceutical industry originally drove much of this research. The terms will be used interchangeably here.

development of PBPK models [30, 31]; however, their usefulness is limited to those parameters with sufficient available data to establish distributions. Parameters such as body weight, blood flow and tissue volumes are subject to variability within the population, however, parameters such as partition coefficients can suffer from a lack of data that makes distributions more difficult to generate.

To our knowledge, at the time of this study, only two models exist which use fuzzy set theory to address uncertainty in PBPK models [28, 32]. Both of these works address uncertainty in partition coefficients for diazepam, a pharmaceutical. The current study represents the first use of fuzzy set theory in a PBPK model for environmental exposures. Here we have chosen to model chloroform because it is the most prevalent of the THMs and therefore has the largest body of literature to provide data.

3 PBPK Model

3.1 PBPK Model Structure

The model structure for this study (Fig. 1) was derived from existing PBPK models [10, 33]. Model compartments include lungs, skin, fat, liver, richly perfused tissues (RPT; e.g. organs) and poorly perfused tissues (PPT; e.g. muscles). Chloroform uptake is through inhalation of room air (to lung compartment) and dermal contact with pool water (to skin compartment). Arterial and venous blood flow are used to distribute chloroform to the compartments. Liver metabolism is considered to be the only relevant metabolic process. Excretion is modeled through exhalation and liver metabolism.

Existing PBPK models [10, 33, 34] were adapted using the concept of fugacity to simplify the expression of model processes [35, 36]. The fugacity method can provide more information about the status of equilibrium and direction of chemical flux between compartments [37].

3.2 PBPK Model Parameters

Physiological (Table 1) and chemical-specific (Table 2) model parameters were used in the deterministic PBPK model. Model differential equations were derived based on previous models [10, 33, 35, 38–40]. Model simulations were conducted using the software Berkeley Madonna [41]. The exposure scenario (i.e. air and water concentrations, time of exposures) was derived from [42]. The concentration in water was assumed to be constant throughout the exposure period.

Fig. 1 PBPK model structure

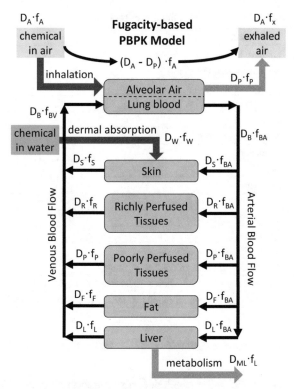

(Model based on Haddad et al. 2006, Catto et al. 2012, Corley 1990, Cahill et al. 2003)

Table 1 Physiological model parameters

Description	Values	Source
General physiological parameters		
Body weight (kg)—BW	70	[33, 43]
Body surface area (cm²)	18,000	[33, 43]
QC—cardiac output (L/h-kg BW)	18	[33]
Alveolar ventilation (L/h-kg BW)	18	[33]
% of QC to each compartment		
Liver	26	[33, 43]
Richly perfused tissues	44	[33, 43]
Slowly perfused tissues	21.6	[33, 43]
Fat	5	[33, 43]
Skin	3.4	[9, 39, 44, 45]
% BW in each compartment		
Liver	2.6	[33, 43]
Richly perfused tissues	5	[33, 43]
Slowly perfused tissues	52	[33, 43]
Fat	21	[33, 43]
Skin	10	[9, 39, 44, 45]

Table 2 Chemical specific model parameters

Description	Values	Source
General chemical properties		
Molecular weight (g/mol)	119.38	
Temperature (kelvin)	298.15	
Ideal gas constant (m³ Pa/K mol)	8.314	
Pa m³/mol	381.43	
Partition coefficients		
Blood:air	10.7	[33, 45, 46]
Fat:air	280	[33, 34]
Liver:air	17	[33, 34]
Richly perfused tissue:air	17	[33, 34]
Poorly perfused tissue:air	12	[33, 34]
Skin:air	19.7	[9, 39, 47]
Absorption and metabolism		
K_p—dermal absorption coefficient (cm/h)	0.1602	[33, 46]
VMAX—capacity of oxidative metabolism (mg/h/kg BW)	12.68	[33]
Km—Michaelis-Menten affinity of oxidative metabolism (mol/m³)	0.448	[33]

3.3   Deterministic PBPK Model Results

The model output included concentration of chloroform in venous blood, arterial blood, and alveolar air (exhaled). The chloroform concentration in alveolar air over time is presented in Fig. 2. Measurements of alveolar air taken by [42] before swimming and after 60 min are presented for comparison. Table 3 shows the measured environmental and biological chloroform concentrations.

Fig. 2 Deterministic PBPK model for chloroform in swimming pools

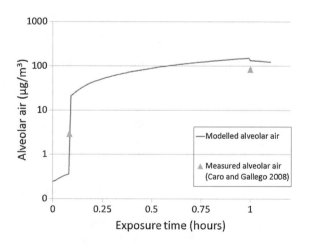

Table 3 Environmental and biological concentrations [42]

Time (min)	Chloroform concentration		
	Water (µg/L)	Air (µg/m^3)	Measured alveolar air (µg/m^3)
5[a]	125	241	2.9 ± 0.2
60			84.1 ± 6.2

[a]Samples were collected before exposure. In the model, pre-exposure alveolar air concentration was used to approximate an ambient concentration of chloroform for the first 5 min of exposure

Table 4 Values of K_P from literature [33]

K_p (cm/h)	Source[a]
0.00438–0.0540	[53]
0.01–0.059	[54]
0.14–0.19	[55]
0.16	[56]
0.16–4.8%	[46]
0.2	[57]
0.22	[9]
0.16–0.42	[58]

[a]Sources were originally listed by [33]

4 Fuzzy PBPK Model

While physiological parameters such as body weight may have sufficient data available to generate distributions for probabilistic analysis, other parameters may lack sufficient data for such characterization. In 1965, Dr. Lotfi A. Zadeh presented fuzzy sets as a new way of representing vague or imprecise information [47]. Since then, there have been many applications in different areas of engineering. In the case outlined here, the imprecise information is the dermal absorption coefficient (K_p, cm/h) for dermal uptake of chloroform.[2]

In spite of the many studies that have evaluated the skin absorption coefficient for PBPK models (Table 4), there is no consensus on which is best. Choosing between the values can be somewhat arbitrary, while fitting the eight values presented here to a probability distribution is impractical.

---

[2]For more detailed discussion of fuzzy sets, fuzzy numbers, fuzzy arithmetic and their use in modeling, see [48, 49–52].

Fig. 3 α-cuts on K_p fuzzy membership functions

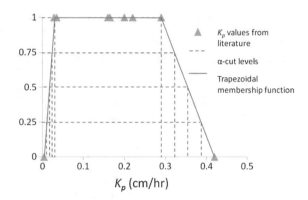

4.1 Membership Functions

A trapezoidal membership function (Fig. 3) was defined using 0.00438 and 0.42 cm/h as the minimum and maximum, respectively. The top of the trapezoid was defined using the other listed values. Where a range was given, the mean was also used [53, 58].

4.2 DSW Algorithm

PBPK model simulations were run using the trapezoidal fuzzy number for K_p according to the DSW algorithm [59]. The algorithm generates fuzzy output for the model using α-cuts and interval analysis. The steps are as follows [48]:

- Select a value α in the membership function.
- Find the intervals in the input membership function corresponding to this α.
- Use interval operations to find the interval for the output membership function for the selected α-cut level.
- Repeat steps for different values of α to build a fuzzy result.

For this analysis, we chose α values of 0, 0.25, 0.5, 0.75 and 1.0 (Fig. 3). The corresponding intervals for K_p (Table 5) were then used in the PBPK model to predict the chloroform concentrations in venous blood, arterial blood and alveolar air. All other model parameters were unchanged.

Table 5 Interval values for K_p

α-cut levels	Left	Right
0	0.00438	0.42
0.25	0.0106	0.388
0.5	0.0168	0.355
0.75	0.0230	0.323
1	0.0292	0.29

Fig. 4 Output of fuzzy
PBPK model

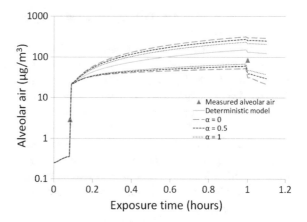

4.3 Fuzzy Model Results

The fuzzy output results for the PBPK model are presented Fig. 4. Results are shown only for α-cut levels 0, 0.5, and 1.0. Results of the deterministic model are represented by the solid line in Fig. 4. The results of the fuzzy model are presented as dashed lines. The measured values of chloroform concentration in alveolar air are also shown.

5 Discussion

In Fig. 2 we see that the deterministic model has an acceptable fit to the measured data. For many PBPK models, the fit between model and experimental data is evaluated only visually while varying the parameters of *Km* and *VMAX* to improve the fit [60]. In fact, classical statistical procedures have been found to not be applicable for the complexity of PBPK models [61]. For the PBPK models presented here, the original *Km* and *VMAX* presented in [33, 34] were used unaltered. In Fig. 4 we see that the fuzzy model (as represented by dashed lines for α-cut levels) shows a similar fit to the deterministic model. While many PBPK models report a single line as concentration over time [62], the fuzzy model has the advantage of showing the uncertainty in the model with the measured data falling within the interval. While the final resulting fuzzy number is highly dependent on the selection of the fuzzy number shape and limits, it nevertheless represents an improvement over a deterministic model.

One benefit of using fuzzy numbers to represent imprecise model parameters is a more complete and realistic estimate of the parameter and therefore of the output of the model. The advantage over probabilistic methods is the ease of defining a membership function as opposed to choosing a probability distribution with so few data. Another benefit is the decrease in computing complexity. Interval operations such as those used in the DSW algorithm are easy to use and require no specialized risk analysis software.

The biggest disadvantage for fuzzy modeling may be the reluctance of engineers and risk assessors to use it [63]. To our knowledge, only two fuzzy PBPK models have been published to date, both being for the pharmaceutical diazepam [28, 32]. Another disadvantage is the possible oversimplification of model parameters in the membership function and the judgement required to choose values for the function limits.

Some environmental models have used probability for aleatory uncertainty (e.g. natural variability in population weight, surface area) combined with fuzzy methods for epistemic uncertainty (for model parameters or equations) [14, 39]. We recommend exploration of this approach in future work for risk assessment for DBPs in swimming pools.

Other future work recommended is

- inclusion of differences in blood flow and metabolism to represent levels of effort in exercise,
- consideration of variability in body weight and metabolism due to age, gender and fitness level,
- use of fuzzy numbers for all partition coefficients,
- use of probabilistic methods where data allows,
- modeling other DBPs, and
- combining the PBPK model with previous exposure model [64] to assess the impact of risk management strategies such as room ventilation, changes in disinfection practices and changes in swimmer hygiene.

While consideration of DBP risk in swimming pools is an important area for future study, we must be mindful of the balance with adequate disinfection. The World Health Organization states:

> The health risks from these byproducts at the levels at which they occur in drinking water are extremely small in comparison with the risks associated with inadequate disinfection. Thus, it is important that disinfection not be compromised in attempting to control such byproducts [19].

This principle should be extended to swimming pools as well.

6 Summary and Conclusions

We have presented a fuzzy-based PBPK model for assessing swimming pool exposures to chloroform using imprecise information. The skin absorption coefficient K_p was assigned a trapezoidal fuzzy membership function. The DSW algorithm was used to generate input for the PBPK model. Model results included chloroform concentrations in venous blood, arterial blood and alveolar air. Comparison to measured concentrations of alveolar air shows that the fuzzy model offers a comparable fit to the deterministic model. Fuzzy parameters have an advantage over deterministic parameters in improving understanding of uncertainty in the model output. Fuzzy parameters also allow the incorporation of uncertainty without assigning probability distributions using insufficient data. We recommend further work on other fuzzy parameters and combinations between fuzzy methods for epistemic uncertainty and probabilistic methods for aleatory uncertainty. Other future work can include refining the model to represent variability between individuals, modeling other DBPs and combining the PBPK model with previous exposure model to assess the impact of risk management strategies.

Acknowledgements This project is part of ongoing research funded under the Canada NSERC Discovery Grant program and UBC Okanagan Internal Research Grants. R.A.D. thanks Dr. Robert Tardif of University of Montreal for expert opinion on parameter estimation.

References

1. C.M. Villanueva, S. Cordier, L. Font-Ribera, L.A. Salas, P. Levallois, Overview of disinfection by-products and associated health effects. Curr. Environ. Health Rep. **2015**(2), 107–115 (2015)
2. S.D. Richardson et al., Occurrence, genotoxicity, and carcinogenicity of regulated and emerging disinfection by-products in drinking water: a review and roadmap for research. Mutat. Res. **636**(1–3), 178–242 (2007)
3. C.M. Villanueva et al., Bladder cancer and exposure to water disinfection by-products through ingestion, bathing, showering, and swimming in pools. Am. J. Epidemiol. **165**(2), 148–156 (2007)
4. M.J. Nieuwenhuijsen, M.B. Toledano, N.E. Eaton, J. Fawell, P. Elliott, Chlorination disinfection byproducts in water and their association with adverse reproductive outcomes: a review. Occup. Environ. Med. **57**(2), 73–85 (2000)
5. M. Goodman, S. Hays, Asthma and swimming: a meta-analysis. J. Asthma **45**(8), 639–647 (2008)
6. J.H. Jacobs et al., Exposure to trichloramine and respiratory symptoms in indoor swimming pool workers. Eur. Respir. J. Off. J. Eur. Soc. Clin. Respir. Physiol. **29**(4), 690–698 (2007)
7. B. Lévesque et al., The determinants of prevalence of health complaints among young competitive swimmers. Int. Arch. Occup. Environ. Health **80**(1), 32–39 (2006)
8. K. Thickett, J. McCoach, J. Gerber, S. Sadhra, P. Burge, Occupational asthma caused by chloramines in indoor swimming-pool air. Eur. Respir. J. **19**(5), 827–832 (2002)
9. B. Lévesque et al., Evaluation of the health risk associated with exposure to chloroform in indoor swimming pools. J. Toxicol. Environ. Health A **61**(4), 225–244 (2000)
10. C. Catto, G. Charest-Tardif, M. Rodriguez, R. Tardif, Assessing exposure to chloroform in swimming pools using physiologically based toxicokinetic modeling. Environ. Pollut. **1**(2), 132–147 (2012)
11. K. Krishnan, T. Peyret, Physiologically based toxicokinetic (PBTK) modeling in ecotoxicology, in *Ecotoxicology Modeling*, ed. J. Devillers, vol. 2 (Springer Science + Business Media, 2009), pp. 145–175
12. US EPA (United States Environmental Protection Agency), *The History of Drinking Water Treatment* (2000)
13. Ontario Sewer and Watermain Construction Association (OSWCA), *Drinking Water Management in Ontario: A Brief History*, no. January, pp. 1–14, 2001

14. K. Olsen, Clear waters and a green gas: a history of chlorine as a swimming pool sanitizer in the United States. Bull. Hist. Chem. **32**(2), 129–140 (2007)
15. J. Rook, Formation of haloforms during chlorination of natural waters. Water Treat. Exam. **23**, 234–243 (1974)
16. US EPA, *Small System Compliance Technology List for the Non-Microbial Contaminants Regulated Before 1996* (1998)
17. J. Beech, Estimated worst case trihalomethane body burden of a child using a swimming pool. Med. Hypotheses 303–307 (1980)
18. S.D. Richardson et al., What's in the pool? A comprehensive identification of disinfection by-products and assessment of mutagenicity of chlorinated and brominated swimming pool water. Environ. Health Perspect. **118**(11), 1523–1531 (2010)
19. World Health Organization, *Disinfectants and Disinfectant By-Products. Environmental Health Criteria 216, Geneva* (2000)
20. Health Canada, *Guidelines for Canadian Drinking Water Quality Ottawa, ON* (1978)
21. H. Gilmour, Physically active Canadians. Health Inf. Res. Div. Health Rep. **18**(3), 45–65 (2007)
22. World Health Organization, *Guidelines for Safe Recreational-Water Environments Vol. 2: Swimming Pools and Similar Environments, Geneva* (2006)
23. Health Canada, *Federal Contaminated Site Risk Assessment in Canada, Part V: Guidance on Complex Human Health Detailed Quantitative Risk Assessment for Chemicals, Ottawa, ON* (2010)
24. L. Erdinger et al., Pathways of trihalomethane uptake in swimming pools. Int. J. Hyg. Environ. Health **207**(6), 571–575 (2004)
25. B. Lévesque et al., Evaluation of dermal and respiratory chloroform exposure in humans. Environ. Health Perspect. **102**(12), 1082–1087 (1994)
26. A.B. Lindstrom, J.D. Pleil, D.C. Berkoff, Alveolar breath sampling and analysis to assess trihalomethane exposures during competitive swimming training. Environ. Health Perspect. **105**(6), 636–642 (1997)
27. International Programme on Chemical Safety, *Characterization and Application of Physiologically Based Pharmacokinetic Models, Geneva, Switzerland* (2010)
28. I.I. Gueorguieva, I. Nestorov, M. Rowland, Fuzzy simulation of pharmacokinetic models: case study of whole body physiologically based model of diazepam. J. Pharmacokinet. Pharmacodyn. **31**(3), 185–213 (2004)
29. R. Tardif et al., Impact of human variability on the biological monitoring of exposure to toluene: I. Physiologically based toxicokinetic modelling. Toxicol. Lett. **134**(1–3), 155–63 (2002)
30. R.S. Thomas, P.L. Bigelow, T.J. Keefe, R.S. Yang, Variability in biological exposure indices using physiologically based pharmacokinetic modeling and Monte Carlo simulation. Am. Ind. Hyg. Assoc. J. **57**(1), 23–32 (1996)
31. C.J. Portier, N.L. Kaplan, Variability of safe dose estimates when using complicated models of the carcinogenic process. Fundam. Appl. Toxicol. **13**, 533–544 (1989)
32. K.-Y. Seng, I. Nestorov, P. Vicini, Physiologically based pharmacokinetic modeling of drug disposition in rat and human: a fuzzy arithmetic approach. Pharm. Res. **25**(8), 1771–1781 (2008)
33. S. Haddad, G.C. Tardif, R. Tardif, Development of physiologically based toxicokinetic models for improving the human indoor exposure assessment to water contaminants: trichloroethylene and trihalomethanes. J. Toxicol. Environ. Health Part A **69**(23), 2095–2136 (2006)
34. R. Corley et al., Development of a physiologically based pharmacokinetic model for chloroform. Toxicol. Appl. Pharmacol. **103**(3), 512–527 (1990)
35. T.M. Cahill, I. Cousins, D. Mackay, Development and application of a generalized physiologically based pharmacokinetic model for multiple environmental contaminants. Environ. Toxicol. Chem. **22**(1), 26–34 (2003)

36. D. Mackay, *Multimedia Environmental Models: The Fugacity Approach*, 2nd edn. (Lewis Publishers, Boca Raton, FL, 2001)
37. M.B. Reddy, R.S.H. Yang, H.J. Clewell, M.E. Andersen, *Physiologically Based Pharmacokinetic Modeling* (Wiley, Hoboken, NJ, 2005)
38. S. Paterson, D. Mackay, A steady-state fugacity-based pharmacokinetic model with simultaneous multiple exposure routes. Environ. Toxicol. Chem. **6**, 395–408 (1987)
39. J. Ramsey, M. Andersen, A physiologically based description of the inhalation pharmacokinetics of styrene in rats and humans. Toxicol. Appl. Pharmacol. **73**(1), 159–175 (1984)
40. M.M. Mumtaz et al., Translational research to develop a human PBPK models tool kit—volatile organic compounds (VOCs). J. Toxicol. Environ. Health Part A **75**(1), 6–24 (2012)
41. R.I. Macey, G.F. Oster, *Berkeley Madonna*
42. J. Caro, M. Gallego, Assessment of exposure of workers and swimmers to trihalomethanes in an indoor swimming pool. Environ. Sci. Technol. **41**(13), 4793–4798 (2007)
43. R. Tardif, G. Charest-Tardif, J. Brodeur, K. Krishnan, Physiologically based pharmacokinetic modeling of a ternary mixture of alkyl benzenes in rats and humans. Toxicol. Appl. Pharmacol. **144**, 120–134 (1997)
44. C.W. Chen, J.N. Blancato, Incorporation of biological information in cancer risk assessment: example—vinyl chloride. Cell Biol. Toxicol. **5**(4), 417–444 (1989)
45. S. Batterman, L. Zhang, S. Wang, A. Franzblau, Partition coefficients for the trihalomethanes among blood, urine, water, milk and air. Sci. Total Environ. **284**(1–3), 237–247 (2002)
46. X. Xu, T.M. Mariano, J.D. Laskin, C.P. Weisel, Percutaneous absorption of trihalomethanes, haloacetic acids, and haloketones. Toxicol. Appl. Pharmacol. **184**(1), 19–26 (2002)
47. L.A. Zadeh, Fuzzy sets. Inf. Control **8**(3), 338–353 (1965)
48. T.J. Ross, *Fuzzy Logic with Engineering Applications*, 2nd edn. (Wiley, West Sussex, England, 2004)
49. R. Sadiq, T. Husain, A fuzzy-based methodology for an aggregative environmental risk assessment: a case study of drilling waste. Environ. Model Softw. **20**(1), 33–46 (2005)
50. D. Dubois, H. Prade, *Fuzzy Sets and Systems: Theory and Applications*, no. Nf. (Academic Press Inc., Boston MA, 1980)
51. K. Zhang, H. Li, G. Achari, Fuzzy-stochastic characterization of site uncertainty and variability in groundwater flow and contaminant transport through a heterogeneous aquifer. J. Contam. Hydrol. **106**(1–2), 73–82 (2009)
52. G.J. Klir, *Uncertainty and Foundations of Generalized Information Theory* (Wiley, Hoboken, NJ, 2006)
53. X. Xu, C.P. Weisel, Dermal uptake of chloroform and haloketones during bathing. J. Expo. Anal. Environ. Epidemiol. **15**(4), 289–296 (2005)
54. R.A. Corley, S.M. Gordon, L.A. Wallace, Physiologically based pharmacokinetic modeling of the temperature-dependent dermal absorption of chloroform by humans following bath water exposures. Toxicol. Sci. **53**(1), 13–23 (2000)
55. J.S. Nakai et al., Penetration of Chloroform, Trichloroethylene, and Tetrachloroethylene Through Human Skin. J. Toxicol. Environ. Health Part A **58**(3), 157–170 (1999)
56. J.N. Mcdougal et al., Dermal absorption of organic chemical vapors in rats and humans. Toxicol. Sci. **14**(2), 299–308 (1990)
57. R.L. Chinery, A.K. Gleason, A compartmental model for the prediction of breath exposure while showering. Risk Anal. **13**(1), 51–62 (1993)
58. T. McKone, Linking a PBPK model for chloroform with measured breath concentrations in showers: implications for dermal exposure models. J. Expo. Anal. Environ. Epidemiol. **3**(3), 339–365 (1993)
59. W. Dong, H. Shah, F. Wong, Fuzzy computations in risk and decision analysis. Civ. Eng. Syst. **2**(4), 201–208 (1985)
60. R.A. Clewell, H.J.I. Clewell, Toxicokinetics, in *Principles of Toxicology: Environmental and Industrial Applications*, ed. S.M. Roberts, R.C. James, P.L. Williams, 3rd edn. (Wiley, Hoboken, NJ, 2015)

61. U.S. EPA, Approaches for the Application of Physiologically Based Pharmacokinetic (PBPK) Models and Supporting Data in Risk Assessment, *Epa/600/R-05/043F*, no. August (2006)

62. N. Tsamandouras, A. Rostami-Hodjegan, L. Aarons, Combining the 'bottom up' and 'top down' approaches in pharmacokinetic modelling: fitting PBPK models to observed clinical data. Br. J. Clin. Pharmacol. **79**(1), 48–55 (2015)

63. H.A. Barton et al., Characterizing uncertainty and variability in physiologically based pharmacokinetic models: State of the science and needs for research and implementation. Toxicol. Sci. **99**(2), 395–402 (2007)

64. R. Dyck, R. Sadiq, M.J. Rodriguez, S. Simard, R. Tardif, Trihalomethane exposures in indoor swimming pools: a level III fugacity model. Water Res. **45**, 5084–5098 (2011)

Mamdani-Type Fuzzy Inference System for Evaluation of Tax Potential

Akif Musayev, Shahzade Madatova and Samir Rustamov

Abstract In the paper, the application of Mamdani-type fuzzy inference method to the expert evaluation of the impact of tax administration reforms on the tax potential is investigated. As input data of the system are taken reforms in tax administration and fuzzified by the triangle, trapezoid, Gaussian and Bell membership functions. It has been shown that the suggested fuzzy approach is one of the effective methods for evaluation of tax potential.

1 Introduction

Taxation problems are one of the most important ones paid attention by experts as well as state institutions at all stages of human history in which economic relations and the state system were formed.

And at the current stage of economic development this problem has been actualized for the economy of any country. Thus, evaluating the tax system impact on the process created by the globalized economic is of great importance for the creation of competitive and innovative national economy. As optimization of taxation is of great importance for economic development, economy of real and public sector, its enough perfection theoretical basis has been created since XIX century and nowadays it is being developed.

A. Musayev (✉)
Institute of Control Systems, Baku, Azerbaijan
e-mail: akif.musayev@gmail.com

S. Madatova
Azerbaijan State University of Oil and Industry, Baku, Azerbaijan
e-mail: shahzademedetova@gmail.com

S. Rustamov
ADA University, Institute of Control Systems, Baku, Azerbaijan
e-mail: srustamov@ada.edu.az

© Springer International Publishing AG, part of Springer Nature 2018
L. A. Zadeh et al. (eds.), *Recent Developments and the New Direction in Soft-Computing Foundations and Applications*, Studies in Fuzziness and Soft Computing 361, https://doi.org/10.1007/978-3-319-75408-6_39

511

But still foreign and local researchers have not been able to create a system, which was adopted unanimously. This is mainly due to the fact that, despite of insufficient grounds economic agents insist on the heavy tax burden trying to avoid taxes and leaning to illegal economies. But in this case, the opportunity given by the economic system (formed) is limited for to realize tax potential.

The researches show that, there is a direct contact between economic development and tax potential formed by it, i.e. tax burden of economy, and perfect tax legislation and administration, knowledge level of the population on economy and taxes and the like factors also impact on this dependence seriously. Therefore, the main purpose of taxation in terms of optimization is not to increase the tax burden of economy, but correct assessment of the potential economic system formed by the tax and maximization of its collection.

The level of tax potential formed by tax depends on tax legislation and administration covering rights and responsibilities, forms and methods of tax control, liability for violation of tax legislation, complaint procedures about the state tax authorities and their officials' activities (inactivity) related to some issues as tax system, general basis of taxation, identification, payment and collection rules of tax, taxation problems of taxpayers and tax authorities, as well as other participants of tax relations. Evaluating the impact of the changes made in the tax administration on tax potential is the most necessary problem to strengthen the financial base of the state development. So, tax is of great importance for the system of financial security of social-economic development. The level of tax rates, tax crimes, tax administration and tax potential have a significant impact on the opportunities and quality of economic development of the state. It must be noted that, as an evaluation method, using fuzzy logic is more advisable.

Generally, implementation of reforms in the tax, administration and inter budget relations area demands objective and reliable evaluation of their impact on tax and it enables to define the volume of real and unconsidered demand on financial resources. The results of such evaluation is necessary for optimization of using financial resources, as well as for maximum financial responsibility, experience, precision from social and other responsibilities of tax administration subjects point.

In the research work referendum and statistic evaluation method will be used for evaluating the impact of the changes made in the tax administration on tax potential, commonly on the state's economy.

The changes made in the tax administration are grounded on the evaluation based on a 100-point scale of employees of tax authorities. Carrying out a survey among independent experts and entrepreneurs in further research works by this method can have positive impact on getting more qualitative and real results of the research. We consider that such evaluation will enable to evaluate social- psychological impact of the changes made in tax administration.

2 Tax Potential

Tax potential consists of maximum sum of tax and payments to be paid by a certain state or a taxpayer within existing tax and other legislation [1–3]. The object of tax potential evaluation may be state, territory, region, district, municipality, customs territory, organization, firm, separate legal and individual entities. At the same time evaluation of the total capacity can be implemented at mentioned objects according to separate and all tax payments concerning each tax and payment proposed in the legislation.

Tax potential is used as synonym to financial potential in many cases, especially in the countries having high developed fiscal system. This is due to the fact that tax forms the major part of budget revenues in these countries. It is no coincidence, some issues related to the tax potential evaluation have been paid attention by scientists and experts in terms of budget regulation in microeconomic level [4, 5].

But these terms differ strictly in financial system of Azerbaijan. So, while the financial potential is defined as a collection of all the financial resources, tax potential is just a part of it. But it should be taken into account that tax is a key part of the financial potential. In this sense, tax potential consists of maximum financial resources which can be mobilized from taxation object through tax. Obviously, tax potential can be defined as the maximum tax inflows involved ideally by a specific object.

In international practice tax potential is presented as maximum income that can be provided in the form of inflows of taxation subject in a particular area under the authority of the government structure. Tax potential consists of per capital potential income of the budget in terms of the last purchaser of tax revenues (in our case the state budget). This income is the maximum one that can be obtained by the government in the conditions of the existing tax legislation in the considered financial year.

Tax potential of the state is determined by the situation of its economic system, and tax legislation of its foreign trade, tax administration and tax policy. In other words, during tax potential evaluation, tax policy that influences on its capacity should be taken into account as one of important elements of the tax relations structure.

Implementation of tax policy in different ways causes significantly different results in the evaluation of tax potential. Therefore, tax potential should be considered as financial resources totality that can be mobilized by taxation system formed by tax policy determined by the legislation in force.

According to the methodological subordination of quantity and quality, analysis and evaluation of tax potential should be determined, first of all, in accordance with quality indicators and then by the determination method of the quantity. From the quality aspect, tax potential means economic relations system formed by the following characteristics [6]:

- Tax potential exists in the economic environment, and its content consists of alienation of total income tax relations' participants (and thus a portion of the

funds) in other words, it is the limitation of raw materials sources of reproduction process in the institutional form in favour of raw materials resources of
reproduction process of the state without prejudice to the reproduction process;

- Tax potential has a feature to organize the mass of tax payments of some tax
  relations participants considering legal territory and economic system of
  taxation;
- Tax potential is a constituent part of total financial resources of tax relations
  objects in certain economic territories; tax potential formation is included by
  realization of economic resources of economic territories; economic resources
  should be determined as the main factor of tax potential;
- Determination of tax potential defines economic resources, reproduction process
  resources and it also means determination of tax potential of further periods.

As quantity, it is expressed by the sum of all payments that can be obtained or
average price of tax and payments sum collected in recent years as well as by tax
potential indicator determined as correlation of tax potential to the number of the
population, territory and other indicators [7–9].

3 Development of Mamdani-Type Fuzzy Inference System

Fuzzy inference is the process of formulating the mapping from a given input to an
output using fuzzy logic [10–13]. There is made a decision about given data at the
end of this mapping. In the work, the problem of evaluating the impact of the
changes and additions made in tax administration on the tax potential by fuzzy
methods is investigated. In the article, we use five-input, five-rule and single-output
fuzzy inference system (FIS) for evaluation of tax reform.

Pre-processing. Input data are collected from 20 experts who evaluated the
reforms in tax administration in 0–100 point scale. Same time these data are verified
by other experts and determined weights for the correctness of evaluations. We
multiplied every input by its weight and found average value (Table 1) at the first
stage.

Table 1 Average value of the expert estimation of tax reform in tax administration

Sign	The reforms in tax administration	Average value
x_1	Some departments have been established on the basis 2,11, 12 of the Regional Tax Offices	9,85
x_2	"ASAN" signature application	59,4
x_3	"MOBILE" signature application	59,4
x_4	E-Audit application	58,9
x_5	The application of unique standards of services to taxpayers	49,6

We define five IF-THEN rules for this problem (Table 2), where y is the output value of the system.

Fuzzy inference process comprises five parts [14]:

1. **Fuzzification of the input variables**. In this step we determine the degree of inputs which they belong to each of the appropriate fuzzy sets via membership functions (MF). Before the rules can be evaluated, the inputs must be fuzzified according to linguistic sets.: For example. How effective is "ASAN" signature application? If this input is estimated by 60 points in (0–100) scale, the membership degree of the "very effective" linguistic set is 0.4.

We use following membership functions for the input data.

The input data x_1 ("Some departments have been established on the basis 2, 11, 12 of the Regional Tax Offices") is fuzzified by a triangle membership function and define three linguistic variables: "irrational", "rational" and "more rational" (Fig. 1). The triangular-shaped membership function defined as follows:

Table 2 The IF-THEN rules of the FIS

Index	IF-THEN rules
1	IF (x_2 is not "sufficient") and (x_3 is not "sufficient") and (x_4 is "effective") and (x_5 is "rational") THEN (y is "good")
2	IF (x_1 is "rational") and (x_2 is "good") and (x_3 is "good") and (x_4 is "more effective") and (x_5 is "rational") THEN (y is "very good")
3	IF (x_1 is "irrational") and (x_2 is "good") and (x_4 is "effectless") and (x_5 is "rational") THEN (y is "sufficient")
4	IF (x_2 is not "sufficient") and (x_4 is "more effective") and (x_5 is "rational") THEN (y is "very good")
5	IF (x_1 is "more rational") and (x_2 is "excellent") and (x_3 is "excellent") and (x_4 is "more effective") and (x_5 is "more rational") THEN (y is "excellent")

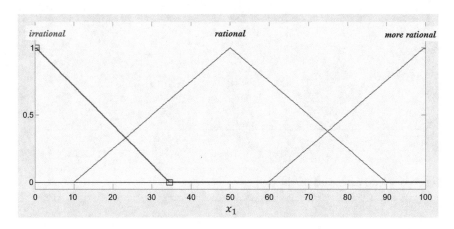

Fig. 1 Graphical illustration of MF of x_1

$$f(x; a, b, c) = \left\{ \begin{array}{ll} 0, & x \leq a \\ \frac{x-a}{b-a}, & a \leq x \leq b \\ \frac{c-x}{c-b}, & b \leq x \leq c \\ 0, & c \leq x \end{array} \right\},$$

where a, b and c are the parameters of the MF.

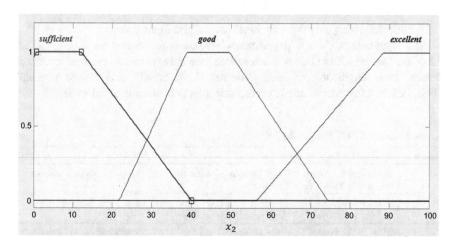

Fig. 2 Graphical illustration of MF of x_2

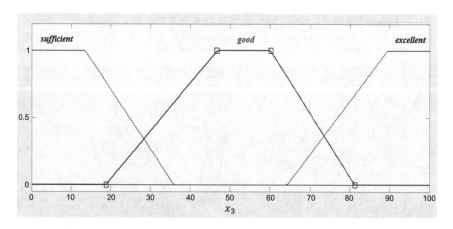

Fig. 3 Graphical description of MF of x_3

The input data x_2 ("ASAN" signature application) and x_3 ("MOBILE" signature application) are fuzzified by trapezoid MF and define three linguistic variables: "sufficient", "good" and "excellent" (Fig. 2 and 3). Trapezoid MF is defined as follows:

$$f(x; a, b, c, d) = \begin{cases} 0, & x \leq a \\ \dfrac{x-a}{b-a}, & a \leq x \leq b \\ 1 & b \leq x \leq c \\ \frac{d-x}{d-c}, & c \leq x \leq d \\ 0, & d \leq x \end{cases},$$

where a, b, c and d are the parameters of the MF.

The input data x_4 ("E-Audit application") is fuzzied by Gaussian membership function and define three linguistic variables: "effect less", "effective" and "more effective" (Fig. 4). Gaussian MF is defined as follows:

$$f(x; \sigma, c) = e^{-\frac{(x-c)^2}{2\sigma^2}},$$

where σ and c are the parameters of the MF.

The input data x_5 ("The application of unique standards of services to taxpayers") is fuzzified by Bell-shaped MF and define three linguistic variables: "irrational", "rational" and "more rational" (Fig. 5). The Bell-shaped MF is defined as follows:

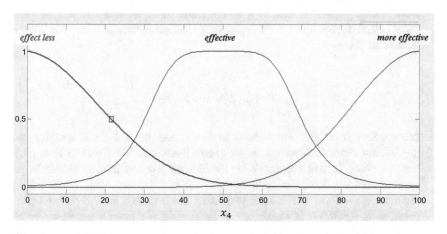

Fig. 4 Graphical illustration of MF of x_4

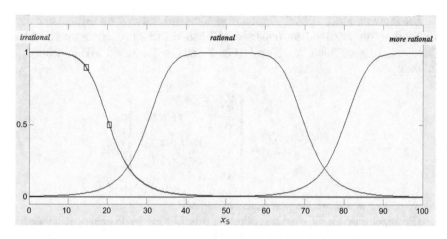

Fig. 5 Graphical illustration of MF of x_5

$$f(x; a, b, c) = \frac{1}{1 + \left|\frac{x-c}{a}\right|^{2b}},$$

where a, b and c are the parameters of the MF.

2. **Application of the fuzzy operator (AND or OR) in the antecedent**. After the inputs are fuzzified, the degree which each part of the antecedent is satisfied for each rule. If the antecedent of a given rule has more than one part, the fuzzy operator is applied to obtain one number that represents the result of the antecedent for that rule [14]. This number is then applied to the output function. The input to the fuzzy operator is two or more membership values from fuzzified input variables. The output is a single truth value. In our system, we used "AND" and probabilistic "OR" operators. The probabilistic "OR" operator is defined as follows:

$$probor(a, b) = a + b - ab$$

3. **Implication from the antecedent to the consequent**. Before applying the implication method, every rule is weighted. Every rule has a weight (a number between 0 and 1), which is applied to the number given by the antecedent. In our system, all rules have the same weight and thus has no effect at all on the implication process. After proper weighting has been assigned to each rule, the implication method is implemented. A consequent is a fuzzy set represented by a membership function, which weights appropriately the linguistic characteristics that are attributed to it. The consequent is reshaped using a function associated with the antecedent (a single number). The input for the implication

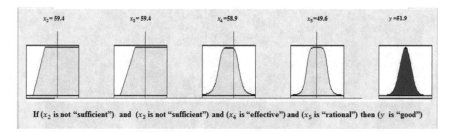

Fig. 6 The Implication process for the first IF-THEN rule

process is a single number given by the antecedent, and the output is a fuzzy set (Fig. 6). The implication is implemented for each rule [14].

4. **Aggregation of the consequents across the rules**. Because decisions are based on the testing of all the rules in a FIS, the rules must be combined in some manner in order to make a decision. Aggregation is the process by which the fuzzy sets that represent the outputs of each rule are combined into a single

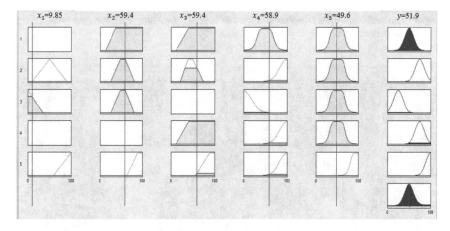

Fig. 7 The Aggregation process of the FIS

Fig. 8 The defuzzification process of the FIS

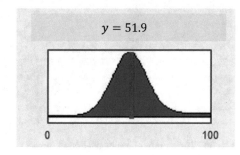

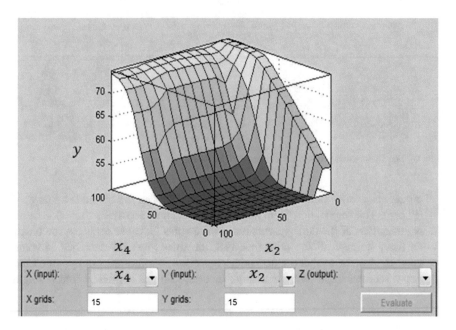

Fig. 9 The surface of the evaluation of impact of tax administration reforms on tax potential that depends on x_4 and x_2

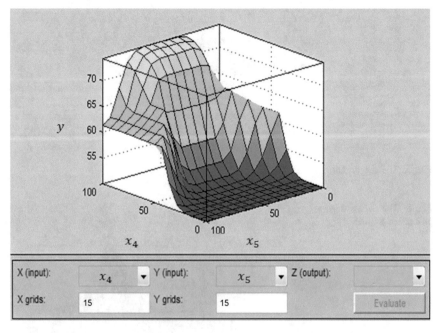

Fig. 10 The surface of the evaluation of impact of tax administration reforms on tax potential that depends on x_4 and x_5

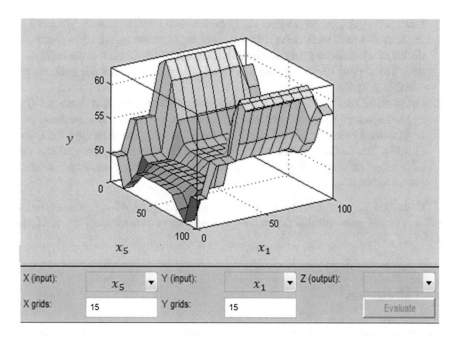

Fig. 11 The surface of the evaluation of impact of tax administration reforms on tax potential that depends on x_5 and x_1

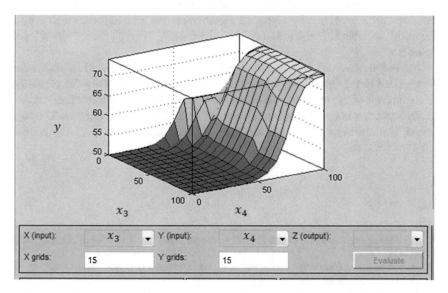

Fig. 12 The surface of the evaluation of impact of tax administration reforms on tax potential that depends on x_3 and x_4

fuzzy set. Aggregation occurs only once for each output variable, just prior to the fifth and final step, defuzzification. The input of the aggregation process is the list of truncated output functions returned by the implication process for each rule. The output of the aggregation process is one fuzzy set for each output variable (Fig. 7).

5. **Defuzzification**. The input for the defuzzification process is a fuzzy set (the aggregate output fuzzy set) and the output is a single number. We use the Center of Gravity Defuzzification (CoGD) method for the defuzzification operation [15–18]. The CoGD method avoids the defuzzification ambiguities which may arise when an output degree of membership comes from more than one crisp output value (Fig. 8).

There are given surfaces of the evaluation of impact of tax administration reforms on tax potential that depends on different input variables in Figs. 9, 10, 11 and 12.

4 Conclusion

The Mamdani-type fuzzy inference method with 5 input, 5 IF-THEN rule and single-output applied to the evaluation of impact of Azerbaijan tax administration reforms in 2013 year on tax potential has been applied in the article. As input data of the system are taken reforms in tax administration and fuzzified by the triangle, trapezoid, Gaussian and Bell membership functions. Fuzzy sets of the output system have been evaluated by Gaussian membership function and determined by 5 linguistic variables: "unsatisfactory", "satisfactory", "good", "very good" and "excellent".

The output value of our suggested fuzzy inference process is 51.9. This value corresponds to the "good" linguistic set in 1–100 point scale that divided regularly.

As a result of the research, the impact of Azerbaijan tax administration reforms in 2013 year on tax potential is evaluated by "good" linguistic variable. The suggested Mamdani-type fuzzy inference system has been realized in the Matlab program package.

References

1. A.F. Musayev, Tax potential and its assessment methods. Tax Magazine of Azerbaijan N5 (119)/2014
2. A.F. Musayev, *Innovation economics and tax stimulation*. (Baku, The University of Azerbaijan, 2014), pp. 184
3. A. Musayev, A. Gahramanov, *Introduction to Econometrics* (Baku, The University of Azerbaijan, 2011), p. 173

4. Tax Reforms in EU Member States, Tax policy challenges for economic growth and fiscal sustainability. Eur. Econom. (2013)
5. A.S. Karatayev, *Instrumentary of Tax Capacity Estimation* (2010)
6. D.N. Slobodchikov, Dissertation. *Tax Potential in the System of Inter-Budgetary Relations.* (Code, HAC- 08.00.10) (2010)
7. J. Yen, R. Langari, *Fuzzy Logic.* Pearson Education (2004)
8. T.J. Ross, *Fuzzy Logic with Engineering Applications.* Wiley (2010)
9. G. Koop, *Analysis of Economic Data* (Wiley, Chichester, 2000)
10. E.H. Mamdani, S. Assilian, An experiment in linguistic synthesis with a fuzzy logic controller. Int. J. Man-Machine Stud. **7**(1), 1–13 (1975)
11. E.H. Mamdani, Advances in the linguistic synthesis of fuzzy controllers. Int. J. Man Mach. Stud. **8**, 669–678 (1976)
12. M. Sugeno, *Industrial Applications of Fuzzy Control*, Elsevier Science Pub. Co., (1985)
13. L.A. Zadeh, Outline of a new approach to the analysis of complex systems and decision processes. IEEE Trans. Syst. Man Cybernet. **3**(1), 28–44 (1973)
14. Mathwork, *Fuzzy Inference Process.* http://www.mathworks.com/
15. S.S. Rustamov, An application of neuro-fuzzy model for text and speech understanding systems./ PCI'2012, in *The IV International Conference "Problems of Cybernetics and Informatics*, vol. I (Baku, Azerbaijan, 2012) pp. 213–217
16. S. Rustamov, E.E. Mustafayev, M.A. Clements, Sentiment analysis using neuro-fuzzy and hidden Markov models of text, in *IEEE SoutheastCon 2013*, (Jacksonville, USA, 2013) inpress
17. S.S. Rustamov, M. A. Clements, Sentence-level subjectivity detection using neuro-fuzzy and hidden markov models, in *Proceedings of the 4th Workshop on Computational Approaches to Subjectivity, Sentiment and Social Media Analysis in NAACL-HLT2013,* Atlanta, USA, 2013, pp. 108–114
18. S.S. Rustamov, On an understanding system that supports human-computer dialogue. PCI'2012, in *The IV International Conference Problems of Cybernetics and Informatics*, vol. I, (Baku, Azerbaijan, 2012), pp. 217–221

Chemical Kinetics in Situations Intermediate Between Usual and High Concentrations: Fuzzy-Motivated Derivation of the Formulas

Olga Kosheleva, Vladik Kreinovich and Laécio Carvalho Barros

Abstract In the traditional chemical kinetics, the rate of each reaction

$$A + \ldots + B \rightarrow \ldots$$

is proportional to the product $c_A \cdot \ldots \cdot c_B$ of the concentrations of all the input substances A, …, B. For high concentrations $c_A, \ldots, c_B$, the reaction rate is known to be proportional to the minimum $\min(c_A, \ldots, c_B)$. In this paper, we use fuzzy-related ideas to derive the formula of the reaction rate for situations intermediate between usual and high concentrations.

1 Chemical Kinetics in Situations Intermediate Between Usual and High Concentrations: Formulation of the Problem

Chemical kinetics: usual formulas. Chemical kinetics describes the rate of chemical reactions. For usual concentrations, the rate of a reaction between two substances A and B is proportional to the product $c_A \cdot c_B$ of their concentrations; see, e.g., [1, 2]. Similarly, if we have a reaction

$$A + \ldots + B \rightarrow \ldots$$

O. Kosheleva (✉) · V. Kreinovich
University of Texas at El Paso, 500 W. University, El Paso, Texas 79968, USA
e-mail: olgak@utep.edu

V. Kreinovich
e-mail: vladik@utep.edu

L. C. Barros
Instituto de Matemática, Estatística, e Computação Científica (IMECC),
Universidade Estadual de Campinas (UNICAMP), Campinas, SP 6065, Brazil
e-mail: laeciocb@ime.unicamp.br

© Springer International Publishing AG, part of Springer Nature 2018
L. A. Zadeh et al. (eds.), *Recent Developments and the New Direction
in Soft-Computing Foundations and Applications*, Studies in Fuzziness
and Soft Computing 361, https://doi.org/10.1007/978-3-319-75408-6_40

with three or more substances, the rate of this reaction is proportional to the products of the concentrations of all these substances $c_A \cdot \ldots \cdot c_B$.

How formulas of chemical kinetics are usually derived. Let us start the explanation of how the *general* formulas of chemical kinetics are derived by first considering the case of *two* substances A and B.

Molecules of both substances are randomly distributed in space. So, for each molecule of the substance A, the probability that it meets a molecule of the substance B is proportional to the concentration c_B. If the molecules meet, then (with a certain probability) they get into a reaction. Thus, the expected number of reactions involving a given molecule of the substance A is also proportional to c_B. The total number of A-molecules in a given volume is proportional to c_A; thus, the total number of reactions per unit time is proportional to $c_A \cdot c_B$.

Similarly, for the case of three or more substances, we can conclude that the reaction rate is indeed proportional to the product $c_A \cdot \ldots \cdot c_B$.

Case of high concentrations. When the concentrations are very high, there is no need for the molecules to randomly bump into each other; these molecules are everywhere. So, as soon as we have molecules of all needed type, the reaction starts. In other words, in this case, the reaction rate is proportional to the concentration of the corresponding tuples– i.e., to the minimum $\min(c_A, \ldots, c_B)$ of all the input concentrations $c_A, \ldots, c_B$.

Example. The formula $\min(c_A, \ldots, c_B)$ can be easily illustrated on the example of a relation which is non-chemical reaction but which is described by the same chemical kinetic-type equations: the relation between predators and prey.

When we have usual (small) concentrations of wolves W and rabbits R in a forest, the probability for a wolf to find a rabbit is proportional to the concentration c_R of rabbits, so the overall amount of rabbits eaten by wolves is proportional to the product $c_W \cdot c_R$.

On the other hand, for high concentrations, e.g., if we throw a bunch of rabbits into a zoo cage filled with hungry wolves, there is no need to look for a prey, each wolf will start eating its rabbit—as long as there are sufficiently many rabbits to feed all the wolves. So:

- When $c_R \geq c_W$, the number of eaten rabbits will be proportional to the number of wolves, i.e., to c_W.
- In situations when there are not enough rabbits (i.e., when $c_R < c_W$), the number of eaten rabbits is proportional to the number of rabbits, i.e., to c_R.

In both cases, the reaction rate is proportional to $\min(c_R, c_W)$.

Empirical evidence for high-concentration reaction rate. The high-concentration reaction rate indeed turned out to be very useful to describe biochemical processes; see, e.g., [3, 4].

Interesting observation: simulations of high-concentration reactions lead to efficient algorithms. It is known that in many cases, difficult-to-solve computational problems can be reduced to problems of chemical kinetics. In such situations of *chemical computing*, we can efficiently solve the original computational problems by either actually performing the corresponding chemical reactions, or by performing a computer simulation of these reactions; see, e.g., [5].

To make the simulations as fast as possible, it is desirable to simulate reactions which are as fast as possible. The reaction rate increases with the concentrations of the reagents. Thus, to speed up simulations, we should simulate high-concentration reactions. This simulation indeed speeds up the corresponding computations; see, e.g., [6, 7].

Main problem. While we know the formulas for the usual and for the high concentrations, it is not clear how to compute the reaction rate for concentrations between usual and high.

What is known. Both formulas $r = c_A \cdot c_B$ and $r = \min(c_A, c_B)$ are particular cases of *t-norms* ("and"-operations in fuzzy logic; see, e.g., [8–10]). This is not a coincidence: there is no reaction if one of the substances is missing, so $c_A = 0$ or $c_B = 0$ imply that $r = 0$—which is exactly the property of a t-norm. Fuzzy t-norms have indeed been effectively used to describe chemical reactions [3, 4].

Remaining problem. The problem is that there are many possible "and"-operations, and it is not clear which one we should select.

What we do in this paper. In this paper, we use the analysis of the corresponding chemical processes to derive the formulas that adequately describe the reaction rate in intermediate situations—and thus, to appropriately select the corresponding "and"-operation.

2   Chemical Kinetics in Situations Intermediate Between Usual and High Concentrations: Analysis of the Problem, Resulting Formulas, and Discussion

Towards formulating the problem in precise terms. Let us start with the case of two substances A and B. As we have mentioned earlier, the two molecules get into a reaction only when they are close enough. When these molecules are close enough, then, within the corresponding small region, the reaction rate is proportional to the minimum $\min(c_A, c_B)$ of their concentrations.

When concentrations are small, then, within each region, we have either zero or one molecule; the probability to have two molecules is very small (proportional to the square of these concentrations) and can, therefore, be safely ignored. In this case, for each region, the reaction occurs if we have both an A-molecule and a B-molecule. The probability to have an A-molecule is proportional to c_A; the probability to have a B-molecule is proportional to c_B. Since the distributions for A and B are independent, the probability to have both A- and B-molecules in a region is equal to the product of these probabilities and is, thus, proportional to the product of the concentrations $c_A \cdot c_B$.

When the concentrations are high, then each region has molecules of both types. The average number of A-molecules in a region is proportional to c_A, i.e., has the form $k \cdot c_A$ for some proportionality coefficient k. Similarly, the average number of B-molecules in a region is equal to $k \cdot c_B$. So the average reaction rate is proportional to $\min(k \cdot c_A, k \cdot c_B) = k \cdot \min(c_A, c_B)$, i.e., is proportional to $\min(c_A, c_B)$.

This analysis leads us to the following reformulation of our problem.

Resulting formulation of the problem in precise terms. Within a unit volume, we have a certain number r of "small regions", i.e., regions such that only molecules within the same region can interact with each other.

We have a total of $N_A = N \cdot c_A$ molecules of the substance A, and we have a total of $N_B = N \cdot c_B$ molecules of the substance B. Each of these molecules is randomly distributed among the regions, i.e., it can be located in any of the r regions with equal probability. Distributions of different molecules are independent from each other. Within each region, the reaction rate is proportional to the minimum $\min(n_A, n_B)$ of the numbers n_A and n_B of A- and B-molecules in this region. The overall reaction rate can be computed as the *average* over all the regions —i.e., in other words, as the mathematical expectation of this minimum.

Analysis of the problem. Based on the above description, the number of A-molecules in a region follows the Poisson distribution (see, e.g., [11]), according to which, for each value k, the probability to have exactly $n_A = k$ A-molecules is equal to

$$\text{Prob}(n_A = k) = \exp(-\lambda_A) \cdot \frac{\lambda_A^k}{k!}. \tag{1}$$

The mean value of the Poisson random variable is λ_A; on the other hand, we have $N \cdot c_A$ A-molecules in r cells, so the average number of A-molecules in a cell is equal to the ratio $\dfrac{N \cdot c_A}{r}$, so

$$\lambda_A = \frac{N \cdot c_A}{r}. \tag{2}$$

In other words, $\lambda_A = c \cdot c_A$, where we denoted $c \stackrel{\text{def}}{=} \dfrac{N}{r}$. Similarly, for the number n_B of B-molecules in the region, we have a probability distribution

$$\text{Prob}(n_B = k) = \exp(-\lambda_B) \cdot \frac{\lambda_B^k}{k!}, \tag{3}$$

with

$$\lambda_B = \frac{N \cdot c_B}{r} = c \cdot c_B. \tag{4}$$

The desired distribution for $n = \min(n_A, n_B)$ can be obtained from the fact that

$$n \geq k \Leftrightarrow (n_A \geq k \,\&\, n_B \geq k).$$

Since A- and B-molecules are independently distributed, the A-related value n_A and the B-related value n_B are also independent. Therefore,

$$\text{Prob}(n \geq k) = \text{Prob}(n_A \geq k) \cdot \text{Prob}(n_B \geq k). \tag{5}$$

Based on (1) and (3), we conclude that

$$\text{Prob}(n_A \geq k) = \sum_{\ell=k}^{\infty} \text{Prob}(n_A = \ell) = \exp(-\lambda_A) \cdot \sum_{\ell=k}^{\infty} \frac{\lambda_A^\ell}{\ell!} \tag{6}$$

and

$$\text{Prob}(n_B \geq k) = \sum_{\ell=k}^{\infty} \text{Prob}(n_B = \ell) = \exp(-\lambda_B) \cdot \sum_{\ell=k}^{\infty} \frac{\lambda_B^\ell}{\ell!}. \tag{7}$$

So, we conclude that

$$\text{Prob}(n \geq k) = \exp(-(\lambda_A + \lambda_B)) \cdot \left(\sum_{\ell=k}^{\infty} \frac{\lambda_A^\ell}{\ell!} \right) \cdot \left(\sum_{\ell=k}^{\infty} \frac{\lambda_B^\ell}{\ell!} \right). \tag{8}$$

The expected value E can be now computed as

$$E = \sum_{k=0}^{\infty} k \cdot \text{Prob}(n = k) = \sum_{k=0}^{\infty} k \cdot (\text{Prob}(n \geq k) - \text{Prob}(n \geq k+1)) =$$

$$0 \cdot (\text{Prob}(n \geq 0) - \text{Prob}(n \geq 1)) + 1 \cdot (\text{Prob}(n \geq 1) - \text{Prob}(n \geq 2)) +$$

$$2 \cdot (\text{Prob}(n \geq 2) - \text{Prob}(n \geq 3)) + \ldots =$$

$$\text{Prob}(n \geq 1) \cdot (1 - 0) + \text{Prob}(n \geq 2) \cdot (2 - 1) + \ldots =$$

$$\sum_{k=1}^{\infty} \text{Prob}(n \geq k). \tag{9}$$

Substituting the expression (8) into this formula, we arrive at the following expression.

Resulting formula for the reaction rate. The reaction rate is proportional to

$$E \stackrel{\text{def}}{=} \exp(-(\lambda_A + \lambda_B)) \cdot \sum_{k=1}^{\infty} \left(\sum_{\ell=k}^{\infty} \frac{\lambda_A^\ell}{\ell!} \right) \cdot \left(\sum_{\ell=k}^{\infty} \frac{\lambda_B^\ell}{\ell!} \right), \tag{10}$$

where $\lambda_A = c \cdot c_A$ and $\lambda_B = c \cdot c_B$ for some constant c.

For a reaction between three or more substances

$$A + \ldots + B \rightarrow \ldots,$$

we similarly get a formula

$$E \stackrel{\text{def}}{=} \exp(-(\lambda_A + \ldots \lambda_B)) \cdot \sum_{k=1}^{\infty} \left(\sum_{\ell=k}^{\infty} \frac{\lambda_A^\ell}{\ell!} \right) \cdot \ldots \cdot \left(\sum_{\ell=k}^{\infty} \frac{\lambda_B^\ell}{\ell!} \right), \tag{10a}$$

where $\lambda_A = c \cdot c_A, \ldots$, and $\lambda_B = c \cdot c_B$ for some constant c.

Towards simplifying the above formula. Let us show that the above formula can be somewhat simplified by expressing it in terms of the upper incomplete Gamma-function. The upper incomplete Gamma function is often used to analyze the Poisson distribution. It is defined as

$$\Gamma(s, x) \stackrel{\text{def}}{=} \int_x^{\infty} t^{s-1} \cdot \exp(-t) \, dt. \tag{11}$$

Its relation to the Poisson distribution comes from the fact that for integer values s, we have

$$\exp(-\lambda) \cdot \sum_{\ell=0}^{s-1} \frac{\lambda^\ell}{\ell!} = \frac{\Gamma(s, \lambda)}{(s-1)!}. \tag{12}$$

Since $\exp(\lambda) = \sum_{\ell=0}^{\infty} \frac{\lambda^\ell}{\ell!}$, we have

$$\exp(-\lambda) \cdot \sum_{\ell=0}^{\infty} \frac{\lambda^\ell}{\ell!} = 1. \tag{13}$$

Subtracting (12) from (13), we conclude that

$$\exp(-\lambda) \cdot \sum_{\ell=s}^{\infty} \frac{\lambda^\ell}{\ell!} = 1 - \frac{\Gamma(s, \lambda)}{(s-1)!}. \tag{14}$$

In particular, for $\lambda = \lambda_A$ and for $\lambda = \lambda_B$, we get the following formulas:

$$\exp(-\lambda_A) \cdot \sum_{\ell=s}^{\infty} \frac{\lambda_A^\ell}{\ell!} = 1 - \frac{\Gamma(s, \lambda_A)}{(s-1)!}; \tag{15}$$

$$\exp(-\lambda_B) \cdot \sum_{\ell=s}^{\infty} \frac{\lambda_B^\ell}{\ell!} = 1 - \frac{\Gamma(s, \lambda_B)}{(s-1)!}. \tag{16}$$

Substituting these expressions instead of the sums into the formula (10), we arrive at the following expression.

Simplified version of the rate formula. The reaction rate is proportional to

$$E = \sum_{k=1}^{\infty} \left(1 - \frac{\Gamma(k, \lambda_A)}{(k-1)!}\right) \cdot \left(1 - \frac{\Gamma(k, \lambda_B)}{(k-1)!}\right). \tag{17}$$

For the reaction between three or more substances, a similar formula takes the form

$$E = \sum_{k=1}^{\infty} \left(1 - \frac{\Gamma(k, \lambda_A)}{(k-1)!}\right) \cdot \ldots \cdot \left(1 - \frac{\Gamma(k, \lambda_B)}{(k-1)!}\right). \tag{17a}$$

Analysis of the above formula. Let us show that in both limit cases—when concentrations are small and when concentrations are large—the formula (10) (and thus, the equivalent formula (17)) leads to the known expressions for the reaction rate.

Indeed, when λ_A and λ_B are small, then $\exp(-(\lambda_A + \lambda_B))$ is approximately equal to 1. Also, terms proportional to λ_A^2 and to higher powers of λ_A are much smaller than the term proportional to λ_A and can, therefore, be ignored. So, in this case, we have $\sum_{\ell=1}^{\infty} \frac{\lambda_A^\ell}{\ell!} \approx \lambda_A$ and $\sum_{\ell=k}^{\infty} \frac{\lambda_A^\ell}{\ell!} \approx 0$ for $k > 1$. Similarly, we have $\sum_{\ell=1}^{\infty} \frac{\lambda_B^\ell}{\ell!} \approx \lambda_B$ and $\sum_{\ell=k}^{\infty} \frac{\lambda_A^\ell}{\ell!} \approx 0$ for $k > 1$. Thus, the formula (10) takes the form $E = \lambda_A \cdot \lambda_B$. Since $\lambda_A = c \cdot c_A$ and $\lambda_B = c \cdot c_B$, this means that in this case, the reaction rate is indeed proportional to $c_A \cdot c_B$.

The estimate for the case when λ_A and λ_B are small was based on the fact that in this case, the terms $\frac{\lambda_A^\ell}{\ell!}$ drastically decrease with ℓ, so we only need to take the into account the largest term—which corresponds to the smallest possible value $\ell = 1$. When λ_A and λ_B are large, the dependence on ℓ is no longer monotonic. The largest value of this term can be estimated if we approximate $\ell!$ by the usual Stirling approximation $\ell! \approx \left(\frac{\ell}{e}\right)^\ell$, reducing each term $\frac{\lambda^\ell}{\ell!}$ to $\left(\frac{\lambda \cdot e}{\ell}\right)^\ell$. This term is the largest when its logarithm

$$L \stackrel{\text{def}}{=} \ell \cdot (\ln(\lambda) + 1 - \ln(\ell))$$

attains the largest possible value. Differentiating L with respect to ℓ and equating the resulting derivative to 0, we conclude that $\ell = \lambda$. For this value ℓ, the term $\left(\frac{\lambda \cdot e}{\ell}\right)^\ell$ turns into $\exp(\lambda)$. Since $\exp(\lambda)$ is equal to the whole sum $\sum_{\ell=0}^{\infty} \frac{\lambda^\ell}{\ell!}$, this means that all other terms in this sum are much smaller—and can thus be, in the first approximation, ignored.

In this first approximation, we can therefore assume that this term is equal to $\exp(\ell)$, while all other terms are 0s. Thus, the sum $\sum_{\ell=k}^{\infty} \dfrac{\lambda_A^{\ell}}{\ell!}$ is equal to 0 when $\ell > \lambda_A$ and to $\exp(\lambda_A)$ when $\ell \leq \lambda_A$. Similarly, the sum $\sum_{\ell=k}^{\infty} \dfrac{\lambda_B^{\ell}}{\ell!}$ is equal to 0 when $\ell > \lambda_B$ and to $\exp(\lambda_B)$ when $\ell \leq \lambda_B$.

So, in the sum (10), the only non-zero terms correspond to cases when $\ell \leq \lambda_A$ and $\ell \leq \lambda_B$, i.e., when $\ell \leq \min(\lambda_A, \lambda_B)$. Each of these $\min(\lambda_A, \lambda_B)$ non-zero terms is equal to

$$\exp(-(\lambda_A + \lambda_B)) \cdot \exp(\lambda_A) \cdot \exp(\lambda_B) = 1,$$

so their sum is indeed approximately equal to $\min(\lambda_A, \lambda_B)$.

Remaining open questions. Formulas similar to chemical kinetics equations are used in many different applications, e.g., in the dynamics of biological species or in the analysis of knowledge propagation. In all these cases, we can consider the product and min operations, and we can also consider intermediate cases.

The above derivation of the formulas for the intermediate "and"-operation uses the specifics of chemical kinetics. It would be interested to provide a similar analysis in other applications areas and see which "and"-operations are appropriate in these situations.

Acknowledgements This work was supported in part by the Brazil National Council Technological and Scientific Development CNPq, by the US National Science Foundation grants HRD-0734825 and HRD-1242122 (Cyber-ShARE Center of Excellence) and DUE-0926721, and by an award "UTEP and Prudential Actuarial Science Academy and Pipeline Initiative" from Prudential Foundation. This work was partly performed when V. Kreinovich was a visiting researcher in Brazil. The authors are greatly thankful to the anonymous referees for valuable suggestions.

References

1. J.E. House, *Principles of Chemical Kinetics* (Academic Press, Burlington, Massachisetts, 2007)
2. G. Marin, G.S. Yablonsky, *Kinetics of Chemical Reactions* (Wiley, Weinheim, Germany, 2011)
3. F.S. Pedro, L.C. Barros, The use of t-norms in mathematical models of epidemics, in: *Proceedings of the 2013 IEEE International Conference on Fuzzy Systems FUZZ-IEEE'2013*, Hyderabad, India, July 7–10, 2013
4. E. Massad, N. Ortega, L. Barros, C. Struchiner, *Fuzzy Logic in Action: Applications in Epidemiology and Beyond* (Springer Verlag, Berlin, Heidelberg, 2008)
5. A.I. Adamatzky, Information-processing capabilities of chemical reaction-diffusion systems. 1. Belousov-Zhabotinsky media in hydrogel matrices and on solid supports. Advanc. Mater. Opt. Electron. **7**(5), 263–272 (1997)
6. V. Kreinovich, L.O. Fuentes, Simulation of chemical kinetics–a promising approach to inference engines, in: J. Liebowitz (ed.), *Proceedings of the World Congress on Expert Systems, Orlando, Florida, 1991*, (Pergamon Press, New York, 1991), vol. 3, pp. 1510–1517

7. V. Kreinovich, O. Fuentes, High-concentration chemical computing techniques for solving hard-to-solve problems, and their relation to numerical optimization, neural computing, reasoning under uncertainty, and freedom of choice, in *Molecular and Supramolecular Information Processing: From Molecular Switches to Logical Systems*, ed. by E. Katz (Wiley, Wienheim, Germany, 2012), pp. 209–235
8. G. Klir, B. Yuan, *Fuzzy Sets and Fuzzy Logic* (Prentice Hall, Upper Saddle River, New Jersey, 1995)
9. H.T. Nguyen, E.A. Walker, *A First Course in Fuzzy Logic* (Chapman and Hall/CRC, Boca Raton, Florida, 2006)
10. L.A. Zadeh, Fuzzy sets. Information and Control **8**, 338–353 (1965)
11. D.J. Sheskin, *Handbook of Parametric and Nonparametric Statistical Procedures* (Chapman and Hall/CRC, Boca Raton, Florida, 2011)

Part XII
Fuzziness and Health Care

Estimating the Membership Function of the Fuzzy Willingness-to-Pay/Accept for Health via Bayesian Modelling

Michał Jakubczyk

Abstract Determining how to trade off individual criteria is often not obvious, especially when attributes of very different nature are juxtaposed, e.g. health and money. The difficulty stems both from the lack of adequate market experience and strong ethical component when valuing some goods, resulting in inherently imprecise preferences. Fuzzy sets can be used to model willingness-to-pay/accept (WTP/WTA), so as to quantify this imprecision and support the decision making process. The preferences need then to be estimated based on available data. In the paper, I show how to estimate the membership function of fuzzy WTP/WTA, when decision makers' preferences are collected via survey with Likert-based questions. I apply the proposed methodology to a data set on WTP/WTA for health. The mathematical model contains two elements: the parametric representation of the membership function and the mathematical model how it is translated into Likert options. The model parameters are estimated in a Bayesian approach using Markov-chain Monte Carlo. The results suggest a slight WTP-WTA disparity and WTA being more fuzzy as WTP. The model is fragile to single respondents with lexicographic preferences, i.e. not willing to accept any trade-offs between health and money.

1 Introduction

Decision making with multiple criteria requires, explicitly or implicitly, making trade-offs between attributes describing decision alternatives. Even if the criteria are quantifiable and expressed as numbers (not only as labels along nominal or ordinal scale, e.g. *ugly*, *mediocre*, and *beautiful*), they may be of very different type, making it difficult to juxtapose them and decide about the exact trade-off coefficient. This is the case when non-market goods, such as health, safety, clean environment, etc., are being valued against money. The present paper focuses on juxtaposing health and financial consequences.

M. Jakubczyk (✉)
Decision Analysis and Support Unit, SGH Warsaw School of Economics, Warsaw, Poland
e-mail: michal.jakubczyk@sgh.waw.pl
URL: http://michaljakubczyk.pl

© Springer International Publishing AG, part of Springer Nature 2018   537
L. A. Zadeh et al. (eds.), *Recent Developments and the New Direction in Soft-Computing Foundations and Applications*, Studies in Fuzziness and Soft Computing 361, https://doi.org/10.1007/978-3-319-75408-6_41

The amount of health gained with a given decision can, in principle, be expressed as a number: an increase in the life expectancy or—more generally—the additional quality-adjusted life years (QALYs). The latter combines the improvements in quality and longevity of life, is formally founded in axiomatic approach [1], and is operationally calculated via assigning numerical values, von Neumann-Morgenstern utilities, to health states [2] defined within some system, e.g. EQ-5D-3L [3]. Still, it is difficult to put a precise monetary value on health, as strikingly visible in systematic reviews of published estimates of the value of statistical life: the standard deviations of published results (e.g. within a given country) are usually of the order of magnitude of the mean values [4–6]. The valuation of non-market goods is also specific in a sense that there is a great difference between the willingness-to-pay (WTP, the amount one is willing to pay for an additional unit of good) and the willingness-to-accept (WTA, the amount one demands to obtain, to accept the loss of one unit) [7–9]. In spite of these difficulties, it is necessary to grasp the preferences quantitatively, in order to support the decision making process and make it transparent.

In social sciences, in order to formally model imprecision, fuzzy sets and fuzzy logic have been used for many years [10]. Therefore, as a solution, it was suggested to also treat the WTP and WTA for health as a fuzzy concept [11], i.e. to define a fuzzy number fWTP (or fWTA) over the universe of $\mathbf{R}_+$ with a non-increasing (non-decreasing, respectively) membership function $\mu_{fWTP}(\lambda)$ ($\mu_{fWTA}(\lambda)$), interpreted as the conviction that it makes sense to pay (accept) λ for an additional QALY (a loss of QALY). It was shown how to support decision making via fuzzy preference relations [11] or choice functions, when multiple alternatives are present [12, 13]. One of the choice functions used the 0.5-cut of fWTP/fWTA, and formal statistical methods how to infer this value based on data collected via surveys in random samples were shown [13]. The choice functions may, however, require knowing the complete shape of the membership function, and learning this shape allows, additionally, a better understanding of the nature of imprecision in the perception of WTP/WTA, e.g. allows comparing the amount of fuzziness between WTP and WTA. Building on the above motivation, in the present paper, I show how to estimate the membership functions using the data collected via Likert-based surveys in random samples.

2 Data

I use the data set previously described in the literature [11]. Briefly, 27 health technology assessment experts in Poland were asked to express their views on how much a society should be willing to pay (should demand to be compensated) for an additional QALY (for a QALY lost). Importantly, the experts were asked to think about the societal value, i.e. the value that should be used in public decision making, not about how much they value the improvement/worsening in their own lives. Therefore, all the respondents were asked about the same thing, while still allowing for a difference in opinions. The respondents were asked about their personal views, not simply to quote the current regulations, which in Poland define the threshold to be precisely three times the annual gross domestic product per capita.

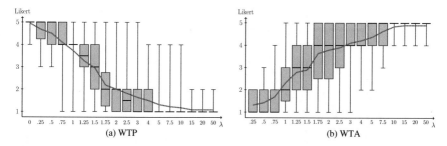

Fig. 1 Data. Values in horizontal axis in hundreds 000s PLN ($\lambda = 0$ was not used in WTA in the survey). Bold line denotes the median. Gray box denotes the first and third quartile. The whiskers denote the min and max. The dotted lines denote the naïvely calculated average response

In the survey, the respondents were asked, inter alia, if they would accept using a technology offering one QALY more (less), if it costed λ more (less); and they declared how much they agreed, using a 5-level Likert scale: 5—totally agree, 1—totally disagree. The raw collected responses are illustrated in Fig. 1. The exact λ values can be read off the horizontal axis ($\$1 \approx 4$ Polish Zloty, PLN).

Three respondents were altogether removed in the WTA part (but included in the WTP analysis and in the sensitivity analysis), as they did not use the *tend to agree* & *agree* Likert answers for any, even strikingly large, λ. Thus, they seemed to disagree with the very possibility of trading off health for money, in this sense rejecting the rules of the game. This paper focuses on estimating the membership function of WTP/WTA, and these three experts reject the very concept of being willing to accept money for health, hence their views cannot be used to quantitatively estimate (crisp or fuzzy) WTA. Still, these opinions are important in constructing a general framework how to decide about public spending in healthcare (perhaps rejecting the very idea of explicit trade-offs), but should be handled separately as they differ qualitatively.

To motivate the present research, let us notice here that—on one hand—it would be, in principle, possible to estimate the membership function in a naïve approach, as follows. We could translate the Likert levels 1–5 into values $\{0, 0.25, 0.5, 0.75, 1\}$, respectively, and average the resulting values (separately for each λ) between the respondents. Then we would simply interpret the resulting average as the value of the membership function, additionally somehow interpolating between available λs. On the other hand, such an approach has several disadvantages. Most importantly, averaging the membership function can result in the outcome being fuzzy, even without any fuzziness in the individual data. This is illustrated in Fig. 2. Assume that four respondents have different, yet crisp, opinions about what values, x, belong to a given set, X, over a universe **R**. In their view, respectively, $X = [0, 1]$, $X = [0, 2]$, $X = [0, 3]$, and $X = [0, 4]$. The membership functions are thus discontinuous (left panel) and take only values $\{0, 1\}$ (and the respondents would only use Likert levels 5 and 1, in their answers). Averaging the results yields a stepwise function (right panel), taking on also values between 0 and 1, and hence implying fuzziness. Thus

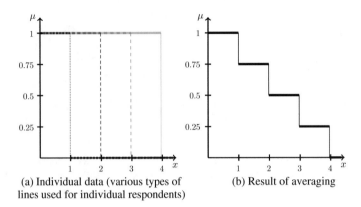

(a) Individual data (various types of
lines used for individual respondents)

(b) Result of averaging

Fig. 2 The artefact of naïve averaging of the membership function, μ. Individuals have crisp, however different, opinions. The average membership function takes on values between 0 and 1, suggesting fuzziness. Discontinuities of μ denoted with thin, vertical lines

a stochastic noise is transformed into fuzziness, while the two types of uncertainty are usually treated as qualitatively different.

Secondly, the naïve approach does not allow for any extrapolation, and so we would only estimate the membership function up to the maximal λ used in the questionnaire. Thirdly, this approach does not allow (easily) to model the heterogeneity of the respondents, so as to estimate the impact of some personal traits on the WTP/WTA. In principle, the modelling approach presented below can include additional explanatory variables characterizing individuals to see how they are associated with the preferences.

Additionally, the naïve averaging might not work if the respondents were presented different λs in the questionnaire; and using various λs might be a good idea if we wanted to collect data for many distinct λs (e.g. to verify the impact of using round numbers), while not asking a single respondent too many questions. In the naïve approach, the averaging would be done over different subsets of respondents and might lead, e.g., to an estimated $\mu_{\mathrm{WTP}}(\cdot)$ being increasing, a logical impossibility.

Finally, explicitly modelling how the individual opinions are translated into Likert levels (as Eq. 2 below) can allow combining various data in a flexible way and may help designing questionnaires in the future (e.g. how many levels to use).

3 Model

3.1 Mathematical Formulation

Below, I present the model for WTP but the idea is analogous for WTA. In what follows, I assume that every individual has their own membership function, $\mu_i(\cdot)$ (WTP suppressed in the subscript, for brevity). I assume that this function is given parametrically by the following equation

$$\mu_i(\lambda) = \frac{1}{1 + a_i \lambda^{b_i}}, \tag{1}$$

where a_i and b_i are strictly positive parameters, specific to a given individual. Hence, I use a standard, decreasing S-shape logistic curve for the natural logarithm of λ. S-shape represents the smooth transition between the full and lack of conviction for greater and greater values (and the location and steepness is given as a function of the parameters, a and b, and so can differ between the respondents). Using the logarithm of λ reflects diminishing sensitivity to equal, absolute increases in cost difference, is motivated by the skewness found in the data [13], and makes the approach robust to considering PLNs per QALY or QALYs per PLN. Still, non-log values are used in sensitivity analysis.

I then assume that, when the respondent faces a survey, $\mu_i(\lambda)$ is translated into the Likert levels, $L_i(\lambda)$, probabilistically, i.e. for each $\mu_i(\lambda)$ there is a probability distribution defined over levels 1–5. Intuitively, the probability of observing a specific level $L_i(\lambda) = k, k = 1, \ldots, 5$, being selected diminishes the farther away $\mu_i(\lambda)$ is from this level's threshold value, θ_k, $\theta_5 = 1$, $\theta_4 = 0.75$, $\theta_3 = 0.5$, $\theta_2 = 0.25$, and $\theta_1 = 0$. More formally,

$$P\left(L_i(\lambda) = k\right) = \frac{\exp(-s_i \times |\mu_i(\lambda) - \theta_k|)}{\sum_{j=1}^{5} \exp(-s_i \times |\mu_i(\lambda) - \theta_j|)}. \tag{2}$$

Parameter s_i, $s_i > 0$, measures the diffusion of probabilities along the Likert-scale. Figure 3 illustrates the above equation for three values of s. Small s_i results in the probabilities being distributed more evenly, and so for any μ_i still all the Likert levels are quite probable (e.g. the respondent is not capable of perceiving own μ_i or is not answering the Likert questions meticulously enough). Increasing s_i results in the level with θ closest to $\mu_i(\lambda)$ being selected with increasing probability. Thus, importantly, the present specification includes as a special case (with $s_i \rightarrow +\infty$) selecting the Likert answer closest to $\mu_i(\lambda)$ (when interpreting Likert answers being transformed to 0, 0.25, 0.5, 0.75, and 1).

Another nice (in my view) feature of the above approach is that, e.g., even if $\mu_i(\lambda) = 0.76$, it is still possible for the middle option (and any other) to be selected, as the numerator in Eq. 2 is always strictly positive. In that sense, the respondent is not entirely able to perceive own membership function being equal to 0.76 and to rule out the middle option completely, for the reason of the middle option (related to $\theta_3 = 0.5$) being blocked by level 4 (having $\theta_4 = 0.75$). Of course, other formulas could be used, e.g. implying randomizing only between two levels, with two θs closest to a given $\mu_i(\lambda)$ from below and above.

Parameters a_i, b_i, and s_i are idiosyncratic for each respondent. They are drawn independently—of each other, and between respondents—from lognormal distributions with the underlying normal distributions with means and inverse variances

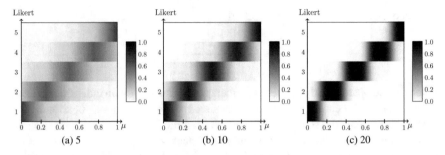

Fig. 3 The conversion of a membership function (horizontal axis) into probabilities of Likert levels (shades of gray for various levels on vertical axis) for three values of diffusion parameter, s_i

$N(m_A, \tau_A)$, $N(m_B, \tau_B)$, and $N(m_S, \tau_S)$, respectively. Point estimates $\widehat{m_A}$, $\widehat{m_B}$, and $\widehat{m_S}$ are taken to define the estimand, i.e. population-level membership function.

The point estimates are taken as medians of posterior distributions. Percentiles 2.5 and 97.5 define the 95% credible interval (95%CI). On technical note, non-informative priors are assumed (normal for means, gamma for inverse variances). The model is estimated with Markov-Chain Monte Carlo method, implemented in JAGS/R. The results come from 20,000 iterations (thinning = 5), with 5000 burn-in iterations.

3.2 JAGS Code

Below, I present the JAGS code used to estimate the model. The following notation is used: nR and nV denote the number of respondents and λs, respectively; the matrix answer[i,j] contains Likert answers, i and k indexing respondents and λs, respectively; l[k] $= [0, 0.25, 0.5, 0.75, 1]$; mA, mB, and mS are the estimated m_A, m_B, and m_S, respectively.

```
model{
#OUTER LOOP OVER RESPONDENTS:
for (i in 1:nR){

  #DRAW INDIVIDUALS' PARAMETERS:
  a[i] ~ dlnorm(mA,tA)
  b[i] ~ dlnorm(mB,tB)
  s[i] ~ dlnorm(mS,tS)

  #MIDDLE LOOP OVER LAMBDAS:
  for (j in 1:nV){
    x[i,j] <- 1/(1+a[i]*pow(v[j],b[i])) #***
```

```
#INNER LOOP OVER LIKERT LEVELS:
for (k in 1:5){
 #PROB. OF LEVEL K (NOT NORMALIZED)
 p[i,j,k] <- exp(-s[i]*abs(x[i,j]-l[k]))
 }

 #PROB. DIST. OF OBSERVABLES:
 answer[i,j] ~ dcat(p[i,j,1:5])
 }
}

#PRIORS
mA ~ dnorm(0,.01)
mB ~ dnorm(0,.01)
mS ~ dnorm(0,.01)
tA ~ dgamma(.01,.01)
tB ~ dgamma(.01,.01)
tS ~ dgamma(.01,.01)
}
```

In order to use the code for WTA, it suffices to replace the ∗ ∗ ∗-marked line with

```
x[i,j] <- 1-1/(1+a[i]*pow(v[j],b[i]))
```

as we expect an increasing membership function, and the sign of parameters is pre-determined by lognormal distributions.

4 Results

4.1 Baseline Results

The estimation yields the following results for WTP:

- $\widehat{m_A} = -17.62$ with 95%CI = $(-20.95; -14.13)$, (notice that we require a to be positive, not m_A)
- $\widehat{m_B} = 1.25$ with 95%CI = $(0.99; 1.44)$,
- $\widehat{m_S} = 2.32$ with 95%CI = $(2.06; 2.61)$.

For the sake of estimating the 0.5-cut, the above implies that $\mu = 0.5$ for ca. 155,700 PLN/QALY, while the official threshold amounts currently to 130,002 PLN/QALY (and 111,381 PLN/QALY at the time of the survey).

The results for WTA are as follows:

- $\widehat{m_A} = -15.43$ with 95%CI = $(-19.58; -11.65)$,
- $\widehat{m_B} = 1.03$ with 95%CI = $(0.74; 1.33)$,
- $\widehat{m_S} = 2.48$ with 95%CI = $(2.28; 2.7)$,

and we get $\mu = 0.5$ for ca. 246,800 PLN/QALY. Hence, there is some disparity in 0.5-cuts. We still have to take into consideration that the current methodology was not focused on estimating the 0.5-cut, but on the whole membership function, and so the estimates of the 0.5-cut may be driven by other regions of the membership function.

The resulting membership functions, based on point estimates, are presented in Fig. 4 (the left panel). Now, having estimated the membership functions allows comparing WTP and WTA quantitatively. Firstly, notice that the 95%CI are wider for WTA (for m_A and m_B, the only ones that impact the shape of the membership function), and so there is more stochastic uncertainty related to WTA than WTP. We can also compare the two with respect to the amount of fuzziness. I use two measures for that purpose. One of them was proposed in the literature and relies on the idea that the more the membership function takes on values close to 0.5, the fuzzier the set is [14]. Formally, the amount of fuzziness is given by

$$\int_{-\infty}^{+\infty} (1 - |2\mu(x) - 1|) \, dx.$$

We obtain, for WTP, 137.98, and for WTA, 284.87 (calculated numerically). I use another approach here, using the particular shape of the membership function (decreasing monotonically from 1 to 0). We can interpret the membership function in the probabilistic fashion, as a flipped vertically cumulative distribution function, and calculate the variance of the variable, λ. Then the more the (upper bounds of the) α-cuts differ, the fuzzier the set is. More formally, the amount of fuzziness is then given by

$$\sqrt{\int_0^1 \sup(WTP_\alpha)^2 d\alpha - \left(\int_0^1 \sup(WTP_\alpha) d\alpha \right)^2}.$$

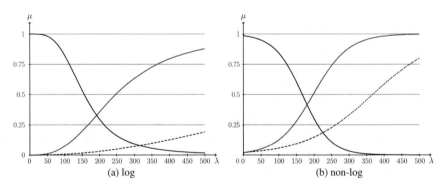

Fig. 4 Membership functions for point estimates (WTP—solid, WTA—dotted). Additionally, results for WTA (dashed) when no respondents are removed

We obtain 108.07 for WTP and 253.44 for WTA, and so again—there is about twice as much imprecision in how WTA is perceived. Probably experts feel more uncomfortably with quantifying trade-offs in the context of selling health, which results in more of both—stochastic noise (differences between the respondents) and fuzziness (inability to precisely locate the threshold).

4.2 Results for WTA for a Complete Data Set

As a sensitivity analysis, I repeat the calculations for the complete data set, i.e. keeping in the three respondents that were removed for WTA analysis. Then, only the results for WTA change, as follows:

- $\widehat{m_A} = -15.13$ with 95%CI $= (-19.94; -10.7)$,
- $\widehat{m_B} = 0.79$ with 95%CI $= (0.33; 1.23)$,
- $\widehat{m_S} = 2.59$ with 95%CI $= (2.39; 2.85)$.

The membership function is also illustrated in Fig. 4 (the left panel, dashed line). As can be seen, including the—qualitatively different-respondents results in the results being completely different, and not too reliable (taking into account the raw data, Fig. 1). This stresses the necessity to handle the data violating the underlying assumptions (accepting some trade-offs) separately.

4.3 No Logarithmic Transformation

For the baseline analysis, the logs of λs were used. Here, I verify the impact of this transformation, redoing the calculations for original values. For that purpose, instead of Eq. 1, we need to use the following one:

$$\mu_i(\lambda) = \frac{1}{1 + e^{a_i + b_i \times \lambda}}. \tag{3}$$

The above expression obviously simplifies to Eq. 1 when we take $\ln(\lambda)$ instead of λ (and denote e^{a_i}, abusing notation, by a_i). We also need to make adequate changes in JAGS code in the line marked with $***$, taking:

```
x[i,j] <- 1/(1+exp(a[i]+v[j]*b[i]))
```

Now, a_i can also take negative values, and so I assume it is normally distributed, with parameters $N(m_A, \tau_A)$.

The estimation yields, for WTP:

- $\widehat{m_A} = -4.2$ with 95%CI $= (-5.3; -3.24)$,
- $\widehat{m_B} = -3.68$ with 95%CI $= (-4.09; -3.28)$,

- $\widehat{m_S} = 2.5$ with 95%CI = (2.29; 2.74),

and for WTA:

- $\widehat{m_A} = -3.92$ with 95%CI = (−5.14; −2.88),
- $\widehat{m_B} = -3.92$ (coincidentally equal to $\widehat{m_A}$) with 95%CI = (−4.42; −3.41),
- $\widehat{m_S} = 2.49$ with 95%CI = (2.21; 2.87).

The results are depicted in the right panel of Fig. 4. As can be seen, this approach reduces the WTP-WTA disparity in terms of 0.5-cuts. The problem with the non-log approach with the $\mu(\cdot)$ given by Eq. 3 is that now necessarily $\mu(0) < 1$ for WTP and $\mu(0) > 0$ for WTA, which does not seem intuitive and desirable. That's why this is not treated as the baseline result.

4.4 Results for WTA for Non-log, Complete Data

For completeness, I present the results of the non-log approach with all the respondents in the WTA analysis. The estimation yields:

- $\widehat{m_A} = -3.84$ with 95%CI = (−5.19; −2.75),
- $\widehat{m_B} = -4.56$ with 95%CI = (−5.48; −3.68),
- $\widehat{m_S} = 2.7$ with 95%CI = (2.33; 3.25).

The results are illustrated with the dashed line in the right panel of Fig. 4. Apparently, the non-log approach is more robust to including respondents strongly opposing to accepting trade-offs resulting in losing some health.

5 Conclusion

Understanding the shape of the membership function of fuzzy trade-off coefficients is important to formally support decision making. When imprecise opinions about possible values of this coefficient are collected via surveys from a group of respondents, these judgements will most certainly differ, and statistical methods must be employed to estimate the joint, average one. Naïve methods may be misleading: e.g. they can artificially suggest more fuzziness than actually present in the raw data. Hence, approaches based on modelling—as one presented in the present paper—may be more useful.

The proposed model clearly distinguishes between the parameterisation of the membership function and the mechanism of its conversion into Likert scale, thus providing flexibility on changing the two independently. Respondents' characteristics could be added to Eq. 1 (or 2, or 3), thus allowing to model heterogeneity and find factors associated with, e.g., accepting greater WTP or larger WTP-WTA disparity.

Further work should be focused on building more robust model, that could handle also respondents with qualitatively different opinions (not crossing the middle Likert option). That may require using models with more flexibility to vary the membership function between the respondents, and so with more parameters: hence, larger data sets are required. As an idea: a scaling parameter could be used to limit the range of values that the membership function can take for a given respondent. Also, perhaps differently shaped functions should be tested for WTP and WTA. That could account for larger fuzziness for WTA, while restoring the equality of 0.5-cuts, detected by different methods [13].

Acknowledgements The research was done during my stay at The Tippie College of Business, The University of Iowa, USA, thanks to the Fulbright Senior Award. This opportunity is greatly appreciated.

References

1. H. Bleichrodt, P. Wakker, M. Johannesson, Characterizing QALYs by risk neutrality. J. Risk Uncertain. **15**, 107–114 (1997)
2. D. Golicki, M. Jakubczyk, M. Niewada, W. Wrona, J. Busschbach, Valuation of EQ-5D Health states in Poland: First TTO-based social value set in Central and Eastern Europe. Value Health **13**, 289–297 (2010)
3. R. Brooks, F. De Charro, EuroQol: the current state of play. Health Policy **37**, 53–72 (1996)
4. F. Bellavance, G. Dionne, M. Lebeau, The value of a statistical life: a meta-analysis with a mixed effects regression model. J. Health Econom. **28**(2), 444–464 (2009)
5. W. Viscusi, J. Aldy, The value of a statistical life: a critical review of market estimates throughout the world. J. Risk Uncertain. **27**(1), 5–76 (2003)
6. L. Hultkrantz, M. Svensson, The value of a statistical life in Sweden: a review of the empirical literature. Health Policy **108**(2–3), 302–310 (2012)
7. W. Dubourg, M. Jones-Lee, G. Loomes, Imprecise preferences and the WTP-WTA disparity. J. Risk Uncertain. **9**, 115–133 (1994)
8. J. Horowitz, K. McConnell, A review of WTP/WTA studies. J. Environment. Econom. Manag. **44**, 426–447 (2002)
9. T. Tunçel, J. Hammitt, A new meta-analysis on the WTP/WTA disparity. J. Environment. Econom. Manag. **68**, 175–187 (2014)
10. M. Smithson, *Fuzzy Set Analysis for Behavioral and Social Sciences.* Springer (1987)
11. M. Jakubczyk, B. Kamiński, Fuzzy approach to decision analysis with multiple criteria and uncertainty in health technology assessment, Ann. Operat. Res. (2015). https://doi.org/10.1007/s10479-015-1910-9
12. M. Jakubczyk, Using a fuzzy approach in multi-criteria decision making with multiple alternatives in health care. Multiple Criter. Decision Mak. **10**, 65–81 (2015)
13. M. Jakubczyk, Choosing from multiple alternatives in cost-effectiveness analysis with fuzzy willingness-to-pay/accept and uncertainty, mimeo (2016)
14. G. Klir, T. Folger, *Fuzzy Sets, Uncertainty, and Information.* Prentice-Hall (1988)

Fuzzy Logic Based Simulation of Gynaecology Disease Diagnosis

A. S. Sardesai, V. S. Kharat, A. W. Deshpande and P. W. Sambarey

Abstract The first step in a knowledge base expert system could be to mathematically evaluate perceptions of the domain experts which are invariably expressed in linguistic terms based on their tactic knowledge followed by the defined steps in differential diagnostic process. We have simulated the process in three stages, especially in gynaecological diseases. Stage I, refers to Type1 Fuzzy Relational Calculus used to arrive at the initial diagnostic labels for gynaecological diseases in patients and to estimate similarity between the domain experts. The case study focused only on the identified gynaecological diseases arrives at comparatively low diagnostic percentage, and therefore termed as Initial Screening Process. The output of the algorithm for patient diagnostic records, considering the variability among the experts, was tested for diagnosing a single disease. After application of '*History*' fuzzy rule base in Stage 2, using Type 1 Fuzzy Inference System, the accuracy was increased to some extent which was further enhanced to high level by Stage III for the prototype of 226 patients diagnosed by the model. The need based research presented will ultimately assist physicians and upcoming gynaecologists.

A. S. Sardesai (✉)
Department of Computer Science, Modern College of Arts,
Science and Commerce, Shivajinagar, Pune 411005, India
e-mail: sardesaicompsci@moderncollegepune.edu.in

V. S. Kharat
Department of Mathematics, Savitribai Phule Pune University,
Ganeshkhind, Pune 411007, Maharashtra, India
e-mail: laddoo1@yahoo.com

A. W. Deshpande
Berkeley Initiative in Soft Computing (BISC)-SIG-EMS, University of California,
Berkeley, California, United States of America
e-mail: ashok_deshpande@hotmail.com

P. W. Sambarey
Department of Gynaecology, B. J. Medical College, Pune, India
e-mail: drsambarey@yahoo.co.in

1   Introduction

In the field of Medicine, efforts on modelling uncertainties using soft computing techniques were initiated in mid-70s. Though there are viable efforts made in the application of fuzzy logic based methods in medical diagnosis [1–7], surprisingly its use in Gynaecology is very limited. Medical diagnosis is a fuzzy or imprecise science.

Gynecology or Gynaecology (science of women) refers to the health of the Female Reproductive System such as: *Uterus*, *Vagina*, and *Ovaries*, while Obstetrics deals with care of women's reproductive tracts and their children during pregnancy (prenatal period), child birth and the postnatal period. The research study presented in this sequel refers only to medical diagnosis in Gynaecology.

In our view, overall approach in gynecological disease diagnosis could be divided in three distinct stages:

Stage 1 refers to Initial Screening Process in order to arrive at a single disease diagnosis for the patients, and based only on the subjective information provided by patients to the physician.

In Stage 2, the patient who has not received a single diagnostic label in Stage 1, is further investigated for single disease diagnosis using parameters like past history, pre-menstrual changes, last menstrual period, marital status, parity, etc. If Stage 2 cannot arrive at a single disease diagnosis for a patient then physical examination and various tests like ultra-sonography, X-ray, blood tests, etc. are conducted and the test results are processed in Stage 3.

The remaining part of the paper is organized as follows: Sect. 2 describes the techniques/methods used in the study. Section 3 demonstrates the case study for stage III. Results and discussions are presented in Sect. 4 followed by concluding remarks.

2   Method

2.1   *System Flow Chart*

Figure 1 presents the overall approach proposed by the authors to arrive at approximately correct diagnosis especially in gynaecological diseases.

FIS (Fuzzy Inference System) defines a nonlinear mapping of the input data vector into a scalar output, using fuzzy rules. Mandani type fuzzy inference system is used in this research which uses centre of gravity as defuzzification method. The details of the method are available in standard literature [8, 9].

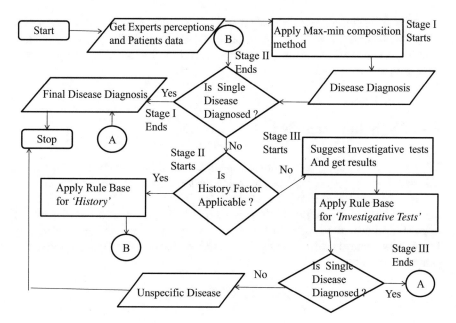

Fig. 1 Overall system approach

2.2 Fuzzy Similarity Measures

Cosine Amplitude Method

It makes use of a collection of data samples, particularly 'n' data samples. If these data samples are collected, they form a data array, $X = \{x_1, x_2, \ldots, x_n\}$. Each of the elements x_i, in the data array X is itself a vector of length m, i.e.

$$X_i = \{x_{i1}, x_{i2}, \ldots, x_{im}\} \tag{1}$$

Each of the data samples can be thought of as a point in m-dimensional space, where each point needs m coordinates for a complete description. Each element of a relation r_{ij}, results from a pair-wise comparison of two data samples, say x_i and x_j, where the strength of the relationship between data sample x_i and data sample x_j is given by the membership value expressing that strength i.e. $r_{ij} = \mu_R(x_i, y_j)$ resulting a relation matrix of size n × n. The relation will be reflexive and symmetric so is a tolerance relation. The cosine amplitude method calculates r_{ij} as [9]:

$$r_{ij} = \frac{\left| \sum_{k=1}^{m} x_{ik} x_{jk} \right|}{\sqrt{\left(\sum_{k=1}^{m} x_{ik}^2 \right) \left(\sum_{k=1}^{m} x_{jk}^2 \right)}}, \text{ Where } 0 \le r_{ij} \le 1 \tag{2}$$

Close inspection of Eq. 2 reveals that this method is related to dot product for the cosine function. When two vectors are collinear (most similar), their dot product is unity. When the two vectors are at right angles to each other (most dissimilar), their dot product is zero.

2.3 Fuzzy Inference System

Consider a multi-input, multi output system. Let the input vector $x = (x_1, x_2, ..., x_n)^T$ and $y = (y_1, y_2, ..., y_m)^T$ be the output vector. The linguistic variable x_i in the universe of discourse U is characterized by: $T(x) = \{T_x^1, T_x^2, T_x^3, ..., T_x^k\}$ and $\mu(x) = \{\mu_x^1, \mu_x^2, \mu_x^3, ..., \mu_x^k\}$ where $T(x)$ is a term set of x, i.e., it is the set of names of linguistic values of x, with each being a fuzzy member and the membership function μx^i defined on U. The linguistic variable y in the universe of discourse V is characterized by:

$$T(x) = \{T_y^1, T_y^2, T_y^3, ..., T_y^k\},$$

where $T(x)$ is a term set of y, i.e., T is the set of names of linguistic values of y, with each Ty^i being a fuzzy membership function μx^i defined on V [9].

Mandani type fuzzy inference system is used in this research which uses centre of gravity as defuzzification method. Software is developed to simulate the Mamdani type fuzzy inference system.

3 Case Study

In the Initial Screening Process, we collected perceptions for symptom-disease relationship from eight different gynaecologists. Patient-symptom relation was defined for 226 Gynaecology patients. This patient's data was collected using personal interview technique from three different hospitals in Pune, Pimpri and Ambejogai. We revisited Fuzzy Relational Calculus and developed the software of disease diagnosis. Table 1 presents the resultant diagnostic after applying patient's data to the software for eight expert's perceptions [8, 10, 11]. The results obtained by the software are cross verified with the diagnosis of three different expert gynecologists.

We believe that some of the patients (Table 1—Row 1) are correctly diagnosed for a single disease. However, the output as: 'Incorrect Diagnosis' (Table 1—Row 4) was discussed with the expert gynecologist and he was of the opinion that, for these patients patient history will be adequate to arrive at a single disease diagnosis.

Table 1 Diagnosis analysis for 8 experts using initial screening model

Diagnosis analysis	Experts							
	E_1	E_2	E_3	E_4	E_5	E_6	E_7	E_8
One disease correct	50	41	50	54	50	53	56	18
Multiple diseases all correct	40	5	2	6	1	4	6	5
Multiple diseases partial correct	107	50	61	59	64	57	54	61
Incorrect diagnosis	29	131 `	113	107	109	112	110	142
Accuracy percentage (Single disease)	22.12	18.14	22.12	23.89	22.12	23.45	24.78	7.96
Overall accuracy percentage	87.16	42.48	50.00	51.77	50.88	50.44	51.33	37.17

3.1 Classification of Experts

After obtaining the result of initial screening as in Table 1, it was observed that there is need to classify experts. The experts were classified using Fuzzy Similarity Measures Cosine Amplitude method [12]. It was concluded that experts $\{E_1, E_2, E_3, E_5, E_7, E_8\}$ agree with each other. This clearly infers that all these gynecologists agree in their diagnostic label for a single disease with 0.981 possibility. We could have considered the perceptions of all the experts. Since expert 1 (gynaecologist) is deeply involved in this research as one of the supervisors, his perceptions were considered for Stage 2 and Stage 3 in this research.

3.2 Stage II (History Rule Base)

After taking a deeper look by the authors, it was inferred that the diagnostic efforts in Gynaecology heavily depend on some of the factors/history such as: Age of the patient, Last Menstrual Periods (LMP), Pre-Menstrual Cycle (PMC) Flow, Irregular/Regular nature of menses, Married/Unmarried status of the patient and Parity. These factors collectively decide some of the diseases. Accordingly 5 fuzzy sets namely, *'Age'*, *'LMP'*, *'PMC_Flow'*, *'Marital Status'*, *'Parity'* are defined and 72 fuzzy rules were formulated and using Type 1 Fuzzy Inference System [12, 13]. Figure 2 shows two fuzzy sets *'Age'* and *'PMC_Flow'*.

Mamdani type Fuzzy Inference System using center of gravity as a defuzzification method is simulated while software development in order to arrive at a single disease as an output.

Using Type 1 Fuzzy Inference System, it was possible to arrive at a single diagnosis for the patients which were shown as unspecified or wrongly diagnosed as depicted in Table 1.

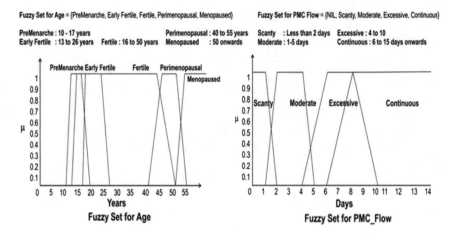

Fig. 2 Fuzzy sets for age and PMC_Flow

Fig. 3 Fuzzy rules

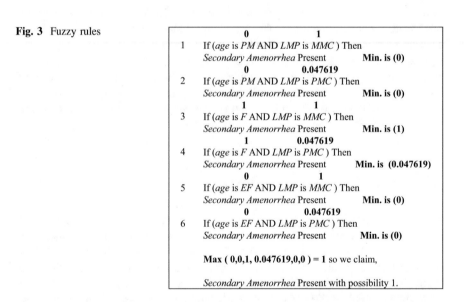

Figure 3 demonstrates the application of fuzzy rules using Mamdani type Fuzzy Inference System, for patient (P_{215}) with history as: Age (35 years), LMP 60 days back. These two factors will fire following 6 rules from the set of 72 rules to yield the output as: "*Secondary Amenorrhea present with possibility 1*".

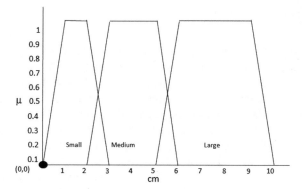

Fig. 4 Fuzzy set for USG_PELVIS_OVARIAN_CYST

3.3 Stage III (Test Rule Base)

If single disease is not diagnosed at Stage II, the investigative tests are suggested, test results is input to Stage III which is a rule base designed for 13 diseases and related imagining, pathology tests and physical examinations. Fuzzy sets are defined after consulting radiologists, pathologists and gynecologists.

Example Fuzzy set for USG_Pelvis_Ovarian_Cyst (absent, small, medium, large) is as shown in Fig. 4.

To test if the *'Ovarian Cyst'* is present, one of test suggested is *'USG Pelvis'*. Figure 4 shows the fuzzy set defined for this test.

We have defined 55 fuzzy sets for investigative tests and are confirmed by six experts. For these 55 fuzzy sets, we defined 715 fuzzy rules.

For patient P_{140}, the diagnosis by Stage I is {*Pelvic Inflammatory Disease (PID), Endometritis, Ovarian Cyst*}, Stage II cannot arrive at final diagnosis so investigative test is suggested "*USG Pelvis*" which resulted in cyst present of size 7.6 cm (Large cyst). So the fuzzy rule fired is as shown in Fig. 5.

The patients which could not get diagnosed by Stage II, are input to Stage III. Out of 226 patients, 50 patients were diagnosed for a single disease and for 11 patients the diagnosis by model was "*Disease Unspecific*". Stage II diagnosed more 29 patients; remaining 138 patients are input to Stage III.

Fig. 5 Fuzzy rule fired for patient P_{140}

(0)   **(0)**
if(*USGPV_Ocyst* is SMALL OR *USGPV_Ocyst* is MEDIUM
(1)
OR *USGPV_Ocyst* is LARGE)
(1)
then *Ovarian Cyst* is Present

4 Results and Discussion

In Stage II, Out of 226 patients, 30 patients are identified using Type 1 Fuzzy Inference System, for whom the 'History' criteria are applicable to arrive at a single disease. We can state in no uncertain terms that 29/30 patients are correctly diagnosed for a single disease. In Stage I, 50 patients were diagnosed correctly using Type 1 Fuzzy Relational Calculus, for single disease diagnosis. In Stage 2 only one patient (P_{100}) was incorrectly diagnosed. The reason for failure is as narrated below:

The history of the patient is: Age: 25 years, fits in fuzzy set *'Age—Fertile'* (F = 1), Marital Status is Unmarried (M_Status = U), Amenorrhea for 4 years which does not fit in any of the LMP fuzzy sets. In the fuzzy sets defined, Amenorrhea is termed when no menses for more than 1.5 months to 4–5 months. Then till 9 months the patient is Gravid, and after the birth of child for next around 1.5 years, it is *Lactational Amenorrhea*. More than 4 years should fall in menopause but menopause is not possible at early age like 25. So none of the rules are fired and the model gives no output. The gynaecologists invariably termed the diseases as *Secondary Amenorrhea* for the symptoms—'age 25 and no menses present'. As it is not a good practice to redefine the fuzzy set for every extreme case, we consider this as exceptional case or may be an outlier. The initial screening model also could result to diagnose this patient as *'Disease Unspecific'*. This patient P_{100} can be considered as an outlier.

In Stage III, 146 patients were diagnosed of which 139 patients got correctly diagnosed and for 9 patients, the model could not give correct diagnosis for known reason one of the reason is as discussed below:

For patient P_{123}, the diagnosis by stage I is {*Endometriosis, Uterine Fibroid, Adenomyosis, Pubertal menorrhagia*}. Test suggested is *Trans-vaginal USG* for *'Endometriosis'* and *'Uterine Fibroid'*. The USG shows no endometrial tissue growth or growth of tumour resulting *'Endometriosis'* and *'Uterine Fibroid'* ruled out. *'Adenomyosis'* is normally confirmed by the test MRI. The result shown no abnormalities found inside the uterus and *'Adenomyosis'* is also ruled out. So the output by model is *'Pubertal Menorrhagia'*. But in this case the output is incorrect as age of the patient is 39 which is not a puberty age. The actual diagnosis for the patient is *'Polycystic Ovarian Syndrome'* (PCOS) which will be diagnosed by the gynaecologist based on his/her experience. In our case, we could not get PCOS as output at possibility value 1 but if we cut down the possibility (α-cut value) to 0.75, we get PCOS as output at initial screening stage.

Similar is the case for patient P_{224}, but the age of the patient is 12 years so she is diagnosed as "*Pubertal menorrhagia*" and is the correct diagnosis.

In the similar manner, further work is in process to diagnose remaining 136 patients using *Tests Rule Base*.

5 Concluding Remarks

The case study demonstrates the utility of Professor Zadeh's fuzzy logic based formalisms in medical diagnosis. The three stage approach developed can be a better diagnostic approach, especially in Gynaecology Disease Diagnosis. The software developed is user friendly. A need for more case studies with huge number of patients is recognized to justify reliability of the software. Smart phone application on android platform is expected to be the final outcome of the applied research.

References

1. A. Mahdi, A. Razali, Al. Alwakil, Comparison of fuzzy diagnosis with K-nearest neighbor and Naïve Bayes classifiers in disease diagnosis. Broad Res. Artif. Intell. Neurosci. **2** (2011). ISSN 2067-3957(online), ISSN 2068-0473(print)
2. A.R. Meenakshi, M. Kaliraja, An application of interval valued fuzzy matrices in medical diagnosis. Int. J. Math. Anal. **5**, 2792–2803 (2011)
3. E. Sanchez, Inverse of fuzzy relations, application to possibility distributions and medical diagnosis. Fuzzy Sets Syst. **2**(1) (1979)
4. F. Steimann, Fuzzy set theory in medicine. Artif. Intell. Med. **11**, 1–7 (1997)
5. L.I. Kuncheva, F. Steimann, Fuzzy Diagnosis. Artif. Intell. Med. **16**, 121–128 (1999)
6. M.A.M. Reis, N.R.S. Ortega, P.S.P. Silveira, Fuzzy expert system in prediction of neonatal rescuscitation. Braz. J. Med. Biol. Res. **37**(2) (2004)
7. P.R. Innocent, R.I. John, Computer aided fuzzy medical diagnosis. Inf. Sci. **162**, 81–104 (2004)
8. G.J. Klir, C. Bo Yuan, *Fuzzy Sets and Fuzzy Logic, Theory and Applications* (Prentice Hall P. T.R., Upper Saddle River, New Jersey, 1995)
9. T. Ross, *Fuzzy Logic with Engineering Applications* (Wiley, Chichester, 2010)
10. A. Sardesai, V. Khrat, A. Deshpande, P. Sambarey, Fuzzy logic based formalisms for gynaecology disease diagnosis. J. Intell. Syst. (2016). https://doi.org/10.1515/jisys-2015-0106
11. A. Sardesai, V. Khrat, A. Deshpande, P. Sambarey, Initial screening of gynecological diseases in a patient, expert's knowledgebase and fuzzy set theory: a case study in India, in *Proceedings of the 2nd World Conference on Soft Computing*, Baku, Azerbaijan (2012), pp. 258–262
12. A. Sardesai, V. Kharat, A.W. Deshpande, P.W. Sambarey, Efficacy of fuzzy-stat modelling in classification of gynaecologists and patients. J. Intell. Syst. (2015). https://doi.org/10.1515/jisys-2015-0001
13. A. Sardesai, V. Khrat, A. Deshpande, P. Sambarey, Fuzzy logic application in gynaecology: a case study, in *Proceedings of the 3rd International Conference on Informatics, Electronics and Vision 2014*, Dhaka, Bangladesh, IEEE explore digital library (2014)

An Ontology for Wearables Data Interoperability and Ambient Assisted Living Application Development

Natalia Díaz-Rodríguez, Stefan Grönroos, Frank Wickström, Johan Lilius,
Henk Eertink, Andreas Braun, Paul Dillen, James Crowley
and Jan Alexandersson

Abstract Over the last decade a number of technologies have been developed that support individuals in keeping themselves active. This can be done via e-coaching mechanisms and by installing more advanced technologies in their homes. The objective of the Active Healthy Ageing (AHA) Platform is to integrate existing tools, hardware, and software that assist individuals in improving and/or maintaining

N. Díaz-Rodríguez (✉) · S. Grönroos · F. Wickström · J. Lilius
Åbo Akademi University, Turku, Finland
e-mail: ndiaz@decsai.ugr.es

S. Grönroos
e-mail: stgronro@abo.fi

F. Wickström
e-mail: frwickst@gmail.com

J. Lilius
e-mail: jolilius@abo.fi

H. Eertink
Novay, Enschede, The Netherlands
e-mail: e.h.eertink@saxion.nl

A. Braun
Fraunhofer IGD, Darmstadt, Germany
e-mail: andreas.braun@igd.fraunhofer.de

P. Dillen
Philips Research, Eindhoven, The Netherlands
e-mail: paul.dillen@philips.com

J. Crowley
INRIA Grenoble, Montbonnot-Saint-Martin, France
e-mail: James.Crowley@inria.fr

J. Alexandersson
DFKI GmbH, Saarbrücken, Germany
e-mail: janal@dfki.de

© Springer International Publishing AG, part of Springer Nature 2018   559
L. A. Zadeh et al. (eds.), *Recent Developments and the New Direction
in Soft-Computing Foundations and Applications*, Studies in Fuzziness
and Soft Computing 361, https://doi.org/10.1007/978-3-319-75408-6_43

a healthy lifestyle. This architecture is realized by integrating several hardware/ software components that generate various types of data. Some examples include heart-rate data, coaching information, in-home activity patterns, mobility patterns, and so on. Various subsystems in the AHA platform can share their data in a semantic and interoperable way, through the use of a AHA data-store and a wearable devices ontology. This paper presents such an ontology for wearable data interoperability in Ambient Assisted Living environments. The ontology includes concepts such as height, weight, locations, activities, activity levels, activity energy expenditure, heart rate, or stress levels, among others. The purpose is serving application development in Ambient Intelligence scenarios ranging from activity monitoring and smart homes to active healthy ageing or lifestyle profiling.

1   Introduction

The ageing of the world population is triggering a large set of projects aiming at the implementation of intuitive Ambient Intelligence homes based on universal design approaches and specifically tailored for elderly and disabled persons [1]. Examples of project using wearables are i2home [1] or EIT Digital Active Healthy Ageing Plat-form.[1] Within this context, we developed an architecture for the aggregation of sen-sor data for Ambient Assisted Living applications [2]. The project platform, among other elements, was composed of (a) a Smart-M3[2] triple storage box that provides an RDF storage [3] for multimedia, interfaces to sensors and actuators, and compu-tational capabilities on an energy-efficient computational platform [4], (b) a unified protocol for interfacing sensors and actuators to the platform, and (c) programming tools for easy generation of ontology libraries that abstract the access to the data store. The ontology proposed in this paper enables the fast application development and interoperability among heterogeneous devices. The M3 storage supports con-necting several M3 boxes together and enables a distributed architecture [5]. M3 provides SSAP (Smart Space Application Protocol) and SPARQL protocols.

One of the features provided by the box is the facilitation of local data processing capabilities for data mining and video processing. M3 storage is a modular and scal-able architecture that can also be integrated into e.g. mobile phones, and it supports distribution to facilitate integration of new applications into the box. The Active Healthy Ageing interface between Smart-M3 storage and PHL data store is in Fig. 1.

The core objective of the Active Healthy Ageing (AHA) project was to design and implement a distributed service-platform in combination with a few applica-tions that are developed in other Health and Well-being (HWB) activities. The AHA-platform is in itself service-independent: it includes general functions, such as data-distribution, service-control, service-configuration, or application-independent

---

[1]https://www.eitdigital.eu/fileadmin/files/HWB-pictures/EIT-Handout-HWB_EoY-1213-spreads-HR.pdf.

[2]Smart-M3 Project: http://sourceforge.net/projects/smart-m3/.

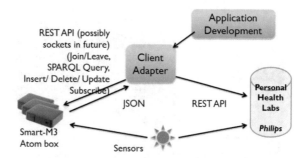

Fig. 1 Active Healthy Ageing interface between Smart-M3 storage and PHL data store

building blocks. The promise of this platform is that it will support services that provide sustainable quality of life improvements. It does this by supporting people to live uncompromised, comfortable, safe, and active even at an advanced age. The AHA-platform can be used by new service-providers, and will lower their barriers for market-entry. The output is of interest for both SMEs and larger companies: it will allow them to focus on their own added value. This will increase innovation in the HWB area due to lower development and exploitation cost.

The project combined several catalysts and carriers, including an Innovation Radar (to highlight the specific state of the art in AHA-platforms), Test bed and Platforms (for the platform architecture and implementation), and Living Lab Tests (for the showcase applications). The activity's aim, in the context of EIT Digital, was to integrate novel technologies to be integrated on a shared platform. The consortium (consisting of EIT Digital IVZW, Novay, Åbo Akademi University, Fraunhofer Gesellschaft, INRIA, Telecom Italia, Engineering, Philips and DFKI) of this activity involves partners from five EIT Digital nodes: Eindhoven, Berlin, Helsinki, Trento, and Paris. In addition, carrier projects include projects like universAAL (with strong focus on data management and privacy), several mobile/cloud combinations, and projects that focus on connectivity.

The potential users of the AHA platform are application/service developers and, of course, agents that favor to age in a healthy manner. The market/user community size is potentially very big; currently there exist platforms for elderly users, with a very limited and narrow focus on telecare and safety.

2 Active Healthy Ageing Platform Architecture

The Active Healthy Ageing Platform architecture is composed from existing software components and/or frameworks. The integration of the AHA Platform Architecture is therefore a bottom-up activity. As a result of all partners contributions, the AHA platform architecture can be seen in Fig. 2. The blue box shows the platform, and the pink box on top shows services to be integrated with the Philips Personal Health Labs data store.

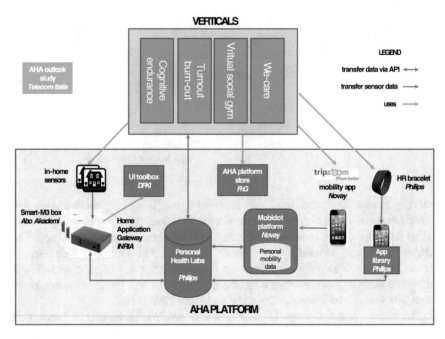

Fig. 2 Active Healthy Ageing Platform architecture

The core element in this architecture is the Personal Health Labs data store (PHL) that stores enriched information coming from other components/services. We distinguish between various types of APIs:

- *Bi-directional data transfer*: This interaction is represented by the blue arrows, that connect directly to the PHL data store (Philips). These are RESTful interfaces, and examples of such services are the Home Application Gateway (INRIA, DFKI), the combination of the Mobidot platform[3] and its mobile app (Novay), and the heart rate bracelet and mobile app.
- *AppStore*: The AHA platform supports an AppStore, provided by FHG. The *Universal Remote Control* is integrated with the *AmiQoLT* Integration Platform, while Inria and DFKI integrated their platforms and data elements in the AHA platform.
- *Verticals*: These are application services that connect directly to the PHL data store. These services are developed in other HWB activities, but integrated with the AHA platform. The AHA outlook study was provided by Telecom Italia.

The interaction between the AHA Platform components happens through synchronous interactions (via RESTful interfaces) provided between the PHL data store and its peers:

(1) *Home Application Gateway* (developed by INRIA-Grenoble/DFKI/Åbo Akademi), based on sensor data interpretation.

---

[3]http://www.mobidot.nl/en/index.php.

(2) *Mobidot MoveSmarter platform* (developed by Novay) interprets mobility data (based on GPS and accelerometer data) and detects individual trips and travel modalities in each trip. The Mobidot platform aggregates sensor data and sends the integrated data towards the PHL data store using a RESTful interface.

(3) The *MoveSmarter* platform free app supports both Android and iPhone handsets.

(4) *HR bracelet* (Philips) monitors heart-rate and derived features like stress levels. It is connected via low-power Bluetooth to a mobile phone that stores the bracelet data in the PHL data-store. The four identified verticals will integrate their sensor data with the PHL data-store.

The definition of the interface between the Home Application Gateway/Smart-M3 Box and the PHL data-store is done via Smart-M3 -based semantic storage box. Using Atom (low-power) board and an RDF store allows for interoperability of heterogeneous information as well as inference reasoning. The AHA platform ontology proposed gathers semantics of different general concepts (users, sensors, physical and physiological variables), while the security/access ontology controls for overall platform data policies [6].

The interface between the Mobidot Platform and the PHL data-store is defined in such a way that the Mobidot platform provides information about the modalities that are used in a single trip. This is the basic mobility information that is stored in the AHA-platform. The Mobidot platform, however, also has support for so-called incentives, a kind-of (virtual and real-world) goodies that can be won by individuals as a reward for proper mobility behavior. The IPersonalMobility interface is responsible for storing and providing so-called trip information (where a trip is an aggregation of modalities used during the specific trip).

The AHA platform application store is a one-stop solution to browse, select, purchase and install new applications that are compatible with the AHA platform. Electronic distribution platforms for software have been around for several decades. While the first systems distributed software without any support for shopping cart or online payment this has become the standard in the last years.

The AHA platform verticals activities produce different software components that either link to the AHA platform from a mobile application or the Home Application Gateway. The mobile systems usually rely on a closed platform and are distributed accordingly. Components developed for the Home Application Gateway should be easily purchasable and distributable over the application server.

3 An Ontology for Wearables Data Interoperability

In order to provide the AHA Platform with sensor data interoperability, we propose an AHA platform ontology that includes variables and features to measure vital signs or other physical and cognitive habilities for personalized health. An ontology is a formal representation of entities, relations and properties of a given domain of knowledge. Ontologies represent the main technology for creating interoperability at a semantic level [7]. Through the development of a formal illustration of the data, it is possible to share and reuse an ontology all over the Web. Ontologies formulate and model relationships between concepts in a given domain [8] and are suitable for adding context-awareness capabilities in sensor network systems. Semantic technologies have shown to be successful in context representation and reasoning, which can serve in object tracking and scene interpretation [9], and human activity recognition [10] among other areas.

3.1 Ontology Entities, Relations and Datatypes for Wearable Devices

By integrating data from different wearable devices and sensors, a set of dimensions or features were selected to be modelled as entities, data properties (literals) and object properties (class-to-class relations). The AHA ontology design consists of a formal specification for the height and weight of the person, the geographical location at which the person is, the logical place at which the person is and logical activity that the person is employing. Other features modelled are the amount of physical activity (exertion) a person is employing, determined at a specific position on the person's body. For each position, maximally one activity count applies at any specific moment. The energy expenditure from physical activity that the person has employed, the heart rate and stress level of the person, the valence of the person,

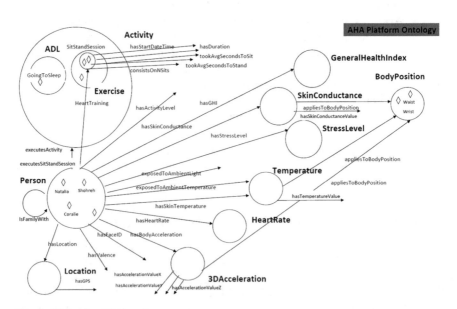

Fig. 3 Wearables AHA ontology

ambient light and temperature that the person is exposed to. More properties are the skin temperature of the person and acceleration at a specific position on the person's body (for each position, maximally one acceleration and skin conductance applies at any specific moment). A summary of the main ontology entities, relations and properties is in Fig. 3.

3.2 Applications

The ontology's vision was designed accounting for the integration into four vertical activities: *Cognitive Endurance, Burn-out turnout, Virtual social gym,* and *We-Care.* Concrete examples on applications of the AHA platform include heart rate unobtrusive motion sensing technologies [11], and different facets of activity recognition in Ambient Assisted Living and health and well-being for healthy ageing lifestyle aspects. Concrete vertical applications in EIT Digital project include *Cognitive Endurance*—which aims at tracking physical activity and heart-rate during the day [12], *Turn out Burnout*—tracking stress, heart rate, activity (lifestyle patterns), sleep [13], etc during daily life [14], and *Virtual Social Gym*—tracking physical fitness (heart rate) and calories burned during a gym session [15].

In general, the applied sensing modalities provided by both sensors and AHA platform are widely applicable and thus enable rapid and low-risk development of applications for third parties in the health and well-being space. Next section shows an application example.

4 Case study: Remote Tracking and Monitoring for Post-surgery Rehabilitation with Kinect

The developed ontology can serve a different range of applications where IoT devices can be connected and information can be shared on a platform independent W3C standard format. In order to test our ontology within the PHL architecture, we developed a remote rehabilitation application. A Kinect for Windows application on remote rehabilitation monitoring/activity recognition, *Rehab@home*, was developed as a case study where the ontology developed served to annotate information useful for both physiotherapists and rehabilitation patients. Rehab@Home encompasses two aspects of health care and well-being: activity monitoring and activity feedback, integrated into everyday lives of (possibly but not uniquely) senior citizens living independently. The application monitors exercise sessions for patients in rehabilitation after shoulder, hip or knee surgery or a simple sit-stand exercise [16]. The final aim is to allow the patient to do the sessions at home giving feedback on the quality and frequency of the exercise to the patient itself and physiotherapist expert,

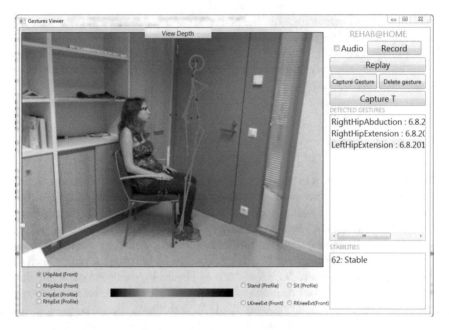

Fig. 4 Kinect remote rehabilitation and monitoring application

remotely and in real time. The current software allows recording new patterns from different users realizing exercises for the system to learn recognizing them.

Figure 4 shows the application for remote rehabilitation monitoring and semantic annotation using Kinect for Windows SDK C# Kinect Toolkit and the developed AHA Ontology. The application records, for instance, a sit-to-stand exercise session amount of repetition times, speed, and joints angles. The demo videos are online.[4,5]

For the integration into PHL store, an exercise activity (recognized by Kinect) is inserted with ontological properties *name, started_at* and *ended_at* as specified in the PHL documentation, for instance SitStandSession example:

```
{"name": "SitStandSession",
"ended_at": "2013-08-13 11:32:29",
"started_at": "2013-08-13 11:20:34"}
```

A client library was developed for the integration of the Activity *SitStandSession* into PHL and the construction of a REST API to access the M3 semantic RDF store [4]. A more tutorized deployed application within the project for elders physical, mental and social activation from home is the Virtual Social Gym [17].

---

[4]Kinect remote rehabilitation demo: https://www.youtube.com/watch?v=XL4JexDNs-Q.

[5]Kinect sit-to-stand demo: https://www.youtube.com/watch?v=g8HOtFTk80c.

5 Conclusion

As a result of the Active Healthy Aging (AHA) Platform project, the definition, design and development of an AHA-platform being the key ambitions of the 2013 Action Line Health and Well-being. The result is a platform that hosts multiple Active Healthy Aging solutions (developed within the action line) on a single platform.

Next to the hosting of the 'in-house' developments, the platform needs to be able in the future to address a number of breakthrough challenges that so far have not been addressed in the fragmented market in Ambient Assisted Living for AHA like technical, legal, cultural and international barriers, mobilization, strengthening and integration of local business communities and international scaling.

The ontology was proposed within the context of EIT Digital partnership among European Union partners within the action line Health and Well-being for the Active Healthy Ageing Platform project (2013). This project's work is part of a starting point, amongst others provided by Philips, INRIA, Novay, Fraunhofer-Gesellschaft and DFKI, to make available a remote monitoring platform also supplied for the Direct Life Labs EIT Digital activity. The future ambition is to create a basic platform with more showcases, higher ambition, and support additional functional and non-functional features.

The wearables Active Healthy Ageing (AHA.owl) OWL 2 ontology aims at serve to both clinical and non-clinical activity tracking and is available online.[6] Reference ontologies developed and used together with AHA ontology are the Kinect Ontology [7][7] and the security and privacy ontology [6].[8]

Acknowledgements We acknowledge the support of EU EIT Digital project no. HWB13070 on Active Healthy Ageing within the Health and Well-being action line and the ICT COST Action IC1303 (European Cooperation in Science and Technology), Algorithms, Architectures and Platforms for Enhanced Living Environments (AAPELE) http://www.aapele.eu. We thank our project partners Marion Karppi (Turku University of Applied Sciences), Antonio De Nigro and Francesco Torelli (R&D Lab—Engineering Ingegneria Informatica), Iman Khaghani Far (University of Trento), Josef Hallberg (Luleå University), Syed Naseh (We-Care), Rafal Kocielnik (TUE) and Marcos Baez (University of Trento).

References

1. J. Alexandersson, i2home-towards a universal home environment for the elderly and disabled. Künstliche Intelligenz **8**(3), 66–68 (2008)
2. N. Díaz Rodríguez, Semantic and fuzzy modelling for human behaviour recognition in smart spaces: a case study on ambient assisted living, Ph.D. dissertation, Åbo Akademi University (Finland) and University of Granada (Spain) (2015)

---

[6]Wearables ontology: https://github.com/NataliaDiaz/Ontologies/blob/master/AHA.owl.

[7]Kinect ontology: http://users.abo.fi/rowikstr/KinectOntology/.

[8]Security and privacy ontology: https://github.com/NataliaDiaz/SecurityAccessControlOntology.

3. P. Karvinen, N. Díaaz Rodrígguez, S. Grönroos, J. Lilius, How to choose a semantic RDF store? an scalability analysis for smart space (2016) (Submitted)
4. F. Wickström, Getting started with Smart-M3 using Python, Technical Report 1071 (2013)
5. A. Berg, P. Karvinen, S. Grönroos, F. Wickström, N. Díaz Rodríguez, S. Hosseinzadeh, J. Lilius, A scalable distributed M3 platform on a low-power cluster, in *Open International M3 Semantic Interoperability Workshop*. TUCS Proceedings, TUCS, ed. by J.-P.S. Soininen, S. Balandin, J. Lilius, P. Liuha, T.S. Cinotti, vol. 21 (2013), pp. 49–58
6. S. Hosseinzadeh, S. Virtanen, N. Díaz-Rodríguez, J. Lilius, A semantic security framework and context-aware role-based access control ontology for smart spaces (2016) (Submitted)
7. N. Díaz Rodríguez, R. Wikström, J. Lilius, M.P. Cuéllar, M. Delgado Calvo Flores, Understanding movement and interaction: an ontology for kinect-based 3D depth sensors, in *Ubiquitous Computing and Ambient Intelligence. Context-Awareness and Context-Driven Interaction*. Lecture Notes in Computer Science, ed. by G. Urzaiz, S. Ochoa, J. Bravo, L. Chen, J. Oliveira, vol. 8276 (Springer International Publishing, 2013), pp. 254–261. https://doi.org/10.1007/978-3-319-03176-7_33
8. M. d'Aquin, N.F. Noy, Where to publish and find ontologies? A survey of ontology libraries. Web Semant.: Sci. Serv. Agents World Wide Web **11**, 96–111 (2012), http://www.sciencedirect.com/science/article/pii/S157082681100076X
9. J. Gómez-Romero, M.A. Patricio, J. García, J.M. Molina, Ontology-based context representation and reasoning for object tracking and scene interpretation in video. Expert Syst. Appl. **38**(6), 7494–7510 (2011). https://doi.org/10.1016/j.eswa.2010.12.118
10. N. Díaz Rodríguez, O.L. Cadahía, M.P. Cuéllar, J. Lilius, M.D. Calvo-Flores, Handling real-world context awareness, uncertainty and vagueness in real-time human activity tracking and recognition with a fuzzy ontology-based hybrid method. Sensors **14**(10), 8 131–18 171 (2014), http://www.mdpi.com/1424-8220/14/10/18131
11. A. Braun, R. Wichert, A. Kuijper, D.W. Fellner, Capacitive proximity sensing in smart environments. J. Ambient Intell. Smart Environ. **7**(4), 483–510 (2015)
12. A. Hedman, J. Hallberg, Cognitive endurance for brain health: challenges of creating an intelligent warning system. KI-Künstliche Intelligenz **29**(2), 123–129 (2015)
13. M. Djakow, A. Braun, A. Marinc, Movibed-sleep analysis using capacitive sensors, in *Universal Access in Human-Computer Interaction. Design for All and Accessibility Practice* (Springer International Publishing, 2014), pp. 171–181
14. R. Kocielnik, N. Sidorova, F.M. Maggi, M. Ouwerkerk, J.H. Westerink, Smart technologies for long-term stress monitoring at work, in *2013 IEEE 26th International Symposium on Computer-Based Medical Systems (CBMS)* (IEEE, 2013), pp. 53–58
15. I.K. Far, M. Ferron, F. Ibarra, M. Baez, S. Tranquillini, F. Casati, N. Doppio, The interplay of physical and social wellbeing in older adults: investigating the relationship between physical training and social interactions with virtual social environments. PeerJ Comput. Sci. **1**, e30 (2015)
16. N. Díaz Rodríguez, S. Grönroos, F. Wickström, P. Karvinen, A. Berg, S. Hosseinzadeh, M. Karppi, J. Lilius, M3 interoperability for remote rehabilitation with kinect, in *Open International M3 Semantic Interoperability Workshop*. TUCS Lecture Notes, ed. by J.-P.S. Soininen, S. Balandin, J. Lilius, P. Liuha, T.S. Cinotti, vol. 21 (2013), pp. 153–163
17. I.K. Far, P. Silveira, F. Casati, M. Baez, Unifying platform for the physical, mental and social well-being of the elderly, in *Embedded and Multimedia Computing Technology and Service*. Lecture Notes in Electrical Engineering, ed. by J.J.J.H. Park, Y.-S. Jeong, S.O. Park, H.-C. Chen, vol. 181 (Springer Netherlands, 2012), pp. 385–392. https://doi.org/10.1007/978-94-007-5076-0_46

Part XIII
Fuzziness in Civil and Environmental Engineering

How to Estimate Resilient Modulus for Unbound Aggregate Materials: A Theoretical Explanation of an Empirical Formula

Pedro Barragan Olague, Soheil Nazarian, Vladik Kreinovich, Afshin Gholamy and Mehran Mazari

Abstract To ensure the quality of pavement, it is important to make sure that the resilient moduli—that describe the stiffness of all the pavement layers—exceed a certain threshold. From the mechanical viewpoint, pavement is a non-linear medium. Several empirical formulas have been proposed to describe this non-linearity. In this paper, we describe a theoretical explanation for the most accurate of these empirical formulas.

1 Formulation of the Problem

Need for estimating resilient modulus. To ensure the quality of a road, it is important to make sure that all the pavement layers have reached a certain stiffness level. To characterize stiffness of unbound pavement materials, transportation engineers use *resilient modulus M_r*.

A material's resilient modulus is actually an estimate of its modulus of elasticity E, i.e., of ratio of stress by strain; the difference from the usual modulus of elasticity if that:

P. Barragan Olague · S. Nazarian · V. Kreinovich (✉) · A. Gholamy
Center for Transportation Infrastructure Systems, University of Texas at El Paso, El Paso, TX 79968, USA
e-mail: vladik@utep.edu

P. Barragan Olague
e-mail: pabarraganolague@miners.utep.edu

S. Nazarian
e-mail: nazarian@utep.edu

A. Gholamy
e-mail: afshingholamy@gmail.com

M. Mazari
Savannah State University, Savannah, GA 31401, USA
e-mail: mazarim@savannahstate.edu

© Springer International Publishing AG, part of Springer Nature 2018
L. A. Zadeh et al. (eds.), *Recent Developments and the New Direction in Soft-Computing Foundations and Applications*, Studies in Fuzziness and Soft Computing 361, https://doi.org/10.1007/978-3-319-75408-6_44

- the usual modulus corresponds to a *slowly* applied load, while
- the resilient characterizes the effect of *rapidly* applied loads—like those experienced by pavements.

A precise definition of the resilient modulus is given, e.g., in [1].

Need to take non-linearity into account. In the usual (*linear*) elastic materials, the modulus does not depend on the stress value. In contrast, pavement materials are usually *non-linear*, in the sense that the resilient stress depends on the stress.

Empirical formulas describing pavement's non-linearity. Several empirical formulas have been proposed to describe this dependence. Experimental comparison [3] shows that the best description is provided by the formula (first proposed in [5])

$$M_r = k_1' \cdot \left(\frac{\theta}{P_a} + 1\right)^{k_2'} \cdot \left(\frac{\tau_{\text{oct}}}{P_a} + 1\right)^{k_3'},$$

where P_a is atmospheric pressure, θ is the *bulk stress*, i.e., the trace

$$\theta = \sum_{i=1}^{3} \sigma_{ii}$$

of the stress tensor σ_{ij} (see, e.g., [6]), and

$$\tau_{\text{oct}} \stackrel{\text{def}}{=} \sqrt{\frac{1}{3} \cdot \sum_{ij} \sigma_{ij}^2 - \frac{1}{3} \cdot \theta^2}$$

is the *octahedral shear stress*.

In terms of the eigenvalues σ_1, σ_2, and σ_3 of the stress tensor,

$$\theta = \sigma_1 + \sigma_2 + \sigma_3$$

and

$$\tau_{\text{oct}} = \frac{1}{3} \cdot \sqrt{(\sigma_1 - \sigma_2)^2 + (\sigma_2 - \sigma_3)^2 + (\sigma_3 - \sigma_1)^2}.$$

What we do in this paper. In this talk, we provide a theoretical explanation for the above empirical formula.

This explanation uses the general idea that the fundamental physical formulas should not change if we simply change the measuring unit and/or the starting point for the measurement scale.

Paper outline. First, in Sect. 2, we briefly explain the general idea, that fundamental physical formulas should not depend on the choice of the starting point or on the choice of the measuring unit.

In Sect. 3, we use this general idea to describe possible dependence of the resilient modulus M_r on, correspondingly, the bulk stress θ and on the octahedral sheer stress τ_{oct}.

Finally, in Sect. 4, we apply similar ideas to combine the two formulas for $M_r(\theta)$ and $M_r(\tau_{oct})$ into a single formula $M(\theta, \tau_{oct})$ that describes the dependence of the resilient modulus on both stresses.

Comment. In our derivation, we are not using physical equation, we are only using expert knowledge—which, in this case, is formulated in terms of invariance. From this viewpoint, this paper can be viewed as a particular case of soft computing, techniques for formalizing and utilizing expert knowledge.

2   General Idea: Fundamental Physical Formulas Should Not Depend on the Choice of the Starting Point or of the Measuring Unit

Main idea. Computers process numerical values of different quantities. A numerical value of a quantity depends on the choice of a measuring unit and—in many cases—also on the choice of the starting point.

For example, depending on the choice of a measuring unit, we can describe the height of the same person as 1.7 m or 170 cm. Similarly, we can describe the same moment of time as 2 pm (14.00) if we use El Paso time or 3 pm (15.00) if we use Austin time—the difference is caused by the fact that the starting points for these two times—namely midnight (00.00) in El Paso and midnight (00.00) in Austin—differ by 1 h.

The choice of a measuring unit is rather arbitrary. For example, we can measure length in meters or in centimeters or in feet. Similarly, the choice of the starting point is arbitrary: when we analyze a cosmic event, it does not matter the time of what location we use to describe it. It is therefore reasonable to require that the fundamental physical formulas not depend on the choice of a measuring unit and—if appropriate—on the choice of the starting point. We do not expect that, e.g., Newton's laws look differently if we use meters or feet.

Of course, if we change the units in which we measure one of the quantities, then we may need to adjust units of related quantities. For example, if we replace meters with centimeters, then for the formula $v = d/t$ (that describes velocity v as a ratio of distance d and time t) to remain valid we need to replace meters per second with centimeters per second when measuring velocity. However, once the appropriate adjustments are made, we expect the formulas to remain the same.

Not all physical quantities allow both changes. It should be mentioned that while most physical quantities do not have any preferred measuring unit—and thus, selection of a different measuring unit makes perfect physical sense—some quantities have a fixed starting point. For example, while we can choose an arbitrary starting point for time, for distance, 0 distance seems to be a reasonable starting point.

As a result, while the change of a measuring unit makes sense for most physical quantities, the change of a starting point only makes sense for some of them—and a physics-based analysis is needed to decide whether this change makes physical sense.

How to describe the change of a measuring unit in precise terms. If we replace the original measuring unit with a new unit which is a times smaller, then all numerical values of the measured quantity get multiplied by a: $x' = a \cdot x$.

For example, if we replace meters with centimeters—which are $a = 100$ times smaller—then the original height of $x = 1.7$ m becomes $x' = a \cdot x = 100 \cdot 1.7 = 170$ cm.

How to describe the change of the starting point in precise terms. If we replace the original starting point by a new one which is b earlier (or smaller), then to all numerical values of the measured quantity the value b is added: $x' = x + b$.

For example, if we replace El Paso time with Austin time—which is $b = 1$ h earlier, then the original time of $x = 14.00$ h becomes $x' = x + b = 14.00 + 1.00 = 15.00$ h.

In general, we can change both the measuring unit and the starting point. If we first change the measuring unit and the starting point, then:

- first, the original value x first gets multiplied by a, resulting in $x' = a \cdot x$, and
- then the value b is added to the new value x', resulting in $x'' = x' + b = a \cdot x + b$.

Thus, in general, when we change both the measuring unit and the starting point, we get a linear transformation $x \to a \cdot x + b$.

3   How Resilient Modulus Depends on the Bulk Stress (and on the Octahedral Shear Stress)

What we do in this section. Let us first use the above idea to describe how the resilient modulus M_r depends on the bulk stress θ.

Which invariances make sense in this case. As we have mentioned in the previous section,

- while the change of a measuring unit makes sense for (practically) *all* physical quantities,
- the change of the starting point only makes physical sense for *some* quantities.

Let us therefore analyze whether the change of the starting point makes sense for the resilient modulus M_r and for the bulk stress θ.

For the resilient modulus, there is a clear starting point $M_r = 0$, in which strain does not cause any stress. So, for the resilient modulus, only a change in a measuring unit makes physical sense.

In contrast, for the bulk stress, we can clearly have several choices of the starting point, choices motivated by the fact that in addition to the external stress, there is also an always-present atmospheric pressure. One possibility is to only count the external stress and thus, consider the situation in which we only have atmospheric pressure as corresponding to zero stress. Another possibility is to explicitly take atmospheric pressure into account and take the ideal vacuum no-atmospheric-pressure situation as zero stress. In the first case, we can select atmospheric pressures corresponding to different heights as different starting points.

What does it mean for the resulting formula to be independent: first approximation. For the dependence $M_r(\theta)$, the requirement—that this dependence does not change if we change numerical values of θ—means the following. For every $a > 0$ and b, the dependence in the new units $M_r(a \cdot \theta + b)$ has exactly the same form as in the old units—if we also appropriately re-scale M_r. So, we should have

$$M_r(a \cdot \theta + b) = c(a,b) \cdot M_r(\theta) \tag{1}$$

for some value c which, in general, depends on a and b.

What are the functions that satisfy this condition: analysis of the problem. Let us find all the functions $M_r(\theta)$ for which, for some function $c(a, b)$, the equality (1) holds for all x, $a > 0$, and b.

From the physical viewpoint, small changes in θ should lead to small changes in M_r, i.e., in mathematical terms, the dependence $M_r(\theta)$ should be continuous. It is know that every continuous function can be approximated, with any given accuracy, by a differentiable function (e.g., by a polynomial). Thus, without losing generality, we can safely assume that the dependence $M_r(\theta)$ is differentiable.

Thus, the function

$$c(a,b) = \frac{M_r(a \cdot \theta + b)}{M_r(\theta)}$$

is also differentiable, as a ratio of two differentiable functions. For $a = 1$, the formula (1) takes the form

$$M_r(\theta + b) = c(1,b) \cdot M_r(\theta). \tag{2}$$

Differentiating both sides of formula (2) with respect to b and setting $b = 0$, we get

$$M_r'(\theta) = c \cdot M_r(\theta), \tag{3}$$

where $f'(x)$ denote the derivative, and c is the derivative of $c(1, b)$ with respect to b for $b = 0$.

The Eq. (3) can be rewritten as

$$\frac{dM_r}{d\theta} = c \cdot M_r,$$

i.e., equivalently, as

$$\frac{dM_r}{M_r} = c \cdot d\theta.$$

Integrating both sides, we get $\ln(M_r) = c \cdot \theta + C_0$ for some constant C_0. Thus,

$$M_r = A \cdot \exp(c \cdot \theta), \tag{4}$$

where $A \overset{\text{def}}{=} \exp(C_0)$.

For $b = 0$ and $a \neq 0$, the Eq. (1) takes the form

$$M_r(a \cdot \theta) = c(a, 0) \cdot M_r(\theta).$$

Substituting the expression (4) into this formula, we conclude that

$$A \cdot \exp(c \cdot a \cdot \theta) = c(a, 0) \cdot \exp(c \cdot \theta). \tag{5}$$

When $c \neq 0$, the two sides grow with θ at a different speed, so we should have $c = 0$ and $M_r(\theta) = \text{const}$.

Thus, the only case when the formula $M_r(\theta)$ is fully invariant is when we have a linear material, with $M(\theta) = \text{const}$.

Since we cannot require all the invariances, let us require only some of them. Since we cannot require invariance with respect to *all* possible re-scalings, we should require invariance with respect to *some* family of re-scalings.

If a formula does not change when we apply each transformation, it will also not change if we apply them one after another, i.e., if we consider a composition of transformations. Each shift can be represented as a superposition of many small (infinitesimal) shifts, i.e., shifts of the type $\theta \to \theta + B \cdot dt$ for some B. Similarly, each re-scaling can be represented as a superposition of many small (infinitesimal) re-scalings, i.e., re-scalings of the type $\theta \to (1 + A \cdot dt) \cdot \theta$. Thus, it is sufficient to consider invariance with respect to an infinitesimal transformation, i.e., a linear transformation of the type

$$\theta \to \theta' = (1 + A \cdot dt) \cdot \theta + B \cdot dt.$$

Invariance means that the value $M(\theta')$ has the same form as $M(\theta)$, i.e., that $M(\theta')$ is obtained from $M(\theta)$ by an appropriate (infinitesimal) re-scaling

$$M_r \to (1 + C \cdot dt) \cdot M_r.$$

In other words, we require that

$$M_r((1 + A \cdot dt) \cdot \theta + B \cdot dt) = (1 + C \cdot dt) \cdot M_r(\theta), \tag{6}$$

i.e., that

$$M_r(\theta + (A \cdot \theta + B) \cdot dt) = M_r(\theta) + C \cdot M_r(\theta) \cdot dt.$$

Here, by definition of the derivative, $M_r(\theta + q \cdot dt) = M_r(\theta) + M'_r(\theta) \cdot q \cdot dt$. Thus, from (6), we conclude that

$$M_r(\theta) + (A \cdot \theta + B) \cdot M'_r(\theta) \cdot dt = M_r(\theta) + C \cdot M_r(\theta) \cdot dt.$$

Subtracting $M_r(\theta)$ from both sides and dividing the resulting equality by dt, we conclude that

$$(A \cdot \theta + B) \cdot M'_r(\theta) = C \cdot M_r(\theta).$$

Since $M'_r(\theta) = \dfrac{dM_r}{d\theta}$, we can separate the variables by moving all the terms related to M_r to one side and all the terms related to θ to another side. As a result, we get

$$\frac{dM_r}{M_r} = C \cdot \frac{d\theta}{A \cdot \theta + b}.$$

Degenerate cases when $A = 0$ can be approximated, with any given accuracy, by cases when A is small but non-zero. So, without losing generality, we can safely assume that $A \neq 0$. In this case, for $x \stackrel{\text{def}}{=} \theta + k$, where $k \stackrel{\text{def}}{=} \dfrac{B}{A}$, we have

$$\frac{dM_r}{M_r} = c \cdot \frac{dx}{x},$$

where $c \stackrel{\text{def}}{=} \dfrac{C}{A}$. Integration leads to $\ln(M_r) = c \cdot \ln(x) + C_0$ for some constant C_0, thus $M_r = C_1 \cdot x^c$ for $C_1 \stackrel{\text{def}}{=} \exp(C_0)$, i.e.,

$$M_r(\theta) = C_1 \cdot (\theta + k)^c. \tag{7}$$

Dependence on the bulk stress: Conclusion. If we represent $\theta + k$ as $k \cdot \left(\dfrac{\theta}{k} + 1\right)$, then we get the desired dependence of M_r on θ:

$$M_r = C_2 \cdot \left(\frac{\theta}{k} + 1\right)^c, \tag{8}$$

where $C_2 \stackrel{\text{def}}{=} C_1 \cdot k^c$.

Dependence on the octahedral sheer stress. Similarly, we can conclude that the dependence $M_r(\tau_{\text{oct}})$ of the resilient modulus M_r on the octahedral sheer stress τ_{oct} has the form

$$M_r = C'_2 \cdot \left(\frac{\tau_{\text{oct}}}{k'} + 1\right)^{c'}, \tag{9}$$

for some constants C'_2, k', and c'.

4 How to Combine the Formulas Describing Dependence on Each Quantities into a Formula Describing Joint Dependence

Idea. We have used the invariance ideas to derive formulas $M_r(\theta)$ and $M_r(\tau_{\text{oct}})$ describing dependence of M_r on each of the quantities θ and τ_{oct}. Let us now use the same ideas to combine these two formulas into a single formula describing the dependence on both quantities θ and τ_{oct}.

Based on the previous analysis, for each pair $(\theta, \tau_{\text{oct}})$, we know the value of the modulus M_r:

- the value $M_1 \overset{\text{def}}{=} M_r(\theta)$ that we obtain if we ignore the octahedral sheer stress and only take into account the bulk stress; and
- the value $M_2 \overset{\text{def}}{=} M_r(\tau_{\text{oct}})$ that we obtain if ignore the bulk stress and only take into account the octahedral sheer stress.

Based on these two values M_1 and M_2, we would like to compute an estimate $M(M_1, M_2)$ for the modulus that would take into account both inputs.

All three values M, M_1, and M_2 represent modulus. Thus, for all three values, only scaling is possible. So, the invariance requirement takes the following form: for every p and q, if we apply the re-scalings $M_1 \to p \cdot M_1$ and $M_2 \cdot q \cdot M_2$, then the resulting dependence $M(p \cdot M_1, q \cdot M_2)$ has the same form as the original dependence $M(M_1, M_2)$—after an appropriate re-scaling by some parameter $c(p, q)$ depending on p and q.

So, for every p and every q, there exists a $c(p, q)$ for which, for all M_1 and M_2, we have

$$M(p \cdot M_1, q \cdot M_2) = c(p, q) \cdot M(M_1, M_2). \tag{10}$$

Analysis of the problem. If we re-scale only one of the inputs, e.g., M_1, we get

$$M(p \cdot M_1, M_2) = c_1(p) \cdot M(M_1, M_2), \tag{11}$$

where $c_1(p) \overset{\text{def}}{=} c(p, 1)$. If we first re-scale by p and then by p', then this is equivalent to one re-scaling by $p \cdot p'$. In the first case, we get

$$M((p \cdot p') \cdot M_1, M_2) = M(p' \cdot (p \cdot M_1), M_2) =$$

$$c_1(p') \cdot M(p \cdot M_1, M_2) = c_1(p') \cdot c_1(p) \cdot M(M_1, M_2). \tag{12}$$

In the second case, we get

$$M((p \cdot p') \cdot M_1, M_2) = c_1(p \cdot p') \cdot M(M_1, M_2). \tag{13}$$

Since the left-hand sides of the equalities (12) and (13) are equal, their right-hand sides must be equal as well. Dividing the resulting equality by $M(M_1, M_2)$, we conclude that

$$c_1(p \cdot p') = c_1(p) \cdot c_1(p'). \tag{14}$$

Differentiating this equality by p' and taking $p' = 1$, we conclude that

$$p \cdot c_1'(p) = c_0 \cdot c_1(p),$$

where $c_0 \stackrel{\text{def}}{=} c_1'(1)$. Thus,

$$\frac{dc_1}{c_1} = c_0 \cdot \frac{dp}{p},$$

so integration leads to $\ln(c_1) = c_0 \cdot \ln(p) + \text{const}$, and

$$c_1(p) = \text{const} \cdot p^{c_0}. \tag{15}$$

For $M_1 = 1$, the formula (11) takes the form

$$M(p, M_2) = \text{const} \cdot p^{c_0} \cdot M(1, M_2), \tag{16}$$

i.e., renaming the variable,

$$M(M_1, M_2) = \text{const} \cdot M_1^{c_0} \cdot M(1, M_2). \tag{17}$$

Similarly, we have

$$M(M_1, M_2) = \text{const}' \cdot M_2^{c_0'} \cdot M(M_1, 1), \tag{18}$$

for some constants const' and c_0'. In particular, for $M_1 = 1$, the formula (18) takes the form

$$M(1, M_2) = \text{const}' \cdot M_2^{c_0'} \cdot M(1, 1). \tag{19}$$

Substituting this expression into the formula (17), we get

$$M(M_1, M_2) = \text{const} \cdot M_1^{c_0} \cdot \text{const}' \cdot M_2^{c_0'} \cdot M(1, 1). \tag{30}$$

Substituting expressions (8) and (9) for M_1 and M_2 into this formula, we come up with the following conclusion.

Conclusion. From the invariance requirements, we can conclude that the dependence of M_r on θ and τ_{oct} has the form

$$M(\theta, \tau_{\text{oct}}) = k_1 \cdot \left(\frac{\theta}{k} + 1\right)^{k_2} \cdot \left(\frac{\tau_{\text{oct}}}{k'} + 1\right)^{k_3},$$

where $k_2 = c \cdot c_0$, $k_3 = c' \cdot c_0'$, and

$$k_1 = \text{const} \cdot \text{const}' \cdot M(1, 1) \cdot C_2^c \cdot (C_2')^{c'}.$$

Thus, we indeed get a theoretical explanation for the empirical dependence.

Remaining open problems. In this paper, we used symmetry ideas to provide a solution to one specific physics-related engineering problem: estimating the resilient modulus for unbound aggregate materials.

While there are not too many papers that use symmetries to solve engineering problems, the use of symmetries is ubiquitous in theoretical physics (see, e.g., [2]), and the use of symmetries can help explain many empirical formulas in soft computing [4].

We therefore hope that our example will lead to future application of symmetry ideas in engineering.

Acknowledgements This work was supported in part by the National Science Foundation grants HRD-0734825 and HRD-1242122 (Cyber-ShARE Center of Excellence) and DUE-0926721, and by an award "UTEP and Prudential Actuarial Science Academy and Pipeline Initiative" from Prudential Foundation. The authors are thankful to the anonymous referees for valuable suggestions.

References

1. American Association of State Highway and Transportation Officials (AASHTO), Resilient modulus of subgrade soils and untreated base/subbase materials, Standard T 292-91, in *Standard Specifications for Transportation Materials and Methods of Sampling and Testing, AAHSTO, Washington, DC* (1998), pp. 1057–1071
2. R. Feynman, R. Leighton, M. Sands, *The Feynman Lectures on Physics* (Addison Wesley, Boston, Massachusetts, 2005)
3. M. Mazari, E. Navarro, I. Abdallah, S. Nazarian, Comparison of numerical and experimental responses of pavement systems using various resilient modulus models. Soils Found. **54**(1), 36–44 (2014)
4. H.T. Nguyen, V. Kreinovich, *Applications of Continuous Mathematics to Computer Science* (Kluwer, Dordrecht, 1997)
5. P.S.K. Ooi, A.R. Archilla, K.G. Sandefur, Resilient modulus models for comactive cohesive soils. Transp. Res. Rec. **1874**, 115–124 (2006)
6. M.H. Sadd, *Elasticity: Theory, Applications, and Numerics, Academic Press* (UK, and Waltham, Massachusetts, Oxford, 2014)

Development of NARX Based Neural Network Model for Predicting Air Quality Near Busy Urban Corridors

Rohit Jaikumar, S. M. Shiva Nagendra and R. Sivanandan

Abstract Accurate prediction of pollutant concentration is very important part in any air quality management program (AQMP). The conventional time series modelling techniques like ARIMA has showing poor prediction and forecasting, as air quality is non-linear and complex phenomenon. Although neural networks have been applied for prediction of air quality data in the previous studies, the model performance was very poor as they don't consider the data as time series in their algorithms. Combining the aspects of both neural networks and time series analysis, Nonlinear Autoregressive models with exogenous input (NARX) based neural networks were found to predict chaotic time series better because of better learning and faster convergence than the conventional neural network algorithms. In the present work, meteorological and traffic parameters near busy urban corridors were used to train NARX based neural network model for the prediction of ambient air quality. Diagnostic analysis between different model variables was done to understand the relationship between one other. The developed model predicted NO_x and SO_2 concentrations with a very good performance over the entire dataset.

1 Introduction

In the recent past, motor vehicles have emerged as one of the critical sources of urban air pollution. Vehicular emissions constitute different pollutants based on the fuel type and quality. The major pollutants released from the vehicles are carbon monoxide (CO), oxides of nitrogen (NO_x), photochemical oxidants, air toxics namely

R. Jaikumar (✉) · S. M. Shiva Nagendra · R. Sivanandan
Department of Civil Engineering, IIT Madras, Chennai 600036, TN, India
e-mail: Biotech.rohit@gmail.com

S. M. Shiva Nagendra
e-mail: snagendra@iitm.ac.in

R. Sivanandan
e-mail: rsiva@iitm.ac.in

© Springer International Publishing AG, part of Springer Nature 2018
L. A. Zadeh et al. (eds.), *Recent Developments and the New Direction in Soft-Computing Foundations and Applications*, Studies in Fuzziness and Soft Computing 361, https://doi.org/10.1007/978-3-319-75408-6_45

benzene, aldehydes, 1-3 butadiene, lead, particulate matter (PM), hydrocarbons (HCs), oxides of sulphur (SO_x) and polycyclic aromatic hydrocarbons (PAHs) [1].

Air quality near urban corridors is highly complex and often represented by chaotic time series. Generally, prediction of time series involves various conventional analytical procedures like linear autoregressive models (AR), moving average models (MA) and autoregressive moving average (ARMA) models introduced by Box and Jenkins [2, 3]. The basic disadvantage with these models is that they are linear and doesn't work when the data is complex like air quality. But, neural networks (NN) are proven to be effective when used to predict non-linear data where the physical process underlying the data is difficult to explain or unclear. To arrive at a solution for such chaotic processes, it is very important to find the relationship between different components of the particular time series. The components are closely related in case of a deterministic process and very minimal relation in a chaotic process. In order to arrive at a better prediction for such chaotic time series, time series component must be integrated into the neural network architecture.

In this paper, a novel neural network architecture is proposed to predict the chaotic air quality time series based upon "Nonlinear Autoregressive models with exogenous input (NARX) based neural network model" [4–6]. NARX based neural network models have been used previously in various studies to predict nonlinear and chaotic time series. The main advantages of NARX based models are (1) training is more effective compare with other algorithms (particularly with gradient descendent method) and (2) the convergence is much faster resulting in better generalization and less computational requirement [4, 5].

2 Methodology

2.1 Study Area

The Fig. 1 shows the corridor under consideration for the current study. It includes S-P Road stretching from Madhya Kailash Temple to Velachery main road junction, OMR from Madhya Kailash Temple to SRP tools bus stop, Tharamani road from SRP tools to velachery bus station, and the Velachery main road till the SP road Junction.

2.2 Traffic Data

Traffic video data at the selected corridors was collected from the ITS laboratory, IIT Madras. The video was manually counted for total number of vehicles every hour.

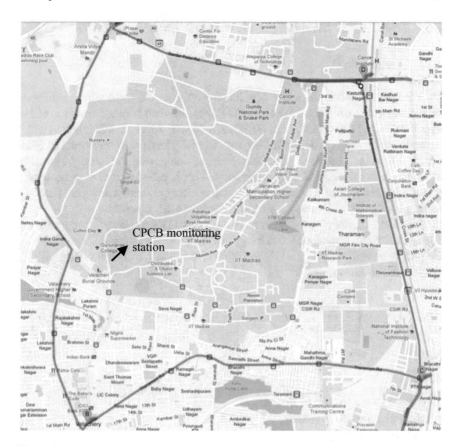

Fig. 1 Study area

2.3 Air Quality and Meteorological Data

Air quality data of different pollutants (CO, NO$_x$, SO$_2$ and RSPM) and meteorological data were collected from the Central Pollution Control Board (CPCB) monitoring station located inside IIT Madras for the entire period of the study.

2.4 Diagnostic Analysis

The relationship between traffic, air quality and meteorological data will be established using different statistical methods such as descriptive statistics, cross correlations, and principal component analysis.

2.5   The NARX Based Neural Network Emission Model

2.5.1   Architecture

The NARX model for prediction of a function Λ can be arrived at using different architectures, but one of the best structure is to use feed forward network with a tapped delay and embedded nodes as shown in Fig. 2. This structure is added with a delayed output layer connection to the inputs. Hence this network is made dependent on the previous output sequence d_u, along with the current input elements to predict the current output sequence. This is called as a time window as it shows a limited part of the time series in the learning algorithm as input. It simply transform a temporal dimension into a different spatial dimension.

Two levels of NN models are used in this work (1) input layer (2) output layer. The general form of the model can be written for next value of time series $y(k + 1)$, the past observation $u(k)$, $u(k-1)$, ..., $u(k-d_u)$ and the past outputs $y(k)$, $y(k-1)$, ..., $y(k-d_y)$ as inputs, as:

$$y(k+1) = \Phi_0 \left\{ w_{b0} + \sum_{h=1}^{N} w_{h0} \cdot \Phi_h \left(w_{h0} + \sum_{i=0}^{d_u} w_{ih} u(k-i) \right) + \sum_{j=0}^{d_y} w_{jh} \cdot y(k-j) \right\}$$

(1)

where,

- $y(k + 1)$—next value of time series,
- $u(k)$—the past observations,
- $u(k-1)$—$u(k-d_u)$—inputs,
- $y(k)$, $y(k-1)$, ..., $y(k-d_y)$—past outputs.

Fig. 2 NARX model with tapped delay line at input

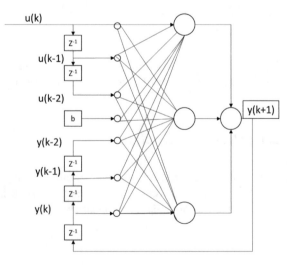

2.5.2 Training Algorithms

A dynamic back propagation algorithm is necessary to find the weights at different nodes and gradients which requires more computation power than a static algorithm and may result in more time. Arriving at a global minima is very difficult as the error surfaces are more complex for a dynamic network [7, 8]. But in the present work, training method used takes advantage of the true real output set thus reducing the training time. Since the network uses decoupled feedback connections, true output can be used. Even general feed forward network with decoupled connections can also be trained using a static algorithm. Also the inputs used in the training process are also real or true inputs and not processed ones, making the overall process very accurate.

But the training process has following limitation. If the number of parameters contained in the network as connections or weights are large, there is a chance of over training the data which may result in a false fit [9]. But in case of NARX neural networks a regularization technique involving a modified performance function to reduce the number of parameters can be obtained. The modified performance function in the form of new MSE$_{reg}$ is given in Eq. 4

$$MSE = \frac{1}{N} \sum_{i=1}^{N} \left(t_j - y_i\right)^2 \tag{2}$$

$$MSW = \frac{1}{n} \sum_{j=1}^{n} w_j^2 \tag{3}$$

$$MSE_{reg} = \xi MSE + (1 - \xi)MSW \tag{4}$$

where

- t_i target and
- ξ performance ratio.

The modified function MSE$_{reg}$ makes the weights and biases of the network very small. This results in a smoother network with less misfits. A modified Levenberg-Marquardt optimization with the new performance function is created and used to train the network. This results in a more generalized form of network solution which can provide widespread application.

3 Results and Discussion

3.1 Descriptive Statistics

Air quality data from January 2012 to March 2014 was collected from the CPCB continuous monitoring station located near the Velachery gate of IIT Madras. The

Table 1 Descriptive statistics of different monitored parameters

Parameter	Mean	Std. deviation
NO_x (ppm)	4.13	4.026
SO_2 (ppm)	10.29133	7.408595
Temp (°C)	29.3464	3.45466
RH (%)	64.9622	16.89799
WS (m/s)	1.7352	0.71749
WD (°)	213.2169	63.07737
VWS (m/s)	0.3570	0.46839
Press (mmHg)	1015.5284	10.44589
Solar (W/m^2)	220.5358	282.26795

air quality parameters namely NO_x, SO_2 and meteorological parameters such as Temperature (Temp), Wind Speed (WS), Wind Direction (WD), Relative Humidity (RH), Vertical Wind Speed (VWS) Barometric Pressure (Press) and solar radiation (Solar) were also collected from the CPCB Monitoring station. The descriptive statistics of the hourly average of air and meteorological parameters are shown in the Table 1.

The weekly variation of the air pollutants is represented in the form of a box plot in Fig. 3. The concentration of all the pollutants is within the National Ambient Air Quality Standards (NAAQS).

3.2 Cross Correlation

Cross correlation analysis was done to study the temporal relationship between the different types of pollutants. Cross correlation measures the lag between two time series. The concentrations of NO_x, SO_2 time series collected from IIT Madras CPCB monitoring station were cross correlated with the total vehicle count of the Velachery main road. The Cross correlation plots of NO_x with TVC are given in the Fig. 4. Also as it can be seen in the figure, there is no significant change in correlation with change in time lag. This indicates there was no lag effect associated with the Pollutant concentration and the traffic count. This can be expected as the monitoring station is located very close to that of the roadside as explained in previous literature [10].

3.3 Principal Component Analysis

In this study, rotated Principal Component Analysis (PCA) is applied to the hourly average concentrations of NO_x, SO_2, RSPM and CO and different meteorological parameters like Temperature (Temp), Wind Speed (WS), Wind Direction (WD),

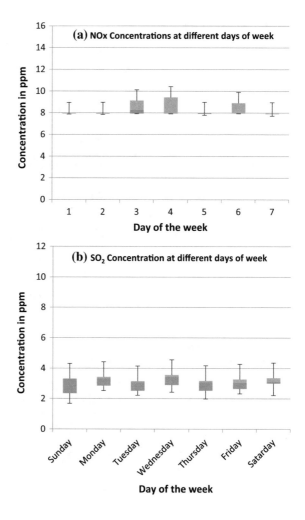

Fig. 3 Weekly variation of pollutant concentration at IITM CPCB monitoring station **a** NO$_x$ **b** SO$_2$

Relative Humidity (RH), Vertical Wind Speed (VWS) Barometric Pressure (Press) and solar radiation (Solar) along with Total vehicular volume (TVC) at Velachery main road. The factors found by PCA do not often have straightforward or unique interpretations. Therefore, in order to clarify the meaning of the components, the most important factors (in terms of variance explained) determined by PCA are subsequently subjected to orthogonal rotation. In this case VARIMAX criterion has been used.

The examined data resulted in three PCs with eigenvalues greater than 1 (shown in scree plot in Fig. 5),

The three components explain 80% of the total variance. Using the values of the respective loadings presented in Table 2, there is a reasonable interpretation for these three components. Only loadings with absolute values greater than 50% of the maximum value are selected for the PC interpretation.

Fig. 4 Cross correlation
between NO$_x$ and TVC

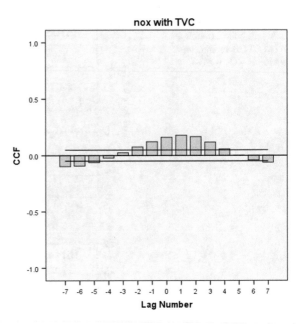

Fig. 5 Principal component
analysis scree plot

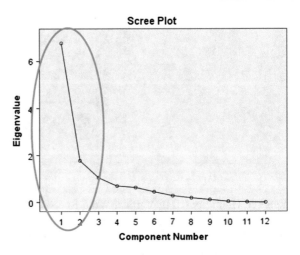

The first component is a measure of Temperature (0.944 Correlation), Relative
Humidity (0.933), wind speed (0.923), wind direction (0.928) and pressure (0.957)
and to a lesser extent NO$_X$ Concentration. This coincides with the previous literature findings which indicate Major meteorological elements affecting the concentration of nitrogen oxides in the air are: wind direction, wind velocity and air
temperature and relative air humidity [11, 12].

TVC (0.776) and RSPM (0.641) are the main contributors in PC$_2$ which con-
trasts meteorological parameters. This can be expected as in cities, motor vehicles

Table 2 Rotated component matrix

	Component		
	1	2	3
No$_x$	0.735	0.062	0.088
So$_2$	0.584	0.226	−0.259
CO	0.253	0.006	0.885
RSPM	0.583	0.641	−0.168
Temp	0.944	0.221	0.141
RH	0.933	−0.194	0.208
WS	0.923	.0272	0.122
WD	0.928	−0.060	0.177
VWS	0.739	0.187	0.496
Pressure	0.957	0.120	0.205
Solar	0.204	0.827	−0.104
TVC	−0.169	0.776	0.220

are the main source of PM emission and it contributes more than 50% of the total PM emissions [13]. Also previous studies have shown a good correlation of PM concentration with TVC [14].

The third PC is a measure of CO (0.885) that might be attributed to be independent of the meteorological parameters. These three independent sources of variance seem to dominate the total data.

3.4 NARX Based Neural Network Prediction Model

Historical Air Quality Data at IITM CPCB Monitoring station from January 2012 to March 2013 was used to develop neural network models. Two models were developed for predicting NO$_x$ and SO$_2$ concentrations. Several types of neural networks were tried. The best fit was observed for neural network model with nonlinear autoregressive exogenous input (NARX). The NARX input based neural network models are a powerful class of models which has been demonstrated that they are well suited for modeling nonlinear systems and specially time series [4]. Major advantages of NARX networks are learning, which is more effective in NARX networks than in other neural network and these networks converge much faster and generalize better than other networks [5, 6]. All the model coding was done in MATLAB 2013b. The best network structure for the model is shown in the Fig. 6.

Here

- x(t) indicates the input time series which comprises of 1 traffic factor (TVC) and **7 meteorological factors**, Wind speed, wind direction, visibility, temperature, Pressure, Solar irradiation and Relative humidity.
- y(t) indicates the the target pollutant time series.

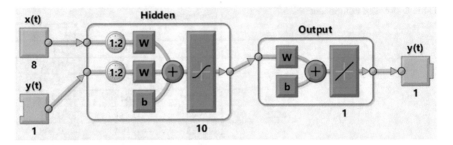

Fig. 6 Best Fit network architecture

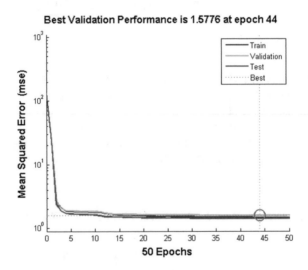

Fig. 7 Training state of SO_2 model

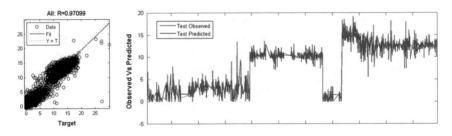

Fig. 8 Model performance for prediction of SO_2 concentration

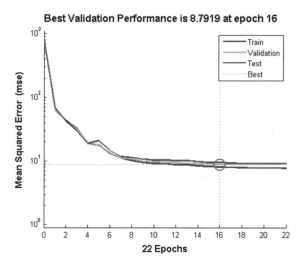

Fig. 9 Training State of NO$_x$ model

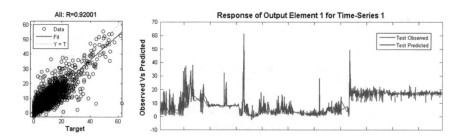

Fig. 10 Model performance for prediction of NO$_x$ concentration

Network with 10 nodes in single hidden layer and sigmoid transfer function is used in the present model [15]. The results for each models are shown in Figs. 7, 8, 9 and 10. The models converged fairly quickly within 50 epochs compared to other architecture from previous literature as shown in Fig. 7. ANN based SO$_2$ model showed a good correlation of 0.971 with fast convergence (Figs. 7 and 8).

ANN based NOx model showed a relatively low correlation of 0.92 and slow convergence as shown in Figs. 9 and 10.

Model performance statistics are given in the Table 3. In comparison with the previous studies involving hybrid neural network approach the current architecture gave better results.

Table 3 Performance statistics of model

Studies	Pollutants modelled	MSE	IA	R
Present study	NO_x, SO_2	8.7919 (NO_X), 1.5776 (SO_2)	0.94 (NO_x), 0.98 (SO_2)	0.92(NO_x), 0.97(SO_2)
Baawain and Al-Serihi [16]	NO_x, SO_2	–	–	0.94, 0.73
Mishra and Goyal [17]	NO_2	26.63	0.85	0.88
Nagendra and Khare [18]	NO_x	69.06	0.76	0.69

4 Conclusion

Urban air quality is a very complex phenomenon with various factors affecting the pollutant concentrations making it nonlinear and stochastic. Conventional Time series models are unable to address the non-linearity. In the present work, NARX based neural network models were developed to predict the air quality near busy urban corridors. The models were able to predict with very good accuracy for the entire datasets. The model results can be used to make early warning systems, regulation and air quality management strategies. Further the model can be integrated into a dispersion model to predict the extreme conditions to give high accuracy hybrid models.

Acknowledgements We would like to thank the Department of Civil Engineering IIT Madras for financial and technical support for the research work.

References

1. CPCB, Air quality monitoring, emission inventory and source apportionment study for Indian cities—India Environment Portal | News, reports, documents, blogs, data, analysis on environment & development | India, South Asia (2010)
2. D.S.G. Pollock, in *Handbook of Time Series Analysis, Signal Processing, and Dynamics* (Elsevier, 1999)
3. T.P.M. Tertisco, P. Stoica, *Modeling and Forecasting of Time Series* (Publishing House of Romanian Academy, 1985)
4. S. Haykin, *Neural Networks: A Comprehensive Foundation*, vol. 13, no. 4 (Macmillan, 1994)
5. T. Lin, B.G. Horne, P. Tino, C.L. Giles, Learning long-term dependencies in NARX recurrent neural networks. IEEE Trans. Neural Netw. **7**(6), 1329–1338 (1996)
6. Y. Gao, M.J. Er, NARMAX time series model prediction: feedforward and recurrent fuzzy neural network approaches. Fuzzy Sets Syst. **150**(2), 331–350 (2005)
7. R.J. Williams, Training recurrent networks using the extended Kalman filter, in *Proceedings of the 1992 IJCNN International Joint Conference on Neural Networks*, vol. 4, pp. 241–246
8. J.A.C.D.P. Mandic, *Recurrent Neural Networks for Prediction* (Wiley, 2001)

9. M.H. Beale, M.T. Hagan, H.B. Demuth, Neural Network Toolbox^{TM} User's Guide (2015)
10. O. Tchepel, C. Borrego, Frequency analysis of air quality time series for traffic related pollutants. J. Environ. Monit. **12**(2), 544–550 (2010)
11. J.P. Shi, R.M. Harrison, Regression modelling of hourly NO_x and NO_2 concentrations in urban air in London. Atmos. Environ. **31**(24), 4081–4094 (1997)
12. R. Kalbarczyk, E. Kalbarczyk, Influence of meteorological conditions on the concentration of NO_2 and NO_x in northwest Poland in relation to wind direction, vol. 94, no. 38, pp. 81–94 (2007)
13. A. Wróbel, E. Rokita, W. Maenhaut, Transport of traffic-related aerosols in urban areas. Sci. Total Environ. **257**(2–3), 199–211 (2000)
14. B. Srimuruganandam, S.M. Shiva Nagendra, Characteristics of particulate matter and heterogeneous traffic in the urban area of India. Atmos. Environ. **45**(18), 3091–3102 (2011)
15. M. Boznar, Pattern selection strategies for a neural network-based short term air pollution prediction model, in *Proceedings Intelligent Information Systems*, 1997 (IIS'97), pp. 340–344
16. M.S. Baawain, A.S. Al-Serihi, Systematic approach for the prediction of ground-level air pollution (around an industrial port) using an artificial neural network. Aerosol Air Qual. Res. **14**(1), 124–134 (2014)
17. D. Mishra, P. Goyal, Neuro-fuzzy approach to forecast NO_2 pollutants addressed to air quality dispersion model over Delhi, India. Aerosol Air Qual. Res. **16**(1), 166–174 (2016)
18. S.M.S. Nagendra, M. Khare, Artificial neural network approach for modelling nitrogen dioxide dispersion from vehicular exhaust emissions. Ecol. Model. **190**(1–2), 99–115 (2006)

How to Predict Nesting Sites and How to Measure Shoreline Erosion: Fuzzy and Probabilistic Techniques for Environment-Related Spatial Data Processing

Stephen M. Escarzaga, Craig Tweedie, Olga Kosheleva
and Vladik Kreinovich

Abstract In this paper, we show how fuzzy and probabilistic techniques can be used in environment-related data processing. Specifically, we will show that these methods help in solving two environment-related problems: how to predict the birds' nesting sites and how to measure shoreline erosion.

1 Formulation of the Problem: Importance of Environment-Related Spatial Data Processing

Importance of environment-related spatial data processing. When analyzing the ecological systems, it is important to study the spatial environment of these systems, and spatial distribution of the corresponding species in this spatial environment; see, e.g., [1, 3, 7].

Studying spatial environment: the importance of studying shorelines. In most locations within an ecological zone, the environmental changes are reasonably slow; it usually takes decades to see a drastic change. However, at the borders between

S. M. Escarzaga (✉) · C. Tweedie
Environmental Science Program, University of Texas at El Paso, El Paso, TX 79968, USA
e-mail: smescarzaga@utep.edu

C. Tweedie
e-mail: ctweedie@utep.edu

S. M. Escarzaga · C. Tweedie · V. Kreinovich
Cyber-ShARE Center, University of Texas at El Paso, El Paso, TX 79968, USA
e-mail: vladik@utep.edu

O. Kosheleva
Department of Electrical and Computer Engineering, University of Texas
at El Paso, El Paso, TX 79968, USA
e-mail: olgak@utep.edu

© Springer International Publishing AG, part of Springer Nature 2018
L. A. Zadeh et al. (eds.), *Recent Developments and the New Direction
in Soft-Computing Foundations and Applications*, Studies in Fuzziness
and Soft Computing 361, https://doi.org/10.1007/978-3-319-75408-6_46

different ecological zones, the changes are much faster. In the border between different types of plants the changes are fast but still gradual: new types of plants appear, their proportion grows, and eventually, they take over the area. However, there are border areas where the change is the most drastic: namely, the shorelines. The shorelines are, in most places, retreating because of the shoreline erosion.

While the overall area of the shorelines is reasonably small in comparison with the areas of the land and the sea areas, shorelines play a large role in ecological systems, since they are a habitat for many species, from birds (like seagulls) to turtles to numerous other creatures.

From this viewpoint, it is important to be able to trace and measure shoreline erosion.

Studying spatial distribution of different species. In addition to tracing and measuring spatial environments which are important for different species, it is also necessary to trace spatial location of these species. This problem is especially important for rare birds. Birds are most vulnerable when they at their nesting sites. It is therefore important to monitor these sites.

Some species use the same nesting sites year after year, but birds from other species vary their sites each year. To be able to monitor birds from these species, it is therefore important to be able to predict their nesting sites.

What we do in this paper. In this paper, we show that fuzzy and probabilistic techniques can help in solving these two environment-related spatial data processing problems.

2 How to Predict Nesting Sites?

Formulation of the environmental problem. We observe nesting sites for a certain bird species. Our goals are:

- to analyze which criteria are important for selecting nesting sites, and
- to come up with formulas that would enable us to predict nesting sites.

Reformulating this problem in precise terms. Let $v_1, \ldots, v_n$ be parameters that may influence the selection of a nesting site: e.g., parameters describing elevation, hydrology, vegetation level, distance form other nesting sites, etc. For each geographical location x, we record the values of these parameters $v_1(x), \ldots, v_n(x)$.

We assume that the birds select a nesting site based on the values of these quantities (at least some of them). In general, this means that a bird tries to maximize the value of some objective function $F(v_1, \ldots, v_n)$ depending on these values v_i.

We do not know the exact form of the dependence $F(v_1, \ldots, v_n)$. However, we can always expand this dependence in Taylor series and keep only terms up to a certain order in this expansion. For example, if we only keep linear terms, this means that we consider objective functions of the type

$$F(v_1, \ldots, v_n) = a_0 + \sum_{i=1}^{n} a_i \cdot v_i$$

for some to-be-determined coefficients a_i. If we also keep quadratic terms, this means that we consider objective functions of the type etc.

$$F(v_1, \ldots, v_n) = a_0 + \sum_{i=1}^{n} a_i \cdot v_i + \sum_{i=1}^{n} \sum_{\ell=1}^{n} a_{i\ell} \cdot v_i \cdot v_\ell,$$

The more terms we keep, the more accurately we describe the objective function and thus, the more accurately we predict the nesting sites.

For each of these approximations, the (unknown) objective function has the form

$$F(v_1, \ldots, v_n) = \sum_{j=1}^{N} A_j \cdot V_j(x), \tag{2.1}$$

where $V_j(x)$ are known values (e.g., $v_i(x)$ and $v_i(x) \cdot v_\ell(x)$) and A_j are the coefficients that need to be determined.

We assume that each year, each of the observed nesting sites x_k has the largest possible value of the objective function among all locations within the corresponding *Voronoi cell* C_k—i.e., among all locations x which are closer to x_k that to any other nesting locations. Under this assumption, we would like to find the weights $A_1, \ldots, A_N$ that best explain the observed nesting sites.

Analysis of the problem. The fact that on the cell C_j, the linear function (2.1) attains its largest value at the site x_j means that

$$\sum_{j=1}^{N} A_j \cdot V_j(x_k) \geq \sum_{j=1}^{N} A_j \cdot V_j(x) \text{ for all } x \in C_k.$$

In other words, we should have

$$A \cdot \Delta(x) \overset{\text{def}}{=} \sum_{j=1}^{N} A_j \cdot \Delta_j(x_k) \geq 0 \tag{2.2}$$

where we denoted

$$A \overset{\text{def}}{=} (A_1, \ldots, A_n),$$

$$\Delta(x) \overset{\text{def}}{=} (\Delta_1(x), \ldots, \Delta_N(x)),$$

and $\Delta_j(x) \overset{\text{def}}{=} V_j(x_k) - V_j(x)$. Similarly, we should have $A \cdot (-\Delta(x)) \leq 0$ for all x.

How can we solve this problem? From the mathematical viewpoint, this problem is similar to the *linear discriminant analysis* (see, e.g., [2]), when we have two sets S and S' and we need to find a hyperplane that separates them, i.e., a vector A such that $A \cdot S \geq 0$ for all $S \in S$ and $A \cdot S' \leq 0$ for all $S' \in S'$. In our case, S is the set of all vectors $\Delta_j(x)$, and S' is the set of all vectors $-\Delta_j(x)$.

The standard way of solving this problem is to compute the mean μ of all the vectors $S \in S$, the covariance matrix Σ, and then to take $A = \Sigma^{-1}\mu$. So, in our case, we should do the following:

- compute all the vectors $\Delta(x)$ with components $\Delta_j(x) = V_j(x_k) - V_j(x)$, where $x \in C_k$; let M be the total number of such vectors;

- compute the average $\mu = \dfrac{1}{M} \cdot \sum_x \Delta(x)$ of these vectors;

- compute the corresponding covariance matrix Σ with components;

$$\Sigma_{ab} = \frac{1}{M} \cdot \sum_x (\Delta_a(x) - \mu_a) \cdot (\Delta_b(x) - \mu_b); \tag{2.3}$$

- compute the desired weights as $A = \Sigma^{-1}\mu$, i.e., as a solution to a linear system $\Sigma A = \mu$.

The above procedure is equivalent to using probabilistic clustering of the vectors $V_j(x)$ and $-V_j(x)$, i.e., clustering based on probabilistic ideas (see, e.g., [6]). Alternatively, we can use *fuzzy clustering* techniques, i.e., clustering based on using fuzzy ideas (see, e.g., [4, 5, 8]).

Once we know the coefficients A_j, we can use the objective function (2.1) to predict the nesting locations as the points x at which the objective function

$$\sum_{i=1}^{N} A_j \cdot V_j(x)$$

attains a local maximum.

How can we gauge the accuracy of the resulting estimate. To gauge the accuracy of this prediction, we can test it against the observed data. Specifically, for each cell C_k, we compute the location c_k at which the weighted combination $\sum_{i=1}^{N} A_j \cdot V_j(x)$ attains its maximum on this cell. The mean square distance between these predicted nesting sites c_k and the actual nesting sites x_k can serve as a natural measure of prediction accuracy.

3 How to Measure Shoreline Erosion?

Formulation of the problem. Many coastal areas are affected by erosion, the sea expands and the shore retreats. A natural way to measure erosion is to observe the shoreline year after year.

In principle, the rate of erosion in each location can be determined as follows: we compute the difference between the observed shoreline locations at two different years, and divide this difference by the number of years between the two observations. In practice, however, observers in different years follow slightly different lines when making their measurement: e.g., lines at a certain distance from water, or at a certain elevation above water, etc. This fact changes the difference between observations and thus, the computed ratio is, in general, different from the actual erosion rate. The difference can be so large that in the areas with known erosion, the computed ratio becomes negative—erroneously indicating the sea retreat.

In short, we have an additional measurement uncertainty. It is desirable to take this uncertainty into account.

How to take this uncertainty into account: First approximation. It is usually assumed that within a few-years period, the rate r of erosion practically does not change. So, if we perform observations at years $t, t+1, ..., t+T$, then we expect the observed coordinates $x_t, x_{y+1}, ...,$ of the shoreline take the form

$$x_{t+i} = x_t + i \cdot r. \tag{3.1}$$

Due to the presence of the above-mentioned observation errors ε_t, the observed coordinate $\widetilde{x}_{t+i}$ has the form

$$\widetilde{x}_{t+i} = x_{t+i} + \varepsilon_{t+i} = x_t + i \cdot r + \varepsilon_{t+i}. \tag{3.2}$$

It is therefore reasonable to use the use the usual Least Squares techniques to estimate the erosion rate r, i.e., to find r as the value corresponding to the following optimization problem:

$$\sum_{i=0}^{T} \left(\widetilde{x}_{t+i} - (x_t + i \cdot r) \right)^2 \to \min_{x_t, r}. \tag{3.3}$$

How can we solve the corresponding minimization problem? Differentiating the expression (3.3) with respect to both unknowns r and x_t, equating the derivatives to 0, dividing both sides of the resulting equation by $T + 1$, and taking into account that

$$\sum_{i=0}^{T} i = \frac{T \cdot (T+1)}{2}$$

and

$$\sum_{i=0}^{T} i^2 = \frac{T \cdot (T+1) \cdot (2T+1)}{6},$$

we get the following system of two equations:

$$\bar{x} = x_t + r \cdot \frac{T}{2};$$ (3.4)

$$\frac{1}{T+1} \cdot \sum_{i=0}^{T} (i \cdot \tilde{x}_{t+i}) = x_t \cdot \frac{T}{2} + r \cdot \frac{T \cdot (2T+1)}{6},$$ (3.5)

where

$$\bar{x} \overset{\text{def}}{=} \frac{1}{T+1} \cdot \sum_{i=0}^{T} \tilde{x}_{t+i}$$ (3.6)

is the arithmetic average of all the observed values $\tilde{x}_{t+i}$.

From these two equations, we can find the estimates r and x_t. Let us first find the estimate r. Multiplying the Eq. (3.4) by $\frac{T}{2}$, we get

$$\frac{1}{T+1} \cdot \sum_{i=0}^{T} \left(\frac{T}{2} \cdot \tilde{x}_{t+i} \right) = x_t \cdot \frac{T}{2} + r \cdot \frac{T^2}{2}.$$ (3.7)

Subtracting (3.7) from (3.5), we get

$$\frac{1}{T+1} \cdot \sum_{i=0}^{T} \left(\left(i - \frac{T}{2} \right) \cdot \tilde{x}_{t+i} \right) = r \cdot \frac{T}{2} \cdot \frac{2T+1}{3} - r \cdot \frac{T}{2} \cdot \frac{T}{2} =$$

$$r \cdot \frac{T}{2} \cdot \left(\frac{2T+1}{3} - \frac{T}{2} \right) = r \cdot \frac{T}{2} \cdot \frac{T+2}{6}.$$ (3.8)

Thus,

$$r = \frac{12}{T \cdot (T+1) \cdot (T+2)} \cdot \sum_{i=0}^{T} \left(\left(i - \frac{T}{2} \right) \cdot \tilde{x}_{t+i} \right).$$ (3.9)

Substituting this expression into the formula (3.4), we get

$$x_t = \bar{x} - r = \frac{1}{T+1} \cdot \sum_{i=0}^{T} \tilde{x}_{t+i} - \frac{12}{T \cdot (T+1) \cdot (T+2)} \cdot \sum_{i=0}^{T} \left(\left(i - \frac{T}{2} \right) \cdot \tilde{x}_{t+i} \right) =$$

$$\frac{1}{T \cdot (T+1) \cdot (T+2)} \cdot \sum_{i=0}^{T} (T \cdot (T+8) - 12 \cdot i) \cdot \tilde{x}_{t+i}.$$ (3.10)

Once we have computed r and x_t from these equations, then, based on a single measurement, we can then estimate the standard deviation σ of the measurement error ε_{t+i} as the mean square difference between the observed and predicted values:

$$\sigma^2 = \frac{1}{T+1} \cdot \sum_{i=0}^{T} (\widetilde{x}_{t+i} - (x_t + r \cdot i))^2. \tag{3.11}$$

In practice, we have several measurements at different spatial locations k, with results $\widetilde{X}_{t,k}$. So, to find σ, we should also average over all these locations:

$$\sigma^2 = \frac{1}{L} \cdot \frac{1}{T+1} \cdot \sum_{k=1}^{K} \sum_{i=0}^{T} (\widetilde{x}_{t+i,k} - (x_{t,k} + r_k \cdot i))^2, \tag{3.11a}$$

where K is the overall number of spatial locations.

Case of $T = 2$. In practice, we often have three consequent years of observation x_t, x_{t+1}, and x_{t+2}, i.e., we have $T = 2$. In this case, the formulas (3.9) and (3.10) take the following form:

$$r = \frac{\widetilde{x}_{t+2} - \widetilde{x}_t}{2}, \tag{3.12}$$

and

$$x_t = \frac{5\widetilde{x}_t + 2\widetilde{x}_{t+1} - \widetilde{x}_{t+2}}{6}. \tag{3.13}$$

Here,

$$x_t + r = \frac{\widetilde{x}_t + \widetilde{x}_{t+1} + \widetilde{x}_{t+2}}{3} \tag{3.14}$$

and

$$x_t + 2r = \frac{\widetilde{x}_t + \widetilde{x}_{t+1} + \widetilde{x}_{t+2}}{3} + r = \frac{-\widetilde{x}_t + 2\widetilde{x}_{t+1} + 5\widetilde{x}_{t+2}}{6}. \tag{3.15}$$

Thus,

$$\widetilde{x}_t - x_t = \frac{\widetilde{x}_t - 2\widetilde{x}_{t+1} + \widetilde{x}_{t+2}}{6}, \tag{3.16}$$

$$\widetilde{x}_{t+1} - x_{t+1} = \frac{\widetilde{x}_t - 2\widetilde{x}_{t+1} + \widetilde{x}_{t+2}}{3}, \tag{3.17}$$

and

$$\widetilde{x}_{t+2} - x_{t+2} = \frac{\widetilde{x}_t - 2\widetilde{x}_{t+1} + \widetilde{x}_{t+2}}{6}. \tag{3.18}$$

Therefore, in this case,

$$\sigma^2 = \frac{1}{3} \cdot \left((\tilde{x}_t - x_t)^2 + \tilde{x}_{t+1} - x_{t+1})^2 + (\tilde{x}_{t+2} - x_{t+2})^2 \right) =$$

$$\frac{1}{3} \cdot \left(\frac{1}{6^2} + \frac{1}{3^2} + \frac{1}{6^2} \right) \cdot (\tilde{x}_t - 2\tilde{x}_{t+1} + \tilde{x}_{t+2})^2 = \frac{1}{18} \cdot (\tilde{x}_t - 2\tilde{x}_{t+1} + \tilde{x}_{t+2})^2. \quad (3.19)$$

By taking the average over all spatial locations, we get

$$\sigma^2 = \frac{1}{18} \cdot \frac{1}{K} \cdot \sum_{k=1}^{K} (\tilde{x}_{t,k} - 2\tilde{x}_{t+1,k} + \tilde{x}_{t+2,k})^2. \quad (3.20)$$

What if positive erosion values are not always within 2-sigma range? The estimated erosion rate r may be negative, but it is OK if within the corresponding 2-sigma interval

$$[r - 2\sigma, r + 2\sigma],$$

we have a non-negative value, i.e., if we have $r + 2\sigma \geq 0$. This means that the difference between the actual (positive) erosion rate and our (negative) estimate r can be explained by the observation uncertainty.

But what if $r + 2\sigma < 0$? This would mean that we need an additional source of error, i.e., that instead of the formula (3.2), we will have

$$\tilde{x}_{t+i} = x_{t+i} + \varepsilon_{t+i} + \delta_{t+i} = x_t + i \cdot r + \varepsilon_{t+i} + \delta_{t+i}. \quad (3.21)$$

In this case, we still determine our estimates x_t and r from the least squares method (3.3). However, now, in addition to the error component (3.11), we have an additional source of error, with some standard deviation σ_δ^2, so the overall variance σ_t^2 now has the form

$$\sigma_t^2 = \sigma^2 + \sigma_\delta^2. \quad (3.22)$$

How can we determine σ_t^2 and σ_δ^2?

For a normal distribution, 95% of the values are within 2 sigma interval. So, for 95% of the estimated erosion values r_k, we should have $r_k + 2\sigma_t \geq 0$, i.e., equivalently, $2\sigma_t \geq -r_k$. If we sort the estimated erosion rates in increasing order, as

$$r_1 < r_2 < \ldots < r_N, \quad (3.23)$$

then this means that the desired inequality should be satisfied for all $k \geq 0.05 \cdot N$, i.e., that we should have $2\sigma_t \geq -r_{0.05 \cdot N}$, $2\sigma_t \geq -r_{0.05 \cdot N + 1}$, etc. Since the sequence r_k is sorted in increasing order, the first inequality implies all the others, so it is sufficient to satisfy the first inequality $2\sigma_t \geq -r_{0.05 \cdot N}$, i.e., $\sigma_t \geq -\frac{1}{2} \cdot r_{0.05 \cdot N}$.

We would like to have the narrowest error bounds, so we choose the smallest $\sigma_t \geq \sigma$ that satisfies this inequality, i.e., we take $\sigma_t = \max\left(\sigma, -\frac{1}{2} \cdot r_{0.05 \cdot N} \right)$.

Resulting algorithm: **Case of general** T. We start with measurements $\widetilde{x}_{t+i,k}$ make at different spatial locations k at years $t, t+1, ..., t+T$.

For each location k, we compute the estimated erosion rate

$$r_k = \frac{12}{T \cdot (T+1) \cdot (T+2)} \cdot \sum_{i=0}^{T} \left(\left(i - \frac{T}{2} \right) \cdot \widetilde{x}_{t+i,k} \right) \tag{3.24}$$

and the estimated initial erosion

$$x_{t,k} = \frac{1}{T \cdot (T+1) \cdot (T+2)} \cdot \sum_{i=0}^{T} (T \cdot (T+8) - 12 \cdot i) \cdot \widetilde{x}_{t+i,k}. \tag{3.25}$$

Then, we estimate the first approximation σ to the corresponding uncertainty as

$$\sigma^2 = \frac{1}{L} \cdot \frac{1}{T+1} \cdot \sum_{k=1}^{K} \sum_{i=0}^{T} (\widetilde{x}_{t+i,k} - (x_{t,k} + r_k \cdot i))^2. \tag{3.26}$$

We then sort the estimated erosion rates in increasing order:

$$r_1 < r_2 < \ldots < r_N, \tag{3.27}$$

and take

$$\sigma_t = \max \left(\sigma, -\frac{1}{2} \cdot r_{0.05 \cdot N} \right). \tag{3.28}$$

This σ_t is the mean square accuracy of the erosion rate estimates r_k.

Resulting algorithm: **Case** $T = 2$. For the case $T = 2$, when we have three consecutive years of measurement, we have simplified formulas

$$r_k = \frac{\widetilde{x}_{t+2,k} - \widetilde{x}_{t,k}}{2}, \tag{3.29}$$

and

$$\sigma^2 = \frac{1}{18} \cdot \frac{1}{K} \cdot \sum_{k=1}^{K} (\widetilde{x}_{t,k} - 2\widetilde{x}_{t+1,k} + \widetilde{x}_{t+2,k})^2. \tag{3.30}$$

Then, we take σ_t as determined by the formula (3.28).

Acknowledgements This work was supported in part by the National Science Foundation grants HRD-0734825 and HRD-1242122 (Cyber-ShARE Center of Excellence) and DUE-0926721, and by an award "UTEP and Prudential Actuarial Science Academy and Pipeline Initiative" from Prudential Foundation. The authors are thankful to the anonymous referees for valuable suggestions.

References

1. S.H. Ackers, R.J. Davis, K.A. Olsen, K.M. Dugger, The volution of mapping habitat for northern spotted owls (Srix occientalis caurina): A comparison of photo-interpreted, Landsat-based, and lidar-based habitat maps. Remote Sens. Environ. **156**, 361–373 (2015)
2. A. Afifi, S. May, *Practical Multivariate Analysis* (Chapman and Hall/CRC, Boca Raton, Florida, 2011)
3. R. Early, B. Anderson, C.D. Thomas, Using habitat distribution models to evaluate large-scale landscape priorities for spatially dynamic species. J. Appl. Ecol. **45**, 228–238 (2008)
4. G. Klir, et al. *Fuzzy Sets Fuzzy Log.* (Prentice Hall, Upper Saddle River, New Jersey, 1995)
5. H.T. Nguyen, E.A. Walker, *A First Course in Fuzzy Logic* (Chapman and Hall/CRC, Boca Raton, Florida, 2006)
6. D.J. Sheskin, *Handbook of Parametric and Nonparametric Statistical Procedures* (Chapman and Hall/CRC, Boca Raton, Florida, 2011)
7. G. Singh, A. Velmurugan, M.P. Dakhate, Geospatial approach for tiget habitat evaluation and distribution in Corbett Tiger Reserve. J. Indian Soc. Remore Sens. **37**, 573–585 (2009)
8. L.A. Zadeh, Fuzzy sets. Inf. Control **8**, 338–353 (1965)

Comparison of Fuzzy Synthetic Evaluation Techniques for Evaluation of Air Quality: A Case Study

Hrishikesh Chandra Gautam and S. M. Shiva Nagendra

Abstract Urban air quality has degraded at an alarming rate due to rapid urbanisation and industrialization in megacities. Therefore, there is an urgent need to assess air quality and suggest risk mitigation measures. In this paper, air quality of Chennai city was evaluated using different Fuzzy Synthetic Evaluation (FSE) techniques i.e. Fuzzy similarity method (FSM) and Simple fuzzy classification (SFC) and the results are compared with the National air quality index (NAQI). In the case of SFC weights for different pollutants were computed using Shannon's information entropy. Seasonal analysis of the criteria pollutants shows highest concentration during the winter season followed by pre-monsoon and summer season. The lowest concentration was observed during Monsoon in most cases. The FSE results are optimistic as compared to the NAQI due to aggregation of pollutant concentration as opposed to maximising function in NAQI which reconfirms the findings of earlier researchers. FSE can be used as a decision making tool to communicate the overall air quality to policy makers/end users (Public) in a simplified qualitative form.

1 Introduction

Urban air quality has degraded at an alarming rate in the recent decades [1]. Chennai being one of the major metropolitan and industrial area has shared the same fate. Chennai regarded as the "Detroit of India" due to its large Automobile industry has been under constant scrutiny of Central pollution control board (CPCB) and Tamilnadu pollution control board (TNPCB) [2]. Chennai city has been classified as a non-attainment area for PM_{10} due to the number of exceedances

H. C. Gautam (✉) · S. M. Shiva Nagendra
Department of Civil Engineering, Indian Institute of Technology Madras, Chennai 600036, India
e-mail: hcgautam.nitk@gmail.com

S. M. Shiva Nagendra
e-mail: snagendra@iitm.ac.in

© Springer International Publishing AG, part of Springer Nature 2018
L. A. Zadeh et al. (eds.), *Recent Developments and the New Direction in Soft-Computing Foundations and Applications*, Studies in Fuzziness and Soft Computing 361, https://doi.org/10.1007/978-3-319-75408-6_47

[3]. This leads to the challenge of informing the residents and policy makers about the air quality of region effectively and in time.

Air quality information dissemination is an integral part of any air quality management system. Air quality index plays a vital role in this purpose. Many air quality indices have been developed in the past few decades [4, 5] to serve the purpose of providing current air quality status for resident's health and safety at the time of exceedances and informed decision making for air quality management strategies of the area.

National Air quality index (NAQI) based on a linear scale working with maximum operator has the problem of eclipsing that is showing lower concentration as compared to the observed concentration and ambiguity showing higher concentration as compared to the observed concentration at or near the class boundaries. These problems occur due to a strict boundary between air quality classes which don't give enough information about the uncertainty and imprecision in air quality data.

The following paper is divided into 7 parts. Followed by the introduction, study area is explained with the location details and types of monitoring location. Conventional air quality indexing techniques are explained in Sect. 3. In the next section the philosophy of fuzzy logic is elaborated with the explanation of Fuzzy similarity method and distance metrics employed in the development of the model. The air quality of the study area is evaluated in the results and discussion section. The paper ends with conclusion describing the main attributes of the model and analysis and future scope of the work.

2 Study Area

Chennai city is located on the south-eastern coast of peninsular India [2]. It has a flat coastal terrain with the average elevation of 6.7 m. It stretches along a length of 25.6 km with an area of 176 Km^2. Chennai is considered as the Indian capital of automobile manufacturing with a share of 30% of automobiles and 40% of auto-parts manufactured in India.

According to Indian meteorological department (IMD) the weather in Chennai is sub-tropical (hot and humid) with 4 seasons i.e. winter, summer, pre-monsoon and monsoon [6]. The season of winter occurs between January to February with an average temperature of 25.6 °C, Summer from March to May with an average temperature of 30.5 °C, Pre-monsoon or southwest monsoon occurs from June to September with an average temperature of 29.7 °C, while Post-monsoon or northeast monsoon occurs from October to December with an average temperature of 26.7 °C. 60% of rainfall occurs during the season of post monsoon or northeast monsoon.

Daily average concentration of Sulphur dioxide (SO_2), Nitrogen dioxide (NO_2) and Respirable suspended particulate matter (RSPM) for 2009–2013 was collected from Tamilnadu pollution control board (TNPCB). TNPCB monitors Air quality data twice a week at 6 locations as depicted in Fig. 1 and Table 1.

Fig. 1 Map showing monitoring locations for air quality monitoring

Table 1 Details of monitoring locations for the study area

Name of location	Type	Location code
Kathivakkam	Industrial	IS-1
Thiruvottiyur	Industrial	IS-2
M C Thiruvottiyur	Industrial	IS-3
Manali	Industrial	IS-4
General hospital	Traffic intersection	TS
Taramani	Residential	RS

3 Air Quality Index

Air quality index has been developed for weighted and aggregated representation of complex environmental data in a simplified and intelligible form. It helps the common public in taking proper precautions in the case of exceedances and decision makers in developing new policies for reducing the harmful effects of air pollution and developing a sustainable society.

The most commonly used air quality index was developed by USEPA earlier named as Air pollution index (API) [7] and later renamed to USEPA AQI. In the above index the air quality is divided into 5 strict linear classes based on concentration breakpoints. A sub-index is calculated for each pollutant using the given Eq. (1)

$$I_p = \frac{I_{HI} - I_{LO}}{BP_{HI} - BP_{LO}} \left(C_p - BP_{LO} \right) + I_{LO} \tag{1}$$

I_p index for pollutant p
C_p Rounded concentration of pollutant p
BP_{HI} Breakpoint that is higher than or equal to C_p
BP_{LO} Breakpoint that is lower than or equal to C_p
I_{HI} AQI value corresponding to BP_{HI}
I_{LO} AQI value corresponding to BP_{LO}

4 Fuzzy Logic

Fuzzy logic is a multi-valued logic formed by extending the binary logic into the realm of partial membership. It was first described by Lotfi A. Zadeh in a seminal paper published in 1965 [8]. Fuzzy logic uses overlapping membership functions to model the uncertainty and ambiguity pertaining in a real world complex system like urban air quality. Fuzzy logic helps in classification using linguistic parameters developed by employing the expert knowledge-base derived by responses from a fuzzy questionnaire.

Fuzzy logic has been applied in a number of papers for the assessment of air quality and development of air quality management tools [9–12]. In assessment of a complex system like air quality fuzzy logic based techniques are able to trap and model the uncertainty arising in data due to error during monitoring, and analysis.

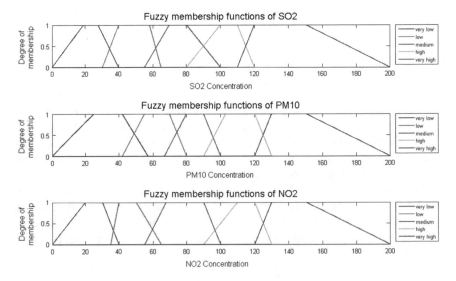

Fig. 2 Fuzzy membership functions for SO$_2$, NO$_2$ and RSPM

The membership functions were determined by employing expert's perception using fuzzy questionnaire duly filled by different air quality experts. The mean of the membership functions developed were used as the final set of membership functions.

The membership functions help in determining the AQI by integrating the health and ecological effects of different pollutants obtained through the expert's experience in the field of air quality into the Fuzzy based model (Fig. 2).

5 Fuzzy Synthetic Evaluation

In Fuzzy synthetic evaluation (FSE) the air quality standards for different pollutants were used to develop linguistic classes which were represented in matrix form with the columns as pollutant and rows as classes [13]. The pollutant concentration value was compared with the standards using the min max approach.

FSE constitutes Fuzzy similarity method (FSM) and simple fuzzy classification (SFC) [14]. In the case of FSM an evaluation matrix R as shown in Eq. (2) is derived from the mean value of each trapezoidal membership function with rows representing the pollutants and columns representing the air quality class and another matrix R' represents the membership value of each pollutant for different air quality class at a given instance [15]. The 2 matrices are compared using the similarity function in Eq. (3) and the maximum value of the similarity function was used to represent the final air quality class for the given case.

$$R = \begin{bmatrix} r_{11}r_{12} & r_{15} \\ r_{21}r_{22} & r_{25} \\ r_{31}r_{32} & r_{35} \end{bmatrix} \tag{2}$$

$$B_j = \frac{\sum_{i=1}^{5} \min\left[r_{ij}(x), r'_{ij}\right]}{\sum_{i=1}^{5} \max\left[r_{ij}(x), r'_{ij}\right]} \tag{3}$$

r_{ij} membership value for pollutant i and class j
i   pollutant 1, …, 3, 1 = SO_2, 2 = NO_2 and 3 = RSPM
j   class 1, …, 5, 1 = very low, 2 = low, 3 = medium, 4 = high, 5 = very high

In case of SFC, weight for each pollutant was derived using Shannon's information entropy function [16]. Information entropy signifies the amount of information contained in a given set of data and represented by the negative log of normalised data [17].

Concentration for jth reading of the ith pollutant was normalised using Eq. (4) from the evaluation matrix. The normalised value r_{ij} is divided by the sum of all the normalised values using Eq. (5) and the final value is used to calculate the

Shannon's entropy for each pollutant in Eq. (6). Weight of a pollutant is calculated using Eq. (7)

$$r_{ij} = \frac{x_{ij} - \min_j\{x_{ij}\}}{\max_j\{x_{ij}\} - \min_j\{x_{ij}\}} \tag{4}$$

$$f_{ij} = \frac{r_{ij}}{\sum_{j=1}^n r_{ij}} \tag{5}$$

$$H_i = -k \sum_{j=1}^n f_{ij} \ln f_{ij} \qquad i = 1, 2, \ldots, m \tag{6}$$

$$w_i = \frac{1 - H_i}{m - \sum_{i=1}^m H_i} \qquad 0 \le w_i \le 1, \qquad \sum_{i=1}^m w_i = 1 \tag{7}$$

Distance metric for each case was calculated from the Eq. (8) utilising the membership value from the evaluation matrix (R) and weight derived using Shannon entropy. The maximum value of the distance metric for each case determines the final air quality category

$$k(j) = \sqrt{\frac{\sum_{i=1}^3 (w_i . \lambda_{ij})^2}{\sum_{j=1}^5 \sum_{i=1}^3 (w_i . \lambda_{ij})^2}} \tag{8}$$

λ_{ij} Membership value for jth value of ith pollutant
k(j)   distance metric for jth value
w_i entropy weight for pollutant I

$$k(f) = \max(k(j)) \tag{9}$$

k(f)   air quality category of the value

6   Results and Discussion

The AQI procedure was applied to air pollutant concentration values obtained from the 6 urban air quality monitoring stations in Chennai. The fuzzy AQI was calculated using FSM and SFC and compared with the conventional National air quality index (NAQI). The results for Kathivakkam and Manali industrial area for 2013 are shown in Tables 2 and 3 respectively.

Table 2 Comparison of Fuzzy synthetic evaluation methods with National air quality index for Kathivakkam for 2013

	NAQI		FSM		SFC	
	AQI	Class	Membership value	Class	Membership value	Class
Jan	163.16	Moderately Polluted	0.28	Unhealthy for sensitive people	0.81	Healthy
Feb	151.70	Moderately Polluted	0.37	Satisfactory	0.78	Healthy
Mar	124.18	Moderately Polluted	0.24	Satisfactory	0.84	Healthy
Apr	90.58	Good (Healthy)	0.25	Healthy	0.85	Healthy
May	88.58	Good (Healthy)	0.26	Satisfactory	0.82	Healthy
Jun	101.5	Moderately Polluted	0.24	Satisfactory	0.79	Healthy
Jul	101.2	Moderately Polluted	0.24	Satisfactory	0.76	Healthy
Aug	92	Good (Healthy)	0.24	Satisfactory	0.79	Healthy
Sep	87.31	Good (Healthy)	0.28	Satisfactory	0.85	Healthy
Oct	61.8	Good (Healthy)	0.18	Healthy	1	Healthy
Nov	86.31	Good (Healthy)	0.29	Satisfactory	0.86	Healthy
Dec	79.98	Good (Healthy)	0.36	Satisfactory	0.94	Healthy

The FSE results are optimistic compared to the NAQI as they show that the air quality is better than air quality determined by the NAQI. This happens because the concentration of all the pollutants are integrated for calculation of index in case of FSE techniques while in the case of NAQI the index is determined only by the concentration of the pollutant with the maximum sub index which causes the NAQI to be less sensitive to the health effect of other pollutants. The FSE techniques give a clearer picture about the air quality as the effect of all the individual pollutants are integrated in the final AQI. The use of membership functions, expert's perception and predetermined Shannon's entropy based weights make the FSE based methods more sensitive to changes in concentration of air pollutants.

The AQI in the winter months of January and February are poor as compared the other months of the year. The best air quality was observed during the Northeast monsoon period of November and December.

Table 3 Comparison of Fuzzy synthetic evaluation methods with National air quality index for Manali for 2013

	NAQI		FSM		SFC	
	AQI	Class	Membership value	Class	Membership value	Class
Jan	–	–	–	–	–	–
Feb	291.25	Poor (unhealthy)	0.33	Unhealthy	0.86	Unhealthy
Mar	304.95	Poor (unhealthy)	0.27	Unhealthy for sensitive people	0.87	Unhealthy forsensitive people
Apr	94.16	Good (Healthy)	0.24	Satisfactory	0.87	Satisfactory
May	192.96	Moderately polluted	0.30	Satisfactory	0.88	Satisfactory
Jun	142.96	Moderately polluted	0.34	Satisfactory	0.71	Healthy
Jul	75	Good (Healthy)	0.34	Satisfactory	0.91	Healthy
Aug	65.62	Good (Healthy)	0.24	Satisfactory	0.88	Satisfactory
Sep	69.28	Good (Healthy)	0.21	Healthy	1	Healthy
Oct	–	–	–	–	–	–
Nov	–	–	–	–	–	–
Dec	–	–	–	–	–	–

Seasonal variation for the 6 monitoring locations for 2009–2012 are shown in Figs. 3, 4, 5 for SO_2, NO_2 and RSPM respectively. No exceedances were observed for SO_2 and NO_2.

In case of all pollutants highest concentration was observed during the winter season due to stable atmospheric conditions which restricts the vertical mixing of air mass and dispersion of pollutants.

In the case of RSPM concentration, exceedances were observed during the years 2009–2011 for some of the industrial locations. All the exceedances were observed during the winter season (Fig. 4).

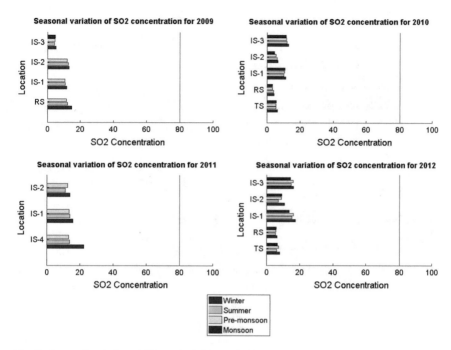

Fig. 3 Seasonal variation of SO$_2$ at TNPCB monitoring locations

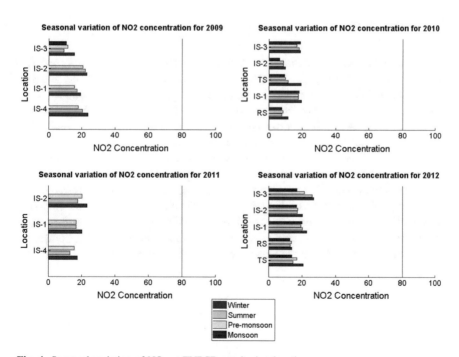

Fig. 4 Seasonal variation of NO$_2$ at TNPCB monitoring locations

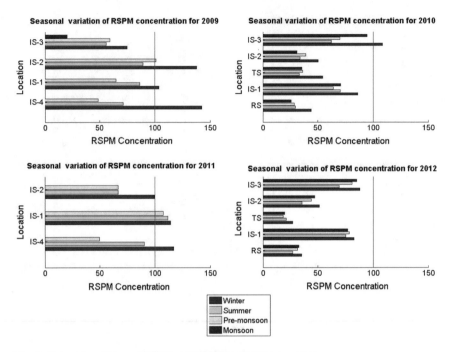

Fig. 5 Seasonal variation of RSPM at TNPCB monitoring locations

7 Conclusion

FSE methods integrates multiple pollutant's concentration and expert's knowledge into the AQI as opposed to conventional NAQI in which a single pollutant with maximum subindex is considered to define the air quality. Urban air quality is not easy to assess and comprehend because of its multi-pollutant characteristics. Contribution of each pollutant can be measured and assessed in the case of FSE.

The given methods are easy to model, less data intensive and the results are comprehensible for local public and decision makers. FSE techniques show more optimistic results as compared to the conventional AQI which can reduce the burden of money and effort on air quality management. The methods overcome the uncertainty of air quality classification near the class boundaries by implying fuzziness in the class boundaries using the overlapping Fuzzy classes and produce an air quality index with higher sensitivity. The above techniques can be used to develop an effective air quality management plan for the urban area of Chennai city.

References

1. J. Fenger, Air pollution in the last 50 years—from local to global. Atmos. Environ. **43**(1), 13–22 (2009)
2. R. Krishnamurthy, K.C. Desouza, Chennai, India. Cities **42**, 118–129 (2015)
3. CPCB, Air quality monitoring, emission inventory and source apportionment study for Chennai (2011)
4. D. Shooter, P. Brimblecombe, Air quality indexing. Int. J. Environ. Pollut. 1–24 (2005)
5. A. Plaia, M. Ruggieri, Air quality indices: a review. Rev. Environ. Sci. Bio/Technol. **10**(2), 165–179 (2010)
6. B. Srimuruganandam, S.M. Shiva Nagendra, Application of positive matrix factorization in characterization of PM (10) and PM (2.5) emission sources at urban roadside. Chemosphere **88**(1), 120–30 (2012)
7. W.-L. Cheng, Y.-S. Chen, J. Zhang, T.J. Lyons, J.-L. Pai, S.-H. Chang, Comparison of the revised air quality index with the PSI and AQI indices. Sci. Total Environ. **382**(2–3), 191–198 (2007)
8. L.A. Zadeh, Fuzzy sets. Inf. Control, 338–353 (1965)
9. B.E.A. Fisher, Fuzzy approaches to environmental decisions: application to air quality. Environ. Sci. Policy **9**, 22–31 (2006)
10. D. Dunea, A.A. Pohoat, E. Lungu, Fuzzy inference systems for estimation of air quality index. ROMAI J. **7**(2), 63–70 (2011)
11. M.H. Sowlat, H. Gharibi, M. Yunesian, M. Tayefeh Mahmoudi, S. Lotfi, A novel, fuzzy-based air quality index (FAQI) for air quality assessment. Atmos. Environ. **45**(12), 2050–2059 (2011)
12. X. Zhao, Q. Qi, R. Li, The establishment and application of fuzzy comprehensive model with weight based on entropy technology for air quality assessment. Procedia Eng. **7**, 217–222 (2010)
13. N.B. Chang, H.W. Chen, S.K. Ning, Identification of river water quality using the fuzzy synthetic evaluation approach. J. Environ. Manage. **63**, 293–305 (2001)
14. G. Onkal-Engin, I. Demir, H. Hiz, Assessment of urban air quality in Istanbul using fuzzy synthetic evaluation. Atmos. Environ. **38**(23), 3809–3815 (2004)
15. L. Abdullah, N.D. Khalid, Classification of air quality using fuzzy synthetic multiplication. Environ. Monit. Assess. **184**, 6957–6965 (2012)
16. Z. Zhi-hong, Entropy method for determination of weight of evaluating indicators in fuzzy synthetic evaluation. J. Environ. Sci. **18**(5), 1020–1023 (2006)
17. S. Al-sharhan, F. Karray, W. Gueaieb, O. Basir, Fuzzy entropy: a brief survey, in *10th IEEE International Conference on Fuzzy Systems* (*Cat. No.01CH37297*), vol. 3 (2001), pp. 1135–1139

Evaluation of Green Spaces Using Fuzzy Systems

M. T. Mota, J. A. F. Roveda and S. R. M. M. Roveda

Abstract Green urban areas have natural attributes that provide important environmental services. However, their impacts depend on the local characteristics, so it is necessary to evaluate the environmental potentials of such areas individually. The lack of methodologies to assess the quality of green urban areas led to the implementation of the present work, which proposes a calculation model for assessing the environmental quality of these spaces. The construction of the model used fuzzy systems capable of handling the subjectivity of the variables, with the creation of fuzzy rule-based systems that permitted working with parameters of different natures. The variables employed were the percentage of vegetation in the area in question and its quality, the latter being determined by analysis of the degree of maturation. Forest formations were considered potentially more advantageous for the provision of environmental services, compared to savannah and grasslands. The model was constructed considering the physical and biological characteristics of the city of Sorocaba, as well as Brazilian standard classifications of vegetation types and their successional stages.

This work is partially supported by Capes Foundation Grant #1846-15-9 and FAPESP Grant #2015/07714-4.

M. T. Mota (✉)
Graduated Program in Environmental Science, Institute of Science and Technology of Sorocaba (ICT), São Paulo State University "Júlio de Mesquita Filho" – UNESP, Sorocaba, São Paulo, Brazil
e-mail: mauricio.mota@posgrad.sorocaba.unesp.br

J. A. F.Roveda · S. R. M. M.Roveda
Institute of Science and Technology of Sorocaba (ICT), São Paulo State University "Júlio de Mesquita Filho" – UNESP, Sorocaba, São Paulo, Brazil
e-mail: jose.roveda@unesp.br

S. R. M. M.Roveda
e-mail: sandra.regina@unesp.br

© Springer International Publishing AG, part of Springer Nature 2018
L. A. Zadeh et al. (eds.), *Recent Developments and the New Direction in Soft-Computing Foundations and Applications*, Studies in Fuzziness and Soft Computing 361, https://doi.org/10.1007/978-3-319-75408-6_48

617

1 Introduction

Intense and chaotic urbanization in Brazil has been responsible for huge reductions of green spaces in cities, which has consequently affected the quality of life of the population.

In 1933, the publication of the Letter of Athens, in the 4th edition of the International Congress of Modern Architecture, provided an in-depth evaluation of the problems plaguing cities and the need to make urban areas more suitable for habitation, proposing an increase in the areas set aside for green spaces.

In Brazil, this new urban design prompted a series of legal provisions that aimed to ensure better planning of the occupation of urban areas. By the end of the 1970s, authorities had a legal obligation to reserve a certain percentage of free space for public use, and in order to prevent disorderly urban expansion, the size of these spaces was related to population density [1, 2]. This strategy was motivated by the need for areas suitable for leisure and recreation; since the areas could not be built on, the presence of vegetation could therefore play an important role in environmental regulation [3].

However, the environmental function of these spaces was only recognized as such in the last decade, with the positive benefits of the associated environmental services including:

- Ecological services related to vegetation, soil permeability, maintenance of biodiversity, climate regulation, and the quality of air, water, and soil [3];
- Psychological aspects related to the exposure of individuals to the natural features of these areas, which contributes to their quality of life. These aspects are associated with leisure and recreation [4–7];
- Health benefits associated with reductions in the incidence of respiratory diseases [8].

It is clear that in order to fulfill the envisaged environmental functions, it is necessary that such areas maintain their physical and biological characteristics as close as possible to the optimum conditions.

Urbanization leads to areas with differing degrees of preservation, notably in terms of vegetation cover, which is a key component in the maintenance of an acceptable microclimate in the urban environment [9–12]. The creation of a methodology for evaluating the quality of green areas would enable comparison of different urbanization projects, in terms of the environmental aspects.

This work proposes a model for analysis of the quality of green areas that could be used to determine the potential of these spaces to provide environmental services. The calculation model uses expressions considering aspects including urban climate, air pollution, preservation of water resources, biodiversity (by providing features such as shelter and nesting spaces), landscape (which influences psychological well-being), and leisure and recreation, amongst others. The use of fuzzy set theory was effective in establishing correlations between variables with different units of measurement, enabling assessment of the quality of an area in terms of its potential to provide environmental services.

2   Methodology

Using fuzzy set theory, an evaluation methodology was constructed that considers the percentage of the area occupied by vegetation, together with the type of vegetation present (forest, savannah, and herbaceous). In this scheme, the higher the percentage of vegetation and the denser the biomass (greatest for forest), the greater the potential benefits in terms of environmental parameters such as temperature, biodiversity, and the provision of leisure and recreation opportunities to the population.

The protocol was based mainly on the quantitative analysis of vegetation, considering the percentage of the green space occupied by vegetation, together with qualitative aspects including the capacity of greater amounts of biomass to improve mood, retain pollutants, and encourage biodiversity.

2.1   Fuzzy Set Theory

Fuzzy set theory, described by Zadeh in 1965 [13], has led to what is known today as fuzzy logic. In this system, abstract forms of reasoning are used, instead of absolute terms. In classical logic, elements are defined as true or false, with assignment of the values 1 and 0, respectively, whereas in fuzzy logic, an element belongs to a fuzzy set where the degree of relevance varies between 0 and 1, with 0 indicating no relevance and 1 indicating maximum relevance. This approach opens up new and more sophisticated ways to analyze a variety of problems. An important feature of this theory is that it enables the expression of uncertainty and subjectivity, because both qualitative and quantitative information can be used. The vagueness associated with the use of linguistic terms (very, little, low, high, etc.), which are traditionally regarded as non-scientific, is admissible in the fuzzy approach, and can enable the system to achieve a representation of the actual reality [14].

This theory has many applications involving the implementation and control of tasks that follow a sequence of linguistic "orders", translated by a set of rules, which are capable of being decoded by a fuzzy controller. Classical fuzzy rule-based systems (FRBS) use fuzzy logic to produce outputs for each input.

In this work, the methodology is FRBS-based, with each input corresponding to an output, governed by a fuzzy controller, which in turn is composed of four main

Fig. 1 Fuzzy controller
scheme. *Font:y authors*

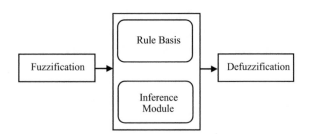

parts: a fuzzification module; a fuzzy controller, which is subdivided according to a rule base and an inference module; and a defuzzification module. A schematic illustration of the functioning of a fuzzy controller is shown in Fig. 1.

2.2 Fuzzification Module

The first stage in the process is fuzzification, in which numerical inputs are converted to fuzzy sets. Each fuzzy set is characterized by its membership function. It is in this step that such functions are built and it is by means of them that individual elements are defined in terms of their degree of belonging to the set. The values used for these functions are obtained by working together with a specialist who has knowledge of the behavior of the variables that compose the model.

2.3 Rule Base

The fuzzy rule base is a set of propositions describing the behavior of the parameters involved in the system. These rules are defined as follows:

$$IF\ (condition)\ THEN\ (action)$$

where the condition and the action are values adopted for linguistic variables such as "excellent", "good", "bad", etc., which were modeled using fuzzy sets. This characteristic of fuzzy logic enables the construction of a model that performs well, even with imprecise variables. This step also requires the knowledge of an expert who determines the action to be taken according to the behavior of the variables.

2.4   Inference Module

This module "*translates*" propositions of the rule base using the techniques of fuzzy logic, defining which t-norms, t-conorms, and inference rules should be used to obtain the fuzzy relation that models the rule base.

2.5   Defuzzification Module

The defuzzification step involves the conversion of a fuzzy set into a numerical value. In the present case, this numerical value is the calculated quality of the area in question. Defuzzification can be performed in several ways, and the centroid method was adopted in this work.

3   Results

The model used to assess the quality of an area and its potential for environmental services in urban environments was constructed using a fuzzy system based on the Mamdani model. The calculation was performed using two input systems, one for the percentage of preserved vegetation cover in the green area, and the other for the quality of this vegetation.

Evaluation of the percentage of vegetation cover employed the direct input of data, by means of a trapezoidal membership function, with values from 0 to 100%, depending on the extent of vegetation cover. The values were classified as follows: very poor (<25%), poor (25%), regular (50%), good (75%), and excellent (>75), as illustrated in Fig. 2. The centroid defuzzification model was used in the calculation.

The system used to evaluate the quality of the vegetation requires previous interpretation prior to the insertion of the data. The input variables correspond to the analysis of characteristics of the vegetation, considering whether the vegetation cover matches one of the existing forest formations in the region, one of the existing savannah (cerrado) formations, or a mixture of both.

The native forest formations in the Sorocaba region are mostly typical of the Atlantic forest and are classified as semi-deciduous seasonal forest. Savannah formations are also very common in the Sorocaba region, and the associated vegetation physiognomies evaluated in this work were the cerrado types: *campo limpo*, *campo sujo*, *campo cerrado*, *cerrado* sensu stricto, and *cerradão*.

In addition to the type of vegetation (forest or savannah), the successional stage was also considered. The methodology adopted for gauging the successional stage was based on CONAMA resolution N° 01/1994.

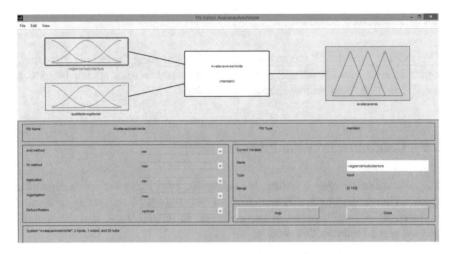

Fig. 2 Illustration of the system constructed using Matlab[®] software for evaluation of the percentage and quality of vegetation cover in a green area

In the case of the input variable related to the different savannah physiognomies, the methodology used was based on São Paulo State Environment Secretariat (SMA) resolution N° 64/2009, which employs the number of tree individuals per hectare to determine the classification.

The membership function for determination of the savannah physiognomy was divided into the five categories described above. The triangulation function was used, because SMA resolution 64/2009 specifies reference numbers for classification.

The construction, from rule base to defuzzification and output, of the system to evaluate the vegetation quality gave emphasis to forest formations and savannah formations with greater density of arboreal individuals, and in the case of mature forests considered the successional stages: advanced, middle, initial, and pioneer. In the case of the savannah formations, prioritization followed the order: *cerradão*, *cerrado sensu strictu*, *campo cerrado*, *campo sujo*, *campo limpo*.

System test

The system was tested by constructing 910 scenarios with different percentages of vegetation cover associated with various possible situations, as well as different percentages of vegetation formations and different successional stages of the formations. Table 1 provides some of the results obtained in the calculations for the different scenarios.

In the 910 different scenarios tested, the vegetation quality grades ranged from 0.82 to 9.20, while the values obtained for the areas ranged from 2.00 to 8.17.

Table 1 Results obtained for evaluation of vegetation and area quality in tests using different scenarios

Forest	Cerrado (savannah)	Successional stage of forest	Successional stage of savannah	Percentage of cover	Quality grade	Area evaluation
20	80	9	100	6	0.82	2.01
20	80	9	200	8	0.88	2.05
20	80	9	300	17	0.95	2.07
20	80	9	500	25	1.12	2.00
20	80	36	1700	83	7.50	7.50
20	80	36	2000	82	9.05	7.62
20	80	36	2300	100	9.20	8.17
50	50	9	100	6	0.82	2.01
50	50	9	200	8	0.88	2.05
50	50	9	1500	75	5.00	5.00
50	50	9	2000	82	7.50	7.50
50	50	9	2300	100	7.50	7.50
50	50	36	1300	67	6.45	6.32
50	50	36	1700	83	7.50	7.50
50	50	36	2300	100	9.20	8.17
80	20	9	100	6	0.82	2.01
80	20	9	200	8	0.88	2.05
80	20	9	300	17	0.95	2.07
80	20	36	1500	75	7.50	7.50
80	20	36	1700	83	7.50	7.50
80	20	36	2000	82	9.05	7.62
80	20	36	2300	100	9.20	8.17

4 Conclusions

The results showed that the model behaved according to expectations, and successfully translated into numerical variables different characteristics related to the quality of vegetation (considering the type of formation, physiognomy, and successional stage) and the percentage of coverage.

The fuzzy approach made it possible to perform a mathematically valid treatment of subjective measures liable to uncertainties. The technique could assist decision makers in identifying areas to be preserved in urban environments, considering attributes important for providing useful environmental services.

The model could be used to evaluate existing areas, as well as areas liable to be urbanized, hence providing the population with valuable comparative information.

In the quality assessment of the vegetation, in some cases identical grades were obtained for different scenarios, indicating that the defuzzification process was not sensitive to certain variations, which was reflected in the final grades for evaluation of the quality of the area. Although several possible modifications of the model were tested, all presented the same difficulty in identifying variability between environments. Nonetheless, despite the difficulty of the model in handling small variations, reflecting a lack of sensitivity in treating the data in certain situations, the model generally performed as expected and was capable of generating satisfactory numerical values appropriate to different environments. The results obtained here demonstrate that this model could be successfully used to evaluate the potential of a defined area to provide environmental services in an urban environment.

Acknowledgements J. A. F. Roveda thanks Capes Foundation for grant #1846-15-9 and S. R. R. M. Roveda thanks São Paulo Research Foundation (FAPESP) for grant #2015/07714-4.

References

1. A.C. Arfelli, Green and leisure areas: considerations for understanding and definition in the activity of land division. In Portuguese: Áreas verdes e de lazer: considerações para sua compreensão e definição na atividade de parcelamento do solo, *Revista de Direito Ambiental*, São Paulo. **9**(33), 37 (2004)
2. M.A. Barreiros, A.K. Abiko, Reflections on the subdivision of urban land. In Portuguese: Reflexões sobre o parcelamento do solo urbano, *Boletim Técnico da Escola Politécnica da USP*, Departamento de Engenharia de Construção Civil, São Paulo (1998), p. 25, http://www.allquimica.com.br/arquivos/websites/artigos/A-00030200652814274.pdf. Accessed: 05 Sept 2013
3. S.M.A. Haq, Urban green spaces and an integrative approach to sustainable environment. J. Environ. Prot. **2**(5), 601–608 (2011), https://doi.org/10.4236/jep.2011.25069
4. P. Guzzo, Green areas. In Portuguese: "Áreas Verdes", http://educar.sc.usp.br/biologia/prociencias/areasverdes.html. Accessed: 15 Apr 2011
5. C. Giuseppe, Go greener, feel better? The positive effects of biodiversity on the well-being of individuals visiting urban and peri-urban green areas. Landscape Urban Plann. **134**, 221–228 (2015)
6. R.C. Loboda, B.L.D. de Angelis, Public urban green areas: concepts, uses and functions. In Portuguese: Áreas verdes públicas urbanas: conceitos, usos e funções. Ambiência - Revista do Centro de Ciências Agrárias e Ambientais, Guarapuava **1**(1), 125–139 (2005)
7. J.C. Nucci, Systemic analysis of the urban environment, densification and environmental quality. In Portuguese: Análise sistêmica do ambiente urbano, adensamento e qualidade ambiental. Revista PUC, Ciências Biológicas e do Ambiente, São Paulo **1**(1), 73–88 (1999)
8. A.M.L.P. Lima, F. Cavalheiro, J.C. Nucci, M.A.L.B. Sousa, N. Fialho, P.C.D. Del Picchia, Use issues in the conceptualization of terms such as open spaces, green areas, and others. In Portuguese: Problemas de utilização na conceituação de termos como espaços livres, áreas verdes e correlatos, in *Proceedings of the Congresso Brasileiro de Arborização Urbana*, São Luiz/MA, vol. 2 (1994), pp. 539–553
9. C. Dacanal, Urban forest fragments and climate interactions at different scales: Study in Campina, SP. In Portuguese: Fragmentos florestais urbanos e interações climáticas em diferentes escalas: estudo em Campina, SP, PhD thesis, Universidade Estadual de Campinas, Faculdade de Engenharia Civil, Arquitetura e Urbanismo (2011) p. 249

10. L.C. Labaki, C.F.B. Bartholomei, L.L.F. Castro, R.F. Santos, Thermal comfort in outdoor spaces: the role of vegetation as a means of controlling solar radiation, in *Proceedings of the PLEA 2000—Architecture, City, Environment*, vol. 1 (Science Publishers, Cambridge. London: James & James, 2000), pp. 501–505
11. T. Honjo, T. Takakura, Simulation of thermal effects of urban green areas on their surrounding areas. J. Energy Build **15**(16), 443–446 (1990). Washington
12. G.M. Heisler, Energy savings with trees. J. Arboric. **12**(5), 113–125 (1986). Washington
13. L.A. Zadeh, Fuzzy sets. Inf. Control **8**(3), 338–353 (1965)
14. L.C. Barros, R. Bassanezi, *Topics of Fuzzy Logic and Biomathematics*. In Portuguese: Tópicos de Lógica *Fuzzy* e Biomatemática, *Coleção IMECC - Textos Didáticos*, vol. 5, Campinas, São Paulo, 2006

A New Methodology for Application of Impact's Identification Using Fuzzy Relation

J. A. F. Roveda, A. C. A. Burghi, S. R. M. M. Roveda, A. Bressane and L. V. G. França

Abstract The Environmental Impact Assessment (EIA) is a mechanism used by governments and private institutions to predict environmental damage and optimize positive impacts. One of the greatest challenges of the EIA process is to achieve an effective integration of the various tools and the analytical procedures. This paper proposes a new methodology for the application of impact's identification and characterization matrices by the use of Fuzzy Logic, using specifically Fuzzy Relations Theory. Three Weight Matrices of State and Association, for each type of expertise (physical, biotic, and anthropic) were created. Fuzzy Relations to aggregate all these matrices were used, and a Weighted Response Matrix was obtained. The developed fuzzy methodology was applied to real cases of EIA, aiming to compare the results. The EIA Mario Covas Road Program—Modified Southern Section was used, and the fuzzy interaction matrix proves to be a successful tool. Because of the results, it was possible to infer that the new methodology, using fuzzy approach is an effective and practical tool in the Environmental Impact Assessment.

J. A. F.Roveda (✉) · S. R. M. M.Roveda · A. Bressane · L. V. G. França
Institute of Science and Technology of Sorocaba (ICT),
UNESP - University Estadual Paulista, Sorocaba, Sao Paulo, Brazil
e-mail: jose.roveda@unesp.br

S. R. M. M.Roveda
e-mail: sandra.regina@unesp.br

A. Bressane
e-mail: adriano.bressane@prograd.sorocaba.unesp.br

L. V. G. França
e-mail: lucirene@posgrad.sorocaba.unesp.br

A. C. A. Burghi
European Joint Master Management and Engineering of Environmental and Energy,
Budapest University of Technology and Economics, Budapest, Hungary
e-mail: carolcisv@hotmail.com

© Springer International Publishing AG, part of Springer Nature 2018
L. A. Zadeh et al. (eds.), *Recent Developments and the New Direction
in Soft-Computing Foundations and Applications*, Studies in Fuzziness
and Soft Computing 361, https://doi.org/10.1007/978-3-319-75408-6_49

1 Introduction

The Environmental Impact Assessment (EIA) is a mechanism used by governments and private institutions that aims the prediction of environmental damage and optimization of positive impacts, simultaneously promoting sustainable development. Increasingly, there is the need to use EIA as a base to establish mechanisms of social control and participatory decision about development projects and economic initiatives.

The EIA is intended to consider environmental impacts even before any decision is taken, that may result in changes in the environmental quality. Therefore, it is necessary to carry out sequential activities to obtain the EIA, involving several participants and focusing on the analysis of environmental feasibility of a proposal.

One of the greatest challenges of the EIA process is to achieve an effective integration of the various tools and the analytical procedures used to investigate the processes and effects of the interactions between human activities and natural and social processes. Therefore, it requires a multidisciplinary approach as it covers many knowledge's areas.

As a result of the wide variety of analysis and topics covered, the environmental impact studies and reports often result in extensive documents, usually with hundreds to thousands of pages, making it difficult the evaluation by the environmental agency. For this reason, the application of Fuzzy Logic [1] could help identifying and emphasizing the aspects of greatest relevance, such as the most significant impacts and the activities that they are related with. The modeling process based on Fuzzy Logic introduces a different, but not complicated approach. Because of that, several researchers have been using it in a variety of environmental applications. For example, Lee [2] created a model to evaluate the water quality of rivers and lakes. Balas [3] classifies beaches and Riedler and Jandl [4] predict the soil and forest degradation, whereas Yildirim and Bayramoglu [5] treated the air pollution, and Conelissen [6] works with sustainable management.

Several tools are used for the identification and evaluation of the impacts, requiring reasonable knowledge of the concepts, apart from detailed understanding of the project and of the environmental dynamics of the place or region. Currently, the most commonly used tools are: checklists, diagrams of interaction and matrices [7].

Matrices are composed of two basic sets: one containing the activities or actions promoted by the evaluated project and the other one with the main components of the environmental system affected [8]. The purpose of the matrix is to identify the possible interactions between the components of the project and the elements of the environment. Matrices are widely used, due to its better representation of the identification and characterization of impacts, when compared to the other methods.

After the selection of the actions and the environmental components, analysts should identify all possible interactions, marking a score of magnitude and importance of each impact, on an arbitrary scale of 0–10. If the magnitude is zero, there is no interaction and the cell is not checked. Impacts can also be identified with linguistic variables or symbols, determined by analysts. After the study of the matrices, it can be concluded that even relating activities and effects, this association is carried out in a simple manner without quantify or qualify the impacts generated.

The Fuzzy Logic proves to have great advantages in its applications, such as its ability to integrate different types of observations, to evaluate inaccurate measurements and to work very well with linguistic terms. Furthermore, this logic enables the work with quantitative and qualitative data [9].

In this context, this project proposes a new methodology for the application of impact's identification and characterization matrices by the use of Fuzzy Logic, using specifically Fuzzy Relations Theory [9]. This logic was applied in an evaluation matrix to reach a new interpretation configuration of the impacts. Therefore, a matrix of fuzzy relations was used in order to indicate not only whether or not there is an association between the two sets but also to identify the magnitude of the relation between these elements. The subjective nature of this theory has proved to be a valuable tool in the determination of environmental conditions and it can also be an effective help in the management and establishment of public policies in order to improve the environment.

2 Methodology

In order to reduce the subjectivity of the identification and assessment of impacts through the interaction matrices, the use of Fuzzy relations was proposed in combination with the common matrix method, aiming a better result on the prediction of the environmental impacts caused by various projects. First of all, it was proposed a fuzzy interaction matrix, based from a matrix proposed by Sanchez [7], which identifies environmental effects and impacts of a limestone mine.

The first step of the methodology is to build the matrices of State (activities of the project vs environmental effects) and Association (environmental effects vs environmental impacts) based on Fuzzy Relations theory, as in Fig. 1, that indicate the degree of relationship between the activities of the project, the effects and environmental impacts. The degree of relation is a value between 0 and 1, as the value 0 represents no relation between the items evaluated and the value 1 represents that they are completely related. The professional who develops the impact identification matrix is the responsible to determine these values.

Once developed both matrices, they are used as a basis for the development of new matrices, which can be filled by other reviewers, including members of the population. These new matrices are filled up only indicating with an "x" the relations that are identified. In the case of this study, three groups of experts were considered, one for each feature of the environment: physical, biotic and anthropic. Therefore, three Weighting matrices were generated, in which the different professionals involved in the EIA process identified the relationships.

	Project's Operation					Support Infrastructure					Environmental Effects	Physical Medium			Biotic Medium			Antropic Medium			
	vegetation removal	organic soil removal	mining	processing of mineral	dam's construction	infrastructure	inputs' supply	inputs' storage	dispatch of products	demolition		change in water quality	change in air quality	change in soil quality	terrestrial habitats destruction	change in animal's populations	change in aquatic ecosystems	visual impact	environmental discomfort	business improvement	increase in tax collection
	0,3	1	1	0	1	0	0	0	0	0,2	topography changes	0	0	0,6	0	0	0	1	0	0	0
	1	1	1	0,8	0	0,8	0	0	0	1	residues generation	0,7	0	0	0	0	0,5	0	0	0	0
	0	0	1	1	0	0,8	0	0,5	0	0	liquid efluents generation	1	0	0	0	0	1	0	0	0	0
	1	1	1	0	0,8	0	0	0	0	0	erosion	0	0	1	0	0	0	0,8	0	0	0
	0,9	1	1	0	0,2	0	0	0	0	0,8	increase in sedimentation	1	0	0	0	0	0,8	0	0	0	0
	0,5	0,5	0,8	0,7	0,5	0,9	0,5	0	0,4	1	noise generation	0	0	0	0	0,5	0	0	0,8	0	0
	0,6	0,3	0,9	0	1	0	0	0	0	0	change in groudwater level	1	0	0	0	0	0	0	0	0	0
	0,2	0,2	0,8	0,8	0,8	0	0,5	0	0,5	0,8	air pollutants generation	0	1	0	0	0	0	0	0,8	0	0
	0,9	1	1	0	0,8	1	0	0	0	0	change in soil characteristics	0	0	1	0	0	0	0	0	0	0
	1	0,6	0	0	1	0,7	0	0	0	0	supression of vegetation and habitats	0	0	0,8	1	1	1	0,3	0	0	0
	0	0	0	0	0	0	1	0	1	0	goods and services demand	0	0	0	0	0	0	0	0	1	1
	0	0	0	0	0	0	0	0	1	0	taxes generation	0	0	0	0	0	0	0	0	0	1
	0,2	0,2	0	0	0,8	0	0,8	0	1	0,8	trucks' traffic	0	0	0	0	0	0	0	0	0	0

(Vertical column labels: "Activities of the Project" and "Environmental Impacts")

Fig. 1 Combined matrices of state (activities of the project versus environmental effects) and association (environmental effects versus environmental impacts)

For the purpose of integrate the Weighting matrices with the matrices of State and Association, the multiplication factor 1 was applied on the interactions where the relation was identified and the multiplication factor 0 where this relation was not identified. So, three Weighting matrices of State and Association were generated, each one different for each type of expertise. It is important to notice that there could be several more of these matrices depending on the expertise considered, such as a fourth matrix filled by representatives of the population.

In order to relate directly the activities of the project with the environmental impacts, a fuzzy composition took place. The composition was made between matrix A (Weighting matrix of State: activities of the project versus environmental effects) and matrix B (Weighting matrix of Association: environmental effects versus environmental impacts), generating a Response matrix C, for each specialist (or expert group). Accordingly, it was possible to define the composition performed:

$$C_{ij} = \max_{1 \leq k \leq n}\left\{\min\left(A_{ik}, B_{kj}\right)\right\} \qquad (1)$$

To reach a common result, the three matrices obtained through the composition were joined by a weighted average of the values. The weights were considered according to the specialty of the professional. E.g. for the impacts of the physical medium, weight 2 was considered to the answers of the expert in this field and weight 1 to the answers of the others. Likewise, the same procedure was performed for the impacts of biotic and anthropogenic medium.

Activities			IMPACTS									
			Physical Medium			Biotic Medium			Antropic Medium			
			change in water quality	change in air quality	change in soil quality	terrestrial habitats destruction	change in animal's populations	change in aquatic ecosystems	visual impact	environmental discomfort	business improvement	increase in tax collection
Activities	Project's Operation	vegetation removal	6,5	0	8	7	7	3	4	2	0,5	0
		organic soil removal	6	1	9	0	2	3	8	2	1	0
		mining	7	4,5	8,25	0	2	5,5	8	6	0	0
		processing of mineral	8	8	0	0	1,5	6	0	8	0	0
		dam's construction	3,5	0	6	0	0	0	7	0	0	0
	Infraestrutura de Apoio	infrastructure	4	0	0	2,5	3,5	3	0	4	0	0
		input's supply	0	5	0	0	1	0	0	5	10	6
		input's storage	2,5	0	0	0	0	2,5	0	0	0	0
		dispatch of products	0	5	0	0	1	0	0	5	8	8
		demolition	5,5	7	1	0	1	3	1	7	0	0

Fig. 2 Weighted response matrix, which lists the activities of the project and its impacts, taking into account the opinion of experts involved in the process

In addition to that, the result was multiplied by 10, so the final answer is a value between 0 and 10. Consequently, it was possible to obtain the Weighted Response Matrix, showed in Fig. 2, which relates the activities of the project with environmental impacts caused, considering the opinion of several specialists.

To summarize the proposed methodology, Fig. 3 shows a scheme of all steps.

Based on the Weighted Response Matrix obtained, the numerical values of each interaction "Activity versus Impact" were classified with linguistic variables, for a better visual communication and understanding of the interactions. The activities were classified according to the degree of relationship with the impact, varying from minimally impacting to extremely impacting.

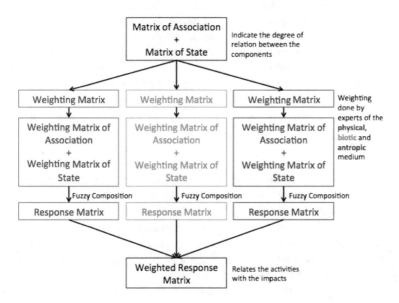

Fig. 3 Scheme of proposed methodology

Table 1 Classification of the degree of relation between activity and impact; and impact and environmental change

Final matrix value	Degree of relation between activity and impact	Degree of relation between impact and environmental change
$0 \leq$ value < 2	Minimally impacting	Minimally significant
$2 \leq$ value < 4	Little impacting	Little significant
$4 \leq$ value < 7	Medium impacting	Medium significant
$7 \leq$ value < 9	Very impacting	Very significant
$9 \leq$ value ≤ 10	Extremely impacting	Extremely significant

According to the CONAMA [10], environmental impact can be defined as any change in the environmental properties. Therefore, the qualification of the impacts was carried out according to the significance of their change to the environment, varying from minimally significant to extremely significant. For this, five ranges of values were considered in the classification, as shown in Table 1. The Final Matrix was then classified according to the linguistic variables.

In order to identify the activities with the most impact degree and the most significant impacts, each activity and impact were analyzed separately. According to the classification of each interaction, the greater degree of relation was selected, due to the fact that if there is only one relation with a greater degree, it is sufficient to classify the activity and impact with this magnitude.

It was possible then to identify the most critical conditions: the activities that generate greater impact and the impacts that generate the most significant changes in the environment. This classification is utterly important for the determination of preventive and corrective actions, in order to investigate the best ways to carry out the activities and to mitigate the impacts caused by them. Hence, the identification of the most impacting activities can support the planning of preventive actions, as well as the forecast of the most significant impacts enables the guidance of corrective measures. Therefore, these actions and measures in combination would integrate an effective environmental management of the project.

3 Results

The developed fuzzy methodology was applied to real cases of environmental impact assessment, aiming to compare the results obtained in the real study with the ones from the fuzzy methodology, thereby to discuss whether the proposed measures would have been the same if the fuzzy application had been used. Therefore, the real environmental study analyzed in this paper is the Environmental Impact Assessment of the Modified Southern Section of the Mario Covas Road Program, a project done in the state of São Paulo, Brazil [11].

The starting point of the analysis of this EIA was the identification of two sets of aspects: the actions of potential impact and the deriving impacts on each environmental component under consideration. It is important to notice the difference between the deriving impact and resulting impact: the first one is the one induced by the actions identified, while the resulting impact is the one that remains even after the adoption of preventive, mitigation or compensation measures. An interaction matrix was used for the identification of the environmental impacts, which relates the actions of the project of potential impacts and the environmental components likely to be affected by them, with the possible impacts to be caused being listed in each interaction.

Environmental control measures were formulated based on the understanding of the potential impacts and the interaction matrix, to be taken as preventive, mitigation or compensatory actions. A detailed impact assessment was carried out in order to qualify and quantify the resulting impact, which is the one that will actually happen even after the effective implementation of the planned measures.

The conclusion of the EIA presents the overall environmental balance of the project, considering the result of the evaluations carried out for each environmental component affected. Regarding the resulting impact of the physical environment, those that were considered most significant occurred during the construction phase, with emphasis on the impact on the relief, in the stability of slopes and in the erosion and its indirect consequences. Among the impacts on the biotic medium, the impact highlighted is the one related to natural vegetation. In addition, impacts in the wildlife were identified such as increasing risk of trampling and hunting. The impacts in the antropic medium were considered diversified, affecting positively or negatively various environmental components. It is common in infra structural projects that components of the anthropic medium are considered as the main recipients of the benefits or positive impacts targeted with the implementation of the project.

After the study of the EIA Mario Covas Road Program—Modified Southern Section, the developed fuzzy methodology was applied in order to analyze its effectiveness in a real study.

The Matrices of Association and State were developed according with the detailed environmental impact assessment carried out in the EIA, which determined the characteristics of each impact and its relation to the activities and aspects of the environment. The value for each relation was determined from this analysis, presenting thus greater fidelity to the project's reality. Furthermore, the matrices were adapted to include the activities of the project, the environmental components and the environmental impacts on the three mediums: physical, biotic and anthropic. The matrices used by the fuzzy methodology also considered other environmental components that had not been analyzed by the EIA study.

Three experts were considered to be involved in the process for the evaluation of the existing interactions, one for each of the mediums: physical, biotic and anthropic. Then, three weighting matrices were obtained.

Following the procedures described in Fig. 3, the matrices were completed by each expert and then integrated to obtain the Response Matrices through the fuzzy composition of the Weighted Matrices of Association and State, according to Eq. 1,

			alteration of relief's morphology, slopes stability and increased erosion	increase of impermeable areas	soil's contamination risk	change in water quality	silting of water courses	change in water courses pluviometric regimen	change in water table's level	change in air quality
			IMPACTS							
			Physical Medium							
ACTIVITIES	Pre Construction	project publicity	0,0	0,0	0,0	0,0	0,0	0,0	0,0	0,0
		contract of the executive project, works and other services	0,0	0,0	0,0	0,0	0,0	0,0	0,0	0,0
		initial mobilization	0,0	3,0	0,0	0,0	0,0	0,0	0,0	0,0
		labour hiring	0,0	6,0	0,0	0,0	0,0	0,0	0,0	0,0
		implementation of provisional administrative facilities	2,0	2,0	2,0	0,0	1,5	0,0	0,0	2,0
		people's and economic activities' replacement	0,0	6,0	0,0	0,0	0,0	0,0	0,0	0,0
		interferences' replacement	0,0	5,0	0,0	0,0	0,0	0,0	0,0	3,8
	Construction	domain's area fencing	0,0	0,0	0,0	0,0	0,0	0,0	0,0	0,0
		provisional interruption in local traffic	5,0	5,0	5,0	0,0	5,0	0,0	0,0	3,8
		system of traffic signals	0,0	0,0	0,0	0,0	0,0	0,0	0,0	0,0
		land cleaning	7,0	7,0	7,0	3,8	5,0	5,0	2,0	6,0
		implementation of access to support works	7,0	7,0	5,0	0,0	3,8	0,0	0,0	0,0
		replacement and/or repair of soft soils	10,0	8,0	7,0	0,0	5,0	0,0	0,0	0,0
		earthwork	8,0	8,0	7,0	1,8	5,5	1,8	0,5	6,0
		activation and use of external support areas to the domain area	3,0	3,0	3,0	0,0	2,3	0,0	0,0	0,0
		material's transportation	0,0	5,0	0,0	0,0	0,0	0,0	0,0	4,3
		stone acquisition and transportation	0,0	6,0	0,0	0,0	0,0	0,0	0,0	6,0
		deviations of waterways	7,0	6,0	0,0	6,0	9,0	8,0	2,0	0,0
		access replacement	2,0	3,0	2,0	1,5	2,0	2,0	2,0	2,3
		final drainage	4,0	4,0	4,0	3,0	4,0	4,0	4,0	0,0
		special works execution	3,0	3,0	0,0	1,5	2,3	2,3	1,5	3,0
		pavimentation	6,0	6,0	6,0	4,5	5,0	6,0	6,0	4,5
		operation of administrative and industrial instalations	2,0	2,0	2,0	0,0	1,5	0,0	0,0	1,5
		slopes stabilization	3,0	3,0	3,0	2,3	3,0	3,0	2,0	0,0
		vertical and horizontal singing	0,0	2,0	0,0	0,0	0,0	0,0	0,0	0,0
		employment demobilization	0,0	0,0	0,0	0,0	0,0	0,0	0,0	0,0
		deactivation of temporary access	2,0	2,0	2,0	0,0	1,5	0,0	0,0	0,8
		deactivation of temporary instalations	2,0	2,0	2,0	0,8	1,8	1,0	1,0	0,8
		domain area recuperation	0,0	1,0	0,0	0,0	0,0	0,0	0,0	0,0
	Operation	road's operation	0,0	6,0	0,0	0,0	0,0	0,0	0,0	9,0
		operational planning and control	0,0	4,0	0,0	0,0	0,0	0,0	0,0	3,0
		daily conservation	0,0	0,0	0,0	0,0	0,0	0,0	0,0	0,0
		road's maintenance	0,0	5,0	0,0	0,0	0,0	0,0	0,0	3,8

Fig. 4 Part of the weighted response matrix obtained from the implementation of the fuzzy methodology in the EIA Mario Covas road program—modified southern section, which relates the activities of the project and possible impacts in the physical medium

in order to find the directly relation between the activities of the project and possible caused impacts, as in Fig. 4.

The Response Matrices were joint into the Weighted Response Matrix and then classified according to the five ranges of values exposed in Table 1, establishing the degree of relationship of the activity with the impact and the degree of the impact and the environment's change.

Figure 5 shows the activities with highest degree of impact and the most significant impacts were identified in accordance with the highest intensity rating of the interactions related to each activity and each impact.

From this final classification, it was possible to identify the activities that generate greater impact and the impacts that cause the most significant changes in the environment. The impacts and their changes can be characterized as positive or negative.

The activities considered to be the ones with most impact are mainly the construction phase: land cleaning, replacement and/or repair of soft soils and deviations

ACTIVITIES	Pre Construction	project publicity	medium
		contract of the executive project, works and other services	medium
		initial mobilization	little
		labour hiring	very
		implementation of provisional administrative facilities	little
		people's and economic activities' replacement	very
		interferences' replacement	medium
	Construction	domain's area fencing	min
		provisional interruption in local traffic	medium
		system of traffic signals	min
		land cleaning	ext
		implementation of access to support works	very
		replacement and/or repair of soft soils	ext
		earthwork	very
		activation and use of external support areas to the domain area	little
		material's transportation	medium
		stone acquisition and transportation	very
		deviations of waterways	ext
		access replacement	little
		final drainage	medium
		special works execution	little
		pavimentation	medium
		operation of administrative and industrial instalations	little
		slopes stabilization	little
		vertical and horizontal singing	little
		employment demobilization	little
		deactivation of temporary access	little
		deactivation of temporary instalations	little
		domain area recuperation	min
	Operation	road's operation	ext
		operational planning and control	medium
		daily conservation	min
		road's maintanence	medium

Fig. 5 Classification of the activities according to their relation with the impacts

of waterways. These activities are considered to cause negative impacts, therefore it is necessary to establish preventive measures and actions. The construction phase was also considered to be the one that causes the most impacts by the real EIA study. In addition to these three activities, the highway's operation was another activity considered to be extremely impacting, during the project's operational phase. The impact caused by this activity can be seen as positive and negative, as although it changes the environmental conditions, it also brings benefits to the urban structure and to the quality of life. Therefore, preventive actions should be taken in order to mitigate the negative impacts, as well as to enhance the positive ones. In a general analysis, the project can be classified as high impacting, since all of the impacts were characterized to be from medium to extreme significance. This characterization was already expected, because only projects with potential for significant environmental degradation is required to be evaluated by EIA in Brazil.

The extremely significant impacts were distributed in the three considered mediums: physical, biotic and anthropic. The impacts of the physical and biotic mediums can be considered exclusively as a negative: modification of the relief's morphology, slopes stability and increased erosion; silting of waterways; change in air quality; and removal of vegetation. For these impacts, it is necessary to plan and develop corrective actions. Potential impacts on road infrastructure, traffic and transport may be considered as negative when related to the construction phase and as positive when related to the operational phase, because during construction there should be temporary changes to a local pattern of traffic distribution in addition to trucks overload in the local road network. During operation, this impact will be positive for assisting local transportation as well as being the main purpose of the project.

Comparing the results obtained by the proposed methodology and the one from the real EIA study, the fuzzy interaction matrix proves to be a successful tool, as the classification of impacting activities and significant impacts was consistent with the results of the case study. Considering its properties, the fuzzy logic allowed to a better treatment of the uncertainties and subjectivities inherent in the EIA.

As another advantage, the developed methodology provides a direct and relatively quick classification of activities and impacts according to their relevance for environmental control.

4   Conclusion

By the analysis of the results obtained using the proposed methodology, it can be concluded that the Fuzzy interaction matrix is an effective tool for the evaluation of environmental impacts, providing a better treatment of the uncertainties and subjectivities inherent in the EIA. The proposed methodology adds value and quality to the data treatment, providing a more accurate result, without a change in the conventional way that experts involved in the process work.

Through the hierarchy process of activities and impacts, the fuzzy logic enabled the acquisition of relevant results for environmental control, in a direct and relatively quick way, replacing the conventionally performed environmental studies that are highly slow and extensive.

In addition, the developed methodology allows the guidance to preventive and corrective measures based on the use of the matrices, promoting the integration of EIA and the environmental management of the project at a stage that would conventionally be only regarding identification of the impacts.

Therefore, the fuzzy approach allows a more realistic decision according to the potential changes, proving to be an effective and practical tool in the Environmental Impact Assessment process.

Acknowledgements Burghi, A. C. A. thanks Sao Paulo Research Foudation (FAPESP) for grants #2011/19634-4. Roveda, J. A. F. thanks Coordenação de Aperfeiçoamento de Pessoal de Nível Superior (Capes) for grants #1846-15-9. This work is partially supported by Capes Foundation Grant #1846-15-9 and FAPESP Grant #2011/19634-4.

References

1. L.A. Zadeh, Fuzzy sets. Inf. Control **8**(3), 338–353 (1965)
2. H.K. Lee, K.D. Oh, D.H. Park, J.H. Jung, S.J. Yoon, Fuzzy expert system to determine stream water quality classification from ecological information. Water Sci. Technol. **36**, 199–206 (1997)
3. C.E. Ballas, A. Ergin, A.T. Williams, L. Koc, Marine litter prediction by artificial intelligence. Mar. Pollut. Bull. **48**, 449–457 (2004)
4. C. Riedler, R. Jandl, Identification of degraded forest soils by means of a fuzzy-logic based model. Plant Nutr. Soil Sci. **165**, 320–325 (2002)
5. Y. Yildirim, M. Bayramoglu, Adaptive neuro-fuzzy based modeling for prediction of air pollution daily levels in city of Zonguldak. Chemosphere **63**, 1575–1582 (2006)
6. A.M.G. Conelissen, J. Van Der Berg, W.J. Koops, M. Grossman, H.M.J. Udo, Assessment of the contribution of sustainability indicators to sustainable development: a novel approach using fuzzy set theory. Agr. Ecosys. Environ. **86**, 173–185 (2001)
7. L.E. Sánchez, Evaluation of environmental impact—concepts and methods. In *Portuguese: Avaliação de Impacto Ambiental – Conceitos e Métodos*, 2nd edn. (Oficina de Textos Ed., 2015)
8. L.B. Leopold, F.E. Clarke, B.B. Hanshaw, J.R. Balsley, *A Procedure for Evaluating Environmental Impact*, vol. 645 (U.S. Geological Survey Circular, Washington, 1971)
9. W. Pedrycz, F. Gomide, *An Introduction to Fuzzy Sets: Analysis and Design* (Mit Press, 1998)
10. CONAMA Resolution, Law n° 01, from Januay 23rd 1986, BRASIL. Provides basic criteria and guidelines for the Environmental Impact Assessment Report (RIMA). DOU Publication, from February 17th 1986, pp. 2548–2549
11. FESPSP (Fundação Escola de Sociologia e Política de São Paulo). Study of Environmental Impact—Mario Covas Higway Program, in *Portuguese: Estudo de Impacto Ambiental—Programa Rodoanel Mario Covas—Trecho Sul Modificado* (2004)

Sustainability Index: A Fuzzy Approach for a Municipal Decision Support System

L. F. S. Soares, S. R. M. M. Roveda, J. A. F. Roveda
and W. A. Lodwick

Abstract Changing the behavior and habits of people and promoting sustainable production and consumption, has caused investment and action by Governments in promoting sustainability as a development model. Although many studies have been developed to evaluate the conditions for sustainability, few take into account the economic, environmental and social aspects altogether. Thus, the aim of this study is the development of an index to measure the degree of sustainability of municipalities using indicators for the social, demographic, economic and environmental dimensions. The methodology employed here is based on the concepts of fuzzy logic which models is subjectivity, uncertainty and imprecision. The index is applied to all 5,565 municipalities in Brazil generating a national sustainability study. After an assessment of the data using our index, a comparison is made with the Municipal Human Development index. Similar results are obtained by both methodologies. However, we argue that a fuzzy logic approach is useful since it uses linguistic variables and is intuitive to implement. Thus, the sustainability index is suitable to aggregate the various indicators and it is an important tool for decision makers. Since it provides information that is useful for the formulation, monitoring and evaluation of public policies.

L. F. S. Soares (✉) · S. R. M. M.Roveda · J. A. F.Roveda
Instituto de Ciência e Tecnologia, UNESP – University Estadual Paulista, Sorocaba,
Sao Paulo, Brazil
e-mail: luifersoares@gmail.com

S. R. M. M.Roveda
e-mail: sandra@sorocaba.unesp.br

J. A. F.Roveda
e-mail: roveda@sorocaba.unesp.br

W. A. Lodwick
Department of Mathematics, University of Colorado, Denver, CO, USA
e-mail: Weldon.Lodwick@ucdenver.edu

© Springer International Publishing AG, part of Springer Nature 2018
L. A. Zadeh et al. (eds.), *Recent Developments and the New Direction
in Soft-Computing Foundations and Applications*, Studies in Fuzziness
and Soft Computing 361, https://doi.org/10.1007/978-3-319-75408-6_50

1   Introduction

Sustainability and sustainable development have become major issues in public policy worldwide. The concept of sustainable development doesn't have a unique definition. One of the most used forms has been the one proposed in the Brundtland Report: "development that meets the needs of the present generation without compromising the ability of future generations to meet their own needs" [1]. Since then sustainability have been promoted as a development model, focusing on sustainable consumption.

There have been many studies concerning the sustainability assessment. Among them we find Ness [2] who describes more than thirty different assessment instruments and Singh [3] who presents a review about the sustainability indices. The sustainability methodologies can be very diverse in their applications to industries [4], energy [5], sustainable management [6], environmental quality [7, 8], and sustainability of ecosystems [9] among others.

However, according to Dahl [10] "much more still remains to be done both to produce indicators of planetary sustainability at the global level, and indicators at the individual level relevant to the changes in personal motivation and behavior essential if a sustainable society is to be built". In this context, models to inform decision makers about municipal sustainability conditions are a useful undertaking. In general, the sustainability conditions are expressed by several indicators and aggregation of them requires consistent ways to deal with information. Besides, the challenge becomes more complex due to difficulty to manage large, diverse, and missing data sets that will be summarized in a single number [11].

An appropriate framework to design models to deal with such challenges is fuzzy logic. Introduced by Zadeh [12], this theory is suitable to analyze subjectivity, uncertainty and imprecision. Various studies have shown that the ability to model expert human knowledge is appropriate to support decision making models [13–17], and a particular case of fuzzy sustainability measures is highlighted in several studies (see Phillis and Andriantiatsaholiniaina [18], Phillis et al. [19], and Gagliardi et al. [20]).

The purpose of this study is to develop a sustainability index to measure municipal sustainability. The methodology employed, is based on a fuzzy inference system, enabling the combination of social, demographic, economic and environmental factors in order to describe the sustainability conditions.

2   Metodology

2.1   Selection of Variables

The model proposed here considers sixteen indicators separated into four dimensions. We use dimensions and indicators for the Sustainability Index

Table 1 Dimensions and indicators considered in this study

Dimensions	Indicators
Social	Life expectancy at birth
	Infant mortality rate
	Schooling rate
	Literacy rate
Demographic	Population density in inhabitants per km^2
	Ratio between the urban and rural population
	Ratio between the male and female population
	% of the population over 25 years
Economic	GDP per capital
	Per-capita income
	Income from work
	Gini index of income distribution
Environmental	Average per capita consumption of water
	Access to the water supply system
	Sanitary sewage type per household
	Access to urban and rural garbage collection

(SI) considered in the study of Sepúlveda [21]. Another consideration in our choice was its free availability. Thus, what we use are the indicators from the Brazilian Institute of Geography and Statistics (IBGE) [22], and they are described below. Table 1 shows the indicators separated into social, demographic, economic and environmental dimensions.

Life expectancy at birth is an indicator that expresses the average number of years that a newborn child, of a particular population group, expect to live.

Infant mortality rate expresses the rate of infant deaths (under one year) in a population, in relation to the number of live births in a given calendar year. It is expressed for every 1,000 children born alive.

Schooling rate of *the adult population* shows the average number of years of study in a population of 25–64 years of age according to groups of years of study (less than 8 years, 8 years, 9 years, 10 years, 11 years, 12 years or more) and the total population of this age group.

Illiteracy rate expresses the percentage of illiterate people of an age group compared to the total population of the same age group. An illiterate person is someone who cannot read and write.

Population density in inhabitants per km^2 is an indicator that displays the number of people per unit of area, (inhabitants/km^2). It is a measure of spatial distribution of the population and allows the study of concentration or dispersion of this population in the geographic area considered.

Ratio between the urban and rural population expresses the ratio between the number of inhabitants in the urban area and the number of inhabitants in the countryside in a population.

Ratio between male and female population expresses the ratio between the number of men and the number of women in a population.

Percentage of the population over 25 years of age is the percentage of persons in the population over 25 years of age.

Gross domestic product (GDP) per capita is an indicator that expresses the gross domestic product (GDP) divided by the number of inhabitants of a locality.

Per capita Income is the ratio between the sum of the incomes of all individuals living in individual households and the total number of these individuals.

Income from work is an indicator that expresses the percentage income from work to the total income.

GINI Index of income distribution is an indicator that measures the degree of inequality in the distribution of individuals according to household per capita income. Its value varies from 0 when there is no inequality (household per capita income of all individuals have the same value), to 1, when inequality is maximum. The universe of individuals is limited to those living in individual households.

Average per capita consumption of water is the water consumption in liters per inhabitant.

Access to the water supply system is the amount of domicile with treated water, i.e. individual permanent domiciles served by piped water from general supply network, with internal distribution to one or more rooms in urban and rural areas.

Sanitation is an indicator that represents the portion of the population served by sanitary sewage system. The variable used is the home sewer connected to the collector (or septic tank), expressed as a percentage, between the urban and rural populations.

Access to urban and rural garbage collection is an indicator that shows the portion of the population living in households served by domestic waste collection services.

2.2 Development of Index

The SI was developed using the concepts of the fuzzy set theory. More specifically, 5 fuzzy rule-based system was used. It has as input and output linguistic variables in the "IF-THEN" rules, allowing for automation and the implementation of an extensive body of human knowledge not possible in quantitative modeling. These concepts have been described in [23–25].

The SI is divided into two levels. In the first one, 4 inference systems were developed to aggregate the indicators into each dimension to evaluate the situation of the considered dimensions: demographic, social, economic, and environmental. The second level, a final system use the results obtained for each dimension as input variables to generate a measure summary, the value of the SI. Thus, SI can provide a measure for the final situation about the sustainability and how is the progress in each one of the dimensions. Figure 1 illustrates the overall architecture.

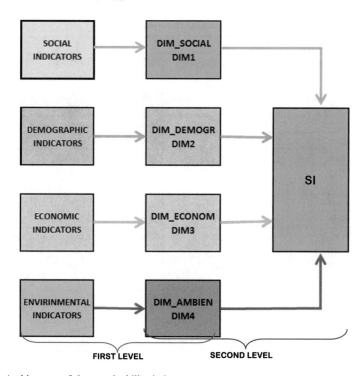

Fig. 1 Architecture of the sustainability index

Both levels use fuzzy membership function to encode the linguistic variables. Each dimension consists of four input variables and each variable was assigned three linguistic terms (low, medium and high). The output variables are assigned seven linguistic terms (very low, low, medium low, medium, medium high, high and very high).

The second level rules are of the type "IF (condition) THEN (answer)" expressing the relationship between the antecedent (condition) and the consequence (answer). For each one of the five systems, an analysis was made of the input and output variables relating them to the IBGE indicators. Each one of the models has 81 rules, since there are four input variables with three fuzzy sets associated with each variable and one output variable. The Table 2 shows some rules used in the social dimension.

Due to facility to incorporate linguistic variables and human expertise, the Mamdani inference method, whose aggregation is performed based on the compositional rule of inference max-min, was used. In the last stage, the data were defuzzified by Center of Gravity method.

The implementation of each of the models was made using the Fuzzy Toolbox MATLAB software, version 7.10.0-R2010a. The final ranking of the SI was

Table 2 Examples of IF-THEN rules in social dimension

Rules	Life expectancy at birth	Infant mortality rate	Schooling rate	Literacy rate	Dim_Social
	Social dimension				
R-1	High	High	High	High	Very high
R-2	High	High	High	Medium	Very high
R-3	High	High	High	Low	Very high
R-4	High	High	Medium	High	Very high
R-5	High	High	Medium	Medium	High
R-6	High	High	Medium	Low	Medium high
R-7	High	High	Low	High	Very high
R-8	High	High	Low	Medium	High
R-9	High	High	Low	Low	Medium high
R-10	High	Medium	High	High	Very high
″	″	″	″	″	″
″	″	″	″	″	″
″	″	″	″	″	″
R-70	Low	Medium	Low	High	Medium low
R-71	Low	Medium	Low	Medium	Low
R-72	Low	Medium	Low	Low	Low
R-73	Low	Low	High	High	Medium high
R-74	Low	Low	High	Medium	Medium low
R-75	Low	Low	High	Low	Medium low
R-76	Low	Low	Medium	High	Medium low
R-77	Low	Low	Medium	Medium	Low
R-78	Low	Low	Medium	Low	Very low
R-79	Low	Low	Low	High	Very low
R-80	Low	Low	Low	Medium	Very low
R-81	Low	Low	Low	Low	Very low

calculated effect considering the effect on the human development index (HDI) [26], and ranges between 0 and 1. In this way, the municipal sustainability was classified into five classes: too high (0.800–1.0); high (0.700–0.799); average (0.600–0.699); low (0.500–0.599); very low (0.0–0.499).

3 Results

A fuzzy rules-based systems was developed to measure the sustainability of different municipalities considering the social, economic and environmental aspects based on the concept of sustainable development.

The MSI was applied to all 5,565 municipalities, distributed in 26 Brazilian States. Only 1% of all municipalities belong to the set of municipalities of Class 1, in which it is considered as having very high sustainability. In Class 2, with high sustainability are 10% of the municipalities and in Class 3, medium sustainability are 7% of the municipalities. Classes 4 and 5, composing 23% and 49% of the municipalities are classified with low and very low sustainability, respectively. In Table 3 it is possible to observe the absolute numbers of municipalities with respect to their sustainability class separated by States of the Federation.

It is important to note that the index proposed here in addition to the general classification can also be partially analyzed, i.e. the analysis of each of the dimensions considered and the value of the SI is useful in defining policy strategies. Table 4 shows the SI and evaluation of each of the dimensions for the municipalities with more than 500 thousand inhabitants. This classification allows policy makers to make a comparison between the municipalities and study the actions developed for each to achieve the higher SI.

The results produced by the SI were compared with those generated by the Municipal human development index (M-HDI). The M-HDI sets the results of HDI for municipalities and is a composite measure of three dimensions indicators of human development: longevity, education and income. Figure 2 shows the ranking of the 20 cities with the highest MSI compared to M-HDI. It is necessary to emphasize that the values of both indices are very close, but the slight variations in each value reflect the environmental assessment which is only seen in the MSI, making it the most appropriate measure to reflect the situation of municipalities in the context of sustainability.

Table 3 Number of municipalities from federative state Brazilian per class by Si

Number of municipalities for sustainability class					
State	Class 1 very high	Class 2 high	Class 3 average	Class 4 low	Class 5 very low	Total municipalities
Acre				2	20	22
Alagoas				4	98	102
Amapá				4	12	16
Amazonas				4	58	62
Bahia		3	5	53	356	417
Ceará		1	1	8	174	184
Espirito Santo	1	1	21	36	19	78
Goiás		29	93	106	18	246
Maranhão				4	213	217
Mato Grosso		2	21	65	53	141
Mato Grosso do Sul			5	44	29	78
Minas Gerais	3	112	209	196	333	853
Pará				9	134	143
Paraíba	1	4	4	9	205	223
Paraná	3	46	141	127	82	399
Pernambuco		2	5	28	150	185
Piaui			2	5	217	224
Rio de Janeiro	1	10	27	41	13	92
Rio Grande do Norte		2	3	16	146	167
Rio Grande do Sul	3	58	113	223	99	496
Rondonia				16	36	52
Roraima			1	5	9	15
Santa Catarina	6	25	69	139	54	293
São Paulo	20	294	194	102	35	645
Sergipe			2	1	72	75
Tocantins		4	19	30	86	139
Distrito Federal	1					1
Total municipalities	39	593	935	1277	2721	5565

Table 4 SI and each dimension results for the Brazilian municipalities with more than 500 thousand inhabitants

Sustainability index

State	Municipality	Population	DIM Social	DIM Demographic	DIM Economic	DIM Environmental	SI
Paraná	Curitiba	1,751,907.00	0.7895	0.6270	0.6268	0.9392	0.9364
São Paulo	Osasco	666,740.00	0.6625	0.7733	0.4113	0.9392	0.8760
Distrito Federal	Brasília	2,570,160.00	0.6500	0.5254	0.6925	0.9392	0.8589
Rio Grande do Sul	Porto Alegre	1,409,351.00	0.6500	0.6500	0.6463	0.9384	0.8589
São Paulo	São Paulo	11,253,503.00	0.6652	0.6446	0.5587	0.9392	0.8426
Paraíba	João Pessoa	723,515.00	0.8300	0.5470	0.3500	0.8869	0.8000
São Paulo	Campinas	1,080,113.00	0.6638	0.6500	0.4478	0.9374	0.7981
São Paulo	Santo André	676,407.00	0.7390	0.6500	0.3500	0.9392	0.7839
Minas Gerais	Belo Horizonte	2,375,151.00	0.6500	0.6500	0.5413	0.9392	0.7839
Rio de Janeiro	Rio de Janeiro	6,320,446.00	0.6500	0.6500	0.4862	0.9387	0.7839
Minas Gerais	Juiz de Fora	516,247.00	0.7082	0.6500	0.3500	0.9392	0.7839
Rio Grande do Norte	Natal	803,739.00	0.8000	0.5220	0.3500	0.6968	0.7683
Pernambuco	Recife	1,537,704.00	0.7520	0.6275	0.3080	0.6500	0.7609
São Paulo	Ribeirão Preto	604,682.00	0.6790	0.6418	0.3674	0.9345	0.7580
Goiás	Goiania	1,302,001.00	0.6611	0.5578	0.4066	0.9392	0.7404
São Paulo	Sorocaba	586,625.00	0.7323	0.5967	0.3500	0.9392	0.7250
São Paulo	São José dos Campos	629,921.00	0.7674	0.5801	0.3500	0.9392	0.7250
Bahia	Salvador	2,675,656.00	0.6500	0.5979	0.3500	0.9392	0.7250

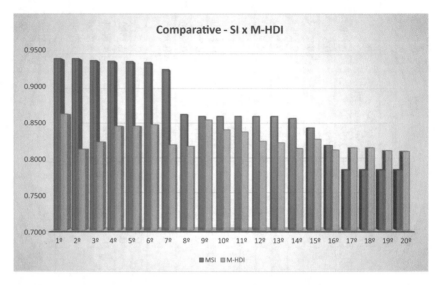

Fig. 2 Ranking of the 20 best cities in SI compared to M-HDI

4   Conclusions

The SI proposed in this study was developed by means of fuzzy logic which incorporates variables of different natures to form a single measure that can point to the conditions of the municipality with respect to sustainable development. The analysis of the results shows that in some states of the country, only the capitals have a higher rating, and in most Brazilian states, the municipalities have a low sustainability index that point to the need to invest in all dimensions.

The variables we chose are easily available which allows the SI to be applied to all Brazilian municipalities establishing a national scenario of the municipalities. In addition, what we developed is an important tool for managers that can create and develop actions in the four dimensions of sustainability, and thus achieve results more effetely. Thus, priorities can be developed for the formulation, monitoring and evaluation of public policies.

The next step in this study will be to measure the accuracy of the model by statistical methods and compared the SI results with another fuzzy methodologies used in sustainability assessment.

Acknowledgements S. R. M. M. Roveda thanks São Paulo Research Foundation (FAPESP) for grant #2015/07714-4. J. A. F. Roveda thanks Capes for grant #1846-15-9. W. A. Lodwick thanks CNPq grant #400754/2014-2. This work is partially supported by FAPESP grant # 2015/07714-4, Capes grant #1846-15-9 and CNPq grant #400754/2014-2.

References

1. G. Brundtland, M. Khalid, S. Agnelli, S. Al-Athel, B. Chidzero, L. Fadika, V. Hauff, I. Lang, M. Shijun, M.M. de Botero, M. Singh, *Our Common Future* (Brundtland report, 1987), p. 144
2. Ness, Barry et al. Categorizing tools for sustainability assessment. *Ecological Economics*, vol. 60, no. 3, pp. 498–508 (2007)
3. R.K. Singh et al., An overview of sustainability assessment methodologies. Ecol. Ind. **15**(1), 281–299 (2012)
4. E.J. Jung, J.S. Kim, S.K. Rhee, The measurement of corporate environmental performance and its application to the analysis of efficiency in oil industry. J. Clean. Prod. **9**(6), 551–563 (2001)
5. F. Begić, N.H. Afgan, Sustainability assessment tool for the decision-making in selection of energy system—Bosnian case. Energy **32**(10), 1979–1985 (2007)
6. G.A. Mendoza, R. Prabhu, Fuzzy methods for assessing criteria and indicators of sustainable forest management. Ecol. Ind. **3**(4), 227–236 (2004)
7. J.H. Popp, D.E. Hyatt, D. Hoag, Modeling environmental condition with indices: a case study of sustainability and soil resources. Ecol. Model. **130**, 131–143 (2000)
8. Y.A. Pykh, E.T. Kennedy, W.E. Grant, An overview of systems analysis methods in delineating environmental quality indices. Ecol. Model. **130**, 25–38 (2000)
9. T. Prato, A fuzzy logic approach for evaluating ecosystem sustainability. Ecol. Model. **187**(2), 361–368 (2005)
10. A.L. Dahl, Achievements and gaps in indicators for sustainability. Ecol. Ind. **17**, 14–19 (2012)
11. O.J. Kuik, A.J. Gilbert, J.C.J.M. van den Bergh, Indicators of sustainable development, in *Handbook of environmental and resource economics* (2002), pp. 722–730
12. L.A. Zadeh, Fuzzy sets. *Information and control* **8**(3), 338–353 (1965)
13. W. Ocampo-Duque, N. Ferré-Huguet, J.L. Domingo, M. Schuhmacher, Assessing water quality in rivers with fuzzy inference systems: a case study. Environ. Int. **32**(6), 733–742 (2006)
14. W. Silvert, Ecological impact classification with fuzzy sets. Ecol. Model. **96**(1), 1–10 (1997)
15. W. Silvert, Fuzzy indices of environmental conditions. Ecol. Model. **130**(1), 111–119 (2000)
16. J.A.F. Roveda, A.C.A. Burghi, S.R.M.M. Roveda, Reconstruction of the environmental quality fuzzy index, in *IFSA World Congress and NAFIPS Annual Meeting (IFSA/NAFIPS), 2013 Joint* (IEEE, 2013). pp. 1086–1089
17. D. Canavese, N.R.S. Ortega, A proposal of a fuzzy rule-based system for the analysis of health and health environments in Brazil. Ecol. Ind. **34**, 7–14 (2013)
18. Y.A. Phillis, L.A. Andriantiatsaholiniaina, Sustainability: an ill-defined concept and its assessment using fuzzy logic. Ecol. Econ. **37**(3), 435–456 (2001)
19. Y.A. Phillis, E. Grigoroudis, V.S. Kouikoglou, Sustainability ranking and improvement of countries. Ecol. Econ. **70**(3), 542–553 (2011)
20. F. Gagliardi, M. Roscia, G. Lazaroiu, Evaluation of sustainability of a city through fuzzy logic. Energy **32**(5), 795–802 (2007)
21. S. Sepúlveda, *Desenvolvimento microrregional sustentável: métodos para planejamento local* (IICA, Brasília, 2005), p. 25
22. I.B.G.E. Brasil, Instituto Brasileiro de geografia e Estatística. Indicadores de Desenvolvimento Sustentavel (Brasil, 2015), p. 351
23. W. Pedrycz, F. Gomide, *An Introduction to Fuzzy Sets: Analysis and Design* (Mit Press,1998)
24. G J. Klir, Y. Bo, Fuzzy sets and fuzzy logic, theory and applications (2008)
25. T.J. Ross, *Fuzzy Logic with Engineering Applications* (Wiley, 2009)
26. S. Anand, A. Sen, Human development index: methodology and measurement (No. HDOCPA-1994-02). Human Development Report Office (HDRO), United Nations Development Programme (UNDP) (1994)

Printed in the United States
By Bookmasters